Offshore Petroleum Drilling and Production

Offshore Petroleum Drilling and Production

Sukumar Laik

CRC Press
Taylor & Francis Group
Boca Raton London New York

CRC Press is an imprint of the
Taylor & Francis Group, an **informa** business

CRC Press
Taylor & Francis Group
6000 Broken Sound Parkway NW, Suite 300
Boca Raton, FL 33487-2742

First issued in paperback 2020

ISBN-13: 978-0-367-57219-8 (pbk)
ISBN-13: 978-1-4987-0612-4 (hbk)

Library of Congress Cataloging-in-Publication Data

Names: Laik, Sukumar, author.
Title: Offshore petroleum drilling and production / Sukumar Laik.
Description: Boca Raton : Taylor & Francis, a CRC title, part of the Taylor & Francis imprint, a member of the Taylor & Francis Group, the academic division of T&F Informa, plc, [2018] | Includes bibliographical references and index.
Identifiers: LCCN 2017038200| ISBN 9781498706124 (hardback : acid-free paper) | ISBN 9781315157177 (ebook)
Subjects: LCSH: Offshore oil well drilling.
Classification: LCC TN871.3 .L285 2018 | DDC 622/.33819--dc23
LC record available at https://lccn.loc.gov/2017038200

Visit the Taylor & Francis Web site at
http://www.taylorandfrancis.com

and the CRC Press Web site at
http://www.crcpress.com

Dedicated to my wife, Jayanti, for her constant

support, cooperation and tolerance.

Contents

Preface

The 1970s was a decade when the quest to understand petroleum drilling and production operations offshore acquired a new impetus. It was a time when global oil production surged on the heels of a rapidly growing world economy. India joined the bandwagon of offshore oil producing nations in 1974 following a momentous discovery of Bombay High. While oil producing nations rejoiced at the acquisition of captive resources, the lack of special techniques and equipment loomed onerously. The problem was further exacerbated by a severe dearth of skilled manpower, an indispensable component for ensuring efficient execution of projects. Nations, however, geared up to the needs of the hour and made suitable investments in building infrastructure while setting directives for improving manpower competency. Workers, directly or indirectly related to the petroleum sector, engineers and young enthusiasts were driven both by need and curiosity to learn about offshore petroleum operations. Hence, the stage for further inquiry and research in this new field (offshore operations) was set.

Academic institutions served as fruitful vehicles of knowledge dissemination thereby generating a steady resource of competent manpower. The Indian School of Mines was a forerunner of such universities in India. Toward this end, a course titled 'Offshore Drilling and Production Practices' was introduced in the undergraduate curriculum of Petroleum Engineering. I had the privilege of teaching this course for many years. One distinguishing and somewhat uncomfortable facet of this course was the lack of a *comprehensive* textbook on E & P operation. Though some books were available, none had covered the subject in its entirety. The source of teaching was mainly the manuals and catalogues of different organisations, journals, conference proceedings and some materials from professional training programmes. Fast and easy access to electronic literature in the late twentieth century, courtesy of the Internet, made 'staying abreast' with the latest more convenient. Technology advancement in deepwater operations has been a significant contributor in current times and deserves a special note.

Even though the digital age made acquiring information a lot easier, there was no 'single' resource that addressed the basics of offshore E & P operations comprehensively. During the 30 years I taught this course, I felt a strong urge to compile and articulate my understanding of the subject. Thus, it seemed providence was at play when the associate editor of Taylor & Francis Group gave a proposal to publish a book on E & P operations. I finally had the required avenue to address the 'missing link'. The book is primarily addressed to undergraduate and graduate students of petroleum engineering, though beginners in the petroleum industry can benefit from its perusal, too. Overall, the book is intended to serve as a 'ready reference' on E & P operations for practitioners and academics alike.

It was my intention to focus exclusively on offshore drilling, completion and production with the 'prepared' reader in mind. Hence, I did not plan on expounding on the prerequisites needed to understand these operations. When a draft content was circulated for review, however, it was suggested that the basics be covered, too. The book in its present form assumes no prior knowledge. It starts with a historical note followed by an introduction in the basics of geology and reservoir engineering. The next section gives a glimpse of the various factors that pose the real challenge to offshore activities. The following section (Chapter 3) introduces the concept of different offshore structures required for carrying out operations like drilling and production. Design aspects have been omitted, though, as there are enough excellent books on 'design of offshore structures' and also because the objective of this book is to focus on operational aspects. So, all relevant topics necessary for a beginner to understand the operations except design are covered. The reader is referred to the references of a particular chapter if indeed the need for further inquiry is felt.

Subsequent sections (Chapters 4, 5 and 6) are dedicated to drilling, completion and production. These chapters initially deal with the basics of the individual operations and then focus on their application offshore. It is hoped that the presentation of topics is such that even a novice can grasp the subject easily. A distinctive feature of the presentation is the systematic coverage of topics in the context of above and underwater systems. Such an ordering of topics was maintained with the hope that a reader proceeding serially would get better clarity. It is worth mentioning here that in Chapter 4 under 'Offshore Drilling', the current topics like UBD, MPD and DGD and drilling through salt to sub-salt layers have been incorporated which are quite relevant in the context of offshore drilling.

Traditionally, storage and pipeline operation have been treated as appendages to the more 'central' production operations. Such a collective treatment blurs the important distinguishing feature of each especially in the case of offshore. These topics are discussed separately in Chapters 7 and 8 for the purpose of avoiding ambiguity in understanding. Furthermore, these two topics have significant relevance for downstream operations of oil and gas which validates a more expansive treatment. It is hoped that such an arrangement would facilitate comprehension.

Any underwater operation requires special assistance from either a diver, an ROV or both. Any offshore operation without these two services cannot be imagined and by not mentioning them this book would have remained incomplete. Hence, the inclusion of Chapter 9 which deals exclusively with divers and ROV.

A very important consideration in the context of offshore operation is safety, particularly for all personnel working away from the shore for extended periods of time. It is absolutely crucial to understand the toll such adverse working conditions have on their health. Environmental pollution is no less important for the environmentalist which of course is equally harmful for human beings and animals. Chapter 10, therefore, is fully dedicated to addressing concerns of health, safety and the environment.

In the process of preparing the manuscript, help was sought from many. In this regard I express special gratitude to two of my colleagues from the Department of Applied Geology, Dr. Syed Tajdarul Hassan and Dr. Ajoy Kumar Bhaumik, who made a significant contribution in drafting the subchapter on 'Background of Geology'. I would like to acknowledge the generous cooperation of all my colleagues in the Department of Petroleum Engineering who provided assistance by providing suitable materials in their area of specialisation. I would also like to acknowledge the contribution of the authors who have been referred to through their publications and are included in the 'References' at the end of each chapter. Some of the illustrations in forms of figures and drawings are taken from the literature or pamphlets of various organisations to whom I express my sincere gratitude. It would be unwise on my part if I did not mention the contribution of two people without whose help and support this work would have not been possible. They are Mr. Rabi Kumar Talukdar, who painstakingly typed out the whole manuscript from my handwritten material, and Mr. Himangshu Kakati, research fellow under my supervision, who drew the figures electronically and organised these files, taken from different sources, in one folder. Toward the last phase, one of my PG students, Mr. Krishan Patidar, helped me in the final compilation and my youngest son, Joyaditya, helped with his valuable criticism and suggestions. My thanks to them as well. Undoubtedly their contribution is commendable. Last but not least I acknowledge the contribution of all my students who by their interactions, seminar/term paper presentations and continuous queries about a textbook on this subject motivated and enriched the endeavour.

I sincerely hope that the book will be as interesting to read as it was to write.

Sukumar Laik

Author

Dr. Sukumar Laik who is presently professor (HAG) of the Department of Petroleum Engineering, Indian Institute of Technology (Indian School of Mines), Dhanbad, India has served as faculty since 1977. He graduated in Petroleum Engg. from Indian School of Mines in 1973 and subsequently obtained his MTech and PhD degrees in Petroleum Engg. from the same institution. As an extra qualification he passed AMIE Sec B (equivalent to BTech) in Mechanical Engg.

Before joining Indian School of Mines, he worked for four years on designing and construction of plant piping and cross-country pipeline in India, Singapore and Indonesia while working with DOSAL Pvt. Ltd. During 1981–1983 he served as visiting faculty in Federal University of Technology, Owerri, Nigeria. His research areas include flow of oil through pipeline, stress analysis of pipeline, offshore drilling and production, hydraulic fracturing simulation and enhanced oil recovery (EOR). Currently he is engaged in a new project on 'Study of Nucleation of Gas Hydrates' where an indigenously fabricated setup (the first of its kind in India) has been made. He has published approximately 60 technical papers in refered journals and has guided many MTech and PhD scholars. He has authored/edited two books, namely, *Dictionary of Petrochemical Engineering* and *Recent Trends in Exploration, Exploitation and Processing of Petroleum Resources*. Presently his major area of research is kinetics of gas hydrates. He is a member of the Society of Petroleum Engineers (USA) and a lifetime Fellow of the Institution of Engineers (India).

He has held several positions in the academic committees outside ISM like member of National Gas Hydrates Programme of the Ministry of Earth Sciences, rules framing committees of the Ministry of Petroleum and Natural Gas and also held administrative positions at IIT(ISM). He was head of the department of Petroleum Engg. from 2000 to 2008 and Dean (Students Welfare) from 2009 to 2012.

1

Introduction

1.1 Historical Background

The history of the petroleum (oil and gas) industry dates to the period before the Christian era when mostly oil and gas shows at or near the ground surface were found. Gradually as the importance of petroleum grew, especially after the invention of the internal combustion engine, efforts were put to go below the ground to find this important mineral and thereafter modern history began. With time as society started to depend on this source of energy, the search also continued and after some period it went beyond the land surface and ventured into water (seas and oceans). Hence, to go into the detailed history of the development of the petroleum industry, we discuss broadly the following: (1) era of surface oil and gas shows, (2) onshore oil and gas industry and (3) offshore oil and gas industry including the genesis of offshore development.

1.1.1 Era of Surface Oil and Gas Shows

Oil in antiquity–Though the modern oil industry is only 160 years old, our early ancestors seem to have taken to collection of crude and asphalt from surface oil seepage from the early days presumably long before the Christian Era. It is well known that in several regions, particularly in the Middle Eastern countries and in Pakistan (formerly North-West India), Egypt, Peru and Mexico, asphalt was used as mortar for water-proofing and building purposes. The archaeological evidence indicates that some of the buildings were constructed about 6000 years ago. There are worldwide records of the heavy residue of bitumen from oil seepages being used for repairing boats and furniture, for treatment of leather and sometimes for embalming. According to the available records, in several cities along the Red Sea oil had been used for lighting during antiquity.

Oil for warfare–There are numerous very ancient instances of the use of burning oil during wars. The use of petroleum products in naval warfare also goes back to antiquity. It was well known that a mixture of naphtha and quicklime was a powerful incendiary when wetted with water. The Arabs had special army units for finding naphtha-filled hand grenades. Crude was also used by some tribes for preparing war paint.

Worshipping at gas shows–There are innumerable instances during the period before the Christian Era of the 'Eternal Fires' from ignition of escaping natural gas becoming objects of worship, particularly in the region surrounding the Caspian Sea; of these, the alters of the fire worshipers in Iran are the most famous. The Jawalamukhi Temple in the Kangra District, India modernised only a few centuries ago, is a more recent instance.

Early methods of collecting oil–It is believed that the technology of collecting oil and bitumen was known before historic times. The Greeks recovered oil by fishing for it in water pools with floating oil; the precious liquid adhering to the branches of some flowering shrubs was collected in cisterns. In other countries, oil was collected from the oil-covered waters either directly in buckets or pots or by dipping blankets, feather bunches, tree branches, linen towels and so on; then, the oil was wrung out into suitable containers.

Oil as medicine–It is not known as to when use of crude oil was extended to medical purposes. Between tenth to sixteenth century, the oil from seepages was extensively used as an ointment for the treatment of rheumatism, burns, sprains and so on, and for treating sores on horses. Marco Polo's writings (thirteenth century) indicate that in the Baku region, the local residents drank oil as medicine. Early in the nineteenth century, a small trade was established in the United States for utilisation of oil to alleviate and cure allergies in the human body.

Oil for lighting and heating–Petroleum (derived from two Greek words *petre* and *elation* [rock + oil] or from the Latin *oleum*) or rock oil, as it was then called, has served humankind as a source of lighting (e.g. Samaritan oil lamp, Figure 1.1) and heating from random occurrences that came naturally to the surface in salt wells or that were expressly obtained from small wells made by hand, in parts of Europe and Asia in the Middle Ages. From the record it is found that a druggist Krier, in Pittsburgh, had an excess production of crude oil beyond that needed for medicinal purposes. He therefore erected a homemade still and refined oil, selling the products for illuminating purposes. Distillation of petroleum was described by the Persian alchemist Mohammad bin Zakariya, Razi (Rhazes). There was production of chemicals such as kerosene in the alembic (al-ambiq), which was mainly used for kerosene lamps. Through Islamic Spain, distillation became available in Western Europe by the twelfth century. It has also been present in Romania since the thirteenth century. In 1745, under the Empress Elisabeth of Russia through the process

FIGURE 1.1
Samaritan oil lamp. (Source: Shutterstock.)

of distillation of the 'rock oil' (petroleum) a kerosene-like substance was prepared which was used in oil lamps by Russian churches and monasteries (although households still relied on candles).

Early records in some parts of the world–The earliest record of any oil well is found to be that in China in 347 AD whose depth was 800 ft (240 m) and was drilled using bits attached to bamboo poles. The ancient records of China and Japan indicate the use of natural gas for lighting and heating. Petroleum was known as 'burning water' in Japan in the seventh century. In the area around modern Baku in Azerbaijan, oil fields were reported to be exploited in the ninth century whose output were reported by Marco Polo in the thirteenth century as hundreds of ship loads.

The earliest mention of petroleum in the Americas occurs in Sir Walter Rayleigh's account of the Trinidad Pitch lake in 1595. Thirty-seven years later, the oil spring of New York finds mention in a visitor's account and in 1753 a map of oil springs of Pennsylvania was published. From 1745 oil sands were mined in Markroiller-Pechelbronn, Alsace which was active until 1970 and was the birthplace of companies like Antar and Schlumberger.

Report of oil production from Yanangyaung was firmly established with several hundred hand-dug wells, yielding up to 400 barrels per day. This production was sent down the Irrawady river in earthenware jars in large boats.

1.1.2 Onshore Oil and Gas Industry

The modern history of petroleum began in the nineteenth century. In 1847, the Scottish chemist James Young noticed natural petroleum seepage in the Riddings colliery at Alfreton, Derbyshire, UK from which he distilled light thin oil suitable for use as lamp oil, at the same time obtaining a thicker oil suitable for lubricating machinery.

In 1848, Young set up a small business refining the crude oil. Eventually, by observing oil dripping from the sandstone roof of the coal mine, he distilled cannel coal at low heat producing a fluid which resembled petroleum. By slow distillation he could obtain many products, one of which he named 'paraffin oil.'

On October 17, 1850, Young patented his method of production of these oils including paraffin oil from coal which was commercialised in 1851 by E.W. Binney & Co. at Bathgate and E. Meldrum & Co. at Glasgow. Around the same time, Canadian geologist Abraham Pineo Gesner developed a process to refine a liquid fuel from coal, bitumen and oil shale which he named kerosene and formed a company called Kerosene Gaslight Company whose activities were expanded to the United States in Long Island, New York by 1854.

Ignacy Lukasiewicz improved Gesner's method to refine kerosene from the more readily available 'rock oil' (petroleum) seeps when the first rock oil mine was built in Poland in 1853. All these developments toward the preparation and use of kerosene from coal as well as rock oil led to the building up of the first modern Russian refinery in the mature oil fields at Baku in 1861, when its production was 90% of the world's total oil production.

It is observed that toward the second half of the nineteenth century, rock oil started gaining in importance and it started getting produced through mining, digging wells by hand and drawing oil from seepages. But it was in 1858 when one of the druggist-promoted American oil companies selling crudely refined oil from seepages engaged Colonel E. L. Drake as superintendent of field operations. He drilled the first well in 1859 at Titusville in Pennsylvania. At a depth of 69 feet, a flow of oil of nearly 800 imperial gallons a day was encountered. Drake's well marks an epoch in the history of the petroleum industry and is regarded as the first commercial oil well (Figure 1.2). Though there was considerable activity around this time in other parts of the world, it is singled out because it was drilled, not dug; it used a steam engine; there was a company associated with it; and it touched off a major boom. By the way, Romania was the first country in the world to have its crude oil output officially recorded in international statistics, nearly 275 tonnes from hand-dug wells; but toward the end of the nineteenth century, the Russian Empire, particularly the Branobel company in Azerbaijan, had taken the lead in production.

During the same period as that of Drake's well, another discovery of oil springs in Ontario, Canada in 1858 touched off an oil boom which brought hundreds of speculators and workers to the area. New oil fields were discovered nearby throughout the late nineteenth century and the area developed into a large refining centre. Early finds like those in Pennsylvania and Ontario led to 'oil booms' in Ohio, Texas, Oklahoma and California in the United States.

With the progress of mechanisation and the appearance of the spark-ignition engine, the industry was obliged to meet the growing demands for energy, transport and lubricants. In 1863, oil was discovered in Peru. The name of John D. Rockefeller became widely known due to his association

FIGURE 1.2
Edwin L. Drake well. (Courtesy of American Oil & Gas Historical Society.)

with cars and oil. He founded Standard Oil Co. in 1870 and created Standard Oil Trust in 1882. In 1885, oil was discovered in Sumatra, Indonesia, and the Royal Dutch Company was launched.

In this context, I would like to make a brief reference to the early history of oil discovery in India, the country of origin of the author, where the first show of oil seepage was observed by geologist H.B. Medlicoff while surveying for coal in Upper Assam near Margherita. In 1866 and 1867 the wells drilled near Jaipur and Namdang were unsuccessful. Later when laying the rail route to Margherita, the engineers of Assam Railways & Trading Company came across the seepages. Then AR&T Co. started drilling in Digboi in 1889 and discovered oil in the first wild-cat-well after drilling to 622 ft (Figure 1.3). Immediately afterward, several development wells were drilled with the help of wooden derricks, thatched with leaves. Later in 1893 oil from Digboi was sent by rail in tank wagons to a 'tea-pot' refinery at Margherita. In 1899, Assam Oil Company (AOC) was formed to take over the management of AR&T Co.'s oil interest in Assam. This is the beginning of the oil industry in India.

FIGURE 1.3
First oil well of India. (From W. B. Metre, *History of Oil Industry, with Particular Reference to India*, Oil India Ltd., publication NOSTALGIA, 1959–2009, pp. 22–23.)

The twentieth century saw the real development in the oil and gas business. At the turn of this century, Imperial Russia's output of oil, almost entirely from the Apsheron Penninsula, accounted for half of the world's production and dominated international markets. Nearly 200 small refineries operated in the suburbs of Baku. As a side effect of these early developments, the Apsheron Penninsula emerged as the world's 'oldest legacy of oil pollution and environmental negligence.' But one of the most important changes that took place in the twentieth century was the discovery of oil in the Middle East, initially in Iran in 1908 and later in Iraq in 1927, then in Bahrain in 1931 and in Saudi Arabia in 1938 along with massive exploration and use.

Since the first discovery of oil by Colonel Drake up to today, two major events took place: World Wars I and II between 1914 and 1918 and 1939 and 1945, respectively, when exploration activities either were halted in many places or slowed down. However, after 1950 again the pace changed and the discovery of oil was reported from Abu Dhabi in 1954, Algeria and Nigeria

FIGURE 1.4
First offshore drilling and production piers at Summerland, California. (Courtesy of American Oil & Gas Historical Society.)

in 1955, Oman in 1957, Groningen gas field in Holland in 1959, Zelten oil field in Libya in 1959 and finally historic OPEC (Organisation of Petroleum Exporting Countries) was formed in 1960 and OAPEC (Organisation of Arab Petroleum Exporting Countries)* in 1968. Thereafter, these two organisations of oil-producing nations played a major role in setting petroleum prices and policy.

1.1.3 Offshore Oil and Gas Industry

The first offshore well producing oil was drilled in 1897, that is, after 38 years of famous Col. Drakes' well in Santa Barbara Channel at Summerland, off the coast of California (Figure 1.4). It was drilled from a 250-ft wooden pier extension. The production of steel piers started in the late 1920s extending 400 m into the sea.

The truly modern offshore industry began in the Gulf of Mexico. The first successful well was in October 1937, in 14 ft of water, a bit over a mile from the Louisiana coast. The well that heralded modern industry, however, was completed by Kerr McGee on November 14, 1947. Although this well was in shallow water (about 20 ft), it was drilled out of sight of land. The rig was

---

* OPEC has 12 member-countries – 6 in the Middle East (Iran, Iraq, Kuwait, Qatar, Saudi Arabia and UAE), 4 in Africa (Algeria, Angola, Libya and Nigeria), and 2 in South America (Ecuador and Venezuela). OAPEC has 11 member Arab countries – Saudi Arabia, Algeria, Bahrain, Egypt, UAE, Iraq, Kuwait, Libya, Qatar, Syria and Tunisia.

movable, and the major patterns of future offshore development were established. The world's largest offshore oil discovery, Saudi Arabia's Safaniya, happened soon afterward in 1951, with the first production in 1957, which was also relatively shallow.

As the industry matured, drilling occurred in progressively deeper waters. By the 1980s 'deepwater' meant about 800 ft of water. By 2008, anything less than 1500 ft was commonly considered shallow. Definitions vary, but 1500 ft and above are now considered deepwater, with depths of more than 7000 ft often called 'ultra deepwater.'

As water depth increased, there was a move away from fixed platforms to cheaper and more mobile submersible and semi-submersible platforms, as well as drill ships. The tallest fixed steel platform was Royal Dutch Shell's 1989 Bullwinkle, in the Gulf of Mexico's Manatee field. The structure is 1386 ft tall in 1326 ft of water and used 44,500 tons of steel on 9500 tons of piling. The first floating drilling vessel was commissioned in 1953, which was capable of drilling in water depths ranging from 400 to 3000 ft. The first versatile deepwater drillship was the 'Glomar Challenger' of the Global Marine Co. USA, commissioned in 1968 and having a water depth rating of 20,000 ft and a drilling depth of 2500 ft with the facilities of satellite navigation and dynamic positioning (DP). The discoveries of Prudhoe Bay in 1968 and Kuparwk in 1969 established the Alaska North Slope, which extended activities in the Arctic offshore continental shelves of the United States and Canada.

The period between 1973 and 1990 saw the discovery and development of giant offshore fields in the North Sea, Mexico, the Caspian, the Arctic and Brazil, as well as the Atlantic near Canada. The period also saw the first deepwater discoveries in the U.S. part of the Gulf of Mexico. Since 1990, there have been giant offshore discoveries in West Africa, Brazil and the U.S. Gulf of Mexico. The overall global success rate since 1985 has been 30% whereas in earlier years it was closer to 10%.

Early major discoveries in the North Sea were Brent in about 460 ft of water and Ekofisk in about 250 ft of water. However, depths as much as 3900 ft (Ormen Lange) have been reached in the North Sea/Norwegian Sea projects. In Brazil's Campos basin, water depths can reach nearly 10,000 ft and in offshore West Africa nearly 5500 ft. A potentially significant recent deepwater discovery is Chevron's 2006 Jack No. 2 well in the Gulf of Mexico. It was drilled not only in 7000 ft of water but also more than 20,000 ft under the sea floor.

The first underwater (subsea) well was completed in 1960 in open sea off the coast of Peru by Peruvian Pacific Oil Company and in the same year Shell Oil Company completed another underwater well in Louisiana after 7 years of design and evaluation. With the advancement in the techniques and with the increased exploratory activities in the deeper continental shelf, the tendency to produce from greater depths using subsea completion method increased manifold.

Modern offshore energy industry benefits come from the hard lessons learned from 60 years of open water experience. Compared to the limits of just a few years ago, today's achievements will no doubt pale in comparison to what the future of offshore exploration will bring.

Offshore India – In India out of 26 sedimentary basins recognised both on land and offshore 10 basins with an area of 0.39 million sq. km are in shallow offshore. In addition, the sedimentary area in deep waters has been estimated to be 1.35 million sq. km. Therefore, the total offshore sedimentary basinal area is about 1.74 million sq. km, most of which remains either unexplored or poorly explored. Directorate General of Hydrocarbon (DGH) carried out surveys in these areas, especially deep waters off the West coast, East coast and Andaman Sea. Based on these surveys, blocks were carved out and offered under various rounds of NELP (New Exploration and Licensing Policy) for exploration of oil and gas. Table 1.1 gives the details of sedimentary basins and basinal areas in the offshore region of India.

The prognosticated hydrocarbon reserves in India are about 18 billion tonnes, out of which about 5 billion tonnes have been converted to geological reserves. This considers up to 200 m water depth in offshore and it is estimated that the hydrocarbon resources may be more than 5 billion tonnes beyond 200 m water depth.

TABLE 1.1

Categorisation of Indian Offshore Sedimentary Basins and Basinal Area

Basin	Basinal Area (sq. km.)
A. *Deep Water*	
Kori-Comorin	1,350,000
85°E	
Narcodam	
Sub-Total	**1,350,000**
B. *Shallow Water*	
Cambay	2500
Bombay Offshore	116,000
Krishna Godavari	24,000
Cauvery	30,000
Kutch	13,000
Andaman-Nicobar	41,000
Saurashtra	28,000
Kerala-Konkan-Lakshdweep	94,000
Mahanadi	14,000
Bengal	32,000
Sub Total	**394,500**
Total (A+B)	**1,744,500**

It was toward the end of 1961 when a Russian seismic ship arrived in Bombay and carried out an experimental seismic survey in the Gulf of Cambay discovering the first offshore structure 'Aliabet' at the estuary of Narmada – Arabian Sea.

The initial reconnaissance survey in Bombay offshore was done in 1964–1965 and the detailed survey in 1972–1973.

The first Indian offshore was spudded on May 19, 1970 in Aliabet. The drilling rig and supporting equipment weighing 500 tonnes were spread out on the 1400 square meter area of the 'Leap Frog' platform consisting of seven blocks each weighing 100 tonnes. Each block was 33 m (long) × 16 m (wide) × 8 m (height). In total, there were 24 legs for 6 platforms. The rig was set on heavy timber planks. The living quarter platform was installed about 100 m away, which was bridge connected. The platforms, which were built in Bombay and installed by ONGC personnel, were designed with the help of Russians to withstand wind velocity of 150 km/h.

The first mobile offshore rig 'Sagar Samrat' arrived in 1973 and moved to Bombay High in January 1974. Oil was struck on February 19, 1974 in the very first well in the L III horizon of Bombay High.

Many fields were discovered after Bombay High like Ratna, Heera, Panna, Bassein, D-18, Neelam, D-1 and so on. A map of the Bombay offshore basin (Figure 1.5) is given for reference.

During the initial stages, developments in offshore were solely done by ONGC. Later some private players like Cairn Energy and Reliance Industries Ltd. developed more offshore areas. Cairn Energy has developed Rawa field of KG Basin and Laxmi & Gouri field of Cambay Basin. Reliance has made sizable gas discoveries in KG Basins in the process of development of this field.

In offshore sedimentary basins, areas under petroleum exploration license (PEL) are close to 0.62 million sq. km with 24 blocks in deepwater and 40 blocks in shallow offshore.

The implementation of NELP has enhanced the exploration activities especially in deep waters. As per this policy, it permits 100% equity stakes by private and foreign companies. Therefore, the blocks are now being awarded under an open acreage system as an ongoing process following international bidding.

1.1.4 Genesis of 'Offshore'

Offshore is simply a particular region of a sea or ocean. The oceans have been in existence for at least 3 billion years – more than half the age of the Earth, yet the origin of the oceans is still in doubt. The topographical features of the planet are changing continuously, albeit slowly relative to the human lifespan. During the main ice ages, the seas were 100–200 m below their present level owing to the presence of large volumes of frozen water; the land masses were consequently larger. Presently, the Earth is emerging from an ice age.

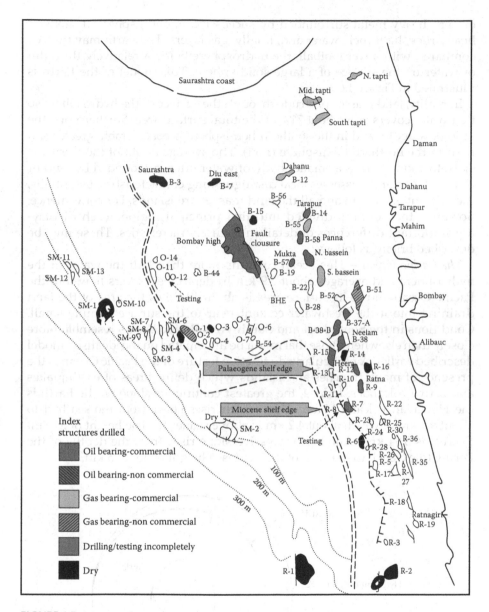

FIGURE 1.5
Map of Bombay offshore basin of India. (Courtesy ONGC.)

To understand the genesis, let us assume that the planet Earth possesses a geometric symmetry. Essentially, Earth would then consist of three major regions: solid material of variable (but generally large) density; liquid of moderate density; and gases of low density. These regions would be held together by gravitational attraction and the substances stratified into a central core of very heavy metal surrounded by successive smooth, spherical shells of heavy rock, light rock, water and, finally, gas layers. The Earth may thus be compared with a wet football, the hydrosphere being a relatively thin film of water on the surface of a large solid sphere. This model of the Earth is illustrated in Figure 1.6.

In reality, land masses protrude through the surface of the hydrosphere so that water covers only about 71% of the total surface area. Furthermore, the ratio of water to land in the southern hemisphere is considerably greater (4:1) than in the northern hemisphere (1.5:1). The average depth of the oceans is close to 4 km. A *sea* is a smaller body of water often surrounded by land or island chains and possesses local distinguishing characteristics. Essentially, there is but one ocean and all the land masses are islands. For convenience, however, the ocean is divided into five principal regions each display-ing important differences of detail in their characteristics. These may be described briefly as follows.

The Pacific Ocean – This ocean contains more than half the water in the hydrosphere and averages about 3.9 km in depth. Few rivers flow into the Pacific Ocean and as its surface area is about 10 times the area of the land draining into it the freshwater content near to the surface is quite small. Conditions in the South-East and South-West Pacific Basins resemble more closely than elsewhere those that might be anticipated on the idealised model described earlier. A particularly distinctive feature of the Pacific Ocean is the presence of many trenches and islands which define areas of earthquakes and active volcanoes. Indeed, the greatest continuous slope on the Earth is the Hawaiian volcano Mauna Kea, which rises sheer from the sea bed to about 9.5 km of which about 4.2 km is above sea level. (The height of Mount Everest is 8.84 km.) Another impressive slope rises from the depths of the Peru-Chile trench to the crest of the Andes, a height of about 5 km.

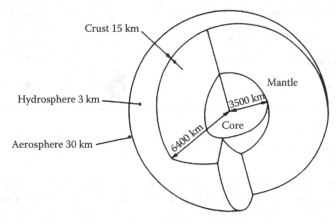

FIGURE 1.6
Model of the Earth.

The Atlantic Ocean – This is a narrow twisted body of water averaging 3.3 km in depth and is adjacent to many shallow seas such as the Caribbean, Mediterranean, Baltic and North Seas, the Gulf of Mexico and so on. Many rivers flow into the Atlantic Ocean including such great ones as the Congo and the Amazon. The surface area of the ocean is only 1.6 times the land area drained into it and so the surface water contains a greater proportion of fresh water than that of the Pacific Ocean. Within the Atlantic Ocean is the so-called Mid-Atlantic Ridge, which extends along the middle of the ocean from north to south and then continues somewhat less definitively through the Indian and South Pacific Oceans. In the Atlantic Ocean, it separates the bottom waters which consequently have different properties on the east and west sides of the Ridge. Moreover, it forms the biggest continuous 'mountain' range on Earth being 16,000 km long and 800 km wide. The total Mid-Ocean Ridge is probably the most extensive single feature of the Earth's topography. The complete length of the ridge is over 48,000 km, its width often over 1000 km and the height relative to the adjacent sea floor varies between 1 km and 3 km. The origin and structure of the Mid-Atlantic Ridge is still in doubt, but many geologists believe it to be the result of upwelling of the mantle material.

The Indian Ocean – This averages 3.8 km in depth and receives fresh water from large rivers such as the Ganges and the Brahmaputra. The surface area of the ocean is 4.3 times that of the land drained into it and is thus intermediate between the Pacific and Atlantic Oceans. An important influence on the Indian Ocean is its susceptibility to radical seasonal variations of atmospheric conditions which are due to the proximity of the large land masses of Africa to the West and India to the North.

The Antarctic Ocean – The three preceding major bodies of water are connected to form this continuous ocean belt known also as the Southern Ocean. This ocean surrounds the frozen Antarctic land mass and induces a deep-ocean circulation.

The Arctic Ocean – Although it is often considered an extension of the Atlantic Ocean, quite recent investigations have shown that the Arctic Ocean is divided into two basins, the Canadian and the Eurasian on either side of the so-called Lomonosov Ridge. The Arctic Ocean surrounds the northern polar ice cap, which permanently covers about 70% of this ocean's surface.

Physiographic features of an Ocean – The main regimes of an ocean, in relation to distance from the land, are the shore, the continental shelf, the continental slope and the deep-sea bottom or Abyssal Plain, which is sometimes called the 'benthic' boundary.

Continental margin – The area of land beneath the sea, from the shore line to the deep ocean or abyssal plain is known as the continental margin which includes the *continental shelf*, the *continental slope* and the *continental rise*.

Typically, an ocean may be about 1000 times as wide as it is deep, a cross-section of it is similar to the one shown in Figure 1.7. The different regions shown here can be described as follows:

Onshore	Inshore or beach	Continental margin			Offshore / Abyssal plain
		Cont shelf	Cont slope	Cont rise	
Distance (average) or width (range)		65–100 km	15–30 km	–	
		1–1200 km	15–100 km	0–600 km	
Depth (Average)		133 m	1830 m	–	Average 3795 m Deepest 11304 m
(Range)		50–550 m	1000–5000 m	1400–5000 m	
Gradient (range)		0°–1°	2°–6°	–	
Area (% of ocean)		6.7% (24 m sq km)	11% (40 m sq km)	3.1% (11 m sq km)	79.2% (285 m sq km)

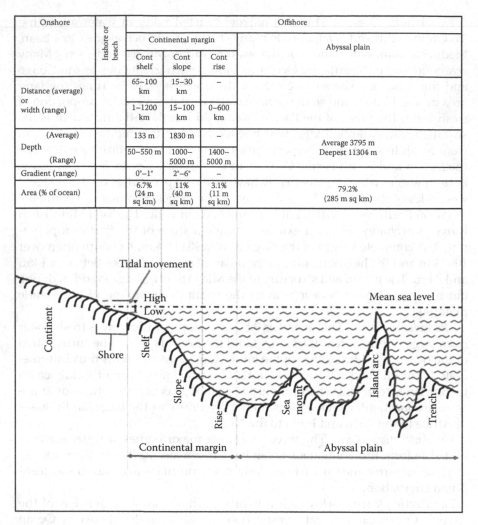

FIGURE 1.7
Physiographic features of the ocean.

Shore – It is defined as that part of the land mass which is closest to the sea and which has been modified by the action of the sea.

Beach – It is the seaward limit of the shore and extends roughly from the highest to the lowest tide levels.

Continental shelf – It is the gently sloping (1 in 500) extension of the land seaward bordering continents which is covered by shallow water, altogether amounting to some 24 million square kilometres or 5.5% of the Earth's surface. Continental shelves vary in width from as little as 1 km to 1200 km and in depth from 50 m to 550 m. The outer edge of the shelf is normally where the inclination of the sea bed suddenly steepens marking the start of the continental slope.

Continental slope – It is a narrow strip of relatively steeply sloping sea bed which extends from the continental shelf. It is seldom more than 80 km wide, often much less, while the water over it generally exceeds 1500 m in depth and sometimes is over 5000 m.

Continental rise – Seaward of the continental slope there is generally another slight slope, called the continental rise, just before the sea bed plunges down to the great depths of the abyssal plain. Water depth ranges from 1400 to 1500 m.

Deep sea bottom/abyssal plain – It is the most extensive area and depths of 3–6 km occur over 76% of the ocean basins. It consists of flat lands although isolated mountain peaks, which occasionally rise high enough to become islands and ridges with valley/trenches along the middle, dot these plains.

1.2 Background of Geology and Reservoir of Oil and Gas

1.2.1 Geology of Rocks

Rocks available in the earth are classified in three major groups: *igneous rock*, *metamorphic rock* and *sedimentary rock*. The first rock generated in the earth by the solidification of the molten magma is known as igneous rock. These rocks are composed of natural crystalline aggregates of minerals at high temperature. These rocks are hard, massive and large bodies, and well abundant on the crust as well as within the deep crust. These may show different structural bodies like batholiths, lopoliths, laccolith, phacoliths, sills, dykes and so on. Sometimes episodic volcanism may show a layered-like structure. The common igneous rocks are granite, basalt, pegmatite, dolerite and so on. Igneous rocks show interlocking texture (mutual relation between the grains) and thus are devoid of porosity. However, post-generation stress may cause development of porosity (e.g., fracture porosity) within the igneous rock. As igneous rocks are formed by the solidification of molten magma (mantle derived) in high temperature, these do not contain any organic carbon within them. Carbon may be present within the igneous rock in the form of elemental carbon (e.g., diamond) or compound (carbonate) but these carbons have never been able to generate any hydrocarbon. Thus, igneous rocks do not have any potentiality to generate hydrocarbon. However, fractured igneous rocks may store hydrocarbon as a reservoir owing to the fracture porosity.

Metamorphic rocks are generated by solid phase transformation of the igneous and sedimentary rocks by the effect of high temperature and pressure.

Pressure and temperature (P-T) changes of the crustal zone are responsible for the changes in the mineralogical composition of igneous and sedimentary rocks in a solid-solid system to generate new metamorphic rocks. The metamorphic reconstitution involves solid state reactions (neocrystallisation) to form new stable minerals and changes in grain boundaries (recrystallisation) as demanded by the new P-T environment. When pre-existing rocks introduced into an environment are different from that in which they were originally formed, they begin to respond to the new physical and chemical conditions through changes in mineralogy, structure and texture. This change can take place isochemically, that is, without any changes in the overall chemical composition or metasomatically, by the gain or loss of chemical constituents, thereby changing the overall chemical composition.

The principal factors leading to the process of metamorphism are temperature, pressure and some chemically active substances (fluids or gases) (Turner and Verhoogan, 1969). The intensity of metamorphism is represented by its grade. Temperature is one of the major factors of metamorphism. Temperature rises with depth below the surface, the rate at which temperature increases with depth varies from place to place but in general it is about 25°–30°C per km and this rate is known as *geothermal gradient*. This rise in temperature is caused by an increase of pressure with depth, and the presence of uncooled magmatic fluids and radioactive substances. Pressure is another important factor. Two types of pressure have been identified below the surface. Load pressure or confining pressure is governed by the load of the overlying rock sequence. A body under such pressure suffers an equal force from all directions thus promoting the formation of minerals of high density and smaller volume. Directed pressure is caused by stress resulting from tectonic movements. This pressure thus causes the formation of rocks and minerals that can withstand high stress and its texture and structure may be modified accordingly.

Sedimentary rocks, also known as secondary rocks, are produced from the degradation of pre-existing igneous or metamorphic rocks. The primary processes involved in the formation of sedimentary rocks are weathering, erosion, transportation, deposition and compaction in the exact sequence (Mukherjee, 2008). Weathering of pre-existing rocks involves both mechanical disintegration as well as chemical decomposition. In the process of disintegration, the rocks are broken up due to the disruptive effects of forces in air and water currents, large changes in the diurnal range of temperature and/or volume changes in the freezing of water. Similarly, pre-existing rocks are decomposed by the solvent action of water, which consists of dissolved gases acquired while falling through the Earth's atmosphere.

The soluble and insoluble materials thus produced by the process of weathering are subsequently transported from its place of origin by rivers, waves, glaciers or currents in air and water. Depending on the agents of transport and its energy condition, the weathered materials are transported in solution, suspension or traction (dragging of heavier sediments along the bed of river or stream channels). This process is followed by the process of deposition, during which the transported sediments are deposited in a basin. The materials carried in suspension by the agents of transport are deposited when (1) the transporting medium is overloaded, and (2) velocity of the transporting agent is reduced. On the other hand, dissolved materials are precipitated when the water current undergoes changes in its chemistry. For example, water containing dissolved materials can mix with water containing different substances leading to chemical reactions and thus precipitation. In the ocean, living organisms extract certain dissolved salts to build their body parts, which are subsequently deposited at the bottom after their death. Until the process of deposition, the loose sediments are accumulated but these sediments will only form a sedimentary rock after the process of compaction. During this process, the loose sediments accumulated in a suitable basin are compacted either by the pressure exerted by their own weight, which is known as welding, or they may be compacted by the deposition of cementing materials forming the bond between the unconsolidated materials. This is known as cementation.

It was mentioned earlier that igneous rocks do not have any potentiality to act as source rock owing to their high temperature origin. Similarly, metamorphic rocks, generated by the process of metamorphism of igneous and sedimentary rocks under high pressure and temperature conditions, also do not have the potentiality to act as a source rock. Some of the sedimentary character may remain in the case of low grade metamorphic rocks, but increase in metamorphic grade destroys the sediment character as a whole. Thus, even if there is any organic carbon present in sedimentary rocks, it is likely to be destroyed owing to the high temperature and lose its potentiality to act as a source rock.

On the other hand, deposition sedimentary rock in basin always contains a variable amount of organic carbon within it. If the organic carbon content in the sedimentary rocks is less than 0.5%, then the potentiality to generate hydrocarbon is considered poor. If the organic carbon content is between 0.5 and 1%, then it is fair; if it is between 1 and 2%, then it is good; if it is between 2 and 3%, then it is very good; and if it is more than 4%, then the potentiality to produce hydrocarbon is considered excellent. Also, the nature of organic carbon plays a major role in determining the type of final product that would be generated after maturation. Terrestrial organic carbon (land derived) is mostly rich in carbohydrates and thus it produces coal. Marine organic carbon, which is mostly algal and sapropelic in nature, produces oil. Thus, for exploration of hydrocarbon in terms of oil, gas or coal, we must concentrate our study on sedimentary rock.

1.2.2 Theory of Origin of Oil and Gas

The name petroleum is coined from the combination of Latin words *petra* (meaning rock) and *oleum* (meaning oil) as it is obtained from rock itself. Petroleum is a naturally occurring yellow to black colour liquid stored in rocks in the subsurface condition and consists of mainly carbon and hydrogen. There are two proposed theories regarding the origin of petroleum with many strengths and weaknesses. These are the *inorganic origin theory* (also known as abiogenic theory or Ukrainian–Russian theory) and the *organic origin theory* (also known as biogenic theory or western theory). Nowadays, the inorganic theory is considered obsolete and people believe in organic origin theory to explain the origin of hydrocarbon.

1.2.2.1 Inorganic Theory

It was first proposed by Georg Agricola during the sixteenth century. Later various other workers added more evidence to strengthen the theory. Among them, Alexander Von Humbolt, Dimitri Mendeleev and Marcellin Berthelot are the most notable. This theory was originated and popularised in and around Russia as most of the publication on this theory was published in the Russian language. This theory has become popular in the west after the publications by Thomas Gold in English. Basically, the inorganic theory again subdivided in two major subdivisions as the Ukrainian–Russian theory and Thomas Gold's Theory. The most interesting consideration of this theory is that petroleum is generated within the rock by an inorganic process and thus the possibility of the estimated reserve may be enormous. As the tectonic activity and magmatic eruption is a continuous process, thus formation of petroleum as per this theory might be started from Precambrian to date.

Ukrainian–Russian Theory – This theory was proposed with the assumption that hydrocarbon may generate by the reaction between iron carbide (or metallic carbide) and water in high temperature (Mendeleev, 1877). This assumption was also supported by the production of acetylene (C_2H_2), which is an organic compound produced by the reaction of water and calcium carbide in natural environment.

$$CaC_2 + 2H_2O \text{ (vapour)} \rightarrow C_2H_2 + Ca(OH)_2$$

Again, Vernadsky (2005) proposed that oxygen content within the deep crust gradually decreases with increasing depth as well as increasing content of hydrogen. The increase in hydrogen as well as decrease in oxygen content leads to the formation of hydrocarbon by the reaction of carbon and hydrogen in reduced environment. Fyfe, Price and Thompson (1978) raised another important point in support of the inorganic origin of hydrocarbon. He questioned the presence of so-called organic porphyrins in petroleum which is also found in meteorites and thus indicates its abiogenic origin. Thus far, more than 80 oil and gas fields in the Caspian district have been explored and developed in crystalline rocks based on this inorganic origin theory.

Gold's Theory – Thomas Gold proposed the deep-seated theory in which the main consideration is that the formation of methane happens in the deep earth, which comes upward through the different fractures developed by earthquake activity (Gold, 1979). Gold and Soter (1980) also prepared a map of oil and gas bearing fields to show the correlation between oil reservoir and past seismicity in support of his theory. He also assumed that mantle-derived methane gets converted to higher hydrocarbons in the upper portion of the crust (Gold, 1993). The presence of helium with hydrocarbons in oil and gas fields also supports a deep source of hydrocarbon (Gold, 1993).

However, major supporting evidence in favour of abiogenic theories is as follows:

- Production of acetylene by the chemical reaction of water vapor with calcium carbide as per the observation of Mendeleev (1877).
- Breakdown of water into H_2 and O_2 and the former would react with hydrogenated graphite and form hydrocarbon (Gold, 1979).
- Hydrocarbons in the form of bitumen are cosmic in origin and precipitated as rain on the Earth from the original nebular materials from which the solar system was formed (Sokoloff, 1890). This theory is also known as extraterrestrial theory.
- Discovery of methane in Titan, the moon of Saturn also suggests the inorganic origin of hydrocarbon (Tomasko et al. 2005; Mitri et al. 2007).
- Presence of organic matter in carbonaceous chondrite (meteorites) (Hoyle, 1999).
- Out-gassing of hydrocarbons below the seafloor through hydrothermal vent even on the surface region thorough volcanoes, mud volcanoes, fumaroles and so on from the deep earth.
- Occurrence of commercial quantities of hydrocarbon in crystalline rocks (Kudryavtsev, 1959)
- Except these, several other supports in favour of abiogenic origin are well described in the papers of Gold (1993) and Glasby (2006).

1.2.2.2 Organic Theory

There are many reasons to consider that hydrocarbons are of biogenic origin. In general, all organic matter like carbohydrate, fat or protein consists of carbon and hydrogen, which are also the essential constituents of hydrocarbons. It is arbitrarily considered that deposition of marine phytoplanktons is responsible for the generation of hydrocarbons by the bacterial as well as thermal cracking. Whereas deposition of terrestrial plant material in aquatic condition leads to the generation of gaseous hydrocarbons and solid coal during diagenesis and metagenesis. Over time, the deposited organic material within the sediments gets decomposed during diagenesis and produces a dark-coloured waxy material known as kerogen. These kerogens are the main constituents for the generation of hydrocarbons. The organic origin theory of hydrocarbons indicates a limited pool of hydrocarbon with respect to inorganic theory.

Evidence that supports the organic origin of hydrocarbons is as follows:

- Close association of hydrocarbon and fossil bearing sedimentary sequence: It is observed that the source rock (shale or limestone) contains a lot of marine fossils (in terms of benthic, planktic and larger foraminifera, diatoms, radiolarians, nanoplanktons, spore and pollens and other invertebrate fossils) within it, which indicates its marine deposits. On the other hand, the chemistry of these organic bodies (in particular the autotrophs) resembles the chemistry of hydrocarbon having common elements as carbon and hydrogen. Thus, these two factors together indicate the contribution of marine organic matter for the generation of hydrocarbons.

- Association of gas and petroleum with coal.

- Presence of some biomarkers: Biomarkers are very stable complex compounds generated in a living body which are made up of carbon, hydrogen and other elements like oxygen, nitrogen and sulphur. Some of these biomarkers (pristane, phytane, steranes, triterpanes and porphyrin) are found in crude petroleum.

- Association with sedimentary rocks: It is observed that all the source rocks are sedimentary rock (shale and limestone) and the generated oil is mostly accumulated in sedimentary rock, which contains many fossils in it.

- Experimental synthesis of oil: Recently experimental synthesis of oil started from the seed of the *Jatrophacurcas* plant (Singh and Padhi, 2009). This also supports the organic origin of hydrocarbon.

- We know that Eskimos use the fats of whales, seals and sea lions to generate heat and light for their daily life. After hunting of animals, they collected these fats and used these as a source of fuel. This event also supports the potentiality of organic matter (in particular, fat) to generate hydrocarbons.

- Rotation of polarised light: Observation of petroleum under petrographic microscope shows that there is a rotation of polarised light when it passes through hydrocarbons. This rotation of polarised light is also observed in all oils which are organically derived. Thus, rotation of polarised light within hydrocarbon strongly shows its organic origin.

- Carbon isotope of hydrocarbons: Organic carbons bear depleted $\delta^{13}C$ values ranging generally in between $-15‰$ and $-30‰$.

Range of $\delta^{13}C$ values for crude oils ranges between $-22‰$ and $-36‰$. The petroleum gases associated with crudes show more depletion in $\delta^{13}C$ to the tune of $-35‰$ to $-65‰$. The bacterial gases (biogenic methane) show even more depletion in $\delta^{13}C$ and range in between $-55‰$ and $-90‰$. Thus, the

$\delta^{13}C$ composition of crude oils and associated hydrocarbons ($-22‰$ to $-90‰$) clearly show that it falls in the range of $\delta^{13}C$ of established biogenic materials and hence acts as strong evidence for its biogenic origin.

1.2.2.3 Criteria and Mechanism for Generation of Petroleum

According to the organic theory, sedimentary rocks are the host of a generation site of hydrocarbons. Many organic carbons (marine and terrestrial) are deposited within the sediments during the deposition of sedimentary rocks in aquatic conditions. The accumulation of organic carbon within the sediments is a syngenetic event and thus the organic carbon buried within the sediments.

Rapid sedimentation/covering – Proper burial and preservation of organic carbon mainly depends on two important factors as sedimentation rate and detachment from atmosphere. Rapid sedimentation helps to quickly bury organic carbon by which the organic carbon gets separated from the dissolved oxygen within the water. The presence of oxygen may oxidise organic carbon and thus the nature and quality of organic carbon may change. Other than oxidation, exposed organic carbon may be consumed by the marine biota as their food by which the quantity of organic carbon may lose. Continuous sedimentation may lead to the increase in burial depth of sedimentary organic carbon by thickening sedimentary sequence. Criteria required for formation of hydrocarbon are as follows:

- *Presence of adequate amount of organic carbon and its quality* – The most important and basic component required for hydrocarbon generation is organic carbon. Thermal and bacterial cracking of hydrocarbon generates liquid and gaseous hydrocarbons. Thus, the quantity of organic carbon is a very important factor.

- *Composition of source rock* – It is observed that high organic carbon concentration bearing zone may not be associated with high organic carbon producing zone. Retaining power of organic carbon in sedimentary succession is also biased by the sediment particle size. The small size particles bear more adsorption capacity to retain organic carbon and thus bear more organic carbon. Also, the porosity of the rock varies proportionally with the grain size. Small grain size bearing sediments have low porosity as well as lower amount of pore water. Thus, the amount of dissolved oxygen becomes less in the pore water as well as oxidation of organic carbon becomes low.

- *Temperature, depth and geologic time* – Temperature plays a key role for the generation of hydrocarbon in terms of gas, liquid or solid. The first requirement for generation of hydrocarbon within a sedimentary rock is the presence of a sufficient amount of organic carbon with its proper quality. Then the sedimentary organic matter undergoes *diagenesis*. Diagenesis is a physico-chemical or biological process operated within the freshly deposited sediments in a low temperature and pressure condition. Diagenesis is the process through which the sediments become consolidated and organic matter degrades. The sedimentary organic matter deposited within the basin degrades during early diageneis by bacterial action and produces methane $(2CH_2O = CH_4 + CO_2)$ (Figure 1.8). Later more sedimentation leads to an increase in the overburden pressure as well as more compaction of sediments and expulsion of pore water to solidify the sediments.

Gradual accumulation of sediment may ultimately increase the thickness of the sediment column and subsidence of the basin may take place. Deep burial of sediments is then suffered by more temperature and pressure conditions owing to the geothermal gradient and overburden pressure. In that way, the sediments gradually enter the *catagenesis* window. Sediments under this window pass through high temperature (60°–130°C) and pressure (300–1500 bars). Presence of high pressure and temperature again makes the condition out of equilibrium and thus compaction of sediments along with the expulsion of pore fluid takes place. Another major change occurs within the organic matter. High temperature is responsible for thermal breakdown (thermal cracking)

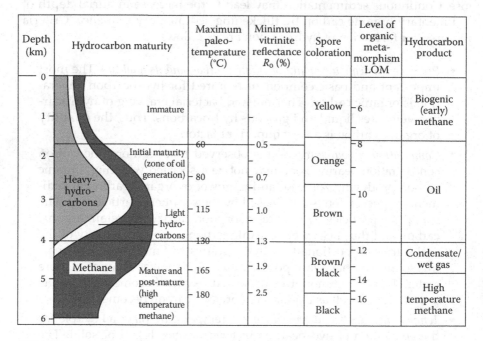

FIGURE 1.8
Schematic representation of diagenesis, catagenesis and metagenesis stages with respect to depth, temperature and products of these stages. (From K. F. North. *Petroleum Geology*. Allen & Unwin, Boston, MA, 1985, pp. 40–41.)

of organic matter to produce methane (wet gas) and liquid hydrocarbon. Thus, this window is also known as the oil window (Figure 1.8).

The evolution of sediments then ultimately reaches its last stage known as *metagenesis* by further burial of sediments in more depth. The temperature and pressure again increase in this stage owing to the geothermal gradient as depth increases. It is the stage of metamorphism and thus mineralogical changes take place along with over-maturation of residual organic carbon or generated liquid hydrocarbon. High temperature burns the organic carbon and thus further carbonisation takes place in the form of either coal or graphite. However, methane (dry gas) is also generated in this window (Figure 1.8).

Depths do not have any direct role on the generation of hydrocarbon. However, increase in depth is responsible for increasing temperature owing to the geothermal gradient which in turn increases the reaction rate for the conversion of organic carbon into hydrocarbon as given by the Arrhenius equation ($k = Ae^{-Ea/RT}$), where k = reaction rate, A = pre-exponential factor which is independent of temperature, Ea = activation energy, R = universal gas constant and T = temperature. Increase in depth is responsible for the compaction of sediments. Complete maturation of organic carbon within the catagenesis window, to produce hydrocarbon, needs some time and is commonly known as cooking time, which is at least 1 Ma. It is important to note that there are mainly six geologic periods during which more than 91% of oil generation has taken place globally and they are as follows: Silurian (9%), Upper Devonian to Tournaisian (8%), Pennsylvanian to Lower Permian (8%), Upper Jurassic (25%), Middle Cretaceous (29%) and Oligocene to Miocene (12.5%).

1.2.3 Migration of Oil and Gas

Hydrocarbons (whether oil, gas or coal) are always generated within a source rock which is either shale or limestone. Shale is a fine-grained rock in which porosity and permeability are almost negligible after compaction. The chemically precipitated limestone is also completely devoid of porosity and thus also permeability. But sometimes fossiliferous limestone may bear some porosity but most of them are intraclastic porosity. As we know that water is always accumulated in porous medium (reservoir), it is likely that oil and gas will also get accumulated in porous medium under subsurface conditions. The presence of oil and gas is always recorded in the highest zone of a reservoir. Also, in the reservoir, gas, oil and water are found in separate phases. Gas occupies the topmost zone whereas the bottom-most zone is occupied by water and oil always occurs as the middle layer, if all the three phases are present in a reservoir. These two simple observations enable us to think about the migration of these three phases in a reservoir system. Thus, it is clear that oil or gas which finally gets accumulated in the reservoir rock must have come out from an impermeable source rock. The movement of fluid and gas in a porous medium can be easily explained whereas the movement of these phases in a nonporous medium is quite difficult to explain. The migration of oil and gas within the nonporous and nonpermeable source rock is called *primary migration* and the movement of oil and gas in a porous medium is called *secondary migration*. We know that formation water present in most of the reservoir rocks is hot and saline in nature and hence denser with respect to oil. This density difference (causes the generation of buoyant force) in between water, oil and gas plays a major role in the movement of oil or gas. The lighter phases move upward owing to the high buoyancy. Thus, buoyant force is the most crucial parameter for the movement of oil. Other than this, overburden pressure, heat flow, thermal conductivity, porosity, permeability, wettability of the medium, capillary pressure and pore throat diameter also play important roles in the migration of oil.

1.2.3.1 Primary Migration

The migration of oil or gas within the source rock is known as primary migration. It is difficult to explain the mechanism of primary migration as it happens within the nonporous and impermeable medium. It is assumed that the clay minerals (kaolinite, montmorillonite, etc.) present within the shale contain a lot of adsorbed water or structure water which becomes separated from the lattice after heating. Vaporisation of these structures or bonded water molecules from the liquid phase requires a lot of space for accommodation. At a greater depth and under high overburden pressure, there will be no scope for increasing volume of rocks to overcome the pressure generated by this water vapour. As a result, the shale itself becomes overpressurised and gradually it becomes fractured. The development of the fractured porosity creates pathways for the movement of oil and gas generated within the shale.

However, several mechanisms are proposed to explain the movement of generated oil and gas from the interior part of the shale to the reservoir rock. These are briefly described as follows:

- Transportation of oil and gas as two immiscible phases in the form of globules and bubbles through the pores in between water wetted grains in compacted source rocks: Here capillary pressure (as the pores are basically very small in diameter and act as a capillary tube) is the main driving force for the movement of oil and gas.

- Colloidal or micellar theory of migration applicable for water-insoluble hydrocarbon molecules at low temperatures: Polar organic molecules having hydrophilic and hydrophobic part may aggregate to form micelle and owing to the hydrophilic nature outside, the hydrocarbon colloid or micelle can form a solution with water which moves away from the source rock.

- Increased temperature in deep earth may be responsible for the increase in the solubility of hydrocarbon molecules in water which leads to migration of hydrocarbon.
- Migration of hydrocarbon by diffusion mechanism: Diffusion is a mechanism of the movement of particles from high density to low density zones. This theory is only applicable for the movement of gas and light oil.
- Hydrocarbon phase migration theory states that the hydrocarbon generated within the source rock migrates as a separate phase with a continuous flow. This is possible where the saturation of oil in terms of total pore fluid in matured source rock is very high and may be as high as 50%.
- Presence of CH_4, H_2S and CO_2 within the water increases solubility of crude in water and migrates as a gas charged solution.

1.2.3.2 Secondary Migration

It is the migration of hydrocarbon within the reservoir rock, which is porous and permeable in nature. Owing to the presence of high porosity and permeability, it is rather easy to explain. Secondary migration may cover a traverse of several metres to hundreds of kilometres. The driving forces for secondary migration are capillary pressure, buoyancy force and hydrodynamic force. A capillary tube is a thin tube having a very small diameter. Capillary action is the ability of a fluid to migrate within the capillary tube without any external force. Here the surface tension and adhesive forces between the fluid and the solid can lift the liquid against the gravity within the capillary tubes. The movement of the fluid within the capillary tube is inversely proportional to the diameter of the capillary tube. Obviously, hydrocarbon molecules with lesser molecular diameter than that of the diameter of capillary tubes will be able to migrate through the tube. Oil and gas are less dense than saline water in the reservoir. Thus, oil and gas move upward owing to buoyancy differences by displacing the same volume of water. Water always flow toward the dip direction of the reservoir bed whereas oil and gas move upward owing to the density differences. Thus, hydrodynamic gradient may create (e.g., hydrodynamic trap) or destroy accumulation of hydrocarbon in traps. Except these, the fracture porosity, which is continuous and larger in size, also helps in the migration of hydrocarbon from one place to another place.

1.2.4 Structure for Accumulation of Oil and Gas

The domain of structural geology deals with the structures present within the rocks. Not all the structures present in the rocks are important to petroleum geology. Structures like fold, fault, unconformity, depositional structure in sediments, diapiric movement and some diagenetic structures are very important to petroleum geology. These structures in the rocks provide the site for accumulation of oil. The presence of nonporous and impermeable rock units above the structure act as a seal. Both seal and structures provide a closure for accumulation of oil and the system together is called a trap. In general, traps are classified into structural traps, diapiric traps, stratigraphic traps, hydrodynamic traps and combination traps (Table 1.2).

Structural traps are mainly of two types, fold trap and fault trap. Folds within the rocks are generated by bending of rock layers from crustal deformation. Folds with upward closing with older rock at the core are known as anticline folds whereas downward closing folds with younger rock toward the core are known as syncline folds (Figure 1.9).

The formation of anticline and syncline folds may be caused by either compressional or compactional or tensional forces acted on rock layers. In petroleum geology, the anticlinal fold has immense importance owing to its upward closing nature. Generally, oil and gas have lower density with respect to formation water, thus owing to the buoyant forces, oil and gas always move up and get

TABLE 1.2

Classification of Different Traps

Structural traps	Fold trap	Compressional fold
		Tensional fold
	Fault trap	Normal fault
		Reverse fault
Stratigraphic trap	Associated with unconformity	Supra unconformity
		Sub unconformity
	Unassociated with unconformity	Depositional
		Diagenetic
Diapiric trap	Salt dome	
	Mud diapir	
Hydrodynamic trap		
Combination trap		

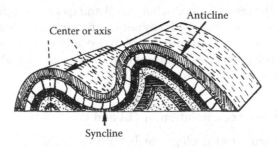

FIGURE 1.9
Anticline and syncline folding in rock layers. (Courtesy of Pearson Scott Foresman/Wikimedia Commons/Public domain.)

accumulated in structural high or at the top of anticlinal folds. Folds may be generated either by compression, compaction, or extension of the rock layers. In case of compressional anticline, the compressional forces are directed toward each other from both sides of rock layers. Thus, owing to this compressional force, the rock layers try to adjust their volume by the formation of folds (Figure 1.10). Sometimes, excess force may exceed the threshold values of the strength of the material and thus rupture along a plane, which is commonly known as a fault.

In the case of compaction, the overburden load plays a major role. Owing to the high overburden pressure, compaction in porous medium takes place and thus vertical subsidence of beds form folded structures. In the case of tensional folds, the rock layers are pulled away from each other by tensional forces. When this tensional force exceeds the threshold values, the gradual crustal thinning and finally faulting takes place in the form of horst and graben. Horsts are the elevated faulted blocks whereas graben are the subsided faulted blocks. Then continuous sedimentation on horst and graben gradually give birth to the folded strata as described in Figure 1.11.

A fault is also a fracture discontinuity structure generated by crustal deformation in which rocks on either side of the fault zone move past each other. If the stress generated during deformation exceeds the threshold strength of a rock, then rupture takes place within the rock layer. Crustal displacement as an effect of tectonic movement can also displace ruptured blocks in different directions. The planar discontinuity surface is called a fault plane.

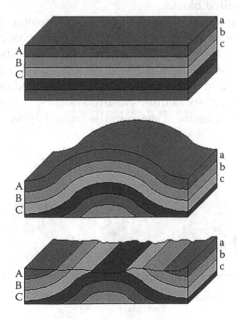

FIGURE 1.10
Development of compressional fold in rock layers. (Courtesy of Actualist/Wikimedia Commons/CC-BY-SA-3.0.)

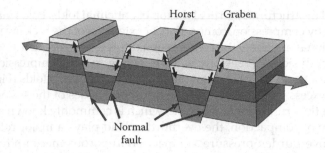

FIGURE 1.11
Development of horst and graben with successive sedimentation and generation of folded structure. (Courtesy of U.S. Geological Survey/Wikimedia Commons/Public domain.)

The displaced blocks on either side of the fault plane are known as foot wall and hanging wall (Figure 1.12).

The block below the fault plane is known as the foot wall whereas the one above the fault plane is known as the hanging wall. A fault can be classified in different ways. Normal fault is a kind of fault where the foot wall relatively moves up with respect to the hanging wall. Reverse faults are those where the hanging wall relatively moves up with respect to the foot wall (Figure 1.13). In addition to these, faults can be classified as dip-slip fault, strike-slip fault and oblique-slip fault based on the direction of relative movement of the faulted blocks.

The entire fault system pertaining to petroleum geology may be classified into eight broad groups based on the potentiality of making closure (Selley, 1998) (Figure 1.14). This classification was done based on the dip direction of the bedding plane and the fault plane, thickness and displacement amount of the bedding plane and relative movement of the faulted blocks. As a whole, there are chances to get six definite closure systems, two having no closure and two with limited closure systems in the faulted zone.

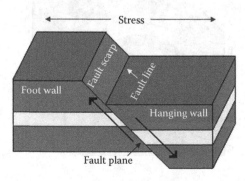

FIGURE 1.12
Normal fault. (Courtesy of Bgwhite/Wikimedia Commons/CC-BY-SA-3.0.)

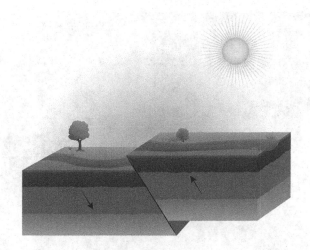

FIGURE 1.13
Reverse fault.

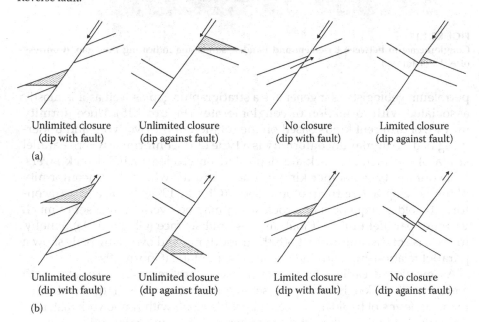

Unlimited closure
(dip with fault)

Unlimited closure
(dip against fault)

No closure
(dip with fault)

Limited closure
(dip against fault)

(a)

Unlimited closure
(dip with fault)

Unlimited closure
(dip against fault)

Limited closure
(dip with fault)

No closure
(dip against fault)

(b)

FIGURE 1.14
Different kinds of fault with possibilities of generation of closure to trap hydrocarbon. (a) Normal fault. (b) Reversed fault.

Stratigraphic traps are mainly classified based on their association with unconformity and the others not associated with unconformity. Unconformity is a time gap (break) in stratigraphy. The presence of unconformity indicates a time period of nondeposition or erosion. The time gap may be in terms of several million years. Unconformity is very important to

FIGURE 1.15
Conglomerate in between basement and Barakar sandstone indicating time gap. (Courtesy of A. K. Bhaumik.)

petroleum geologists as it generates a stratigraphic gap as well as it is mostly associated with formation of conglomerates (Figure 1.15). Unconformity may be of different kinds based on the relation of the overlying and under-lying beds. Angular unconformity is a type of unconformity where parallel strata of sedimentary rock are deposited on eroded inclined rock layers. Nonconformity is another kind of unconformity where the unconformity plane is lying above the plutonic body (Ghosh, 1993). In case of paracon-formity, older sequences of a rock strata may be overlain by a set of much younger parallel units without an erosional surface (Ghosh, 1993). Finally, in the case of disconformity both the underlying and overlying beds show a parallel relationship with the unconformity plane (Ghosh, 1993).

As mentioned earlier, unconformity planes are generally associated with conglomerates. Conglomerate is a sedimentary rock where larger rock frag-ments in terms of boulder, cobble and pebble along with framework materials are embedded within the cemented materials. Thus, unconformities generally lack any pore space and act as a good seal to arrest migration of hydrocarbons. On the other hand, sometimes strike valleys are associated with unconformi-ties. Strike valleys are generated by the differential erosion of different rock bodies owing to the competency differences. For example, in the case of alter-nate shale and limestone layers, shales are soft and get quickly eroded with respect to the limestone bodies. This leads to the development of valleys which are parallel to the strike of shale layers. The unconformity plane will thus lie on the eroded surface of shale and limestone layers. Now, if these strike valleys are filled by sandstone and then above the sandstone if there is development of shale layers, then the embedded sandstone within the shale and limestone layers will produce a good reservoir with a complete trap (Figure 1.16).

Sometimes in angular unconformity, if a sandstone body is covered by shale layers below and above the unconformity then it also may be able to produce a complete closure (Figure 1.17).

Stratigraphic traps not associated with unconformity are mainly *depositional traps*. They may be different kinds of sedimentary bedding like sheet

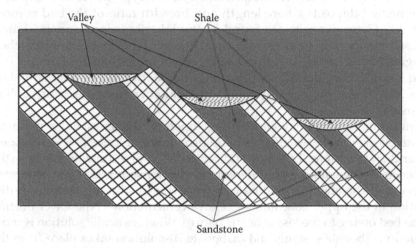

FIGURE 1.16
Supra unconformity trap.

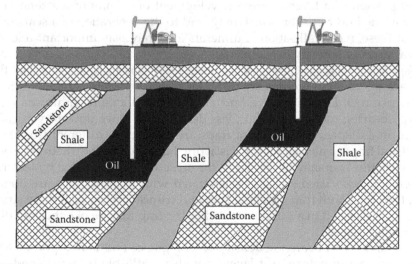

FIGURE 1.17
Subunconformity trap.

deposits, elongated deposits (belt, dendritic, ribbon or shoestring, pods, etc.), pinch out structure, channel deposition, bars and reefs.

Sheet type bed – This is a large tabular sedimentary bed having almost equal length and breadth (length:breadth = 1:1). This sort of bed is formed in turbidite fans, braided alluvial plains and so on.

Elongated deposit – Here the sedimentary deposits are elongated in shape with varied length:breadth ratio. It may be of *belt type* deposits where the length:breadth ratios of the bed are more than 3:1. These are generated mainly in a barrier bar environment. *Dendritic* type deposits are a form of elongated deposits where length and breadth ratio of the bed is more than 3:1. These deposits are found in fluvial tributary channels, deltaic and submarine fan areas. *Ribbon or shoestring* deposits are nonbranched elongated beds having length:breadth ratio of more than 3:1. *Pods* are isolated sedimentary deposits where length:breadth ratio of the bed is less than 3:1. These are generally found in tidal current deposits or aeolian deposits.

Stratigraphic traps may also be developed by diagenesis of sediments. However, the diagenesis effect is more prominent in limestone (carbonates) rather than sandstone. Leaching and cementation of sediments owing to the presence of active fluid with different pH play a major role in the process of diagenesis. The alkaline solution is more active in siliceous system for dissolution in the upper zone and successive cementation in the lower portion of the bed owing to the loss of acidic nature. Whereas acidic solution is more active in carbonate systems and carbonate dissolution takes place from the upper zone owing to the acidic nature of the solution. After reaction, the solution loses its acidity and the cementation takes place in the lower area of the bed. Thus, as a whole, chances of increase in porosity are there in the upper portion of a layer whereas development of a non-porous zone may occur in the lower portion which may lead to the generation of a seal zone. Except these, recrystallisation of minerals also may play important roles in the variation of porosity in sediments.

Diapiric trap – Diapiric movement within rocks is mainly controlled by the buoyant forces owing to density differences of the different rock layers. A less dense rock layer has more buoyancy force underlain within the high density bearing rock layers. Thus, less dense rocks move upward vertically and form a diaper. These sorts of traps are genetically different from true structural traps as no tectonic force is acting here. However, structure associated with this trap may be similar to the structure as originated by tectonic movement. Associated structures generated with diapiric traps are domal trap, field trap, fault trap, pinchout trap and truncation trap. All these structures are described in a salt diaper. A diapiric trap may be of either salt diaper or mud diaper.

Salt diaper – Salt bears the density of 2.03 gm/cc and is non-porous and nonpermeable in nature. Salt layers are also malleable in nature and the density of the salt never changes during compaction as it is non-porous.

Freshly deposited clay or sand layers above salt layers have initial lower density. But as successive sedimentation goes on, owing to the compaction effect, the net volume of clay and sand layer gradually decreases and thus these become denser with respect to the salt layer. Then, finally the density of the salt layer gradually decreases to values less than the density of the overlying rock layers. In this situation, owing to less density, the salt layer starts moving upward owing to buoyancy force and as a whole all the rock layers above the salt also move up. As the rocks are more brittle in nature, thus during this upward movement some of the rock layer may get folded owing to the sidewise contraction of rock length, or may get dome shaped as a result of the overall upward movement of the system, or faulted owing to the brittleness of the rock, or pinching out or truncation of layers may take place against the salt layers. As a whole, all these structures give birth to a new set of closures where the accumulation of hydrocarbon may take place. As described earlier, salt has no porosity and permeability, thus a salt dome itself may act as a seal against truncation or pinching out of other rock layers against it. The upward movement of salt is also known as halokinesis. The entire trap system is illustrated in Figure 1.18.

Mud diapir – Alternate formation of shale and sandstone is observed in the delta region and the sandstone body gradually pinches out toward the open ocean where the prodelta starts with the deposition of clay. Also, the delta and prodelta regions show a higher rate of sedimentation. Thus, in the prodelta region the clay bed is overlain by shale layers. Rapid sedimentation allows the shale to shed off pore water quickly and become non-porous and impermeable with respect to the underlying mud. Thus, as a result, underlying mud remains with a lot of pore water under high pressure. This cause build-up of overpressure within the underlying mud layer as well as the

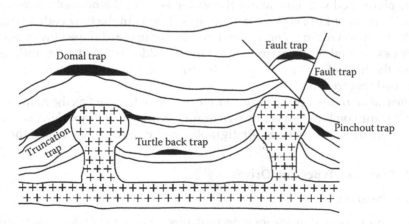

FIGURE 1.18
Diapiric movement of salt and associated traps generated during the diapiric movement.

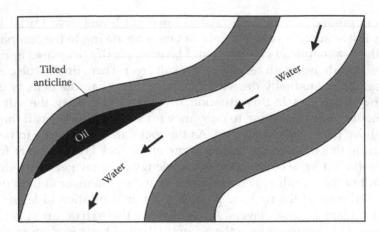

FIGURE 1.19
Hydrodynamic trap.

density of the mud layer becomes lower with respect to the overlying shale
layer. As a result, the upward movement of the mud layer starts owing to the
buoyancy force and leads to the formation of a mud diapir and associated
structures as described in the salt diapir.

Hydrodynamic trap – This is generated by the hydrodynamic movement of
water and subsequent generation of trap within the rock layer. In a folded
or bent layer of porous and permeable rock of which one end is exposed
to the atmosphere, the inflow of surface water may take place owing to the
gravitational movement of water downward. At the same time if oil is there
in the down part of the layer, then oil will try to move up as the density of oil
is less than the density of water. Thus, as a whole, a downward flow of water
and a upward flow of oil will take place. During the movement of both the
liquid phases, oil will flow above the water layer in the opposite direction.
Thus, if a bend or kink is present in upper side within the flow path, then oil
will be trapped within it and will not go further upward if the flow of water
continues. The oil will remain trapped in the folded or bend zone until and
unless the flow of water is stopped. In this way, the hydrodynamic trap is
generated (Figure 1.19).

Combination trap – It is the most common type of trap and as the name suggests it is the combination of any two or more varieties of trap. Here, we may
get combination of structural, stratigraphic and diapiric traps all together.

1.2.5 Reservoir Types and Drives

1.2.5.1 Reservoir

A reservoir is formed of one (or more) subsurface rock formations containing
liquid and/or gaseous hydrocarbons, of sedimentary origins with very few

exceptions. The reservoir rock is porous and permeable and it is bounded by impermeable barriers which trap the hydrocarbons.

The main reservoir rocks are made up of sandstone and/or carbonates. These are normally stratified in successive beds. Based on the type of rock, they can be classified as sandstone reservoirs and carbonate reservoirs.

1.2.5.1.1 Sandstone Reservoirs

These are by far the most widespread, accounting for 60%–80% of all reservoirs. The rock is formed of grains of quartz (silica, SiO_2). If the grains are free, they form sand. If the grains are cemented together, they form sandstone.

Sandstones are very often stratified in a simply superimposed pattern, or with intersecting beds. A vertical cross-section generally shows alternating deposits of sandstone, shale, shaly sand, silts, clays and so on. The majority of the oil and gas reservoirs are of this type.

1.2.5.1.2 Limestone/Carbonate Reservoirs

Carbonate rocks are of varied origins:

1. Detrital: formed of debris (grains of limestone, shells, etc.)
2. Constructed: of the reef type
3. Chemical: formed by the participation of bicarbonate and originating in marine muds

They consist of limestone ($CaCO_3$) and/or dolomite ($CaCO_3$, $MgCO_3$) and often display reservoir characteristics. The two are often intermingled, however, in various ways. Dolomitic and calcitic rocks may be interbedded, or dolomite may form irregular bodies that stand out in slight relief on weathered surfaces that gives them a patchy appearance.

Though sandstone reservoirs are in the majority, limestone and dolomites contain nearly half of the world's petroleum reserves. The shift in importance from predominantly sandstone reservoirs to carbonate reservoirs has been brought about by the discovery and development of the great oil fields of the Middle East, chiefly Iran, Iraq, Kuwait and Saudi Arabia, western Texas and western Canada. Much of the production is obtained from limestone reservoir rocks, which are folded into large anticlines.

1.2.5.2 Reservoir Characterisation

For a rock to form a reservoir:

1. It must have a certain storage capacity: this property is characterised by the *porosity*.
2. The fluids must be able to flow in the rock: this property is characterised by *permeability*.

3. It must contain a sufficient quantity of hydrocarbons, with a sufficient concentration: the impregnated volume is a factor here, as well
as the *saturation*.

1.2.5.2.1 Porosity

Porosity is a measure of the void space within a rock expressed as a fraction
(or percentage) of the bulk volume of that rock.

The general expression for porosity is

$$\varnothing = \frac{V_b - V_g}{V_b} = \frac{V_p}{V_b} \qquad (1.1)$$

where

$\varnothing$ = porosity
V_b = bulk volume of the rock
V_g = net volume occupied by solids (also called grain volume)
V_p = pore volume = the difference between bulk and grain volumes.

In actual rocks porosity is classified as

A. *Absolute porosity* – total porosity of a rock, regardless of whether the
individual voids are connected, and

B. *Effective porosity* – only that porosity due to voids which are
interconnected.

The effective porosity is of interest to the oil industry, and all further discussion will pertain to this form. In most petroleum reservoir rocks, the
absolute and total effective porosity are, for practical purposes, the same.
Geologically, porosity has been classified into two types, according to the
time of formation:

1. *Primary porosity* (intergranular): Porosity formed at the time the
sediment was deposited. The voids contributing to this type are the
spaces between individual grains of the sediment.

   The sedimentary rocks that typically exhibit primary porosity
   are the clastic (also called fragmental or detrital) rocks which are
   composed of erosional fragments from older beds. These are normally classified by grain size, although much looseness of terminology exists. Typical clastic rocks that are common reservoir rocks are
   sandstones, conglomerates and oolitic limestones.

2. *Secondary porosity*: Voids formed after the sediment was deposited.
The magnitude, shape, size and interconnection of the voids bear no
relation to the form of the original sedimentary particles.

Porosity of this type has been subdivided into three classes based on the mechanism of formation.

1. Solution porosity – voids formed by the solution of the more soluble portions of the rock in percolating surface and subsurface waters containing carbonic and other organic acids. This is also called vugular porosity and the individual holes are called vugs. Unconformities in sedimentary rocks are excellent targets for zones of solution porosity. Voids of this origin may range from small vugs to cavernous openings. An extreme example is Carlsbad Cavern.

2. Fractures, fissures and joints – voids of this type are common in many sedimentary rocks and are formed by structural failure of the rock under loads caused by various forms of diastrophism such as folding and faulting. This form of porosity is extremely hard to evaluate quantitatively due to its irregularity.

3. Dolomitisation – This is the process by which limestone ($CaCO_3$) is transformed into dolomite, $CaMg\ (CO_3)_2$. The chemical reaction explaining this change is

$$2CaCo_3 + MgCl_2 \rightarrow CaMg(CO_3)_2 + CaCl_2$$

It has long been observed that dolomite is normally more porous than limestone. This is the reverse of what might normally be expected since dolomite is less soluble than calcite. The reason being the porosity formed by dolomitisation is due to solution effects enhanced by a previous chemical change in limestone.

It should be realised that primary and secondary porosity often occur in the same reservoir rock.

1.2.5.2.1.1 Typical Porosity Magnitude A typical value of porosity for a clean, consolidated and reasonably uniform sand is 20%. The carbonate rocks (limestone and dolomite) normally exhibit lower values with a rough average near 6% to 8%. These values are approximate and certainly will not fit all situations. The principal factors that complicate intergranular porosity magnitudes are

1. Uniformity of grain size – The presence of small particles such as clay, silt and so on, which may fit in the voids between larger grains greatly reduces the porosity. Such rocks are called dirty or shaly.

2. Degree of cementation – Cementing material deposited around grain junctions reduces porosity.

3. Packing – Geologically young rocks are often packed in an inefficient manner and are highly porous.

4. Particle shape – Various formations with different particle shapes
exhibit considerable porosity data.

1.2.5.2.2 Permeability

In addition to being porous, a reservoir rock must be permeable; that is, it must
allow fluids to flow through its pore network at practical rates under reasonable pressure differentials. Permeability is defined as a measure of a rock's
ability to transmit fluids. The quantitative definition of permeability was first
given in an empirical relationship developed by the French hydrologist Henry
D'Arcy, who studied the flow of water through unconsolidated sands.

This law in its differential form is

$$\upsilon = -\frac{k}{\mu}\frac{dp}{dL} \qquad\qquad (1.2)$$

where

υ = apparent flow velocity
μ = viscosity of the flowing fluid
dp/dL = pressure gradient in the direction of flow
k = permeability of the porous media

Consider the linear system of Figure 1.20.

The following assumptions are necessary to the development of the basic
flow equations:

1. Steady state flow conditions exist.

2. The pore space of the rock is 100% saturated with the flowing fluid.
Under this restriction k is the absolute permeability.

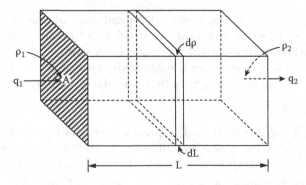

FIGURE 1.20
Linear flow system illustrating Darcy's equation.

3. The viscosity of the flowing fluid is constant. This is not true since $\mu = f(p, T)$ for all real fluids. However, this effect is negligible if μ at the average pressure is used, and if conditions 4–6 hold.
4. Isothermal conditions prevail.
5. Flow is horizontal and linear.
6. Flow is laminar.

With these restrictions in mind, let

$$v = \frac{q}{A} \tag{1.3}$$

where
q = volumetric rate of fluid flow
A = total cross-sectional area perpendicular to flow direction

This is a further assumption since only the pores, and not the full area, conduct fluid. Hence, v = an apparent velocity. The actual velocity, assuming a uniform medium, = v/ϕ.

Case I: Linear Incompressible Fluid Flow
Substitution of Equation 1.3 in Equation 1.2 gives

$$\frac{q}{A} = -\frac{k}{\mu} \times \frac{dp}{dL} \tag{1.4}$$

Separation of variables and insertion of the limits gives

$$\frac{q}{A} \int_0^L dL = -\frac{k}{\mu} \int_{p_1}^{p_2} dp \tag{1.5}$$

which, when integrated, is

$$q = \frac{kA(p_1 - p_2)}{\mu L} \tag{1.6}$$

or

$$k = \frac{q\mu L}{A\Delta p}$$

Equation 1.6 is basic and the following units serve to define the darcy.

If

$q = 1$ cc/sec

$A = 1$ cm^2

$\mu = 1$ centipoise

$\Delta p/L = 1$ atmosphere/cm

Then

$k = 1$ darcy

A permeability of 1 darcy is much higher than that commonly found in reservoir rocks. Consequently, a more common unit is the millidarcy, where

$$1 \text{ darcy} = 1000 \text{ millidarcys}$$

Case II: Linear Compressible Fluid Flow

Consider the same linear system of Figure 1.20, except that the flowing fluid is now compressible; then q will not be constant but a function of pressure. It can be shown mathematically that the same expression as that of Equation 1.6 will hold good in this case also if we express q at the average pressure in the system.

Similar expression can be derived for radial flow of compressible and incompressible fluids also.

1.2.5.2.2.1 Dimensional Analysis of Permeability The physical concept of the darcy is enhanced by a dimensional analysis of Darcy's equation. This is accomplished by resolving all quantities into their basic dimensions of mass M, length L and time t, and solving for the dimensions of k.

$$k = \frac{q\mu L}{A\Delta p}$$

where

$q =$ volume/time $= L^3/t$

$L = L$

$A =$ area $= L^2$

$$\Delta p = \frac{force}{area} = \frac{mass \times acceleration}{area}$$

$$\Delta p = \frac{ML/t^2}{L^2} = \frac{M}{Lt^2}$$

$$\mu = M/Lt$$

$$\therefore k = \frac{\left(\dfrac{L^3}{t}\right)\left(\dfrac{M}{Lt}\right)(L)}{(L^2)\left(\dfrac{M}{Lt^2}\right)} = L^2$$

The fact that permeability is not a dimensionless quantity, but has the units of area, gives an insight into its nature. Its magnitude is primarily dependent on pore size. Consequently, a large void, such as a vug or fracture, may have far more permeability than thousands of small voids. The pore size of a clastic rock is largely dependent on its grain size. Hence, a fine-grained rock will have a lower permeability than a coarse-grained rock of the same porosity if other factors, primarily cementation, are constant.

1.2.5.2.2 Typical Permeability Magnitudes   In general, rocks having a permeability of 100 md or greater are considered fairly permeable, while rocks with less than 50 md are considered tight. Values of several darcys are exhibited by some poorly cemented or unconsolidated sands. Many productive limestone and dolomite matrices have permeabilities below 1 md; however, these have associated solution cavities and fractures which contribute the bulk of the flow capacity. Current stimulation techniques of acidising and hydraulic fracturing allow commercial production to be obtained from reservoir rocks once considered too tight to be of interest.

1.2.5.2.3 Saturation

It is common to have more than one phase present in the pores of the reservoir rock. Saturation of a fluid phase defines relative amount of the particular phase present in the pore space.

$$Saturation = \frac{Volume\ of\ particular\ fluid\ phase\ in\ pore\ space}{Total\ pore\ space}$$

If a rock with a pore volume V_p contains V_w volume of water, V_o volume of oil and V_o volume of gas, when $V_p = V_o + V_g + V_w$, then the oil, gas and water saturations are $S_o = \dfrac{V_o}{V_p}$, $S_g = \dfrac{V_g}{V_p}$ and $S_w = \dfrac{V_w}{V_p}$, which are expressed in fraction (or percentage), so that $S_o + S_g + S_w = 1$ (or 100%).

To know the volume of oil and gas in place, it is necessary to find out the values of oil saturation and gas saturation. The distribution of various phases in pore space depends on saturation of the phase and their *wettability* to the rock.

Wettability is the property of a fluid to adhere to a solid surface. It is measured by the contact angle, which is the angle made by the fluid with the rock surface in the presence of some other fluid.

If the contact angle is less than 90°, the fluid is wetting and if it is more than 90°, then the fluid is said to be nonwetting. The fluids making contact angle of about 90° are said to have intermediate wetting characteristics.

A typical distribution of two phases in porous media for oil and water at different saturations is shown in Figure 1.21.

There are other saturation and wettability dependent parameters like effective and relative permeability and capillary pressure of the rock.

1.2.5.2.3.1 Effective Permeability   The rock may have more than one fluid phase present in the pores. Effective permeability is the permeability of a rock to one particular fluid phase in the presence of other phase/ phases.

Effective permeability may be denoted as K_o, K_w and K_g for oil, water and gas, respectively. It is a function of saturations of fluid phases.

1.2.5.2.3.2 Relative Permeability   It is ratio of effective permeability of a particular phase to the absolute permeability of the rock. It is expressed in fraction or percentage: $K_{ro} = K_o/K$, $K_{rw} = K_w/K$, $K_{rg} = K_g/K$.

Relative permeability of a rock to a fluid phase is a function of saturation, wettability characteristics of the phase to the rock, degree of consolidation of the rock and so on.

Some typical two-phase relative permeability curves are shown in Figure 1.22.

1.2.5.2.3.3 Capillary Pressure   It is a measure of the tendency of the rock to suck in the wetting phase to repel the nonwetting phase. It depends on contact angle of fluid phases and their interfacial tension. The capillary

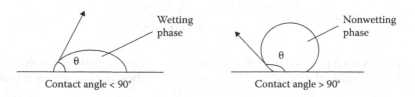

FIGURE 1.21
Typical distribution of two phases in porous media for wetting and nonwetting phase.

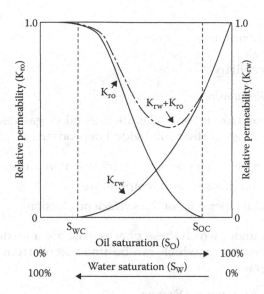

FIGURE 1.22
Typical two-phase relative permeability curve.

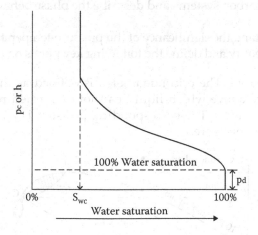

FIGURE 1.23
Typical capillary pressure curve.

pressure changes with change in saturation. It may help or retard movement of oil toward wellbore depending on whether oil is nonwetting or wetting phase to the rock. A typical capillary pressure curve is shown in Figure 1.23. Mathematically, capillary pressure (P_c) is expressed as

$$P_c = \frac{2\gamma \cos\theta}{r}$$

where
    γ = interfacial tension
    θ = contact angle
    r = radius of capillary

1.2.5.3 Types of Reservoirs

Petroleum reservoirs are broadly classified as oil or gas reservoirs. These broad classifications are further subdivided depending on:

- The composition of the reservoir hydrocarbon mixture
- Initial reservoir pressure and temperature
- Pressure and temperature of the surface production

The conditions under which these phases exist are a matter of considerable practical importance, which can be understood from the pressure-temperature diagram.

1.2.5.3.1 Pressure-Temperature Diagram

Figure 1.24 shows a typical pressure-temperature diagram of a two-component system which is essentially used to classify reservoirs, classify the naturally occurring hydrocarbon systems and describe the phase behaviour of the reservoir fluid.

To fully understand the significance of the pressure-temperature diagrams, it is necessary to identify and define the following key points on these diagrams:

- *Cricondentherm* – The cricondentherm is defined as the maximum temperature above which liquid cannot be formed regardless of pressure (point N). The corresponding pressure is termed the cricondentherm pressure.

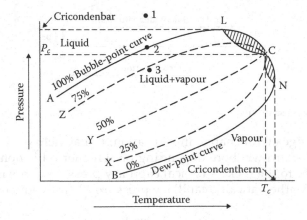

FIGURE 1.24
Typical pressure-temperature diagram for a two-component system.

- *Cricondenbar* – The cricondenbar is the maximum pressure above which no gas can be formed regardless of temperature (point L). The corresponding temperature is called the cricondenbar temperature.

- *Critical point* – The critical point for a multicomponent mixture is the state of pressure and temperature at which all intensive properties of the gas and liquid phases are equal (point C). At the critical point, the corresponding pressure and temperature are called the critical pressure P_c and critical temperature T_c of the mixture.

- *Phase envelope* (two-phase region) – The region enclosed by the bubble-point curve and the dew-point curve (line ACB), wherein gas and liquid coexist in equilibrium, is identified as the phase envelope of the hydrocarbon system.

- *Quality lines* – The dashed lines within the phase envelope are called quality lines. They describe the pressure and temperature conditions for equal volumes of liquids. Note that the quality lines converge at the critical point (point C).

- *Bubble-point curve* – The bubble-point curve (line AC) is defined as the line separating the liquid-phase region from the two-phase region.

- *Dew-point curve* – The dew-point curve (line BC) is defined as the line separating the vapor-phase from the two-phase region.

In general, reservoirs are conveniently classified on the basis of the location of the point representing the initial reservoir pressure p_i and temperature T with respect to the pressure-temperature diagram of the reservoir fluid. Accordingly, reservoirs can be classified into two types. These are

- *Oil reservoirs* – If the reservoir temperature T is less than the critical temperature T_c of the reservoir fluid, the reservoir is classified as an oil reservoir.

- *Gas reservoirs* – If the reservoir temperature is greater than the critical temperature of the hydrocarbon fluid, the reservoir is considered a gas reservoir.

1.2.5.3.2 Oil Reservoirs

Depending upon initial reservoir pressure p_i, oil reservoirs can be subclassified into the following categories:

1. *Undersaturated oil reservoir* – If the initial reservoir pressure p_i (as represented by point 1 of Figure 1.24) is greater than the bubble-point pressure p_b of the reservoir fluid, the reservoir is labelled an undersaturated oil reservoir.

2. *Saturated oil reservoir* – When the initial reservoir pressure is equal to the bubble-point pressure of the reservoir fluid, as shown in Figure 1.24 by point 2, the reservoir is called a saturated oil reservoir.

3. *Gas-cap reservoir* – If the initial reservoir pressure is below the bubble-point pressure of the reservoir fluid, as indicated by point 3 on Figure 1.24, the reservoir is termed a gas-cap or two-phase reservoir, in which the gas or vapor phase is underlain by an oil phase. The appropriate quality line gives the ratio of the gas-cap volume to reservoir oil volume.

1.2.5.3.3 Gas Reservoir

In general, if the reservoir temperature is above the critical temperature of the hydrocarbon system, the reservoir is classified as a natural gas reservoir. Based on their phase diagrams and the prevailing reservoir conditions, natural gases can be classified into four categories:

- Retrograde gas-condensate
- Near-critical gas-condensate
- Wet gas
- Dry gas

Retrograde gas-condensate reservoir – If the reservoir temperature T lies between the critical temperature T_c and cricondentherm of the reservoir fluid, the reservoir is classified as a retrograde gas-condensate reservoir. This category of gas reservoir is a unique type of hydrocarbon accumulation in that the special thermodynamic behaviour of the reservoir fluid is the controlling factor in the development and the depletion process of the reservoir. When the pressure is decreased on these mixtures, instead of expanding (if a gas) or vaporising (if a liquid) as might be expected, they vaporise instead of condensing.

Near-critical gas-condensate reservoir – If the reservoir temperature is near the critical temperature, the hydrocarbon mixture is classified as a near-critical gas-condensate. Since all the quality lines converge at the critical point, a rapid liquid build up will immediately occur below the dew point as pressure is reduced.

Wet-gas reservoir – Such a situation occurs when reservoir temperature is above the cricondentherm of the hydrocarbon mixture. Because the reservoir temperature exceeds the cricondentherm of the hydrocarbon system, the reservoir fluid will always remain in the vapor phase region if the reservoir is depleted isothermally.

As the produced gas flows to the surface, however, the pressure and temperature of the gas will decline and if the gas enters the two-phase region, a liquid phase will condense out of the gas and be produced from the surface separators.

Dry-gas reservoir – The hydrocarbon mixture exists as a gas above cricondentherm both in the reservoir and in the surface facilities. That means even with the depletion of pressure and temperature, the gas does not enter the two-phase region. The only liquid associated with the gas from a dry-gas reservoir is water. Usually a system having a gas-oil ratio greater than 100,000 scf/STB is considered to be a dry gas.

1.2.5.4 Reservoir Drives

Reservoir drive mechanisms refer to the energy in the reservoir that allows the fluids to flow through the porous network to the wellbore. In its simplest definition, reservoir energy is always related to some kind of expansion.

Several types of expansions take place inside and outside the reservoir, as a consequence of fluid withdrawal. Inside the reservoir, the expansion of hydrocarbons, connate water and the rock itself provides energy for the fluid to flow. Outside the producing zone, the expansion of a gas cap and/or an aquifer may also supply a significant amount of energy to the reservoir. In this case, the expansion of an external phase causes its influx into the reservoir and will ultimately result in a displacement process.

There are five basic types of mechanisms that are commonly used to classify the dynamic behaviour of a reservoir:

1. Fluid expansion
2. Solution gas drive
3. Water drive
4. Gas cap drive
5. Compaction drive

Very few reservoirs belong completely to one of these categories. In fact, in most cases the main producing mechanism for a given reservoir may change during the exploitation of the field. Typically, for example, an undersaturated oil reservoir produces under fluid expansion conditions in the initial period of the exploitation, until the reservoir pressure falls below the bubble point pressure. At this stage, the solution gas drive mechanism becomes predominant. In addition to that, in most cases, more than one mechanism is active at any time for a given reservoir. Therefore, the most common producing mechanism could be referred to as a *combination drive* (Figure 1.25).

1.2.5.4.1 Fluid Expansion

Fluid expansion occurs as the reservoir undergoes a pressure depletion. In such conditions, when no external influx is present, the reservoir fluid essentially displaces itself.

In gas and gas condensate reservoirs, fluid expansion is often the predominant drive mechanism and accounts for the recovery of a significant part of

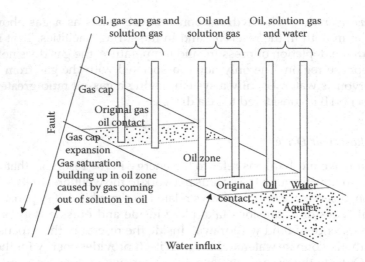

FIGURE 1.25
Cross-section of a combination drive reservoir.

the hydrocarbon originally in place. On the contrary, in the case of under-saturated oil reservoirs, the liquid phase expansion contributes only a little to oil recovery, since oil compressibility is usually very low, especially in medium to heavy gravity oils.

In undersaturated oil reservoirs producing by fluid expansion, the pressure declines very rapidly, while the producing gas-oil ratio (GOR) remains constant and equal to the original solubility ratio, Rs_i.

1.2.5.4.2 Solution Gas Drive/Depletion Drive

When the reservoir pressure falls below the saturation pressure, gas is liberated from the hydrocarbon liquid phase. Solution gas drive, or dissolved gas drive, indicates the process of expansion of the gas phase, which contributes to the displacement of the residual liquid phase.

Initially, the liberated gas will expand but not flow, until its saturation reaches a threshold value, called critical gas saturation (refer to the saturation vs. relative permeability curve in Figure 1.22). Typical values of the critical saturation ranges between 2% and 10%. When this value is reached, gas starts to flow with a velocity that is proportional to its saturation. The more the pressure drops, the faster the gas is liberated and produced, thus lowering further the pressure, in a sort of chain reaction that quickly leads to the depletion of the reservoir.

At the surface, solution gas drive reservoirs are characterised in general by rapidly increasing GORs and decreasing oil rates, while the pressure decline tends to be less severe than in the liquid expansion phase. Generally, no or little water is produced. The ideal behaviour of a field under dissolved gas drive depletion is illustrated in Figure 1.26.

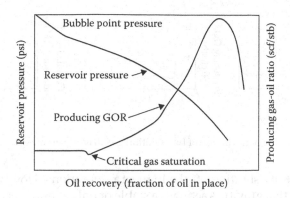

FIGURE 1.26
Production behaviour of a solution gas drive reservoir.

As can be observed, the GOR curve has a peculiar shape, in that it tends to remain constant and equal to the initial Rs_i while the reservoir pressure is below the bubble point; then it tends to decline slightly, until the critical gas saturation is reached. This decline corresponds to the existence of some gas in the reservoir that cannot be mobilised. After the critical saturation is reached, the GOR increases rapidly and finally declines toward the end of the field life, when the reservoir approaches the depletion pressure. The final recovery factor in this kind of reservoir is normally rather low, ranging approximately from 7% to 35% of the original oil in place (OOIP).

For this reason, the availability of a reliable set of gas-oil relative permeability curve is mandatory when evaluating a solution gas drive reservoir. Whenever possible, the curves determined in the laboratory should also be compared with field derived K_{rg}/K_{ro} values.

Other rock and fluid parameters influence the performances of a solution gas drive reservoir to a lesser extent.

1.2.5.4.2.1 Gravity Segregation One important point when dealing with solution gas drive reservoirs is the influence of gravity. It is generally assumed that in these kinds of reservoirs the impact of gravity is negligible. The important implication is that the recovery is not rate-sensitive and that these reservoirs can normally be produced at the highest sustainable rates. However, in some favourable cases, the effect of gravity may act so as to induce the segregation of phases in the reservoir. When this happens, gas is not produced but rather migrates toward the top of the structure to form a secondary gas cap which in turn acts to maintain the pressure in the reservoir.

The occurrence of a gravitational segregation in the reservoir can often be inferred from the production behaviour of the field. For example, when gravity is acting, wells located updip in the structure will produce preferentially with higher GOR with respect to the wells located downdip. Other important information comes from the rate of success of workovers aimed at reducing the

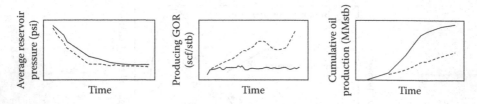

FIGURE 1.27
Impact of gravity in the performance of two solution gas drive reservoirs.

producing GOR by shutting off the highest perforations. However, the most important sign that gravity is acting is a stable or only slightly increasing GOR.

Figure 1.27 shows the production profiles relative to two oil fields, both producing by solution gas drive. The practically constant GOR profile indicates that a gravitational segregation process is taking place in one of the two reservoirs (solid line). The two fields have roughly the same oil in place, but, as it can be noted, the cumulative production and consequently the total recovery are very different in the two cases.

1.2.5.4.3 Water Drive

Many hydrocarbon reservoirs are connected down-structure with natural aquifers, which can provide an important source of producing energy. As oil is produced, pressure declines in the reservoir and when the pressure disturbance reaches the aquifer, water starts to expand and to flow into the reservoir. Therefore, in this case, the producing mechanism is related to a displacement process, since the expansion takes place mostly outside the reservoir.

From a geometrical point of view, water drive fields may be described as bottom or flank/edge water drive, depending on the relative configuration of the aquifer and the reservoir. In bottom drive reservoirs, the oil zone is completely underlain by water, while in flank/edge water drive reservoirs, the oil is in contact with the aquifer only in the peripheral parts of the field. These different configurations pose distinct problems, when production is concerned, because the former are more prone to water coning problems, while the latter will in general experience water fingering.

The efficiency of a water drive mechanism depends in the first place on the volume of interconnected water. In fact, since water compressibility is very low, on the order of 5×10^{-6} vol/psi, several thousand barrels of water must be present and able to expand, in order to produce a single barrel of oil.

Another important parameter of water drive systems is aquifer permeability. High permeability is essential for a water drive to be effective, since the pressure gradients must propagate relatively rapidly, to allow for a sufficient volume of water to be involved. It could be argued that a lower threshold of around 100 mD is necessary for a water drive system to behave efficiently.

The production performance of a water drive reservoir is quite different from those observed for solution gas drive. When the aquifer volume is large enough, the reservoir will in general show a fairly low pressure decline and, furthermore, this decline may become smaller with time, since the aquifer response is often delayed. When the water influx rate equals the fluid production rate, the reservoir pressure may eventually stabilise to a constant value, which is somewhat lower than the initial pressure. In extreme but not infrequent examples, for very large aquifers with high transmissibility, pressure does not show any decline with time and remains equal to the initial pressure, even in the presence of high withdrawal rates.

The producing GOR will remain constant and equal to the initial solution GOR, if the pressure remains above the bubble point.

As far as the oil production is concerned, a slow but steady decline is generally observed, which is related to the progressive water invasion of the structurally lowest wells. Therefore, unless wells are abandoned and worked over, water drive fields are characterised by a progressive increase of the water production. Moreover, if water and oil have approximately the same mobility, the total fluid production of the field remains fairly constant. The ideal production performance of a water drive field is illustrated in Figure 1.28.

In those cases where the aquifer supply is not strong enough to maintain the pressure above the bubble point at the desired production rates, the pressure may decline below the saturation pressure and some gas is liberated in the reservoir. To some extent, this may be beneficial, since the expansion of the gas phase provides the reservoir with an additional source of energy, which may reduce the rate of pressure decline. However, if the pressure keeps declining, the producing GOR may rise significantly and the solution gas drive process may prevail, thus impairing the final recovery.

Water drive reservoirs usually exhibit the highest recovery efficiency. Reported values of recovery factors in these reservoirs range from 30% to 80% of the original OIP (oil-in-place) with an average in the vicinity of 50%. Significantly, the highest figures have been reported for high permeability reservoirs.

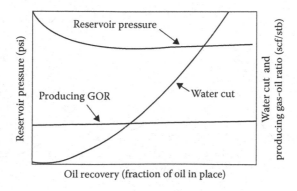

FIGURE 1.28
Production behaviour of a water drive reservoir.

1.2.5.4.3.1 Unstable Displacements   The recovery of a water drive reservoir may or may not be rate sensitive, depending on the stability of the displacement process. The existence of a stable process, in turn, depends upon many reservoir and fluid properties.

In general, two types of unstable displacements occur in water drive reservoirs, that is, coning and fingering. Although the two processes may occur in the same reservoir, the former is typical of bottom aquifers and the latter of flank aquifers.

The first of these processes, water coning (or *cusping*, as it is often called for horizontal wells) defines the movement of the fluid contact surface (water-oil contact, in this case) toward a producing well, due to the presence of viscous pressure gradients established around the wellbore by the production itself. This movement is counterbalanced by gravity forces; therefore, the existence and the magnitude of a coning process can be defined, at any time, by the resulting potential gradients around the wellbore. Figure 1.29 shows a simplified example of a water coning in a vertical well.

Water fingering, on the other hand, defines the formation of water tongues that under-run the oil zone, because of excess production off-takes. This is a highly unwanted situation in all cases because it not only causes early water

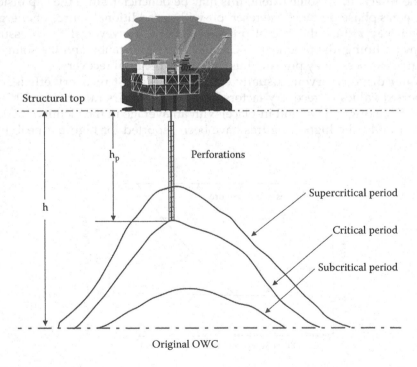

FIGURE 1.29
Water coning in a vertical well.

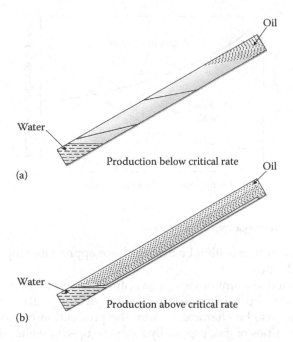

FIGURE 1.30
Stable displacement (a) and water fingering (b).

breakthrough in structurally high wells, but it also poses serious problems to the recovery of the oil left above or below the water tongue (Figure 1.30).

1.2.5.4.4 *Gas Cap Drive*

A gas cap drive is the producing mechanism whereby a volume of free gas in the upper part of the structure of a reservoir expands into the oil zone to displace oil downdip, toward the producing wells.

Where an original (primary) gas cap exists, the oil phase is saturated and the pressure at the gas-oil contact is equal to the saturation pressure. Therefore, a small pressure drop in the reservoir, related to fluid withdrawal, causes some solution gas to evolve from the oil phase. In other words, a gas cap drive is always accompanied by some degree of solution gas drive.

The relative impact of the two mechanisms basically depends on the size of the gas cap. The larger the gas cap (with respect to the oil volume), the smaller the pressure drop in the reservoir necessary for the gas cap to expand. Therefore, the larger the relative size of the gas cap, the smaller will be the impact of the solution gas drive process in the reservoir.

A gas cap can be either primary or secondary. In the latter case, a gravitational segregation mechanism must work in the reservoir, which allows for the gas to migrate up structure.

The rock and fluid characteristics necessary for this process to happen are high vertical permeability, favourable oil mobility and low flow velocity.

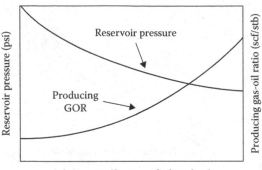

FIGURE 1.31
Production behaviour of a gas cap drive reservoir.

Also, the reservoir must either be thick or have appreciable dip, to provide a high vertical closure.

From a production point of view, a gas cap drive reservoir is usually characterised by a slow but fairly constant pressure decline with cumulative production. It may also be characterised by the production of substantial and increasing quantities of gas, especially from the up-dip wells. However, this is an unwanted situation; therefore, wells are progressively worked over or shut-in, in order to prevent withdrawal of gas from the gas cap and preserve reservoir energy. Water production depends upon the presence and the activity of a natural aquifer, but is generally negligible. Figure 1.31 shows the ideal production behaviour of a gas cap drive reservoir.

1.2.5.4.5 Compaction Drive

Compaction drive is the producing mechanism related to the decrease in pore volume that in some reservoirs is a consequence of fluid withdrawal. To understand the mechanism, consider that the effective pressure P_E acting on the rock at a given depth corresponds to the difference between the total pressure P_T (corresponding to the weight of the overburden formation) minus the fluid pressure P_F (what we ultimately refer to as formation pressure). Therefore, the following simple relationship holds:

$$P_E = P_T - P_F$$

Whenever pressure depletion is observed in the reservoir, this implicitly generates an increase in the effective pressure acting over the rock framework. Depending on the compressibility of the formation, this increase may induce a decrease in the pore volume and therefore provide some energy to the system.

In most cases, the pore volume compressibility of the reservoir formations (not to be confused with the bulk volume compressibility) is of the order of magnitude of 5×10^{-6} vol/psi, which is a small value compared to fluid compressibility, especially gas. For this reason, in most cases the contribution of compaction to the total recovery is small and often negligible. In addition, it is often considered that compressibility is constant with pressure.

However, many significant exceptions have been documented in the literature, where abnormally high values of compressibility provide the reservoirs with an important source of energy. The most well-known examples are the oil fields along the eastern flank of the Maracaibo Lake, Venezuela, the offshore fields near Long Beach, California, and most importantly the Ekofisk Field in the Norwegian sector of the North Sea. In all these fields, compaction of the reservoir formation is associated with a significant subsidence at the surface, which is the most spectacular expression of the underground rock compaction.

The production behaviour of a typical compaction drive reservoir is difficult to define, since it depends upon, among other factors, the particular rock framework and its mechanical properties. In general, however, the pressure and production behaviour is similar to that of water drive reservoirs, with the difference that only little water is produced, related in this case to the expulsion of some connate water or water trapped in shaly layers. In fact, as they are related to overpressure, typical compaction drive reservoirs are often isolated systems, disconnected from regional aquifers.

1.2.6 Oil and Gas Reserve Estimation

1.2.6.1 Reserve

Before estimating the reserve, it is necessary to know what is meant by reserve and how it differs from other similar terms like *resources* and *in situ hydrocarbons*, as they all have very close connotations.

The volume of hydrocarbon *in situ* that is present in a reservoir rock is commonly known in the oil industry as original oil in place (OOIP) or HIIP (hydrocarbon initially in place). From this OOIP, it is only possible to extract a fraction of hydrocarbons, known as *reserves*. So, a factor called *recovery factor* or *recovery efficiency* is used which when multiplied with HIIP, the value of hydrocarbon reserve or reserve is obtained. Thus, reserve may be defined (exact definition given by Society of Petroleum Engineers (SPE)/World Petroleum Congress (WPC) mentioned below) as OOIP volume which is economically feasible to recover when an oil development project is implemented. *Resources*, on the other hand, are all the hydrocarbon volumes still to be discovered (nondiscovered resources) or that are discovered, but lack economic feasibility. The diagram in Figure 1.32 (known as a McKelvey diagram) graphically demonstrates these.

Due to the many forms of occurrence of petroleum, the wide range of characteristics, the uncertainty associated with the geological environment and the constant evolution of evaluation technologies, a precise definition and classification system was found to be nonpractical and hence in 1994 during the World Petroleum Congress, a task force was formed with members

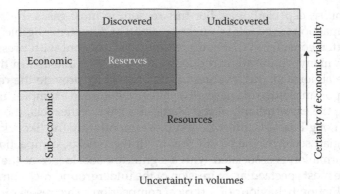

FIGURE 1.32
McKelvey diagram. (From J. S. Gomes and F. B. Alves, *The Universe of the Oil & Gas Industry –
From Exploration to Refining*, Partex Oil & Gas, 2013.)

drawn from organisations like SPE and WPC to draft a common definition of
reserve and their classification which would be acceptable worldwide.

As per SPE/WPC, reserve classification can be summarised as follows.

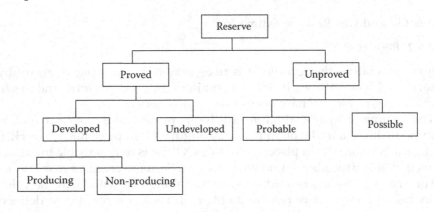

According to SPE/WPC, 'Reserves are those quantities of petroleum which
are anticipated to be commercially recovered from known accumulations from
a given date forward.' All reserve estimates involve some degree of uncertainty. The uncertainty depends chiefly on the amount of reliable geologic and
engineering data available at the time of the estimate and the interpretation
of these data. The relative degree of uncertainty may be conveyed by placing
reserves into one of two principal classifications, either *proved* or *unproved*.

Unproved reserves are less certain to be recovered than proved reserves
and may be further subclassified as *probable* and *possible* reserves to denote
progressively increasing uncertainty in their recoverability. Identifying
reserves as proved, probable and possible has been the most frequent classification method and gives an indication of the probability of recovery.

Because of potential differences in uncertainty, caution should be exercised when aggregating reserves of different classifications.

Reserves estimates will generally be revised as additional geologic or engineering data becomes available or as economic conditions change.

Reserves may be attributed to either natural energy or improved recovery methods. Improved recovery methods include all methods for supplementing natural energy or altering natural forces in the reservoir to increase ultimate recovery. Examples of such methods are pressure maintenance, cycling, water flooding, thermal methods, chemical flooding and the use of miscible and immiscible displacement fluids. Other improved recovery methods may be developed in the future as petroleum technology continues to evolve.

1.2.6.1.1 Proved Reserves

Proved reserves are those quantities of petroleum which, by analysis of geological and engineering data, can be estimated with reasonable certainty to be commercially recoverable, from a given date forward, from known reservoirs and under current economic conditions, operating methods and government regulations. Proved reserves can be categorised as developed or undeveloped.

1. *Developed* – Developed reserves are expected to be recovered from existing wells including reserves behind pipe. Improved recovery reserves are considered developed only after the necessary equipment has been installed, or when the costs to do so are relatively minor. Developed reserves may be subcategorised as *producing* or *nonproducing*.

   Producing – Reserves subcategorised as producing are expected to be recovered from completion intervals which are open and producing at the time of the estimate. Improved recovery reserves are considered producing only after the improved recovery project is in operation.

   Nonproducing – Reserves subcategorised as nonproducing include shut-in and behind-pipe reserves.

2. *Undeveloped reserves*. Undeveloped reserves are expected to be recovered: (1) from new wells on undrilled acreage, (2) from deepening existing wells to a different reservoir, or (3) where a relatively large expenditure is required to (a) recomplete an existing well or (b) install production or transportation facilities for primary or improved recovery projects.

1.2.6.1.2 Unproved Reserves

Unproved reserves are based on geologic and/or engineering data similar to that used in estimates of proved reserves; but technical, contractual, economic or regulatory uncertainties preclude such reserves being classified as proved. Unproved reserves may be further classified as probable reserves and possible reserves.

1. *Probable reserves* – Probable reserves are those unproved reserves which analysis of geological and engineering data suggests are more likely than not to be recoverable. In this context, when probabilistic methods are used, there should be at least a 50% probability that the quantities actually recovered will equal or exceed the sum of estimated proved plus probable reserves.

2. *Possible reserves* – Possible reserves are those unproved reserves which analysis of geological and engineering data suggests are less likely to be recoverable than probable reserves. In this context, when probabilistic methods are used, there should be at least a 10% probability that the quantities actually recovered will equal or exceed the sum of estimated proved plus probable plus possible reserves.

1.2.6.2 Reserve Estimation

Commonly used reservoir performance and reserve estimation techniques are

1. Volumetric method
2. Material balance method
3. Decline curve analysis method
4. Reservoir simulation technique

Generally, hydrocarbon volume is estimated by both volumetric and material balance method, then an attempt is made to determine the reason for discrepancy between the two calculations. If the material balance should give a larger reservoir than the volumetric calculation, it is possible that either part of the oil reservoir has not yet been located or an unrecognised influx of water exists.

1.2.6.2.1 Computation of Reservoir Volume by the Volumetric Method

The bulk volume of a hydrocarbon bearing reservoir is determined by its physical boundaries as given by the reservoir structure map and position of oil-water contact (OWC). The structure map is a projection of the surface of a reservoir boundary on a horizontal reference plane. It displays lines of equal elevation (contours), relative to sea level, of boundary surface. The oil or gas-bearing net pay volume for a reservoir may be computed in several different ways.

1. From subsurface data, a geological map is prepared (Figure 1.33), contoured on the subsea depth of the top of the sand (solid lines)

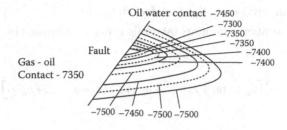

Oil water contact −7450
−7300
−7350
−7350
−7400
−7400

Fault

Gas - oil
Contact - 7350

−7500 −7450 −7500 −7500

Geological contour map on top
(—) and base (····) of reservoir

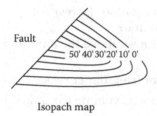

Fault

50' 40' 30' 20' 10' 0'

Isopach map

FIGURE 1.33
Contour map and isopach map.

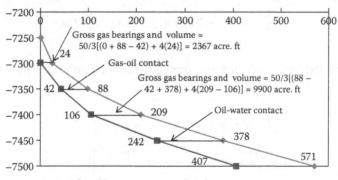

Area enclosed by contour: Acre-feet diagram

FIGURE 1.34
Volume calculation using Simpson's rule.

and on the subsea depth of the base of the sand (dashed lines). The total area enclosed by each contour is then planimetered and plotted as abscissa on an acre-feet diagram (Figure 1.34) against the corresponding subsea depth as the ordinate. Gas-oil and oil-water contacts as determined from core, log or test data are shown as horizontal lines. After connecting the observed points, the combined gross volume of oil and gas-bearing sand may be

A.   Planimetered from the acre-feet diagram
B.   If the number of contour intervals is even, computed by Simpson's rule.

$$\text{Volume} = \frac{h}{3}\left[a_0 + 4(a_1 + a_3 + ... + a_{n-1}) + 2(a_2 + a_4 + ... + a_{n-2})\right] + t_n a_n$$

where
h = contour intervals (feet)
a_0 = area enclosed by zero contour (at oil-water contact-acres)
a_1 = area enclosed by first contour (acres)
a_2 = area enclosed by second contour (acres)
a_n = area enclosed by nth contour
t_n = average formation thickness above the top contour

Simpson's rule is accurate for irregular curves.

C. With somewhat less accuracy, volume calculated by the trapezoidal rule.

$$\text{Volume} = 1/2 \ h \ (a_0 + 2a_1 + 2a_2 + ... + 2a_{n-1} + a_n) + t_n a_n ...$$

$$V = \frac{h}{2}\left[a_0 + a_n + 2(a_1 + a_2 + ... + a_{n-1})\right] + t_n a_n$$

D. Computed by means of the somewhat more complicated pyramidal rule

$$V = \frac{h}{3}\left(a_0 + a_n + 2[a_1 + a_2 + ... + a_{n-1}]\right.$$

$$\left. + \left[\sqrt{a_0 a_1} + \sqrt{a_1 a_2} + ... + \sqrt{a_{n-1} a_n}\right]\right) + \frac{h_n}{3} a_n$$

For frustum of pyramid with area a_0 and a_1

$$V_{0-1} = \frac{h}{3}\left(a_0 + a_1 + \sqrt{a_0 a_1}\right)$$

For the best accuracy, the pyramidal formula should be used. However, because of its simpler form, the trapezoidal formula is commonly used which introduces an error of <2% when the ratio of successive area of contours is 0.5. If the ratio of areas of any two successive isopach lines is less than 0.5, the pyramidal formula is applied.

From a study of the individual well logs or core data it is then determined what fraction of the gross sand section is expected to carry and produce hydrocarbons. Estimating this net-pay fraction is often aided greatly when micro- or contact logs are available. In shaly sand sections, the area under the SP curve above the shale line is sometimes used as a yardstick for the net-pay fraction. Multiplication of this net-pay factor with the gross sand volume yields the *net pay volume*.

1.2.6.2.1.1 Volumetric Method of Reserve Estimation The volumetric calculation of oil or gas reserve consists primarily of determining the volume of rock holding hydrocarbons, the total voids in such rock, the volume percentage of the voids containing hydrocarbons and the percentage of these hydrocarbons economically recoverable in the stock tank. Reserves are normally expressed in stock tank barrels (STB) or stock tank cubic meter for liquid, or million standard cubic feet (MMSCF) or million standard cubic meter (MMSCM) for gas.

(a) Oil in reservoir (no free gas present in oil-saturated portion)

$$N = \frac{7758 V_o \emptyset_R (1 - S_w)}{B_o}, \ STB$$

where 7758 is the number of barrels per acre-foot and ϕ is the porosity, S_w is the interstitial water saturation, B_o is the formation volume factor, $\left(FVF = \frac{ReservoirVolume}{StocktankVolume} \right)$ V_o is net pay volume of the oil bearing portion of reservoir, acre-foot.

(b) Free gas in reservoir or gas cap (no residual oil present)

$$G = \frac{43560 V_g \emptyset (1 - S_w)}{B_g}, \ SCF$$

where 43,560 is the number of cubic feet per acre-foot and V_g = net pay volume of the free-gas bearing portion of a reservoir, acre-foot, B_g = gas formation volume factor.

(c) Solution gas in oil reservoir (no free gas present)

$$G_g = \frac{7758 V_o \emptyset_R (1 - S_w) R_s}{B_o}, \ SCF \ of \ solution \ gas$$

where R_s = gas solubility factor.

1.2.6.2.2 *Material Balance Method*

This method is used only after the hydrocarbon production has started and the production and reservoir data are gathered. The material balance is one of the basic means of predicting reserves, as well as future reservoir performance. Theoretically, it contains no errors, practically speaking though, it is never exact because of errors in the data used and the assumptions necessary to express it in terms of measurable properties. This method is based on the premise that the pore volume of a reservoir remains constant or changes in a predictable manner with the reservoir pressure when oil, gas and/or water are produced. This makes it possible to equate the expansion of the reservoir voidage caused by the withdrawal of oil, gas and water minus the water influx. The material balance equation (MBE) in its most general form reads:

$$N = \frac{N_P \left[B_t + (R_P - R_{Si}) B_g \right] - (W_e - W_P B_W)}{(B_t - B_{ti}) + \dfrac{mB_{ti}}{B_{gi}} (B_g - B_{gi})}$$

where
    N = reservoir oil in place, STB
    N_p = cumulative oil produced, STB
    m = ratio between initial reservoir free gas volume and initial
  reservoir oil volume $= \dfrac{GB_{gi}}{NB_{oi}}$

    G = Initial gas cap volume, SCF
    B_t = two-phase formation volume factor for oil = $B_o + (R_{si} - R_s)B_g$
    B_g = gas formation volume factor
    B_o = oil formation volume factor
    B_w = water formation volume factor

    R_p = cumulative gas-oil ratio, SCF/STB $= \dfrac{G_P}{N_P}$
    R_s = gas-solubility factor
    W_e = cumulative water influx, bbl
    W_p = cumulative water produced, bbl
    G_P = Cumulative produced gas, SCF

The calculation of N is one of the basic uses of balances. The simplest, rough approach is to assume that m and W_e are zero.
    Thus, MBE reduces to

$$N = \frac{N_P \left[B_t + (R_P - R_{si}) B_g \right]}{(B_t - B_{ti})}$$

A more realistic approach is to determine m volumetrically.

The assumption that W_e equals zero initially is not bad, for no large amount of water influx should be expected during the early stages of depletion. The validity of this assumption may be checked in two ways. One is to compare the N thus found with the volumetric value of N. Another is to calculate N over successive intervals. If N remains substantially constant, we may reasonably deduce that no water drive exists. Where a water drive does exist, values of N determined may be plotted against some time function such as N_p. Extrapolation back to $N_p = 0$ then yields a good probable value.

Regardless of the method used, the final value of N chosen should be one that is consistent with all data. In the very early stages of depletion, the volumetric method is usually the most accurate, whereas the material balance gains preference in the later stages of depletion.

In the study of reservoirs which are produced simultaneously by the three major mechanisms, for example, depletion drive, gas cap drive and water drive, it is of practical interest to determine the relative magnitude of each of these mechanisms which contribute to the production. This MBE can be rearranged to obtain three fractions, whose sum is one, which is called the depletion drive index (DDI), the segregation (gas cap) index (SDI) and the water drive index (WDI).

That is,

$$DDI + SDI + WDI = 1.$$

1.2.6.2.3 *Production Decline Curve Analysis*

Estimates of ultimate recovery by extrapolation of a performance trend fundamentally all follow the same platform. The two quantities one usually wishes to determine are either remaining oil reserves or remaining productive life. Cumulative production and time are therefore normally selected as independent variables and are plotted as abscissas. A varying characteristic of the well performance which can be easily measured and recorded is then selected as a variable to produce a trend curve.

By plotting values of this continuously changing dependent variable as ordinates against the values of the independent variable (cumulative production or time) as abscissas, and graphically extrapolating the apparent trend until the known end point is reached, an estimate of the remaining reserves can be obtained. Amongst the many dependent variables which can be used in estimates based on performance trends, the rate of production is by far the most popular when production is not restricted. In that case one commonly refers to production decline curves. The two main types are rate-time and rate-cumulative curves for each of the two independent variables. Rate of oil production as the dependent variable has the advantage of always being readily available and accurately recorded.

Gradual changes in the production rate of a well may be caused by:

1. Decreasing efficiency or effectiveness of the lifting equipment.
2. Reduction of productivity index, or completion factor, or increase in the skin effect due to physical changes in and around the wellbore such as deposition of wax, salt or asphaltenes from the produced fluids or the accumulation of loose sand, silt, mud or cavings.
3. Changes in bottom-hole pressure, gas-oil ratio, water percentage or other reservoir conditions.

Unless defective conditions of the wellbore are detected or cured, the reserve estimates obtained by decline-curve analysis will be limited to those recoverable under existing and sometimes only partially effective wellbore conditions.

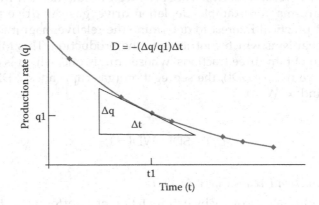

FIGURE 1.35
Rate of decline definition (normal plot).

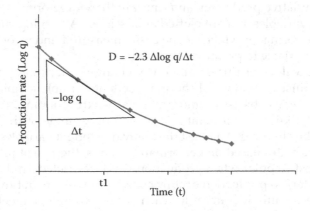

FIGURE 1.36
Rate of decline definition (semi-log plot).

1.2.6.2.3.1 Nominal and Effective Decline There are two types of decline. The nominal decline rate D is defined as the negative slope of the curve (Figures 1.35 and 1.36) representing the natural logarithm of the production rate q versus time t, or

$$D = -\frac{d(\ln q)}{dt} = -\frac{\frac{dq}{dt}}{q}$$

Nominal decline, being a continuous function, is used mainly to facilitate the derivation of the various mathematical relationships.

The effective decline rate De, being a stepwise function and therefore a better agreement with actual productive recording practices, is more commonly used in practice. It is the drop-in production rate from q_i to q_1 over a period of time equal to a unit (1 month or 1 year) divided by the production rate at the beginning of the period.

$$D_e = \frac{q_i - q_1}{q_i}$$

The time period may be 1 month or 1 year for effective monthly or annual decline, respectively.

1.2.6.2.3.2 Different Types of Production-Decline Curves Three types of production–decline curves are commonly recognized (Figure 1.37):

1. Constant-percentage decline (Exponential decline):

    With constant percentage decline, the nominal decline rate D does not change with time

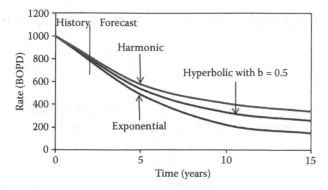

FIGURE 1.37
Three types of production decline curves.

$$D = \frac{-\dfrac{dq}{dt}}{q}$$

which after integration leads to the rate-time relationship,

$$q = q_i e^{-Dt}, \text{ Or, } \ln\left(\frac{q_i}{q}\right) = Dt$$

After integrating a second time, the cumulative production at time t is obtained as expressed by the rate-cumulative relationship

$$N_P = \frac{q_i - q}{D}$$

where q_i is the rate at the start of the decline and q is the rate at the end of the period.

2. Hyperbolic decline:

With hyperbolic decline, the nominal decline rate D is proportional to a fractional power n of the production rate, this power being between 0 and 1

$$D = \frac{-\dfrac{dq}{dt}}{q} = bq^n$$

in which the constant 'b' under initial conditions $= \dfrac{D_i}{q_i^n}$

After integration, the following rate-time relationship is obtained:

$$q = q_i(1 + nD_i t)^{-\frac{1}{n}}$$

After a second integration, the cumulative production at time t is obtained as expressed by the rate-cumulative equation

$$N_P = \frac{q_i^n}{(1-n)D_i}\left(q_i^{1-n} - q^{1-n}\right)$$

3. Harmonic decline (n = 1):

With harmonic decline, the nominal decline rate D is proportional to the production rate q. Hence, the rate-time equation is

$$q = q_i(1 + D_i t)^{-1}$$

and the cumulative production equation is

$$N_P = \frac{q_i}{D_i} \ln\left(\frac{q_i}{q}\right)$$

1.2.6.2.4 *Reservoir Simulation Technique*

Reservoir simulation is no more than a methodology, a valuable tool to predict the behaviour of oil reservoirs that relates the changes in reservoir pressure to the production and injection levels, to the reserves and to the properties of rock and fluids, such as porosity, saturation and permeability.

This relationship is also no more than a mere material balance that is so familiar to engineering students, that is, what comes in minus what goes out of a defined volume is equal to what accumulates in that volume. In the case of reservoir engineering, the flows entering or leaving the volume are given by the Darcy equation, which models the flow of fluids in porous media. On the other hand, what accumulates is a function of the fluid volume, the compressibility of the medium (rock and fluids) and the pressure variation in the volume in question.

As can be seen, the basic principles of reservoir simulation are simple. The complexity begins when the reservoir is described in geological terms and in terms of the characterisation of the variations in rock and fluid properties, both laterally and in depth. To be able to model this situation mathematically, the three-dimensional physical model of the reservoir properties must be broken down into small intercommunicating blocks forming a three-dimensional mesh or grid (Figure 1.38). Each grid block is processed in accordance with the above principles. The result is a system of equations that model the flow of fluids between each block and its adjacent blocks, and the pressure variation of each block due to the compressibility and accumulation of fluids.

Since the number of grid blocks of a model is usually very large, computers are needed to resolve this matrix, using software specifically developed for this purpose, the so-called *reservoir simulator*. With the advent of increasingly powerful computers, these reservoir models can be refined to more than a million grid blocks.

Reservoir simulation is used not only in analysing the behaviour of the reservoir under different production mechanisms, but also in planning the development of an oilfield or in planning monitoring campaigns.

FIGURE 1.38
Three-dimensional reservoir simulation model. (From J. S. Gomes and F. B. Alves, *The Universe of the Oil & Gas Industry – From Exploration to Refining*, Partex Oil & Gas, 2013.)

Reservoir simulation serves essentially to help engineers to understand the production and pressure mechanisms of a reservoir and, therefore, to predict production profiles as a function of time, which, in the end, correspond to the reserves extracted. When production estimates are evaluated in economic terms, they enable the oil and gas company to define not only the economically recoverable reserves, but also the more profitable levels of production. When properly used, reservoir simulators can in fact provide managers with valuable tools for decision-making regarding the development of oil and gas fields.

A reservoir simulator is software that has been developed to construct mathematical models of the physical reality of the reservoirs and of the dynamic processes underlying production. The basic equations of the reservoir simulator are described below and applied to each grid block of the reservoir model.

- Material balance equation (input – output = what is produced/injected + what is accumulated)
- $Q = k. A. \Delta P/\mu. L$ (Darcy's equation), which calculates the flow rate of fluids between a grid block of a reservoir model and any adjacent block through their common interface as follows:

  Q = Fluid flow rate

  k = Permeability of the porous medium

  A = Section available for the fluid to flow, usually the contact area between the blocks

ΔP = Pressure difference between the block in question and the adjacent block

μ = Fluid viscosity

L = Distance between the centres of the blocks

The simulator calculates the flow of fluids in the reservoir. The principles underlying the simulation are simple. The equations of multiphase flow between each block and neighbouring blocks are expressed in partial differential equations for each phase (oil, gas and water) over time. As already mentioned, these equations are obtained from the conventional equations that describe the behaviour of fluids, such as Darcy's equation for the flow of fluid in porous media. In addition to this equation, the mass and energy conservation principle makes it possible to calculate the accumulation in each block of the simulator, resulting in a variation in pressure or temperature.

The partial differential equation of each phase is then formulated in the form of finite differences; the reservoir volume is related to a vast number of blocks in the simulation model and the reservoir's production period is divided into several time steps. Mathematically speaking, the problem is *discretised* in space and time and can be solved using numerical methods.

Figure 1.39 shows the equation that is the basis of a simulator. In a simplified situation in two dimensions and one phase with identical blocks, the flow is established between blocks i,j and the four adjacent blocks with which it has contact surfaces and through which fluids can flow.

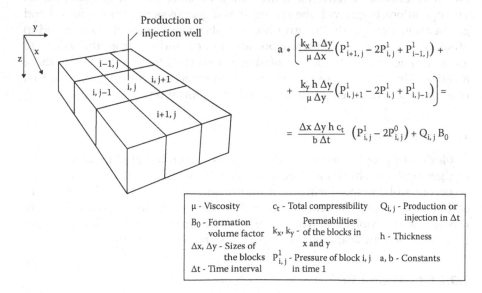

$$a * \left[\frac{k_x\, h\, \Delta y}{\mu\, \Delta x} \left(P^1_{i+1,\,j} - 2P^1_{i,\,j} + P^1_{i-1,\,j} \right) + \frac{k_y\, h\, \Delta y}{\mu\, \Delta y} \left(P^1_{i,\,j+1} - 2P^1_{i,\,j} + P^1_{i,\,j-1} \right) \right] =$$

$$= \frac{\Delta x\, \Delta y\, h\, c_t}{b\, \Delta t} \left(P^1_{i,\,j} - 2P^0_{i,\,j} \right) + Q_{i,\,j}\, B_0$$

μ - Viscosity	c_t - Total compressibility	$Q_{i,\,j}$ - Production or injection in Δt
B_0 - Formation volume factor	Permeabilities k_x, k_y - of the blocks in x and y	h - Thickness
Δx, Δy - Sizes of the blocks	$P^1_{i,\,j}$ - Pressure of block i, j in time 1	a, b - Constants
Δt - Time interval		

FIGURE 1.39
The basis of the formulation of a simulator. (From J. S. Gomes and F. B. Alves, *The Universe of the Oil & Gas Industry – From Exploration to Refining*, Partex Oil & Gas, 2013.)

The solution method is therefore based on a system of finite difference equations generated for all the model blocks. These are large matrices that can be solved by easily programmable numerical methods.

Simulators are divided into three main categories. The first category includes *three-phase simulators*, which simulate three phases – oil, gas and water – where the gas can enter or leave the oil stream, depending on the pressure. Models of this type are usually characterised as *black oil*, since they treat the oil as a single phase, which is usually associated with its dark colour.

Compositional and thermal simulators fall into the second category, which applies to reservoir models involving special development processes. The third category includes *stream line simulators*, often used in preliminary studies, especially in the preselection and ranking of static models before the dynamic models are built. In terms of formulation, these various types of simulators involve different aspects.

1.3 Offshore Oil and Gas Operations

Oil and gas operations cover a vast spectrum of activities starting with the establishment of the existence of this natural resource in the mother Earth up to utilising it in the best possible way. These operations when carried out in an offshore environment are called offshore oil and gas operations. The operations begin with the geologist and geophysicist hunting for oil and gas bearing strata under the ground in onshore and below the ocean bed in offshore and ends with the transportation of oil and gas from the source to its destination or in the case of offshore from the offshore site to the shore location. These operations can be broadly divided into four main areas, that is, exploration, drilling, production and transportation.

1.3.1 Exploration

Exploring or prospecting for hydrocarbons is an integrated process involving geology, geophysics and geochemistry with the objective of finding a structure suitable for storing an economically sufficient quantity of hydrocarbons (oil or gas). On land, sometimes shallow *in situ* testing is used in conjunction with geophysical work whereas offshore prospecting almost always relies on a geophysical method – primarily the seismic method.

1.3.1.1 Geological Method

Geology is the science of rocks. The geologist carries out a survey either at the ground surface aerially or through satellite with the aim to identify the formations, study the stratigraphy, sedimentology and the structures like fold, fault and so on. For establishing the geometry of the structures, core samples are also drilled out. In offshore, when an area is found to be favourable from geophysical studies, coring is done from core-drilling ships.

These specialised ships can remain dynamically positioned on location and drill in seas with 30 ft (9 m) waves and in depths approaching 4000 ft (1200 m).

1.3.1.2 Geophysical Method

The different geophysical methods that can be applied to oil and gas exploration are gravimetric, magnetic, electromagnetic, electrical (resistivity) and seismic out of which the seismic method is the most commonly used in oil and gas exploration. It is based on the principle of the propagation of seismic waves underground or under the ocean bed.

A seismic survey is a program for mapping geologic structures by creating seismic waves and observing the arrival time by the waves reflected from acoustic-impedance contrasts or refracted through high velocity layers.

1. Seismic refraction – This method is essentially used for shallow investigations. Sometimes, it is applied as a reconnaissance tool in petroleum exploration.
2. Seismic reflection – This method is used as a day-to-day exploration tool to map shallow as well as deep subsurface geologic structures.

In most cases, the seismic reflection technique is used extensively for hydrocarbon exploration to map subsurface structures suitable for oil/gas entrapment in the sedimentary basins in onshore and offshore areas.

Seismic exploration consists of three stages of operation:

1. Data acquisition
2. Data processing
3. Data interpretation

1.3.1.2.1 Seismic Data Acquisition

Seismic data acquisition begins with a seismic source creating seismic waves and ends with the recording of the echoes from beneath the surface that result from the source disturbances. There are several techniques for generating artificial seismic waves. The most common are chemical explosives such as dynamite, mechanical vibrators (vibrator trucks) and compressed air guns (Figure 1.40), which are used at sea.

The Earth's motion is detected by a transducer called a geophone on land and by pressure detector called a hydrophone at sea. Hydrophones are pressure detectors that are sensitive to variation in pressure. Normally, hydrophones are kept a few feet below the water surface in marine operations. Piezo-electric and magnetostriction types are common in hydrophone.

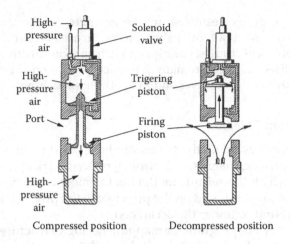

High-pressure air
Solenoid valve
High-pressure air
Trigering piston
Port
Firing piston
High-pressure air

Compressed position   Decompressed position

FIGURE 1.40
Compressed air guns.

The sensing element in piezo-electric crystal is barium titanate and lead zirconate.

From the detectors, that is, geophones or hydrophones, the signal is transmitted either by cable or by telemetry to a seismic recording unit, where the signals are combined and recorded on magnetic tape (Figure 1.41). Modern seismic data is acquired by digital recorders capable of recording the entire seismic signal encountered in oil exploration. The most common feature of any digital recorder (digital seismograph) is the *dynamic range* of the system which is the ratio of the largest recoverable signal (for a given distortion level) to the smallest recoverable signal. The smallest recoverable signal is often taken as the noise level of the system. The dynamic range of any system is expressed in decibels as

$$db = 20\log_{10}\frac{A}{B}$$

where A and B are the largest and smallest recoverable signal.

Before digital recording, the signals pass through an analogue-to-digital converter. The graph that records geophone/hydrophone vibration is commonly called a *trace*. For marine operations (Figure 1.42), digital streamers of 240 channel capacity are used. Due to fast technological advances, new seismic recording systems are now available for data acquisition to overcome some existing problems and can be used in difficult areas.

Until very recently, most seismic data acquisition campaigns were processed in two dimensions (2D campaigns) and it was not until 1972 when with the advent of new technology and powerful computers capable of processing and interpreting data in three dimensions, 3D data acquisition

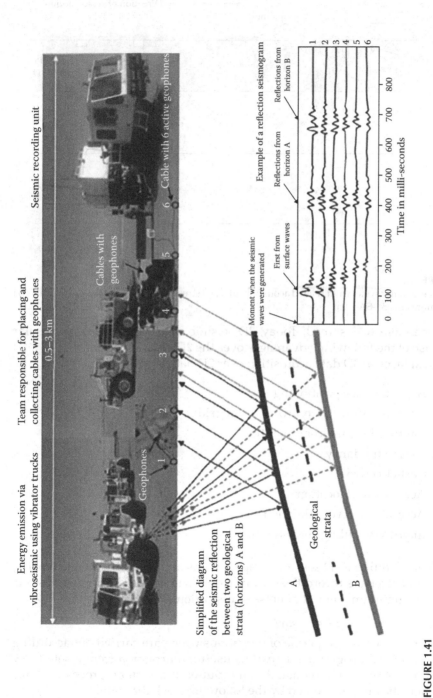

FIGURE 1.41
Seismic acquisition process. (From J. S. Gomes and F. B. Alves, *The Universe of the Oil & Gas Industry – From Exploration to Refining*, Partex Oil & Gas, 2013.)

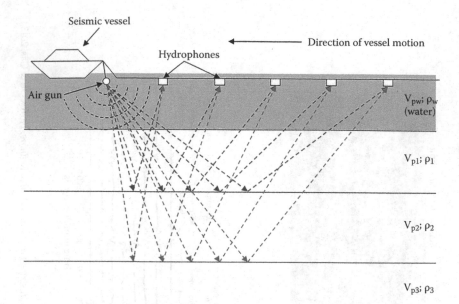

FIGURE 1.42
Marine streamer system and related acoustic sources of streamer noise. (Courtesy of Wikimedia Commons/CC-BY-SA-3.0.)

became more widespread. Today, 3D seismic campaigns are very popular because of the following advantages over the 2D technique.

Advantages of 3D data acquisition over 2D data acquisition:

1. Accurate trace positioning
2. Improved velocity control (1/2 km grid)
3. Improved signal-to-noise ratio
4. Data redundancy
5. Cost-effective
6. Accurate well location
7. Accurate reserve estimation
8. Improved drilling success ratio

When planning a 3D seismic campaign, it is important that the geometry, shots and receivers record the seismic signal with the widest range possible and the minimum amount of noise or distortion.

1.3.1.2.2 Seismic Data Processing

Data processing is a sequence of operations which are carried out according to a predefined programme to extract useful information from a set of raw (normally observational) data. As an input-output system, processing may be schematically represented by the following block diagram.

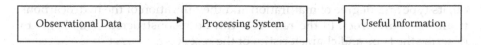

The type of processing is mainly controlled by the particular software, whereas the actual execution, size of the processed data and speed of computation are mainly governed by the system hardware, that is, the computing equipment.

Referring to the block diagram, the observational data which are records made in the field, act as an input to the processing system and the seismic stacked sections representing as truly and clearly as possible the subsurface geology of the area act as the output. The input may be raw data, either 2D or 3D needs suitable processing sequences to produce a stacked section.

1.3.1.2.3 Seismic Data Interpretation

The interpretation of seismic data in geological terms is the objective and final product of seismic exploration with the purpose of locating structures which provide excellent traps containing commercial quantities of oil or gas. However, many such structural traps do not contain oil or gas in economic quantities. Since drilling of an oil/gas well is very costly, the effort is made to derive maximum information about the geological history of the area and about the nature of rocks in an effect to form an opinion about the probability of encountering petroleum in that structure.

Seismic data are usually interpreted by geophysicists or geologists with a strong background of data acquisition and data processing operations. The object of seismic interpretation is usually to prepare two-way reflection time contour (isochron) maps showing the depth to a series of reflectors which have been picked on the stacked seismic sections.

To carry out this process of seismic survey in offshore, a ship equipped with all of the above facilities on board known as a seismic vessel is deployed.

1.3.1.3 Geochemical Method

In this method, the aim is to find traces of hydrocarbons in the air and in the soil, as a result of their migration to the surface through a system of fractures. When hydrocarbons migrate to the surface, they leave traces of hydrocarbons in the soil.

When properly analysed from a geochemical point of view (spectrography of the heavy components C15 – C30), the type and degree of maturation of the hydrocarbons become known. When samples are collected in depth, near the source rock, it is also possible to assess the type and degree of maturation with the help of petrographic analyses (reflectivity of organic compounds such as vitrinite).

When the geochemical and petrographic studies are properly integrated with migration models, conceptual models can be drawn up on the origin of oil, its type and degree of maturation and the migration of the hydrocarbons from the source rock to the reservoir can be reconstructed. This helps to describe the type and characteristics of the reservoir.

1.3.2 Drilling

Once the indication is received from geologists and geophysicists about the existence of hydrocarbon-bearing structures within an economic range, to validate their conceptual model an exploratory well needs to be drilled to provide confirmatory evidence about the presence of oil and gas and their commerciality. Thus, to discuss briefly drilling (discussed in detail in a later chapter), depending on the purpose and the timing, it can be broadly divided into two types:

1. Exploratory drilling
2. Development drilling

1.3.2.1 Exploratory Drilling

To confirm or deny the presence of hydrocarbons, exploratory wells are drilled. This is the first well drilled in an unknown structure, so it is sometimes called a wild-cat well. During such drilling, a lot of information is gathered regarding different formations encountered with depth and simultaneously logging, coring, testing and so on are also done. These data or information help in studying the commercial viability of the new field. A few exploratory wells other than the wild-cat wells are drilled to delineate the reservoir boundary which helps in planning for further development of the field in a most economical way so that the life of the field can be expanded. Since the investment made in drilling a single well is very high, it is a matter of concern for the investor or the owner. As far as onshore exploratory wells are concerned, the decisions do not depend much on the type of equipment except their capacities but in offshore, the deployment of the type of platform is to be made judiciously.

Here it is not advisable to deploy a *fixed platform* because in the case of non-commerciality, the platform has to be abandoned. Therefore, in this case, it is the *mobile units* that need to be used like a jack-up rig or submersible unit or floating units, for example, *semi-submersible* or *drill ship*. Even if no commercial hydrocarbon is found, these mobile units can be shifted to some other drilling site and used.

1.3.2.2 Development Drilling

Once the results from exploratory wells are positive, then a detail development planning is done so that exploitation is made in the most profitable way. The most important is to mark the position of development wells on the structural map and this depends on whether it is onshore or offshore. In offshore generally development drilling is done from a fixed platform preferably self-contained from where many wells are drilled to reach different locations in the reservoir by using the method of *directional drilling*. This form of drilling has the advantage of having the flow from many wells converge to one surface location for production. Every self-contained platform can permit drilling a minimum of 8 to 10 wells; there are other types of fixed platforms which are also used for development drilling, details of which are discussed in later chapters.

1.3.3 Production

Immediately after development drilling during which different casings are lowered to reach up to the target formation, production tubing is lowered from the surface which becomes the main communication path for the well fluid to flow up to the surface. Before starting actual production of well fluid, which is nothing but a mix of oil, gas and water, the well needs to be completed in all respects which is known as well *completion*. Basically, it consists of certain operations like perforation, setting packers, installing a 'Christmas tree' at the top of the wellhead and so on, the details of which are covered in subsequent chapters. In offshore, depending on whether the Christmas tree is placed above the water surface or below, these are termed *deck level completion* or *subsea completion*.

Broadly speaking, the production operations that are carried out during the producing life of a well can have two broad divisions: (1) down the hole operations and (2) surface operations. Those operations that are carried out through the wellbore during its producing history like workover operations, wireline operations, stimulation, artificial lift and so on, fall under the first category whereas those operations that are carried out at the surface like separation, treatment, storage and transport are commonly termed surface operations.

As far as down-the-hole operations are concerned, there is no difference between onshore and offshore so long it is deck level completed but in the case of subsea completed wells a conduit from the subsea well up to the deck of a floater is necessary.

When surface operations which are mainly processing of well fluid are to be carried out the arrangement is totally different in the case of offshore as compared to onshore though conceptually they are similar. Here many wells are completed in a single platform, if deck level completed. These platforms are called *well platforms*, which normally house 4–8 wells and are normally unmanned, that is, there is no provision for living quarters for the working personnel. Throughout the offshore field, many such well platforms are visible.

From such well platforms well fluids flow to a large self-contained platform known as a *process platform*, which are multi-decked and have separators needed for separation of well fluids into oil, gas and water, treatment facilities for oil, gas and water, storage (not very large) facility for storing oil, flares for burning gas, auxiliary equipment like pumps, compressors, power generators, material handling equipment like cranes and so on, safety devices, telecommunication devices, other utilities, helipad living quarters and so on.

In deep waters, instead of a fixed process platform, a floating production system (FPS) is used where mostly floater ships are used whose decks consist of all the surface facilities required for separation, treatment and so on, including facilities for living quarters, safety devices, utilities and other auxiliary support systems. Sometimes the hull of these ships is used for storage of oil and contains facilities for offloading the crude as well. In such cases, they are termed floating, production storage and offloading (FPSO).

These FPSs receive the well fluid from subsea completed well through different production risers which are flexible pipes of different configurations and sizes. For offloading also other risers are used which are commonly termed export risers. An elaborate discussion on these have been made in subsequent chapters.

1.3.4 Transportation

Whenever transportation is talked about in the oil and gas industry it either refers to liquid and gas transportation or worker transportation. For liquid and gas transportation, it starts from the well itself. Referring to offshore, each well platform is connected to a process platform by production risers and a flowline that carries well fluid. From the process platform the separated oil and gas separately move through individual pipelines either to the shore or any storage devices. Sometimes ships are also used to transport the separated oil in the absence of any underwater pipeline. So, flowlines, risers, ships or barges are the major modes of transportation in the first case.

In the second case, worker transportation from shore to platform or from one platform to the other is accomplished either by boats or helicopters. High-speed crew boats transport work crews when time is available and the distance is less than about 50 miles (80 km). Helicopters transport crews and other personnel over long distances and/or when time is important. Simultaneous to the worker transportation, materials (consumables/nonconsumables) and equipment transportation is also accomplished with work boats. These boats are generally about 30 ft (9 m) wide and 140 ft (43 m) long. They are versatile, high-powered and essential to offshore operations. These are sometimes referred to as OSV (offshore supply vessel) also. Thus, all fixed platforms where boats are to anchor must be provided with mooring bitts, bumpers, cranes, stairs and so on.

Bibliography

Ahmed, T. *Reservoir Engineering Handbook*. Elsevier, Inc., 2006.
Burcik, E. J. *Properties of Petroleum Reservoir Fluids*, John Wiley & Sons Inc., 1957.

Clayton, B. R. and R. E. D. Bishop, *Mechanics of Marine Vehicles*, Gulf Publishing Company, 1982.

Cole, F. W. *Reservoir Engineering Manual*, Gulf Publishing Company, 1969.

Cosentino, L. *Integrated Reservoir Studies*, Editions Technip, Institut Francasis Du Petrole Publications, 2001.

Cosse, R. *Oil and Gas Field Development Techniques – Basics of Reservoir Engineering*, Editions Technip, Institut Francais Du Petrole, 1993.

Craft, B. C. and M. F. Hawkins, *Petroleum Reservoir Engineering*, Prentice-Hall Inc., 1959.

Fyfe, W. S., N. J. Price, W. S. Thompson. *Fluids in the Earth's Crust*. Elsevier, Amsterdam, 1978.

Gatlin, C. *Petroleum Engineering – Drilling & Well Completions*, Prentice-Hall Inc, 1960.

Ghosh, S. K. *Structural Geology: Fundamentals and Modern Developments*. Pergamon Press, Oxford, 1993, pp. 494–501.

Glasby, G. P. Abiogenic origin of hydrocarbons: An historical overview. *Resource Geology*, 2006, 56, 85–98.

Gold, T. and S. Soter. The deep – earth-gas hypothesis, *Sci. Amer* 1980, 242(6), 130–137.

Gold, T. Terrestrial sources of carbon and earthquake outgassing. *Journal of Petroleum Geology*, 1979, 1(3), 3–19.

Gold, T. The origin of methane in the crust of the earth. USGS, Prof. Paper, 1993, 1570, 57–80.

Gomes, J. S. and F. B. Alves. *The Universe of the Oil & Gas Industry – From Exploration to Refining*, Partex Oil & Gas, 2013.

Graff, W. J. *Introduction to Offshore Structures*, Gulf Publishing Company, 1981.

History of the Offshore Industry 2014. http://www.offshore-mag.com/index/about -us/history-of-offshore.html.

History of the Petroleum Industry 2014. http://en.wikipedia.org/wiki/history-of -the petroleum-industry.

Hoyle, T. *Mathematics of Evolution*. Acron Enterprises, LLC, Memphis, TN, 1999.

Kudryavtsev, N. A. Oil, Gas and Solid Bitumen in Igneous and Metamorphic Rocks, *Vses. Nauchno – Issled. Geol. Razed*. 1959, Inst. No. 142, 263p.

Levorsen, A. I. *Geology of Petroleum*, W. H. Freeman and Company, 1954.

Mendeleev, D. L'Origine du petrole. Rev. Sci., 2e Ser. 1877, 8, 409–416.

Metre, W. B. *History of Oil Industry, with Particular Reference to India*, Oil India Ltd., publication NOSTALGIA, 1959 – 2009, pp. 22–23.

Mitri, G., A. P. Showmana, J. I. Luninea and R. D. Lorenza. Hydrocarbon lakes on Titan, *Icarus* 2007, 186, 385–394.

Mukherjee, P. K. *A Textbook of Geology*, The World Press Private Ltd., Kolkata, 2008.

North, K. F. *Petroleum Geology*. Allen & Unwin, Boston, MA, 1985, pp. 40–41.

Offshore Petroleum History – American Oil & Gas History 2014. http://aoghs.org /offshore-history/offshore-oil-history.

Selley, R. C. *Elements of Petroleum Geology*. Academic Press, 1998, pp. 1–470.

Singh, R. K. and S.K. Padhi. Characterisation of Jatropha oil for the preparation of biodiesel. *Natural Product Radiance*, 2009, 8(2), 127–132.

Sokoloff, N. V. Kosmischer Ursprung der Bituminas', *Bull. Soc. Imp. Nat. Moscou*, nuov. Ser. 1890, 3, 720–739.

SPE/WPC Reserves Definitions Approved, *Journal of Petroleum Technology*, May, 1997, 527–528.

Tomasko, M. G. et al. Rain, winds and haze during the Huygens probes descent to Titan's surface. _Nature_ 2005, 438, 765–778.

Turner F. J. and J. Verhoogan. _Igneous and Metamorphic Petrology_, McGraw-Hill, 1969.

Vernadsky, V. I. _The History of Minerals in the Earth's Crust_. Vol. 2, Part-I, Moscow-Leningrad, 1933.

Whitehead, H. _An A – Z of Offshore Oil & Gas_, Gulf Publishing Company, 1983.

2

Ocean Environment/Sea States

Working offshore is more complex than working onshore. This extra dimension of complexity is created by the environment. The ocean environment, which comprises meteorological and oceanographic data other than earthquake, environmental conditions, affects everything offshore, from the design of offshore platforms, submarine pipelines and terminals to the planning, installation and maintenance procedures. Experienced specialists working in this area are generally consulted while defining the pertinent meteorological and oceanographic conditions affecting a platform site.

The level of oceanographic and meteorological information needed during the design and installation of a platform depends on the amount of knowledge already available concerning the platform site and its general surroundings. Normal or operating meteorological and oceanographic environmental conditions, which are expected to occur frequently during the life of the structure, are needed to plan field operations such as installation and to develop the operational environmental load. Extreme environmental conditions, defined as those conditions which recur with a return period of typically 100 years, are needed to develop the extreme environmental load.

2.1 Meteorology

Meteorology is the science concerned with atmospheric phenomena especially in relation to weather and climate. Broadly it is concerned with

1. The observation of weather elements
2. The communication of observations to forecast centers
3. The analysis of observations using available techniques
4. The preparation of forecasts of weather and sea conditions
5. The analysis of the forecasted conditions on offshore operations
6. The communication of forecasts and the operational recommendations
7. Preparation of climatological studies

Hence, in a nutshell, it deals with weather to a large extent and with climate and monsoons to a lesser extent.

Weather is defined as the state of the atmosphere at a specific time with respect to heat or cold, wetness or dryness, calm or storm and clearness or cloudiness. Often the term is used to identify rain, storm, or other unfavorable atmospheric conditions.

Climate is defined as the average course or condition of the weather in a given area over a period of years as exhibited by temperature, wind velocity and precipitation. Climate is spoken of as being hot, cold, wet or dry. Sometimes it is referred to as harsh or mild. A basic relationship exists between climate and latitude (distance north or south of the equator).

Because offshore drilling crews usually stay in one region during a drilling operation, changes in climate rarely affect them, unless and until movement of marine vessels takes place from one part of the world to another.

On the other hand, changes in weather often occur in minutes or hours as rapid-moving weather systems approach and pass a given point. Even though in certain areas of the world the weather can change radically in a short span of several hours, it is often possible to predict these sudden weather changes to the hour as much as 24 hours in advance.

Since meteorology or weather is concerned with atmosphere, let us discuss what atmosphere is and how air or, for that matter, wind plays a big role in generation of weather elements.

2.1.1 Atmosphere

The atmosphere of Earth is defined as a layer of gases surrounding the planet that is retained by Earth's gravity. The atmosphere protects life on Earth by absorbing ultraviolet solar radiation, warming the surface through heat retention (greenhouse effect) and reducing temperature extremes between day and night (the diurnal temperature variation).

The common name given to the atmospheric gases used in breathing and photosynthesis is *air*. Dry air contains about 78% nitrogen (by volume), 21% oxygen (by volume) and about 0.95% argon, 0.04% carbon dioxide and small amounts of other gases. Air also contains water vapour, on average around 1% at sea level and 0.4% over the entire atmosphere.

Earth's atmosphere can be divided into four primary layers. Starting from highest to lowest, there is the thermosphere (50 to 440 miles), mesosphere (31 to 50 miles), stratosphere (7 to 31 miles) and troposphere (0 to 7 miles). Although air content and atmospheric pressure vary at different layers, air suitable for the survival of terrestrial plants and terrestrial animals currently is only known to be found in Earth's troposphere.

When the surface of the Earth is heated due to radiation of the sun, called *insolation*, it in turn heats the air above it. The unequal heating of Earth causes the air in the atmosphere to be heated unequally. The differential heating of Earth is caused by the angle at which the sun's rays strike the Earth's surface and the varying capability of the regions of the Earth to absorb heat. Land surfaces have faster rates of heating and cooling than do water surfaces.

Thus unequal heating of the atmosphere results in a mixture of warm and cool air. Warm air in the atmosphere expands, becomes less dense and rises. Cool air contracts, become more dense and moves downward. Warm air lying over cool air will eventually cool and move downward and new warm air will replace it. The warm and cool air in the atmosphere exert pressure over the surface of the Earth, which is known as *atmospheric pressure*. A *low-pressure area* is formed by the warm air expanding and rising in the atmosphere. A *high-pressure* area is formed by the cool air contracting and moving downward in the atmosphere. The rotation of the Earth causes these low- and high-pressure areas to move and circulate. This circulation of air is responsible for the creation of wind and other weather phenomena.

2.1.2 Wind

Horizontal movement of air in the Earth's atmosphere is known as wind. Other than heating and cooling of the atmosphere, there is also another factor known as *Coriolis force* which is simultaneously responsible for producing *prevailing winds* (Figure 2.1).

Coriolis force is an apparent force caused by the rotation of the Earth about a vertical axis which affects the wind direction in both the northern and southern hemispheres. Wind or water in motion will move clockwise in the northern hemisphere and counter-clockwise in the southern hemisphere (Figure 2.2). Typically in northern hemisphere, air circulates clockwise around a high-pressure area (Figure 2.3a) and counter-clockwise around a low-pressure area (Figure 2.3b). Because of the Coriolis force, air circulation is opposite in the southern hemisphere, that is, moving counter-clockwise around a high-pressure area and clockwise around a low-pressure area.

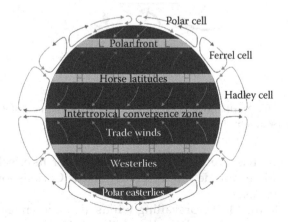

FIGURE 2.1
Circulation of prevailing winds. (Courtesy of Dwindrim/Wikimedia Commons/CC-BY-SA-3.0.)

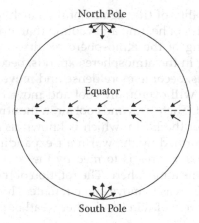

FIGURE 2.2
Effect of the Coriolis force on wind.

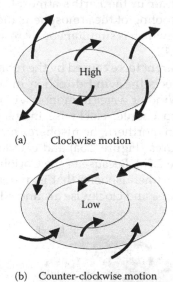

FIGURE 2.3
Circulation of air in the Northern hemisphere. (a) Clockwise around a high-pressure area. (b) Counter-clockwise around a low-pressure area.

The Coriolis force has no effect on wind at the equator. Its effect increases as it moves from middle latitude toward poles and the Coriolis force increases as wind speed increases.

Referring to Figure 2.1, the prevailing winds also known as *surface winds* can be measured at a surface observing station. It is important to note that *wind direction* is always reported as *from*, for example, a northeast wind indicates it blows from the northeast, a southwest wind means it blows from the southwest direction.

Similarly, it also is to be noted that *wind speed* is conventionally reported in *knots* (1 knot = 1 nautical mile/h = 1.15 statute mile/h = 1.852 km/h). Wind speeds are categorised as Beaufort numbers (devised in 1805 by an Irish Royal Navy officer named Rear Admiral Sir Francis Beaufort and adopted in 1830), which are presented in tabular form known as the *Beaufort Scale* (Table 2.1). Beaufort numbers between 7 and 10 are known as *gales* in which 7 indicates *moderate gale* (28–33 knots), 8 indicates a *fresh gale* (34–40 knots), 9 indicates a *strong gale* (41–47 knots) and 10 indicates a *whole gale* (48–55 knots).

The common prevailing winds from North pole to South pole (as shown in Figure 2.1) are Polar Northeasterlies (near 60° N latitude), prevailing Westerlies in north (between 30° N and 60° N latitudes), northeast and southeast trade winds (from 30° N and 30° S latitudes), prevailing Westerlies in south (between 30° S and 60° S latitudes) and polar Southeasterlies (near 60° S latitude).

The prevailing winds are affected by *monsoons*, large seasonal wind systems that reverse direction in the winter and summer months. Monsoons are caused by large temperature differences between land and sea. Land surfaces have faster rates of heating and cooling and greater variation of temperatures than water surfaces. Land areas are high in pressure in the winter and low in pressure in the summer.

In the winter months, high-pressure air intensifies as it moves over land areas and eventually meets low-pressure air over the warm oceans. Strong winds, which frequently reach speeds of 35 knots, result. These winter monsoons exhibit little or no rainfall and are sometimes referred to as dry monsoons.

In the summer months, warm winds reverse and travel back over the oceans, meeting low-pressure air on land. The low-pressure air is intensified by contact with the warm air. Summer monsoons are usually characterised by heavy rainfall and are sometimes referred to as wet monsoons. High humidity and large differences in temperature between and surrounding oceans may cause a monsoon fog to occur.

Monsoons have little or no effect in polar regions and in areas surrounding the equator because temperatures of land and water are similar in these areas. Monsoons are common in Southern Asia, Northern Australia, North and South America and parts of Africa. The greatest monsoons occur in Southern and Southeastern Asia.

2.1.3 Weather Phenomenon

The major factor in the development of weather is *air masses.*

Air masses are bodies of air that remain for extended periods of time over large land and sea areas (usually 1600 km or more across) with uniform

TABLE 2.1

Beaufort Wind Scale with Corresponding Sea State Codes

Beaufort Number	Wind Velocity (knots)	Wind Description	Sea State Description	Sea State	
				Term and Height of Waves (ft)	Condition Number
0	Less than 1	Calm	Sea surface smooth and mirror-like	Calm, glassy 0	0
1	1–3	Light air	Scaly ripples, no foam crests		
2	4–6	Light breeze	Small wavelets, crests glassy, no breaking	0–0.3	1
3	7–10	Gentle breeze	Large wavelets, crests begin to break, scattered whitecaps	Smooth, wavelets 0.3–1	2
4	11–16	Moderate breeze	Small waves, becoming longer, numerous whitecaps	Slight 1–4	3
5	17–21	Fresh breeze	Moderate waves, taking longer form, many whitecaps, some spray	Moderate 4–8	4
6	22–27	Strong breeze	Larger waves, whitecaps common, more spray	Rough 8–13	5
7	28–33	Moderate gale	Sea heaps up, white foam streaks off breakers	Very rough 13–20	6
8	34–40	Fresh gale	Moderately high, waves of greater length, edges of crests begin to break into spindrift, foam blown in streaks		
9	41–47	Strong gale	High waves, sea begins to roll, dense streaks of foam, spray may reduce visibility		
10	48–55	Storm/whole gale	Very high waves, with overhanging crests, sea white with density blown foam, heavy rolling, lowered visibility	High 20–30	7
11	56–63	Violent storm	Exceptionally high waves, foam patches cover sea, visibility more reduced	Very high 30–45	8
12	64 and over	Hurricane	Air filled with foam, sea completely white with driving spray, visibility greatly reduced	Phenomenal 45 and over	9

Source: Modified from Nora Sheppard, *Wind, Waves and Weather*, Petroleum Extension Service, 1984; and B. B. Clayton and B. E. D. Bishop, *Mechanics of Marine Vehicles*, Gulf Publishing Company, 1982.

heating and cooling properties. Air masses acquire their distinctive temperature and moisture characteristics from the surface over which they originate, known as *air mass source regions* (Figure 2.4).

In Figure 2.4, lowercase c and m refer to continental and maritime air masses developing over land areas and water areas, respectively. The capital letters A, P and T refer to Arctic, Polar and Tropical air masses development over the poles, upper latitudes (cold regions) and low latitudes (warm regions), respectively. Thus, cP refers to Continental Polar air mass source region and mT refers to Maritime Tropical air mass source region.

These different air mass source regions have different temperatures and moisture content. For example, tropical air masses resulting from heating of land surface (cT) have high temperatures and low water vapour content. Whereas, maritime air masses (mT) have lower temperatures and much higher water vapour content. It is this water vapour content in the air masses that contributes to the energy that produces hazardous weathers like frontal storms, thunder storms, hurricanes and related disturbances. Whenever two air masses meet, a *front* is formed.

Fronts are surfaces of discontinuity or interfaces between unlike masses of air. Fronts occur when masses of warm and cold air (such as mT and mP) meet. Changes in temperature, humidity and cloud formation and wind speed and direction occur. Fronts are associated with most of the severe weather conditions on Earth.

There are four types of fronts, namely, (1) cold fronts, (2) warm fronts, (3) stationary fronts and (4) occluded fronts (Figure 2.5).

A *cold front* exists when cold air replaces warm air. The warm air is lifted to ride up over the cold air. When sufficient moisture is present in the warm air and unstable layers of air exist, thunderstorms may be triggered. At times,

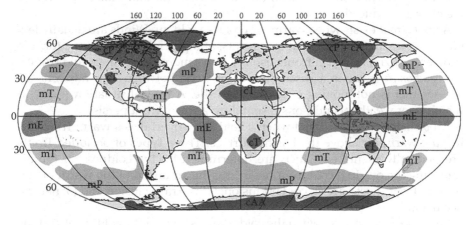

FIGURE 2.4
Air mass source regions. (Courtesy of Wikimedia Commons/Public domain.)

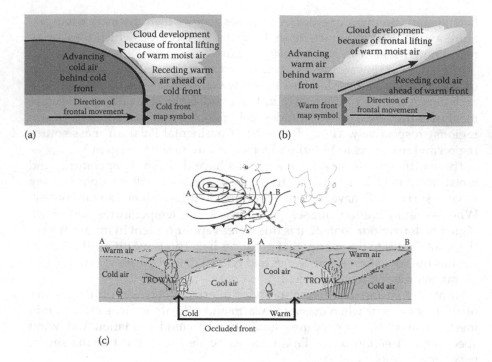

FIGURE 2.5
Cross-section of fronts: (a) Cold front; (b) warm front; (c) cold-type occluded front. (Courtesy of Pierre_cb/Ravedave/Wikimedia Commons/CC-BY-SA-3.0.)

squall (a wind of 16 knots or higher, sustained for at least 2 minutes) lines containing thunderstorms travel ahead of the front and violent prefrontal weather is produced.

A *warm front* exists when warm air replaces cold air. Winds are usually less violent in warm fronts than in cold fronts, and precipitation is in the form of showers or drizzle. Warm fronts usually move more slowly than cold fronts.

A *stationary front* is a discontinuity or zone between two adjacent air masses that is not moving or is moving little. A typical situation is one in which poor weather conditions resembling a warm front persist.

An *occluded front* is formed when a cold front overtakes a warm front lifting the warm air above the Earth's surface. Three types of occluded fronts are generally identified which are determined by the coldness of the air replacing the original cold front regarding the air ahead of the warm or stationary front. These are (1) cold occlusion, (2) warm occlusion and (3) neutral occlusion.

A *cold occlusion* occurs when the coldest air is behind the cold front and the warm front is forced aloft.

A *warm occlusion* occurs when the coldest air is ahead of the warm front and the cold front is forced aloft.

A *neutral occlusion* occurs when the temperatures of the cold air masses of the cold and warm fronts are the same.

When the interface between cold and warm air masses becomes unstable (in the case when polar easterlies and westerlies converge or plunge down) as a result of the effects of friction and the differences in density of two air streams, a wave-like flow pattern is created as is shown in Figure 2.6. Low pressure develops at the crest of the wavelike front because the warm, humid, low-density air rides over and is replaced by the cold, dry, dense air. The warm, moist air rises to progressively heavier cloud (defined later) until saturation is reached and intense precipitation accompanied by strong winds occurs (Figure 2.6). These strong winds blow into the low pressure centre and thus develop a 'low' or potential storm centre – a *cyclone*, which may travel along an occluded front. Such wavelike cyclones can be very destructive to offshore operations. The rate of progression of a cyclone varies, but on average it is about 10–15 m/sec and so long as the difference in temperature and moisture content is maintained, the cyclone persists. If the cyclonic area is broad and shallow, it is called a trough of low pressure.

The term *cyclogenesis* is defined as the development of a low-pressure system. Air around low-pressure (cyclonic) systems moves in a counter-clockwise direction in the Northern hemisphere (Figure 2.7) and a clockwise direction in the Southern hemisphere. Although the term cyclone is sometimes used to describe severe storms occurring in the atmosphere, scientists use the term to describe any low-pressure system.

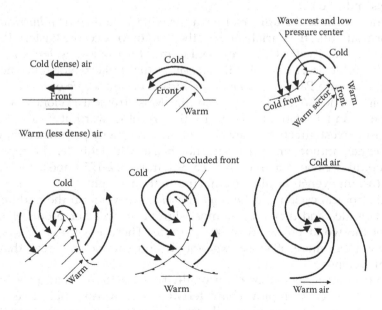

FIGURE 2.6
Development of a cyclone (Radial view).

FIGURE 2.7
Wind movement around a cyclonic system in the Northern hemisphere.

In a tropical country like India, tropical cyclones are found, which fall in the region of the intertropical frontal zone and occur commonly in the summer months. A storm with winds of 65 knots (Beaufort number 12) or greater is known as a *tropical cyclone*. Owing to the lack of great contrasts in temperature, the tropical cyclone tends to be less specific in size, shape and wind system. Nevertheless, over tropical oceans, relatively small and consequently extremely intense cyclones do develop. These are variously called *hurricanes, typhoons* and so on.

Tropical cyclone is the universal term for what is known as a *hurricane* in the North Atlantic, the Caribbean Sea, the Gulf of Mexico, the Eastern Pacific and the Central Pacific. Other regional names for tropical cyclones include *cordonazo* in Mexico; *taino* in Haiti, *typhoon* in the Western North Pacific and most of the south Pacific; *cyclone* in the Indian ocean; *baguio* or *baruio* in the Philippines; and *Willy-Willy* in Australia. Most tropical cyclones develop between 5° and 15° latitude. They usually travel forward at 9 to 13 knots, from east to west direction. They also follow a northerly path when they are in the higher latitudes and turn to the northeast at 40° latitude. The approach of a tropical cyclone is marked by heavy winds (112–175 knots), and rising waves (height over 60 ft) and currents (more than 4 mph).

Beside tropical cyclones, *extra-tropical cyclones* develop in the middle and upper latitudes and they are the most significant weather producer in all parts of the world poleward of 30° latitude. These low-pressure systems probably affect more offshore exploration and production areas than do tropical cyclones.

Extra-tropical cyclones may occasionally travel to the low latitudes, where they lose their extra-tropical characteristics over warm tropical seas and become tropical. This situation will result in a tropical cyclone that is not developed from a tropical wave.

Another hazardous weather disturbance is known as a *waterspout*. Most common in the spring and early summer, the waterspout (Figure 2.8) develops when cold air and warm, humid air meet. Waterspouts are closely associated with thunderstorms; a waterspout funnel forms from a cumulonimbus cloud (see Section 2.1.4 'Cloud'). The high speed of the whirling funnel causes its centre to be extremely low in pressure. This accounts for the explosive action created when an object (e.g., a drilling platform) at sea and waterspout meet. Hail, heavy rain and other disturbances are usually experienced before and after a waterspout's passage. Waterspouts usually accompany tropical cyclones. Because of their high wind speeds, waterspouts are known as the most violent of all storms at sea. However, they are usually less destructive than tropical and extra-tropical cyclones because they cover a small area.

Two types of waterspouts exist. The first type is known as a *tornado* that ventures out to sea. This has cyclonic rotation and travels at a speed from 23 to 35 knots. Its funnel has a speed of from 86 to more than 260 knots and has an average width of 1000 ft.

The second type of waterspout develops at sea. It is difficult to classify because it develops in contrasting areas under varied conditions and exhibits no set pattern of behaviour. This second type can form in good or bad weather conditions. They can form in tropical regions or in the north and

FIGURE 2.8
Waterspout. (Courtesy of Dr. Golden Joseph, NOAA/Wikimedia Commons/Public domain.)

south temperate zones. Such waterspouts can develop from both a cumulonimbus cloud and the ocean's surface. In reality, the dark funnel, produced by condensation circling the low pressure centre, descends from the cloud toward the ocean surface. The seawater begins to circulate around the low-pressure point above it. The cloud and water funnels eventually meet. Such waterspouts seldom last for more than an hour. They may be very fierce or very weak.

Thus far, the major weather disturbances have been discussed for which air masses are the major factors. Beside this, water droplets in different form also give some warning to the offshore personnel, particularly those involved in the transportation of manpower and materials to the rig site. For them it is necessary to recognise various types of cloud and fog formations.

2.1.4 Cloud

A cloud is defined as a mass of visible, suspended water droplets or ice crystals that are formed by condensation. There are about 10 types of cloud formations depending upon their height in the sky from near the ground and their composition (i.e. either water droplets or ice crystals). These are illustrated in Figure 2.9 and brief descriptions of each type from near the ground to a height of 45,000 ft are given below by grouping them in four parts: (1) to (4) from near the ground to 6500 ft, (5) from ground to 45,000 ft, (6) to (7) from 6500 to 23,000 ft and (8) to (10) from 16,500–45,000 ft.

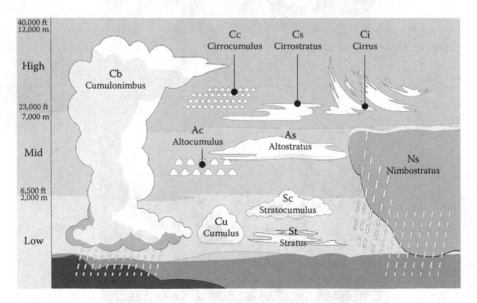

FIGURE 2.9
Types of clouds. (Courtesy of Valentin de Bruyn/Wikimedia Commons/CC-BY-SA-3.0.)

1. *Stratus (St) clouds* are low level clouds composed of water droplets, which cover the sky in a grey sheet.

2. *Nimbostratus (Ns) clouds* are thick, dark grey, shapeless, low level clouds that are composed of water droplets. They block out the sun and are associated with continuous rains or snow, high winds and hazardous sea conditions.

3. *Stratocumulus (Sc) clouds* are grey or white clouds that appear in rolls (in cold seasons) or as a continuous sheet broken into irregular parallel bands, and are composed of water droplets. These clouds may produce light rain or snow that will hamper visibility.

4. *Cumulus (Cu) clouds* are white and puffy clouds with a dark base. The shape of this type of cloud constantly changes. Prominent in the summer months, cumulus clouds generally cover only 25% of the sky. Two different cumulus formations are significant. (a) Cu clouds lying horizontally with little vertical development which are known as fair weather cumulus. (b) Cu clouds showing pronounced vertical development, known as vertical cumulus causing heavy rain and strong winds.

5. *Cumulonimbus (Cb) clouds* are also called thunderheads. They form from cumulus clouds that have reached great vertical development. A well-developed cumulonimbus cloud has a fibrous top that is drawn out in the direction of motion, resulting in an anvil shape. The top of the cloud is composed of ice crystals, and the bottom of the cloud is composed of water droplets. Thunderstorm conditions like waterspouts, high winds, heavy rain and hail can be expected from such clouds.

6. *Altostratus (As) clouds* are bluish or greyish layers of uniform mid-level (from 6500 ft to 23,000 ft) clouds that cover large portions of the sky. These clouds may obscure the sun or give it a watery appearance. These are composed of either ice crystals or water droplets. At sea, they are usually the first indication of an approaching storm that may bring large waves.

7. *Altocumulus (Ac) clouds* are white or grey mid-level (6500–23,000 ft) clouds that appear as cloud rolls arranged closely together. This type of cloud is composed of either ice-crystals or water-droplets.

8. *Cirrus (Ci) clouds* are high level (from 16,500 to 45,000 ft) white feather-like clouds that appear in all seasons. These are composed of ice crystals.

9. *Cirrostratus (Cs) clouds* are very thin, white, high level (16,500 to 45,000 ft) clouds that produce halos visible around the sun or moon. Such type of cloud is composed of ice crystals and appears as a thin veil when seen in the sky.

10. *Cirrocumulus (Cc) clouds* have a thin, white and rippled appearance. They sometimes form a banded arrangement known as mackerel

sky. Cirrocumulus are the rarest of all cloud types and are often formed from cirrus and cirrostratus clouds. These are high level (16,500 to 45,000 ft) clouds composed of ice-crystals.

Fog is a cloud of minute water droplets or ice crystals suspended in the air so that the cloud bottom rests upon the Earth's surface, either ground or water. The formation of fog is related to the 'relative humidity,' which is measured by knowing the 'dew-point spread.' The difference between the actual air temperature and the dew point is known as the dew-point spread and when this spread is 5°F or less, fog occurs. There are three types of fog, which are described below.

1. *Sea fog or advection fog* is caused by the movement of warm, moist air over a cold surface having temperature less than the dew point of the air. Four-fifths of all maritime fogs are sea fogs.
2. *Steam fog or sea smoke* is created when cold, dry air moves over a much warmer body of water. The warm water begins to evaporate, adding water vapour to the cold air which becomes saturated quickly and a light 'steam' is formed just above the water's surface.
3. *Ice fog*, which is common in Arctic regions, occurs in moist air during cold, calm conditions in winter. Ice fog, made up of ice crystals, may be produced artificially by the combustion of aircraft fuel in cold air.

2.2 Oceanography

Oceanography is concerned with the prediction of wave and swell conditions from forecasted wind conditions or the determination of past wave and swell conditions from known past wind or pressure conditions. Concurrently, ocean current and tide are also a matter of concern for oceanographers. In higher latitudes, sea ice is also an added parameter for calculating its effect on offshore structures for which information is also required to be gathered.

2.2.1 Wave

Ocean waves may be formed by several forces, including the frictional contact between wind and ocean, the tidal attraction of the sun and the moon and the vibration of the Earth.

Wind waves are formed by the contact of wind with an ocean's surface, and have the single greatest effect on offshore operations. As wind blows over an ocean, friction exists between the wind and the ocean's surface. The energy transferred from wind to water via friction results in the motion of

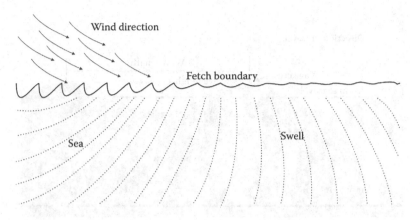

FIGURE 2.10
Area of wave development.

the ocean's surface. Wind energy is transferred to only the very top surface of the water. The top layer of water imparts energy to the water directly beneath it and so on (energy decreasing as the process proceeds). This transfer of energy through friction causes wind waves to form.

Wind waves develop and grow where a certain wind prevails from the same direction for a period of time. This area of wave development is called the *fetch* (Figure 2.10). Fetch is also defined as the distance over which the wind blows to generate the observed waves at a given position or point. A fetch may be hundreds of miles long. Maximum wave heights exist at the leeward (or downwind) boundary of the fetch. Waves that are still in their fetch, under the influence of its wind, are called *sea*. Waves that have moved out of their fetch and into weaker winds are called *swell*. Swell decreases in height and has regular movement. In comparison to sea, which appears choppy, swell appears round-topped.

Tidal waves are caused by the action of the sun and the moon on the oceans of the Earth. They have the longest wave periods of all, some lasting for over 24 hours. Tidal waves travel at speeds of approximately 400 knots.

Tsunamis or seismic sea waves, often confused with tidal waves, are produced by submarine earthquakes or explosions. When a seismic disruption occurs, great volumes of water are displaced, transferring energy from the ocean floor to the surface. This energy transfer results in huge breakers that inundate the shoreline, destroying boats and installations. Tsunamis are virtually unnoticeable in deep water. They have deep water wave heights of approximately 3 ft, wave lengths of hundreds of miles, and wave periods of 15 to 60 min.

2.2.1.1 Wave Description

The wind-generated waves are mostly known as surface gravity waves. For its description, let us consider a wave form that is a simple harmonic so that at any time it has a sinusoidal shape as shown in Figure 2.11.

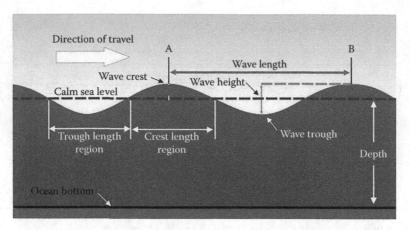

FIGURE 2.11
Wave description. (Courtesy of NOAA/Wikimedia Commons/Public domain.)

Wave height (H) is measured from the crest (the top of the wave) to the trough (the lowest point of the water surface level between crests). The heights of wind waves rarely exceed 55 ft, but they have been known to exceed 100 ft. Wave heights and sea characteristics can be reported in terms of the Beaufort scale (Table 2.1). The scale ranges from Beaufort number 0, which indicates calm, mirror-like seas with no waves, to Beaufort number 12 indicating a height of more than 45 ft and a sea that is completely white with blowing foam and spray.

Four factors – wind speed, duration, fetch and water depth – determine wave height. Wind speed is the most important. The faster or stronger the wind blows, the higher the waves will be although wind speed affects wave height only up to a certain extent. Duration of wind speed means that the longer the strong wind blows, the higher the waves will be. For example, if a 30-knot wind blew over an area of unlimited fetch for 5 hours, 7½ ft waves would develop. If that same 30-knot wind blew over the same area of water for 30 hours, the waves would reach 16 ft. Similarly, the longer the fetch, the higher are the waves. For example, if a 35-knot wind was blowing across a 1-mile area (short fetch), waves might reach only 2 ft in height. If that same 35-knot wind blew across an area of unlimited fetch, waves of at least 30 ft in height would eventually develop. Water depth is significant only in shallow seas and bays, where the shallowness causes waves to be relatively small. In an open ocean, water depth is sufficient for maximum wave development.

Wavelength (λ) is defined as the horizontal distance between two successive wave troughs or crests. Wind waves or surface gravity waves have wavelengths that are usually 12 to 35 times their heights. Waves in swell have wavelengths that are 35 to 200 times their heights. *Amplitude* (A) is a measurement of the vertical distance from the wave axis or the still water level to the top of the wave peak (crest) and bottom of the wave trough. Depending on the type of wave, the wave height (H) may be equal to twice the amplitude (i.e., H = 2A) or may not, because in certain waves crest amplitude is greater than the trough amplitude. The *wave period* (T) is the time interval between the passage of two successive wave troughs or crests. This can be determined by measuring the time taken by two peaks to pass a certain point.

Frequency (f or ω) of a wave is a measurement of how often a wave recurs in a measured amount of time. One completion of the repeating pattern is called cycle. The frequencies of progressive waves or those that move forward indicate how fast a wave moves forward in units of cycles per minute. Wave frequency is related to wave period by Period $\alpha \dfrac{1}{\text{Frequency}}$. Most wind waves have short wave periods of 3 to 30 sec.

Wave *interference* is caused when sea and swell, or two or more systems of swell, converge. Irregular waves are formed in this meeting of different wave systems. Wave height may be increased when the crests combine, and flat zones may be caused when crests and troughs meet. When these different wave systems cross each other at an angle, the irregular wave patterns produced are called cross sea.

Breakers, sea-surface waves that have become too steep to be stable, form offshore when the wave height increases beyond a height-to-length ratio of approximately 1:7. The water at the crest moves faster than the rest of the wave, causing the crest to fall forward. These breakers appear as *whitecaps*, waves with white, foamy tops. Whitecaps can also be generated when strong winds blow the tops off waves. When viewed from above, these whitecaps form a white, streaking pattern.

Wave movement or *wave velocity* expressed in feet per second is called *celerity (c)*.

2.2.2 Current

An ocean current is defined as the predominantly horizontal movement of the ocean waters. Ocean currents may be crudely divided into two types. The first type has a surface flow, which usually travels in a cyclical pattern (Figure 2.12). The second type has a deep, slow-moving flow, which usually travels in a meridional direction. Many currents have associated with them a reverse flow. These counter-currents flow adjacent to the accompanying current.

Ocean currents are classified as warm or cold and act as a heat distribution system for the Earth. Cold currents from the high latitudes flow toward each other and circulate. Horizontal differences in temperature are equalised; thus, temperature extremes between different areas are reduced. The meeting of warm and cold currents may cause fog or frontal activity to occur.

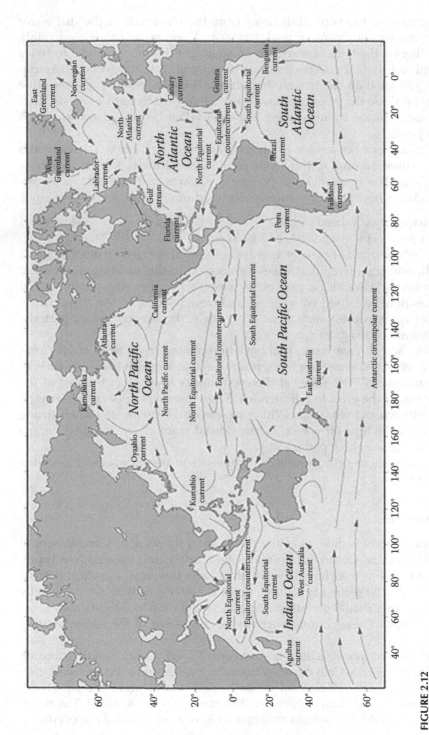

FIGURE 2.12
Ocean surface currents. (From B. B. Clayton and B. E. D. Bishop, *Mechanics of Marine Vehicles*, Gulf Publishing Company, 1982. Copyright American Meteorological Society. With permission.)

A current's direction of motion is known as its *set*. A current's speed of motion is called *drift*. The speed at which a current moves depends on the density of the water and the speed and direction of the wind.

2.2.3 Tides

Tides, the periodic rising and falling of the surface of the oceans and water bodies connected to the oceans (gulfs and bays*), are primarily caused by the unequal gravitational attraction of the sun and the moon on the Earth. Tides are also caused by other forces; the arrangement of ocean bottom contours and land masses; wind action; the motion of ocean currents; the exit of large river floods at the mouth of streams; the transport of water by high waves; and the development of low pressure centres.

Lunar tides are caused by the moon. *Solar tides* are caused by the sun. Tides caused by the sun and those caused by the moon are essentially the same, although the moon's impact on the Earth's tides is greater. The forces of the moon and of the sun on the Earth are dependent on their masses and distances from the Earth. Though the moon has 27 million times less mass than the sun, it is 390 times closer to the Earth. The moon's tidal force on the Earth is roughly twice that of the sun; therefore, the tides caused by the moon are approximately twice the size of those caused by the sun. The period of time between successive local applications of the sun's maximum tidal force is 12 hours. The moon has a tidal force period that is slightly longer, approximately 12½ hours.

The behaviour of tides is affected by the moon's revolution about the Earth and the sun's and moon's planes of motion that lie at an angle to the equator. Also, the Coriolis force causes most tides to move clockwise in the Northern hemisphere and counter-clockwise in the Southern hemisphere.

The horizontal movement of the tide as it rises and falls and moves toward and away from the coast is known as the *tidal current*. When the tidal current rushes toward the shore, it is called the *flood current*. When it travels back toward the sea, it is called the *ebb current*. In open sea, tidal currents rotate continuously 360° once or twice a day. Near coastal areas, the direction of tidal currents depends on the region's topography. Near the shore, tidal currents often reach 1.5 to 2.5 knots.

Ordinary tides are completely predictable. Barring forces such as storms or high winds, the tide will take approximately 6 hours to fall to its low point. The high point of the tide is called *high tide*, and the low point of the tide is called *low tide*.

The range of the tide refers to the difference in height between consecutive high and low tides at a given place. Tidal range is affected daily by the positions of the moon and the sun (Figure 2.13). The range of the tide will be at its highest

---

* Bay – Part of the sea filling a wide-mouthed opening of a land (e.g. Bay of Bengal). Gulf – Piece of sea like a bay but less open at the mouth (e.g. Gulf of Cambay).

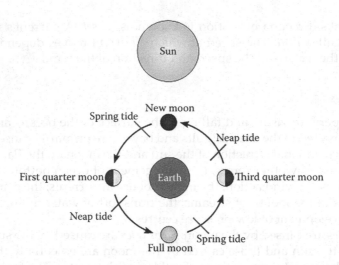

FIGURE 2.13
Position of sun, Earth and moon in tidal force system. (Courtesy of KVDP/Wikimedia Commons/Public domain.)

when the moon and the sun are aligned (during new and full moons). When this moon–sun alignment occurs, the tide is known as a *spring tide*. When the moon and the sun lie at right angles to each other (during first and third quarters), the tidal range will be at its lowest. This tide is known as a *neap tide*. Average tidal range varies from under 2 ft in areas of the Mediterranean, the Arctic and the South Pacific to over 50 ft in the Bay of Fundy, Canada. The range of the tide is very much dependent on the formation of the area in which it travels. Tides may increase in height when they move through narrow passages.

The frequency with which tides occur is dependent on the depth, size and shape of the oceans. Most of the world's oceans and seas, including the Atlantic Ocean, have tides which have two high water levels and two low water levels per day. This type of tide is known as the *semidiurnal tide*. Tides in the waters surrounding Alaska and the Gulf of Mexico have only one high water level and one low water level per day. This type of tide is known as a *diurnal tide*. Characteristics of both semidiurnal and diurnal tides can be found in some islands in the Pacific Ocean. This *mixed tide* exhibits discrepancies in the heights of the high tides or low tides that occur each day.

A *storm surge*, or *storm tide*, is a high tide near the shore that is significantly higher than normal. A storm surge often occurs when a cyclonic storm or a *squall* (a wind of 16 knots or higher, sustained for at least 2 minutes) occurs. Cyclonic storms cause tides to grow as high as 20 ft over their predicted heights. The size and behaviour of a surge depends on the atmospheric pressure at the storm's centre, the storm's speed and direction, the horizontal scale and the topography of the ocean bottom and coastline. When a cyclonic storm meets the coastline, the surge is usually greatest to the right of its path.

2.2.4 Sea Ice

Ice in the sea comes mainly from the freezing of sea water and is thus called 'sea ice' or in other words sea ice is frozen seawater. When seawater freezes, crystals of pure ice form first and increase the salinity of the remaining liquid. However, some of the concentrated salt solution (brine) becomes trapped within the open structure of crystals.

When ice thickens, or when rafting occurs, the brine gradually trickles down from the elevated ice to the water level and eventually leaves almost saltless, clear, old ice.

Sea ice must therefore be considered as a material of variable composition with properties that depend significantly on its previous history.

Whereas pure water freezes at 0°C (32°F), the freezing point of sea water with a salinity of 3.5% is –1.91°C (28.6°F).

2.2.4.1 Types of Sea Ice

Incipient freezing in the sea is indicated by a *greasy* appearance on the surface arising from the presence of *flat ice crystals*. As freezing continues, individual plates of ice develop in quality *frazil ice* and these tend to aggregate into *slush* ice. Further, aggregation results in rounded sheets of *pancake* ice, which then freeze together to form *flow* or *sheet* ice.

Various other terms are used to define different types of sea ice depending on whether it is attached (or frozen) to the shoreline (Figure 2.14). If attached, it is called *land fast ice, coast ice,* or more often *fast ice* (from fastened) and unlike fast ice, *drift ice* (or *pack ice*) occurs further offshore in very wide areas and encompasses ice that is free to move with currents and winds. The physical boundary between fast ice and pack ice is the *fast ice boundary*. Pack ice covers more than half of visible sea surface and the pack ice zone is further divided into *shear zone, marginal ice* zone and *central pack*.

Pack ice is formed into a mass by rafting of pans, floes and brash. A *pan* is a drifting fragment of the flat thin ice that forms in bays or along the shore.

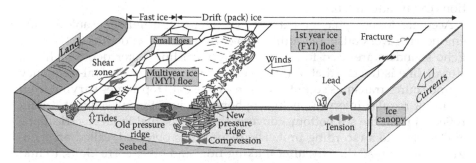

FIGURE 2.14
Sea-ice features. (Courtesy of Lusilier/Wikimedia Commons/CC-BY-SA-3.0.)

A *floe* is floating ice formed in a large sheet on the surface of a body of water. *Brash* is a mass of ice fragments.

At an air temperature of –30°C some 30 mm of ice can form in 1 hour and 300 mm in 3 days, the rate decreasing in thickness because ice is a poor conductor of heat.

Sea ice thickness can vary considerably, for example, *floes*, floating ice fields, grow to thicknesses of 10 ft over portions of the Arctic Ocean. When floes are driven together, they form *ice ridges*. Some ice ridges extend 20 ft above and 60 ft below the water surface. *Icebergs* are large, floating masses of ice detached from glaciers. The average iceberg is calculated to weigh 1 million tons.

2.3 Sea Bed Soil

The problems encountered in the design of offshore structures in relation to soil mechanics are principally those regarding the ability of the foundation unit to resist vertical or lateral forces, problems of scour and of pulling out the legs of mobile exploration units. Prediction of soil behaviour is necessary before a structure can be located in an offshore site.

Basically, there are two general types of commercial investigations currently available. The first group relies on ship-based geophysical methods which yield data without even touching the sea-bed. These methods use equipment like echo-sounders, sub-bottom profilers, side-scan sonars, television cameras and so on, to determine the general sea-bed topography. With these surveys, it becomes possible to prepare contour maps of the sea bottom and determine the kinds and number of soil layers below the sea-bed up to a limited depth.

The second group obtains soil samples or carries out *in situ* tests to determine the soil properties of the sea-bed at the site. With the assumption that this sample is characteristic of the soil layer in the area, the design of foundations can be attempted.

Accurate topographic, or bathymetric, charts are indispensable to any engineering work offshore for they depict the exact water depths in the area. Echo-sounders are used for continuously recording water depths while a survey line is traversed at sea in a vessel. Here a single or dual-frequency sound source transmits sound waves in a cone with top angles varying from 5°–15° at regular intervals. A hydrophone receives the subsequent echoes reflected from the sea bottom, converts the sonic signal to an electrical one and feeds it to a recording unit. The record thus obtained, called an echogram, can be read directly, after making tidal and other corrections, to find the depth of the sea-floor.

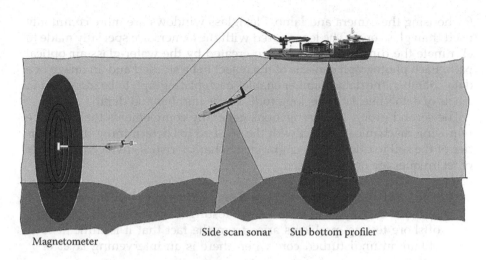

Magnetometer · Side scan sonar · Sub bottom profiler

FIGURE 2.15
Sub-bottom profiler and side scan sonar.

The sub-bottom profiler (Figure 2.15) is an instrument not unlike the echo sounder, but operating on an electrical frequency of 3.5–7 kHz as compared to 30–210 kHz of the echo-sounder. The lower frequency of the sub-bottom profiler provides a sea-bed profile up to 70 m penetration with a resolution of 76 cm at best, for strata with different reflective properties, mainly of a difference in density. The sparker and boomer systems emit high energy sound pulses at regular intervals.

While a sparker system emits electrical sparks underwater to produce sound pulses, the boomer system uses a vibrating membrane. The sound reflections are picked up by hydrophones and in this case a sea-bed penetration of 100 to 300 m is possible.

The side scan sonar is an instrument similar to the sparker and boomer profiler except that it maps the sea-bed sideways. It produces right- and left-hand records which look like a set of pictures of the sea-bed on both sides of the towed sound source called the 'fish.' The records show bottom irregularities, rock outcrops, sunken objects, pipelines, well heads, sand waves, faulting areas and seeps. The 'fish' has a transmitter and receiver mounted on it and is towed some 12 m off the sea bottom at towing speeds of about 6–8 knots.

Next best to active personal observations or remote television viewing is photographic camera recording of the sea-bed. Even when a person makes visual observations he or she requires a camera because of its ability to capture a scene for later and repeated study.

The chief characteristics of an underwater photography unit are its ruggedness and the ability of the camera-and-lamp housing to withstand high pressure at oceanic depths. Long aluminium and steel cylinders are widely used for housing the camera and lamp. Plexiglass windows are more commonly used than glass ones. The lenses used with the camera are specially made to eliminate the dimensional distortion created by the water-glass-air optical path. Each photograph consists of the object to be studied and an image of a data chamber; the data chamber enables each photograph to be identified in terms of data, time, latitude, longitude, frame number and depth.

The second group of investigations generally complements the first type requiring mechanical contact with the sea-bed for determining the properties of the soil to calculate its engineering characteristics. Two general types of techniques are normally used:

1. Undisturbed samples of soil are taken and tested in the laboratory, in the same way as in onshore soil testing. However, in the case of offshore testing, problems arise from the fact that it is difficult to obtain an undisturbed core when there is an intervening layer of water.

2. In situ testing of soils by suitable modification of the existing instruments to operate under water.

Underwater soil sampling techniques are broadly of two types, shallow or deep penetration sampling. Shallow sampling may be done using corers (gravity and piston type corers), snapper or grab sampler and dredges. Deep penetration sampling operations are used to obtain from considerable depths below the sea-bed repetitive samples which are relatively undisturbed. They require a means of holding the drilling equipment in a fixed position while the boring penetrates the sea-floor and sampling or testing operations are carried out.

The problem of sampling disturbance is serious with all types of coring devices. Extensive studies have been carried out on the influence of the size and shape of the sampler on disturbances and the influence of disturbance and stress change on the strength and compressibility of the sea-floor. Sometimes, in situ testing of soil may be carried out in conjunction with the deep penetration sampling operations. Instruments conventionally used are: the Dutch-cone apparatus, the shear-vane or the pressure meter. The major problem with all in situ testing devices is that of correlation of data. The results obtained 100 m below the sea-bed have to be correlated with empirical data collected and analysed in the past so as to enable the designer to predict the soil's properties and characteristics.

Normally the soil properties that are tested in the laboratory are density, porosity, that is, volume of voids in the sample, and void-ratio, that is, the ratio of volume of voids to the volume of soil particles. Beside these key properties, other parameters of interest to the designer are cohesiveness, mineralogical composition, permeability, shear strength (i.e. resistance of soil to lateral movement) and compressibility (i.e. rate at which the settlement of soil occurs under load). Behaviour of a soil under cyclic stresses, its liquefaction potential, is of interest. While conventional soil mechanic techniques permit a reasonably accurate assessment of the required soil parameters, there are a few areas where special methods must be utilised for offshore soil. One of these is the soil-to-pile structure interaction problem.

Foundation investigations are required to see if suitable anchor-holding capacity can be derived from the soil on the ocean surface to permit anchoring of floating equipment like exploration vessels, floating tankers and so on.

For mobile jack-up drilling rigs, soil boring tests are generally required to a depth of approximately 16 m below the anticipated leg penetration at the selected site. Rig failures have often occurred where harder layers such as sand overlay softer layers such as clay. In this case, even test loading of the legs to a load over the anticipated maximum does not always preclude leg-bearing failure because the behaviour of clay under load varies with time. Leg loads are also variable and when drilling is in progress, leg loads are vibratory in character. These vibrations can sometimes cause the hard sand layers to weaken locally, enabling the leg to punch through, causing subsidence at leg loads lower than the test load. A high-resolution profiler survey will indicate boulders, shallow gas pockets and the homogeneity of the soil. Since profiler surveys do not adequately show differences in soil properties but only soil densities, it is essential to carry out soil borings parallel with profiler surveys.

For laying submarine pipelines, a bathymetric route survey of water depths and sea-bed slopes along the anticipated route is required. Further gravity cores or vibro-cores to a penetration of up to 6 m need to be collected to assess the sea-bed and the presence of sand waves.

For the design of offshore structures requiring deep penetration piles, the work begins by carrying out a bathymetric survey which will provide the exact water depth, the general sea-bed slopes and the likelihood of moving sand waves. Sparker or boomer surveys are carried out to map the area and soil layers so that if the anticipated platform location is changed, some correlation is available.

Platform strength and stability depend on the actual and lateral piling capacity which can only be determined by a detailed knowledge of the soil under the sea-bed. Adequate soil information can usually be obtained from one soil boring per platform location drilled to a penetration in excess of the pile penetration anticipated. Soil properties required for actual and lateral loads in the pile design can be derived from various field and laboratory tests.

The above environmental parameters are amongst the key inputs for design, construction, installation and life time use of offshore structures and marine vessels. While the state of knowledge today permits assessments to be made with a limited degree of accuracy or certainty, a rigorous evaluation of environmental data that are available is one of the most important steps in the design of offshore structures.

Bibliography

Malhotra, A. K. *An Introduction to Ocean Science and Technology*, National Book Trust, India, 1980.

Clayton B. B. and B. E. D. Bishop, *Mechanics of Marine Vehicles*, Gulf Publishing Company, 1982.

Wilson, J. F. *Dynamics of Offshore Structures*, John Wiley & Sons, Inc., 1984.

Sheppard, N. *Wind, Waves and Weather*, Petroleum Extension Service, 1984.

Sea ice. http://en.wikipedia.org/wiki/sea_ice, January 2, 2015.

3

Offshore Drilling and Production Platforms/Units

To drill an oil/gas well in the middle of sea or ocean, it is essential to have some platform on which necessary drilling equipment and other accessories and ancilliaries can be placed. As discussed in Chapter 1, depending on whether the type of well is exploratory or development in nature, the type of platform can be fixed or mobile. The term platform signifies some type of structure and in between these two major types of structures there is one more category known as a compliant type platform, which has its own application. Thus, depending on the number of wells to be drilled, the type of facilities to be installed on the topside, for example, oil rig, living quarters, helipad and so on, the depth of water, that is, shallow (<3 m), deep (350–1500 m), or ultradeep (>1500 m), practicality of methods of construction, fabrication and installation, meteorology and oceanography and finally the economics which vary globally from place to place, offshore platforms can be broadly categorised into three types:

1. Fixed platform
2. Compliant platform
3. Mobile platform/units

Elaborate discussions on all these three types have been made below.

3.1 Fixed Platforms

Fixed type platforms are those whose leg or base is fixed at the seabed forever either by way of penetration to the seabed or by its own weight. Moreover, they remain fixed at one place and cannot be moved to any location.

Different types of fixed platforms are

1. Jacketed platform:
   a. Well-protector platform
   b. Tender platform

    c. Self-contained template platform

    d. Self-contained tower platform

  2. Gravity platform

3.1.1 Jacketed Platform

The term 'jacket' here means a steel structure in the shape of a truncated pyramid that rises from the seabed to above the water line. This is built from tubular steel members and not structural irons. The cylindrical shapes of the tubular members create less resistance to waves and currents, reducing the amount of steel and thus its weight and cost. Other than the jacket, the jacketed platform consists of the 'decks' or the 'superstructure' on which the drilling, production and all other equipment rest, and 'piles,' which are driven through the leg of the jacket for securing the platform to the seabed. The superstructure generally comprises three decks, namely, (1) drilling deck or main deck, (2) wellhead/production or mezzanine deck, and (3) cellar deck. These decks are supported on a gridwork of girders, trusses, and columns in two rows (A & B) of four columns (1, 2, 3 and 4) each in a typical 8-leg jacket platform (Figures 3.1 and 3.2).

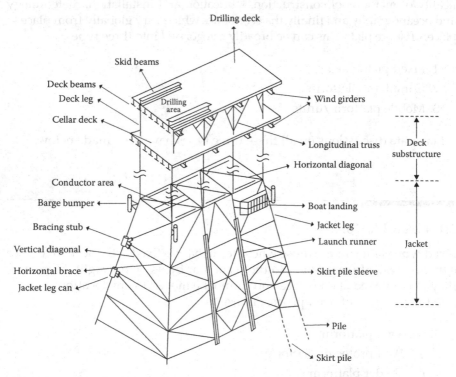

FIGURE 3.1
Principal elements of an eight-leg platform.

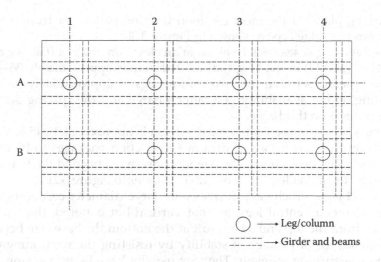

FIGURE 3.2
Plan view of the deck showing arrangement of legs and girders in eight-leg jacket platform.

The bottom ends of the deck columns are joined with the piles which protrude from jacket legs after being driven through them.

The jacket consists commonly of eight large-diameter tubular legs framed together by many relatively smaller sized tubular members known as 'braces.' These braces can be of three types, that is, (1) horizontal bracing, (2) diagonal bracing both in the vertical and horizontal plane and (3) X- or K-type bracing (Figure 3.3a). The designer from his experience decides on

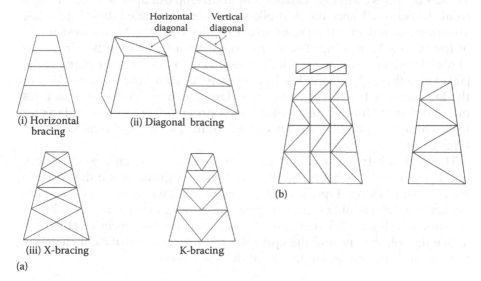

FIGURE 3.3
(a) Types of bracing. (b) Warren bridge type structure.

the framing plan but the most common framing pattern or truss work is of the Warren bridge type as shown in Figure 3.3b.

The leg spacing at sea deck level (i.e. at an elevation of 10–14 ft above mean water level [MWL]) in the longitudinal direction is approximately 36–45 ft, and the spacing between the two central legs is approximately 45–60 ft depending on the availability of a launch barge and the spacing between launch runners on this barge.

The leg spacing in the transverse direction is approximately 45 ft, which depends on the dimensions of drilling/production packages to be placed on the deck. The decks normally extend or overhang beyond the boundary outlined by the jacket legs by about 12–15 ft (refer to Figure 3.2).

The jacket legs, which vary in sizes with large-diameter corner legs and smaller diameter central legs, are not vertical but battered, that is, they taper out from the top and as a result at the bottom the base size becomes larger giving the advantage of stability by resisting the environmentally induced overturning moment. The legs usually have batter varying from 1:7 to 1:12.

Other than three main components of an 8-leg platform, that is, jacket, superstructure and piles, there are some auxiliary elements in it. Those are skirt-pile sleeves, drilling area deck substructure, deck modules, drilling area, conductor/riser area, boat landing facility, barge bumpers, launch runner, bracing stub and jacket leg can.

Skirt-pile sleeves are used for driving skirt piles in between the main piles around the periphery of a jacket leg. These skirt piles are used to increase the capacity of the structure in resisting the overturning moment. These skirt piles do not get extended up to the top but up to a certain height from the bottom. Hence the skirt-pile sleeves or pile guide tubes (Figure 3.4) are incorporated into the jacket structure between the two lowest levels of the horizontal bracing. The sleeves must be sufficiently offset from the plane of the jacket side to permit the piles to pass parallel to the plane of the jacket side through pile guides. In deepwater designs, the lower portion of the legs are constructed of very large diameter tubes so that several piles may be driven through pile guide tubes provided in the large diameter legs. This enlargement of the lower ends of the jacket legs is called a *bottle* (Figure 3.4).

The deck substructure in between two decks or between the cellar deck and the top of the jacket is usually made up of a group of parallel Warren trusses with Warren type bracing to form the planar trusses into a space frame. The web members and the upper and lower portions may be tubular. As shown in Figure 3.5, there are two deck levels – the lower level is high above the splash zone and the upper deck is placed at a sufficient elevation to place and operate the equipment at the lower level.

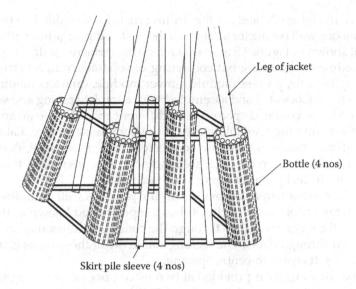

FIGURE 3.4
Skirt pile sleeves and bottles at the bottom of jacket leg.

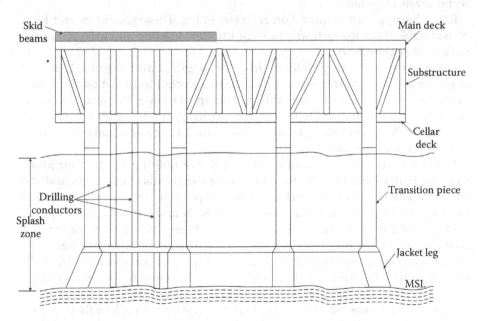

FIGURE 3.5
Deck substructure.

As shown in Figures 3.1 and 3.5, the drilling rig is to be skidded on two skid beams from one well conductor location to the next using large hydraulic jacks. The layout shown in Figure 3.5 is an exclusive arrangement for drilling wells.

Deck modules are nothing but consisting of separate compact structures like drilling module, process module, power module, quarters module and so on, which are fabricated and assembled with complete piping and so forth on shore within a confined space. These different modules are arranged in a planned way into the available space on the deck substructure. Later after placing different modules these are connected using pipe joints, flanges or welding and wherever on the deck areas some spaces are vacant, those will be covered with steel plates or removable hatch plates.

The conductor/riser area is provided on the jacket as well as on the decks where an array of conductor or riser tubes are positioned known as the well bay because the wells are drilled through the conductor tubes and later wells are produced through risers. The rectangular array of these tubes is usually set up on a 6–8 ft centre-to-centre spacing.

A boat landing facility is provided at two places, one on each longitudinal side of the jacket between columns 2 and 3. Each boat landing should have platforms for the safe embarking and disembarking of the crew personnel on or from the boats at two different elevations because of the variation in water levels at different times.

Barge bumpers are required on each jacket leg. These should extend for a considerable distance vertically to accommodate loading and unloading in a variety of sea conditions.

Launch runners are provided either on legs A2 and A3 or B2 and B3 depending on which transverse side is to be selected to lie flat on the transportation barge or launch barge during transportation of the jacket from the onshore construction yard to the offshore drilling site. These launch runners move over the skid beams placed on the launch barge and make the whole jacket slide into the water.

Jacket leg cans or chord and bracing stub are nothing but tubular joints in a particular framing plan for connecting the tubular jacket legs and the smaller diameter brace tubes (Figure 3.6). Depending on the position of the braces, tubular joints are designated as T, Y, K, N and so on (Figure 3.7).

Depending on the usage, the jacketed platform can either be of smaller size containing the necessary equipment for drilling or other purpose only or may have adequate strength and sufficient space to contain an entire drilling rig with all auxiliary equipment along with crew quarters and other utilities. Such large size platforms are normally referred to as self-contained platforms. So, following are the detailed descriptions of each of these types.

3.1.1.1 Well-Protector Platforms

Platforms built to protect the risers on producing wells in shallow water are called well-protectors or well jackets. Usually a well jacket serves from

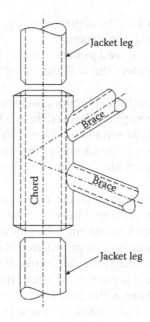

FIGURE 3.6
Leg cans and bracing stub.

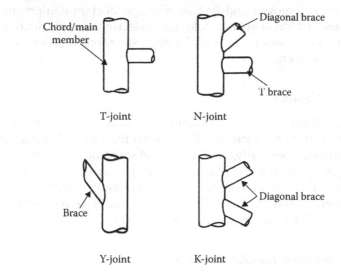

FIGURE 3.7
Tubular joints.

1–4 wells. Such a platform may be either one large pipe or caisson,* or an open lattice truss template/jacket structure.

There are two main types of well jackets: the slip-over type and the development type. Both types protect the well (or wells) from ship collisions and environmental forces.

The slip-over jacket is used for exploratory wells in 50–100 ft water depths. After the well is drilled through a caisson driven a few feet into the sea bottom, a well jacket is slipped around the caisson. This jacket is a four-legged structure with one side open meant for slipping around the well. The open side may later be added with additional bracing. It normally serves one well only.

On the contrary the development jacket may accommodate several wells and this is installed prior to drilling. This is used in conjunction with mobile drilling units like jack-up or semisubmersible, so that after the mobile unit leaves the site the well conductors or production risers are protected by the four-legged jacket. While drilling, the slot in the drilling vessel or the cantilevered portion of the jack-up is positioned above the top open portion of the truncated pyramidal structure of the jacket.

Sometimes at a greater depth (>1000 ft) these jackets can act as drilling platforms which contain only equipment like derrick, substructure, mud tank, mud pumps and power units. The living accommodations and other facilities are housed in another adjoining platform. After drilling, the drilling equipment is removed and in that place production equipment like a Christmas tree, production manifold for collecting the production from wells and so on are placed. Thus, under such situations it acts as a drilling well-protector platform.

3.1.1.2 Tender Platform

The tender platform is not used as commonly now as it was 40 years ago. In terms of size and operation, it falls between the well jacket and the self-contained platform. Generally, the derrick and substructure, primary power supply and mud pumps are placed on the platform. The drilling crew quarters, remaining equipment and supplies are located on the tender ship moored adjacent to the platform. The two are usually connected by a long walkway. Figure 3.8 shows such a type of platform.

3.1.1.3 Self-Contained Template Platform

The self-contained platform is a large, usually multiple-decked, platform which has adequate strength and space to support the entire drilling rig

---

* The dictionary meaning of 'caisson' is 'water-tight case used in laying foundations under water.'

FIGURE 3.8
Tender platform. (Courtesy of energy.gov; Wikimedia Commons/Public domain.)

with its auxiliary equipment and crew quarters, and enough supplies and materials to last through the longest anticipated period of bad weather when supplies cannot be brought in. The bad weather period is normally three to four days so the platform should accommodate supplies and materials for approximately twice this time.

There are two types of self-contained platforms: the template type and the tower type. The tower platform is also a template structure; however, the piles are driven in a different manner.

The most common type of offshore structure in service today is the template (jacket) structure (Figure 3.9). The term template is derived from the function of these structures to serve as a guide for the piles. The piles after being driven are cut off above the templates, and the deck is placed on top of the piles. The template is prevented from settling by being welded to the pile tops with a series of rings and gussets. Hence, the template carries no load from the deck but merely hangs from the top of the piles and provides lateral support to them.

Self-contained template platforms have been designed and constructed in many sizes and shapes. Early template structures had many legs and a

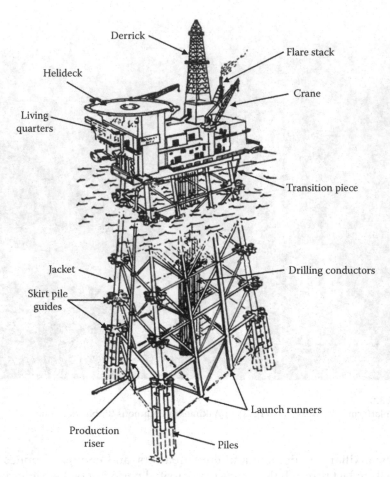

FIGURE 3.9
Template platform.

multiplicity of horizontal and diagonal braces. More recently, the trend is toward eight-legged jackets about which detailed description has already been made.

In some cases, 10-legged jackets are also used where an additional two legs are used (one on each side) in the narrow direction at the centre of the three legs for piling in addition to the four legs on each side.

Jackets constructed in more recent years frequently have larger-diameter corner legs; typical pile sizes of eight-pile jackets are 60-in. (1.5-m) OD for corner legs and 48-in. (1.2-m) OD for in-between legs. Allowing 1-in. annular clearance between the pile and the inside of the leg means that the jacket legs have internal diameters of about 62 in. (1.6 m) and 50 in. (1.3 m), respectively.

3.1.1.3.1 *Installation of Template Platform*

Installation of a platform consists of transporting the various components of the platform to the installation site, positioning the platform on the site, and assembling the various components into a stable structure in accordance with the design.

The installation procedure is classified as below:

a. Transportation

b. Removal of jacket from transport barge

c. Erection

d. Pile installation

e. Super structure installation

f. Grounding of installation welding equipment

All the operations are discussed below in detail, as is being done in Bombay High Offshore.

a. Transportation

For transportation of a template type of platform which is being used in the Bombay High field, consideration has been given to the jacket, piles and superstructure for various equipment to be placed on the platform like the test separator, fire water pump, generator and so on.

The jacket is normally taken to the site on a launch barge itself which is properly balanced before sailing and sea fastenings are done at several places. By sea fastening, the jacket almost becomes an integral part of the barge.

The jacket is provided with a launch truss on the side of it laying on the barge. The jacket is supported on these two launch trusses on the barge with a thick layer of grease in between to provide a smooth launching of the jacket into the sea.

The other platform parts should be loaded on the barge in such a manner to ensure a balanced and stable condition. Ballasting of the barge as required is done to provide a good balance. Adequate sea fastening is designed and provided in case of transportation of super structures and piles, so that they remain intact in severe weather to be encountered during the voyage.

From the fabrication yard to the installation site, a safe route is fixed based on ocean survey maps, to ensure a safe journey in the sea. Accordingly, a number of tow vessels are also determined.

b. Removal of Jacket from Transport Barge

When the launch barge carrying the jacket reaches the location of installation, the jacket is launched into the sea around 1½ miles away from the exact location of installation.

The launch barge consists of launch ways, rocker arms, controlled ballast and dewatering unit and a power unit (hydraulic ram, etc.) to assist the jacket to slide down the ways.

So, for launching the jacket, the barge is first anchored in the opposite direction of launching. Then the hydraulic rams are placed on the launch ways, sea fastenings are cut and a winch is connected to the jacket via a pulley so that pulling the sling on the winch helps in launching the jacket. After that the ballast tanks of the launching barge are flooded so that sufficient angle is provided to launch the jacket; that is, the barge tilts toward the side of launching. Then by giving hydraulic pressure in rams, the winch sling is simultaneously pulled to assist the launching. As the jacket starts sliding, the winch takes care of launching and after some time the jacket goes by its own weight. Rocker arms take care of the easy dropping of the jacket into the sea. A jacket that is to be launched has to be water-tight and buoyant.

c. Erection

Erection is accomplished with the help of a 'derrick barge,' which has a high capacity derrick (crane) that can rotate in any direction, living facilities for about 100–200 people, a helipad, all necessary diving facilities, NDT facilities and all sorts of facilities for anchoring, power, steam and so on.

So, when the jacket floats in the sea, with the help of two boats, it is brought to the site of erection where the derrick barge is already anchored. Anchors should hold the barge tight enough so that at the time of erection, the barge should not move from its position. The process of erection of the template has been schematically presented in Figure 3.10.

At the site of installation, four slings are attached to the top portion of the jacket for which lifting eyes are provided and the jacket is slowly upended. In this process, the jacket comes to the position of installation. It is kept in this position (orientation) with the help of slings and tugs. Then flooding valves are opened slowly by which water starts entering into the legs and buoyancy tanks and the jacket starts sitting down. But before upending, divers survey the sea bottom for any possible collection of debris and other things that may create problems in the installation. When the jacket is put on the sea bottom, the level of the top portion is measured and levelled so that after piling proper correction could be avoided.

d. Pile Installation

Proper installation of piling is vital to the life and permanence of the platform and requires each pile to be driven to or near design penetration, without damage, and for all field-made structural connections to be compatible with the design requirements.

The jacket columns and pile sleeves are provided with rubber diaphragms so that they remain waterproof at the time of floating and when piles are put inside, they get ruptured. Each pile has stabbing guides so that at the time

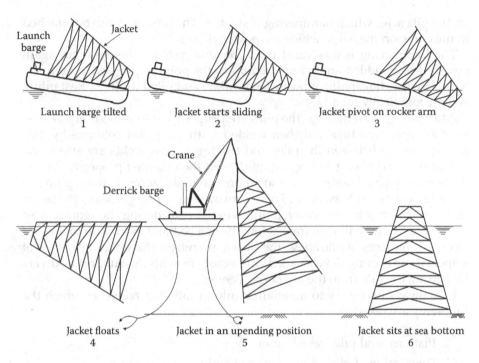

FIGURE 3.10
Process of erection of a template.

of stabbing a guide in the other one, we get a good joint for welding. Lifting eyes help in lifting each pile add-on section and when a pile has been driven, these eyes rest the pile on the jacket columns.

Pile driving refusal is the term encountered at the time of pile driving when it is found hard to drive a pile. This is defined as 'the point where pile driving resistance exceeds either 300 blows per foot (BPF) for five consecutive feet, or 800 BPF of penetration.' If there has been a delay in pile driving operations for 1 hour or more, the refusal criteria stated above shall not apply until the pile has been advanced at least 1 foot following the resumption of pile driving.

In a pile driving operation, first the cover plate of the jacket column is removed by gas cutting, then with the help of a barge crane the lower piece of the pile is lifted, lowered into one column until it rests on the lifting eyes, then another piece of pile is brought nearer to the previous one and stabbed with it after which these two are joined by welding and the joint is tested ultrasonically (NDT). Next by removing the eyes of the lower pile, it is lowered along with all other piles which are subsequently joined one after the other in a similar way until it punctures the lower diaphragm in the column and pierces into the sea bed. Then, accordingly, the hammer is lifted and put on the pile after which hammering is started. The hammers can be attached to the crane on the barge which are power-driven.

The hammering is continued until the pile rests on the eyes and again more piles are added and hammered up to the point of pile driving refusal. Similarly, the diagonally opposite leg is taken and piled through. Rest piling is done identically for all legs.

After piling of all the legs, the piles are properly cut to provide equal level for the super structure and then welded with the jacket columns by putting spacers in between the piles and the legs. These welds are also ultrasonically tested after grinding smoothly and then painted properly. Despite these main piles, if design demands more strength for foundation, piling in the pile sleeves has to be done. There is provision of pile guides in the jacket through which piles are lowered and welded considering the requirement of the length that will penetrate below the sea bed and extend up to the pile sleeve. Rest pieces of pile above the sleeve are retrievable either by means of a special connecting device called connectors or with the support of divers by cutting the pile from the top of the sleeves.

It is always necessary to maintain a pile installation record in which the following are noted:

1. Platform and pile identification
2. Penetration of pile under its own weight
3. Blow counts throughout driving with hammer identification
4. Penetration of pile under its own weight and weight of hammer
5. Unusual behaviour of hammer or pile during driving
6. Interruptions in driving, including 'setup' time
7. Lapsed time for driving each section
8. Elevations of soil plug and internal water surface after driving
9. Actual length of each pile and cut offs
10. Pertinent data of a similar nature covering driving, drilling, grouting, or concreting of grouted or belled piles

Grouting is the next step that is performed after piling. For this there is a grout line for each leg and sleeve which is connected to the bottom of the leg. Cement slurry is pumped into this and from the top return water present in the annular space comes out through an outlet connection. After some time when the cement is fully filled in the annular space, cement starts coming out. The samples of cement slurry going in and coming out of each leg and sleeve are taken and cubes are made out of them by placing in wooden frames which are later sent for hardness tests to get an idea of cementation in the annular space. Divers survey the cementation of sleeves.

e. Superstructure installation

Normally four decks are installed on an unmanned platform. These are

1. Cellar deck
2. Main deck
3. Mezzanine deck
4. Helideck

The main deck and the cellar deck are fabricated in one piece and the mezzanine deck and helideck are in another piece. They are transported on a cargo barge and are directly lifted and stabbed over the piles. But before stabbing a transition piece is added which makes slanted piles vertical. The main deck with the cellar deck is lifted with the help of a derrick and set with proper care being exercised to ensure proper alignment and elevation. The deck elevation shall not vary more than ±3 in. (76 mm) from the design elevation shown in the drawing. The finished elevation of the deck shall be within ½ in. (13 mm) of level.

Once the super structures are installed, all stairways, hand rails, boat landings, rub strips and barge bumpers and other similar systems are installed.

After this all piping hook up work takes place. Simultaneously, all electrical connections are made, various pumps and generators are tested and commissioned. After hook-up of all piping systems, they are hydraulically tested according to the codes and contract specifications.

3.1.1.4 Self-Contained Tower Platform

The self-contained tower platform is characterised by relatively few large-diameter, non-battered legs and fewer diagonal braces of larger size than those used in regular template type structures. The tower type jacket was conceived to eliminate the need to launch the structure from a barge; it can be floated to location using the buoyancy of its larger-diameter legs. The tower platform was originally designed to supply deepwater structures for the Pacific coast and for Cook Inlet, Alaska.

The regular template structure has many cross braces, horizontal and diagonal. The tower structure has relatively few braces, and none extend into the splash zone above the mean low-water level. This reduces the lateral resistance the structure offers to the passage of large storm waves and eliminates having bracing members in the path of ice floes in cold climates.

The pile foundation for a tower platform usually consists of several groups of piles, frequently four. Each group of piles is driven through one of the large-diameter legs with each pile subsequently serving as a conductor through which a well may be drilled.

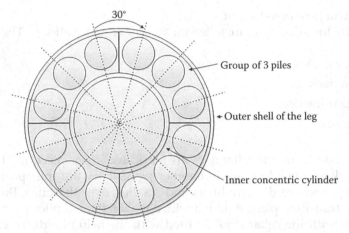

30°

Group of 3 piles

Outer shell of the leg

Inner concentric cylinder

4 groups of 3 piles each through one leg

FIGURE 3.11
Arrangement of piles in one leg of tower type platform.

FIGURE 3.12
Tower type platform. (Courtesy of Offshore Exploration Conference, 1968.)

There are from 8 to 12 cylindrical tubular piles per leg. The piles are arranged in a circle around the inside of the large-diameter leg (Figure 3.11). The guide tubes for the piles are structurally positioned using bulkheads with an inner cylinder inside that is concentric with the outer shell of the leg. Cement grout is used to fill the annulus between the outer cylindrical shell and the inner cylinder. This large usage of grout keeps the outer cylindrical legs from failing by local buckling and bonds the piles to the inner and outer cylinders to achieve composite action.

The four-legged tower platforms in Cook Inlet have legs ranging in diameter from 14–17 ft (4.3–5.2 m). The bracing member, all of which are below the level of formation of maximum ice thickness, range in diameter from 48–74 in. (1.2–1.9 m) (Figure 3.12).

The piping and valves required for flooding the legs for upending the tower are located inside the legs, as are the conduits for cathodic protection anodes, the flowline risers, pumps and instrumentation. Upending and positioning of the tower structure in the water can be accomplished without the use of a derrick barge, although one is needed for placing the various segments of the deck structure on the upended tower.

3.1.2 Gravity Platform

Concrete gravity structures rest directly on the ocean floor by virtue of their own weight. These structures offer an attractive alternative to piled steel template platforms in hostile waters like the North Sea. The advantages are that the structures can be constructed onshore or in sheltered waters, towed semisubmerged to the offshore location and installed in a short time by flooding with seawater (ballasting). Other advantages are the elimination of steel piling, the use of traditional civil engineering labour and methods, less dependence on imported building materials, a structure tolerant to overloading and the high durability of the materials.

There are various designs of gravity platform, namely condeep platforms, seatank platforms, andoc platforms and CG Doris platforms. Figures 3.13 through 3.16 depict these four types. The basic components of the first three types consist of cellular concrete foundation, called *caisson*, with two, three or four hollow-concrete steel *shafts* or *towers* to support the *deck structure* and the *base or skirt*.

3.1.2.1 Condeep Platform

Condeep (abbr. concrete deep water structure) refers to a make of gravity-based structure for oil platforms constructed by Norwegian contractors. Details of its main component are as follows:

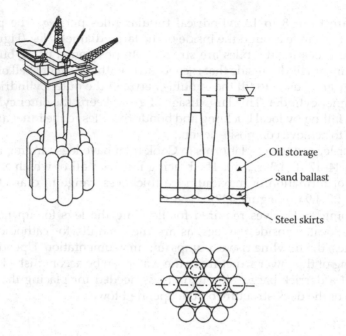

FIGURE 3.13
Condeep gravity platform.

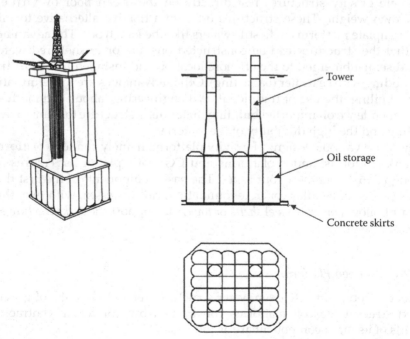

FIGURE 3.14
Seatank gravity platform.

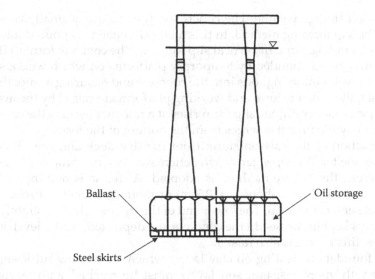

FIGURE 3.15
Andoc gravity platform.

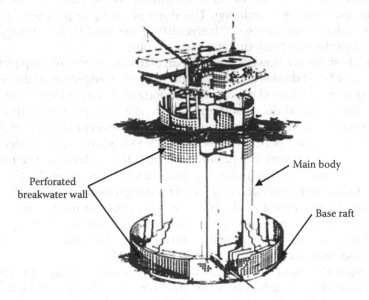

FIGURE 3.16
CG Doris gravity platform.

Caisson: In the condeep platforms, the caisson which remains submerged in water is hexagon-shaped and usually consists of 19 vertical, interconnected cylindrical cells with spherical domes at each end (Figure 3.13). Each cell has an outer-shell diameter of 66 ft (20 m) and is more than 164 ft (50 m) in height. The cells provide buoyancy during construction and towing and later serve as oil storage volume. The concrete cell walls are normally constructed by the slip forming method. In this method, concrete is poured into a continuously moving form. In vertical slip forming, the concrete form (cell in this case) may by surrounded by temporary platforms on which workers stand, placing steel reinforcing rods into the concrete and ensuring a smooth pour. Together, the concrete form and working platform are raised by means of hydraulic jacks. Generally, the slip form rises at a rate that permits the concrete to harden by the time it emerges from the bottom of the form.

The construction of the caisson starts from the dry dock and then it is floated to the sheltered deepwater construction site. For the proper placement of concrete, the 'tremie method' is adopted. A tremie is nothing but a downpipe or vertical tube about 1 ft (0.3 m) or more in diameter through which concrete enters the form. The lower end of the tremie, which is slightly flared, is always kept immersed in the concrete just deposited. As the level of concrete rises, this tremie is also raised.

Skirt: For foundations resting on clay layers which often show sufficient slipping strength, more resistant soil layers must be reached with skirts installed beneath the base raft. These skirts must be thin enough to penetrate the ground and yet be capable of transmitting the vertical and horizontal forces imposed on the foundation. The skirts serve three purposes, namely penetration to the weaker soils, transmitting the load to the stronger soil below and protecting the foundation from scour.

Figure 3.17 shows the dowels* and skirts in plan view of the condeep platform. The location of steel dowels and the general vertical arrangement of dowels and skirt compartments formed beneath the caisson are shown in Figure 3.18.

Below the spherical dome bottoms of the cells the cylinder walls extend downward as 3 ft (0.95 m) thick concrete skirts to a level of 1.6 ft (0.5 m) below the tip of the domes. Beneath these concrete skirts, corrugated steel skirts extend 11.5 ft (3.5 m) from the 12 outer cells and the centre cell. The purpose of using the skirts is to improve the foundation stability, make the structure less vulnerable to erosion and facilitate the grouting process. The skirts prevent lateral movement of the platform on the ocean floor. They contain and facilitate evacuation of water from beneath the structure during penetration and serve as containment walls for the grout. The skirt area is usually divided into several compartments. (See Figure 3.17.)

To prevent the platform from sliding and causing damage to the steel skirts as touchdown is achieved, three steel pipe dowels are provided. These extend down 13 ft (4 m) below the level of the steel skirts and penetrate the ocean floor first.

The dowels are positioned outside the cylindrical cells as shown in Figure 3.17. The acceptable sea-state conditions for touchdown (more specifically, the initial contacting of the ocean floor with the steel dowels) are based on the axial and bending strengths of the dowels.

---

* A 'dowel' is a solid cylindrical rod or peg used for holding together components of a structure.

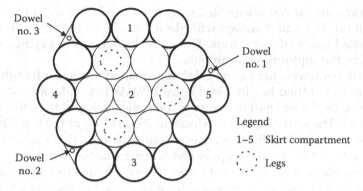

FIGURE 3.17
Skirt compartments and dowel location.

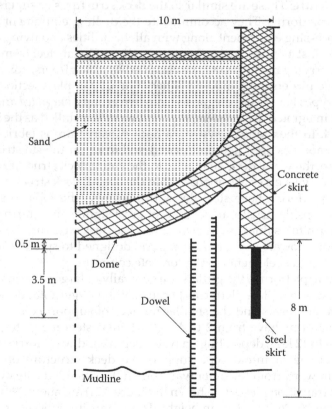

FIGURE 3.18
Concrete and steel skirt.

Towers/columns/shafts/legs: Referring to Figure 3.17, three of the caisson cells (cell no. 1, 3 and 5 shown with dotted circle) in the condeep design are extended into shafts or towers. These are slender, tapered cylinders constructed by the slipforming technique.

The shafts or towers have a conical shape for about the lower two-thirds to three-quarters of their height. (See Figure 3.19.) As the shafts are extensions of the cells, the lower shaft diameter is the same as that of the cells, that is, 66 ft (20 m). The shaft diameter reduces to 39 ft (12 m) at the top. The wall thickness of the conical shafts decreases from 30 or 41 in. (75 or 105 cm) at the bottom to 20 in. (50 cm) at the upper end of the conical section.

The concrete legs serve as enclosures for well conductors by protecting those from wind and waves. Steel guides with drilled plugs are cast into the lower concrete domes so that conductor tubes may be inserted past the plugs with little effort.

Deck structure: These are similar to the deck structures or superstructure of template platforms. They accommodate the drilling equipment and production processing equipment along with all the utilities and living quarters.

All the deck structures are made of three-dimensional steel elements. The main load-carrying members may be classified as plate girders, box girders or trusses. Truss girders are composed either of tubular or plate sections (Figure 3.20). Deck types have been categorised into two groups: *integrated* and *modularised*. In the integrated deck system, the equipment is installed as the structure is fabricated. In the modular system, a deck substructure is fabricated, and equipment modules are lifted and placed within or on the substructure. In common use, the word 'deck' may refer to the entire deck structure or to any one of the many levels or elevations within the overall deck structure.

Concrete platforms have steel decks to minimise the topside structural weight. The topside structural weight is of great significance during tow-out of the platform to its final offshore location. The deck structure is connected to the platform towers using *transition pieces* designed to spread the applied loads from the steel elements to the concrete columns.

For concrete platforms, the girders are generally of larger dimensions than girders for steel template platforms because of the larger total deck weights for concrete platforms and the smaller number of support columns. Tower concrete platforms have from 2 to 4 legs whereas steel template platforms have from 4 to 12 legs, depending on water depths and wave forces. Concrete platforms of the manifold type support the deck structure on columns extending upward from the outer wall and on the extended core cylinder.

The steel transition pieces (as shown in Figure 3.21) are about 39 ft (12 m) in diameter and 20–26 ft (6–8 m) in height. These transition pieces transfer the loads from the rectangular deck substructure to the circular concrete towers. This load transfer requires curved steel panels forming, approximately, an inverted truncated cone. The vertical stiffeners shown in Figure 3.21 may be channels, angles, tees or combinations of these sections. Horizontal stiffeners often encircle the inside of the periphery also.

FIGURE 3.19
Slipforming of condeep tower. (Courtesy of Bluemoose/Wikimedia Commons/CC-BY-SA-3.0.)

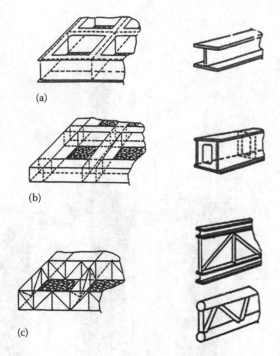

FIGURE 3.20
Types of girders. (a) Plate girder. (b) Box girder. (c) Trusses made of beams and tubulars.

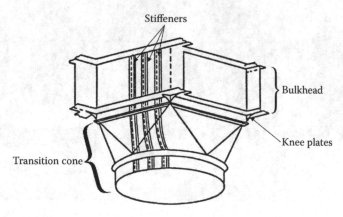

FIGURE 3.21
Transition piece to join deck substructure with concrete tower.

3.1.2.2 Andoc and Sea Tank Platform

Andoc is named after its construction firm, namely Anglo-Dutch Offshore Concrete a joint venture of Dutch and British firms. An example of an andoc platform is Dunlin A (UK). Sea tank is designed by Sea Tank Company Paris and its examples are Brent C (UK) Frigg TPI (UK) and Comorant A (UK). These platforms are quite similar to the condeep platform having the tower design with the same four components as described there. Further there is a lot of similarity between andoc and sea tank as well (Figures 3.14 and 3.15). Both these platforms have four towers and have caissons built-up of square cells for oil storage. The outer cell walls of the caissons are constructed as circular cylindrical shell panels and the caisson cell roofs consist of a combination of conical shells and flat slabs of variable thicknesses.

One major difference between the andoc and sea tank platforms is the mounting of the deck structure. Like the condeep platform, the sea tank platforms have the steel deck structure fastened to the tops of the concrete towers by a specially designed steel transition joint, whereas the towers on the andoc platform change from concrete shells to steel shells just below the lower astronomical tide water level.

3.1.2.3 CG Doris Platform

The first concrete gravity platform in the North Sea in Norwegian waters was a CG Doris platform, named Ekofisk 1. The structure has a shape like a marine sea island and is surrounded by a perforated breakwater wall. The original proposal of the French group CG Doris (Compagnie General pour les Developments Operationelles des Richesses Sous – Marines) for a pre-stressed post-tensioned concrete 'island' structure was adopted on cost and operational grounds, and DORIS was the original contractor for the structural design. Further examples of the CG Doris designs are the Frigg CDP1 (UK), Frigg MP 2 (UK) and Ninian (UK).

This platform is different from the previous three platforms in a way that this is known as a *manifold type platform* whereas those three are categorised as *concrete tower type platforms*. There is one more type of manifold platform known as Petrobras (abbr. PetroleoBrasileiro S. A) gravity platform, whose examples are Pub 3, Pub 2 and Pag 2 in Brazil.

The CDP 1 platform, which is a CG Doris type, rests on a 331 ft (101 m) diameter base raft. The bottom slab of the raft is from 2.3–2.6 ft (0.7–0.8 m) thick and is stiffened by a system of radial and circumferential walls 49 ft (15 m) high. The two outer circumferential walls are perforated and serve as antiscour walls.

As shown in Figure 3.16, the main body of the platform is approximately a cylinder made of six vertical cylindrical shell panels the radii of which are less than the radius of the overall body so that the panels form a lobed shape as shown in Figure 3.22. This lobed cylinder has a 203 ft (62 m) outside diameter. For a height of 223 ft (68 m) from the base slab, the lobed cylinder

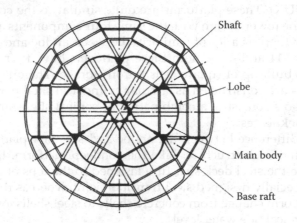

FIGURE 3.22
Horizontal cross-section of CG Doris platform.

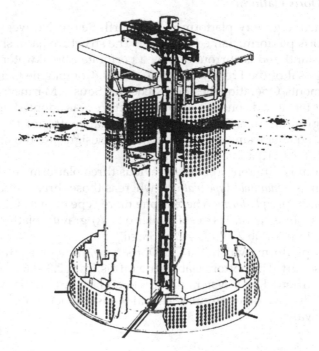

FIGURE 3.23
Frigg MCP-01 platform. (Courtesy of Total Marine Ltd.)

is stiffened by 6 radial diaphragms 1.8 ft (0.55 m) thick. The upper part of this lobed cylinder is a 4 ft (1.2 m) thick breakwater wall stiffened at the top (elevation 351 ft or 107 m) by radial precast prestressed beams 13 ft (4 m) deep. The innermost cylinder, or shaft, has an inside diameter of 30 ft (9 m) and encloses the gas pipelines. Twenty-four vertical tubes are installed through the ballast inside the lobed wall for well drilling and oil production. The deck structure is supported by fourteen 6.5 ft (2 m) diameter steel columns filled with concrete; these bear upon the precast, prestressed radial concrete girders at the top of the shell walls. The deck structure is mounted on a system of 45 parallel deck beams outlining an area 210 ft × 207 ft (64 m × 63 m) and located 403 ft (123 m) above the ocean floor. The beams are between 13 and 16 ft (4 and 5 m) deep and weigh about 303 tons (275 tonnes) each. The total deck area is 10.8×10^4 ft^2 (10,000 m^2).

Figure 3.23 shows a cut-away of the Frigg MCP-01 platform. The structural features of this platform are the same as for the CDP 1 platform except that the height of the perforated breakwater wall is 6.5 ft (2 m) lower because the water depth is 6.5 ft (2 m) less, and there is only one deck.

3.1.2.4 Installation of Gravity Platform

The installation of a tower type gravity platform (condeep) involves various steps starting from construction of the bottom structure up to the shafts or towers with intermittent movement from dry dock to sheltered water and then erection of the deck and the final settling of the platform. All these steps are depicted in Figure 3.24. Similarly, the construction steps for manifold type (Ekofisk) platform are shown in Figure 3.25. As the two figures indicate, the construction steps are basically the same.

1. Construction begins in a dry dock excavated close to the sea and closed by a steel sheet pile wall. The base slab is cast of reinforced concrete, and using slip forming, the base raft is formed. In the case of manifold type the holes in the antiscour perforated wall are closed with removable steel plugs before floating out of dry dock.
2. The dry dock is filled with water and the sheet pile wall (dock gate) is removed to permit towing of the raft to the construction site.
3. Next the construction of all the structures like cell walls, upper domes, shafts and so on except the deck is accomplished in sheltered deepwater in floating condition using a slipforming technique.
4. Now for installing the deck substructure in the case of the tower type platform (condeep), the constructed structure is temporarily immersed by ballasting to a deep draft and then the completed deck substructure is floated to the top of the shafts and then the whole unit is deballasted to lift it.

   In the case of the manifold type platform (Ekofisk), the deck substructures are made of prestressed and reinforced concrete.

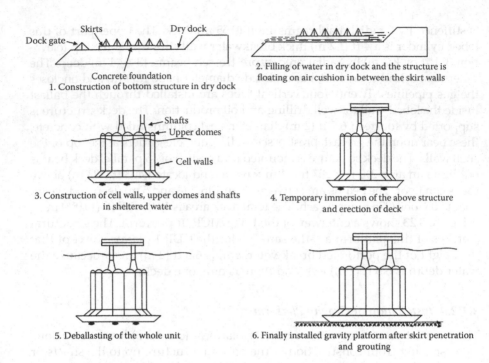

FIGURE 3.24
Sequence of installing a concrete tower type platform.

5. After lifting the unit, deck modules are joined with the deck substructure through transition pieces, wherever applicable, by welding.

6. Then the completed platform is deballasted to a minimum drag draft, and towed using several tugboats from the sheltered deepwater construction site to the final offshore location. During towing, the tops of the caisson cells for the tower type platforms should be submerged 33 ft (10 m) below water level because then regardless of the condition of the sea, the structure remains stable. This takes on average 3–5 days. Several precautions are taken during this towing and normally a model test is done to decide upon the number of tugboats, towing speed, towline forces, minimum draft and the dynamic response of the structure to waves. The economical towing speed is about 2 knots (3.7 km/h). It is also necessary to do this operation when forecast of favourable weather is made for at least 3–4 days.

7. Finally, the installation of the platform on the ocean floor by ballasting is done. This is the most complex and represents the critical phase in the life of the structure.

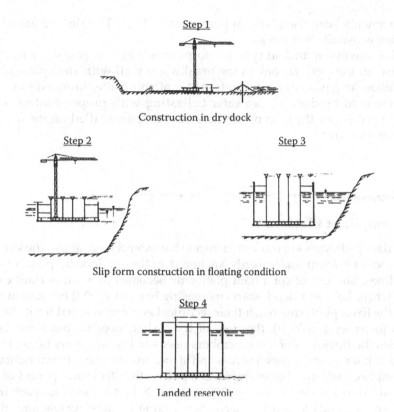

Step 1

Construction in dry dock

Step 2

Step 3

Slip form construction in floating condition

Step 4

Landed reservoir

FIGURE 3.25
Sequence of installing a concrete manifold type platform.

For a tower type structure (condeep), the following steps are followed for its installation:

  i. Touchdown
 ii. Dowel insertion
iii. Steel skirt penetration
 iv. Concrete skirt penetration
  v. Dome seating
 vi. Water extraction and grouting inside the skirt compartments

It is important to have control over the ballasting at different steps so that the platform is always kept level. The ballast in various compartments provides vertical downward forces whereas there is water pressure beneath the base raft, so by controlling both, the chances of tilting the platform can be avoided. Water extraction and cement grouting injected inside the skirt compartments keep the platform permanently level. The time required for grouting is usually 3–5 weeks.

In the case of a manifold type platform (Ekofisk), it is necessary to close and seal all the perforations in the breakwater wall with steel plate plugs and neoprene gaskets and second, at the offshore site, immersion of the platform is to be done by seawater ballasting with proper control of its decent speed with the help of submersible pumps installed on the outside of the main wall.

3.2 Compliant Platforms

Compliant platforms are those structures that extend from above the surface to the ocean bottom and directly anchored to the seafloor by piles and/or guidelines. The use of compliant platforms becomes more important especially when the water depth starts increasing beyond 1000 ft because at that point the fixed platforms reach their technical and commercial limit. As the depth increases (>1000 ft), the base of fixed structures becomes too large and also the thickness of steel members needs to become more because the dynamic interaction between waves and structures reaches critical limits for the template platform. For example, comparing the structural period of the jacket at two water depths, that is, 300 ft and 1000 ft, it is found to reach from less than 3 seconds to nearly 6 seconds resulting in considerable amplification of the energy transmitted to the structure by the operating sea state (i.e. wind and waves). Under these conditions, fatigue aspects may be critical and additional steel may be required to reinforce the structural joints and stiffen the members – thus reflecting on fabrication costs and operational/installation difficulties.

Common types of compliant platforms are

1. Compliant tower
2. Guyed tower
3. Tension leg platform

3.2.1 Compliant Tower

These are tall structures built of cylindrical steel members but slender in shape. Pilings tie it to the sea bed but in a small footprint with a narrow base. These structures do not have the same stability as the fixed platforms; rather they sway with the current, waves and winds to the extent 10–15 ft off centre in extreme cases, but during normal operating conditions, the motions are much less under environmental load.

To help it sway, several rigid steel sections of which the C.T. is made of are joined together by hinges. It is fairly transparent to waves and is designed to flex with the forces of waves, wind and current. It uses less steel than a conventional jacket platform which becomes obvious when comparison is made between the structural weights of the world's deepest fixed platform (Bullwinkle) and the world's tallest compliant structure (Petronious) in Table 3.1.

Compliant towers (Figure 3.26) are designed to have considerable 'mass' and 'buoyancy' in their upper regions and as a result they have a sluggish response to any forces because its mass and stiffness characteristics are tuned such that its natural period would be much greater than the period of waves in the extreme design environment. The typical 10–15 second cycle wave passes normally through the structural frame without any response just as water reeds behave when waves pass through them. In fact, an early version of compliant towers was named *Roseau*, meaning reed in French.

The construction of compliant towers is more or less similar to jacketed structures except that the requirement of steel (or tonnage) is less in this case.

A compliant tower structure can be divided into four basic structural components:

- The foundation piles.
- The base section.
- The tower section(s). Depending on the water depth and the means of transport, the tower can be made in one or more sections.
- The deck.

The base and the tower sections are lattice space structures fabricated from tubular steel members and thus termed the jacket base and the jacket tower sections. Normally the tower section is much larger than the base section.

TABLE 3.1

Structural Weights of Fixed Platform and Compliant Towers

Platform	Water Depth, ft (m)	Topside Weight (Tons)	Structural Weight (Tons)
Bullwinkle (fixed)	1353 (413)	2033	49,375
Baldpate (CT)	1650 (503)	2400	28,900
Petronious (CT)	1754 (535)	7500	43,000

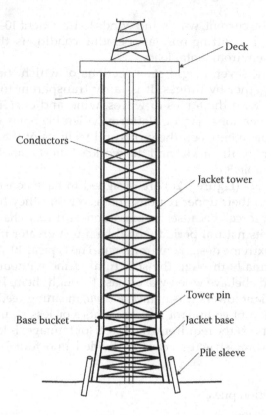

FIGURE 3.26
Compliant tower.

3.2.1.1 Installation of compliant tower

The jacket base section is transported on and launched off the deck of a launch barge at site. The top of the jacket is connected to a derrick barge and the bottom to its assisting tugs. Once in the water, the jacket base section is upended by the derrick barge assisted by the jacket buoyancy. Once vertical, the jacket is lowered and manoeuvred into position often with the guidance of a preinstalled docking pile.

Piles are transported to the site on cargo barges, lifted off and upended, using the cranes of the derrick barge, lowered, stabbed through the jacket base pile sleeves and driven to target penetration. Pile driving is similar to that described in the case of the template platform.

After the verticality and orientation of the jacket base are achieved, piles are grouted to the pile sleeves. The base structure is now safely secured to the seabed and ready to receive the next tower section.

Then, the tower section is transported on the deck of a launch barge and launched into the water. Due to the large weight and height of the tower section, it is designed such that it is self-upending after separating from the launch barge and going into the water. Once vertical, the tower section is ballasted to the required float-over draft. The tower section is then towed and positioned over the preinstalled jacket base section. With assistance from the attending derrick barge and position-holding by tugs, ballasting continues until the pins at the base of the tower section engage one by one in their respective receiving buckets at the top of the preinstalled base section. Grout is then injected into the gap between the pin and the bucket, which provides the structural continuity and the integrity of the entire subsurface structure (base and tower sections). The tower is now ready to receive the topside deck. The topside deck can then be lifted by the derrick barge and set onto the tower structure.

3.2.2 Guyed Tower

A guyed tower (Figure 3.27) is a slender structure made up of truss members, which rests on the ocean floor and is held in place by a symmetric array of

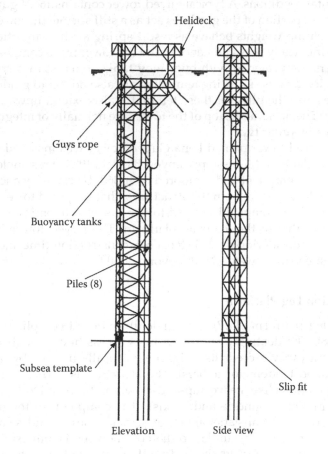

FIGURE 3.27
Guyed-tower platform.

catenary guy lines. A guyed tower may be applicable in deep hostile waters where the loads on the gravity base or jacket-type structures from the environment are prohibitively high.

The tower is normally supported on the seabed on closely located ungrouted pipe piles to simulate the behaviour of a hinged base and is supported laterally by anchored guy lines attached to the structure near the water surface. The guy lines typically have several segments. The upper part is a lead cable which acts as a stiff spring in moderate seas. The lower portion is a heavy chain with clump weight. The upper part of the guy lines is attached to the deck in wedge-like clamps and then passes vertically down the structure to fairleads approximately 50 ft below the water line. From the fairleads, the lines extend radially at an angle of approximately 30° to clump weights on the ocean floor. From the clump weights, the guy lines further extend to anchor piles. The number and size of the guy lines are dictated by several factors, including the size of the structure, water depth, degree of redundancy desired and environmental conditions. A typical guyed tower contains 16–24 guy lines.

As the upper portion of the guy lines act as a stiff spring, the lower portion along with clump weights behave as a soft spring as they are lifted off the bottom during heavy seas and hence make the tower more compliant.

The towers are provided with buoyancy tanks to assist in carrying part of the deck load. Subsea drilling templates are also added to guide the conductor pipes and the piles for these templates also extend upward into the tower's legs. The decks at the top of the tower are normally of integrated type rather than a modular type.

The first guyed tower named Lena Guyed Tower was installed in 1000 ft (300 m) water depth in Mississippi Canyon Block in 1983. It resembles a jacket structure but is compliant and is moored over 360° by catenary anchor lines.

In a water depth where both the structures, that is, guyed tower and template platforms are compatible, guyed tower has an economic advantage in comparison to that of the template platform. The guyed tower has a cost advantage, by considering overall steel cost, construction time and installation time, especially in water depth beyond 1000 ft.

3.2.3 Tension Leg Platform

A tension leg platform (TLP) is a vertically moored compliant platform (Figure 3.28). The floating platform with its excess buoyancy (because the platform's buoyancy exceeds its weight) is vertically moored by taut mooring lines called tendons (or tethers). The structure is vertically restrained precluding motions like 'heave' (up-and-down motion) and 'roll' and 'pitch' (rotational motion around x- and y-axis). It is compliant in the horizontal direction permitting translational motions, that is, 'surge' and 'sway.'

The TLP is structurally similar to the common semisubmersible drilling structure (discussed later). Its size is directly related to the payload required and the environmental forces it has to withstand. This normally indicates

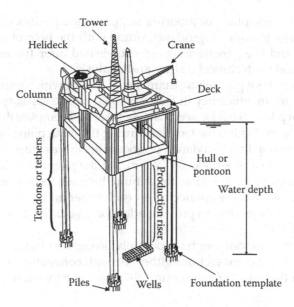

FIGURE 3.28
Tension leg platform.

that the buoyancy will be between 2 and 5 times that of a semisubmersible structure.

The main components of the TLP are the hull structure, the deck structure, the foundation with piles and the tendons/tethers.

The *hull structure* mainly comprises pontoons, corner and centre nodes and columns or legs. Its shape can be square, triangular or hexagonal, although square is more common.

The *deck structure* is made up of bulk head girders and structural deck pallets on which all necessary equipment and accessories for drilling and production are placed.

The construction of the hull and deck can be made in two ways:

1. *Separate construction*: Here after constructing the hull and deck separately, the mating is done at the site following the same sequence similar to the gravity platform.
2. *Single piece construction*: Here the deck is built on to the hull and equipment is placed on the completed platform.

The most difficult and uncertain part of TLP construction is the placement of the seafloor template as foundation and piling.

Foundation consists of four separate steel octagonal templates. Each template is anchored to the seabed by 8 tubular steel piles, 72-in. diameter, stabbed through the sleeves in the template and driven to a minimum penetration of 58 m.

The foundation templates or mooring templates are positioned on the seabed using a large temporary guidance frame with the help of a floater. The pile diameter and the penetrations are established after the extensive site investigation and geotechnical testing and analysis.

The pile group has a pitch circle diameter of 16.5 m, which ensures that the group has maximum efficiency in terms of axial tensile capacity.

An additional pile sleeve is located at the centre of each template through which a pin-pile is used to stabilise the template during the main piling operation and to ensure that the required foundation positional tolerances are maintained. The pin–pile is driven to a penetration of 24.5 m. All the piles are connected to the template injecting cement grout into the annulus between the pile and the sleeve.

The mooring sleeves are connected to each other and to the pile sleeves by 4.5 m. deep plate girders to provide additional stability and a more even distribution of load.

Tendons/tethers: Tendons can be separately brought to the site by joining/assembling their different sections either through connectors or by welding. Depending on their way of assembling they can be either towed dry or wet to the site.

1. *Dry tow of tendon section*: A typical tendon string is made up of a bottom section, several main body sections and a top section made of a length adjustment joint (LAJ). The bottom connects through a mechanical connector to the pile or foundation template. The individual sections are joined together with a mechanical connector such as the Merlin connector. The main body sections are typically fabricated in sections of 240–270 ft lengths, shipped on a cargo barge to the installation site, where they are lifted and upended by a crane barge. During the tendon assembly process, the weight of the tendon string that has already been assembled is supported on a tendon assembly frame (TAF), which is a purpose-built structure that is installed over the side of the derrick barge. The maximum length of individual sections is determined by the available hook height of the derrick barge. Tendon strings with longer sections require fewer mechanical connectors but a larger installation crane boom.

2. *Wet tow of complete tendon*: As an alternative to using tendon connectors, tendon strings are assembled by welding individual sections together. The tendons are subsequently launched and wet towed to the site in the same way as the pipe bundles. Buoyancy modules may be strapped onto the tendons to provide additional buoyancy and control stresses during the wet tow. Once at the site, the tendons can be upended with the help of winches or cranes and controlled removal of the buoyancy modules. This method saves the cost of mechanical connectors. The tow operation must be designed carefully to ensure that failure of any component during the wet tow does not lead to the total loss of the tendon string.

3.2.3.1 Installation of TLP

The TLP can be installed in two ways: (1) Tendons are preinstalled before the arrival of hull or the platform and (2) tendons and hull are installed simultaneously.

1. Tendons can be installed prior to the hull arrival to site. To ensure that the tendon and its components remain taut, upright and to keep the stresses within design allowables, temporary buoyancy modules are provided in the form of steel cans which connect to the tendon at its top. This temporary buoyancy module (TBM) is clamped to the tendon after the tendon assembly is complete. The tendon and the TBM assembly are lifted from the side of the derrick barge and manoeuvred until the tendon bottom connector is stabbed into the foundation and latched in position. The TBM is then deballasted such that it applies sufficient tension to the tendon until it is hooked up to the TLP. Figure 3.29 shows a schematic of a preinstalled tendon. When the platform arrives on site, it is ballasted until the tendon connector engages the LAJ teeth at which point the connector is locked off. Once the connector is locked off, the connector allows the downward movement of the platform under wave action but prevents any upward movement. This is known as 'ratcheting.'

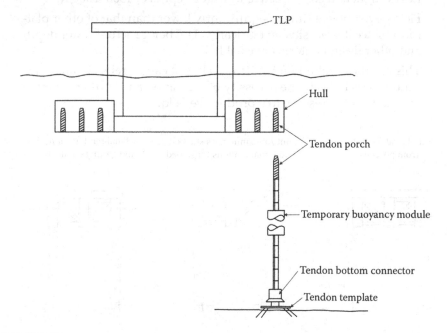

FIGURE 3.29
A pre-installed tendon before hook-up to TLP.

The ballasting operation continues in parallel with the ratcheting motions, until the desired draft is reached. At that point, the ballast water is pumped out causing the tension in the tendons to increase while the hull draft only reduces marginally by the amount of tendon-stretch. The de-ballasting operations are considered complete when the desired pretension is reached in the tendons.

2. In the second method of installation, once the tendon is assembled on site, the derrick barge hands it over to the platform/hull where it is hung from the tendon porches. Once all the tendons are hung from their respective porches, their bottom connectors are stabbed into their piled foundations. Tendon pretensioning is achieved using mechanical tensioners similar to chain jacks. The pretensioning operation can proceed in several stages with only one group of tendons being tensioned during each stage in order to limit the number of the tensioning devices required.

The tendon porches in this type of installation have to be open on one side to allow the tendon to be inserted laterally. This restriction does not apply to the preinstalled tendons. Figure 3.30 shows a schematic of the complete mooring sequence.

Some salient features of the TLP are

1. Less weight and cost-effective in water depths of 1500–2000 ft.

2. Field development time is significantly lower than that of other platforms since it is possible to fabricate a TLP before actual water depth and other design criteria are established.

3. This has the capability for shifting. Thus, it can be relocated if delineation wells indicate the necessity of relocation and could be moved from a depleted reserve to a productive field.

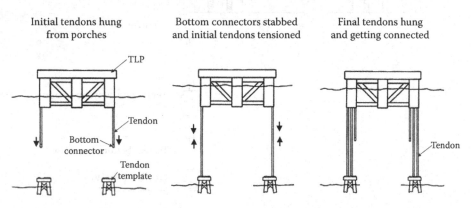

Initial tendons hung   Bottom connectors stabbed   Final tendons hung
from porches   and initial tendons tensioned   and getting connected

FIGURE 3.30
Sequence of mooring a TLP.

4. Wells are required to be completed as subsea. But it allows the extension of the wellhead to the surface which permits easy workover and maintenance of the wells.

5. The sway and surge motions are not critical to equipment and personnel.

3.3 Mobile Units

As discussed in Chapter 1 (Sections 1.3.2 and 1.3.3), in the case of exploratory wells and deep waters the mobile units for both drilling and production are being used. These mobile units can be classified as follows:

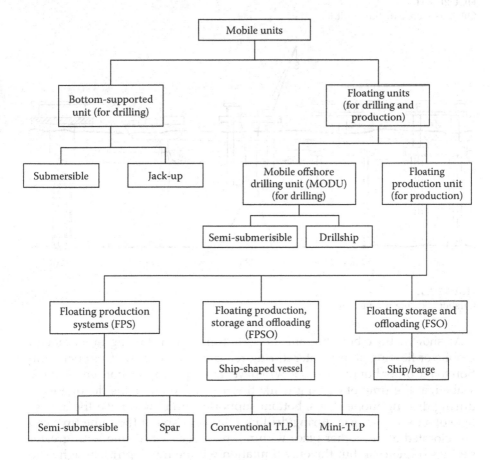

Figures 3.31 and 3.32 show schematic classifications of mobile units separately for drilling and production units.

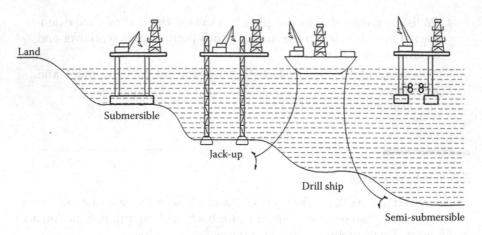

FIGURE 3.31
Offshore mobile drilling unit.

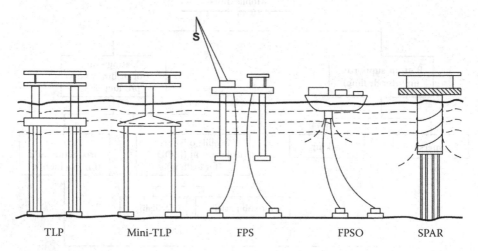

FIGURE 3.32
Offshore floating production unit.

As shown above both bottom-supported units and floating units remain mobile or floating above water during relocation or shifting from one location to the other but the first one will have its support at the bottom, or at the seabed, at the time of drilling while the second one remains floating even during drilling mode. Thus, bottom-supported units will have the advantage of a fixed platform during drilling operation and at the same time can be relocated at any other place within a relatively short time with greater savings of the cost. But the only limitation will be the depth. Though jack-ups are quite handy and popular up to a water depth of 350–400 ft (120 m), submersibles have restriction up to a very shallow depth, that is, about 50 ft of water. The jack-ups constitute about 50% of the world's offshore drilling fleet.

3.3.1 Bottom-Supported Unit

3.3.1.1 Submersible

The concept of using a mobile unit to have a bottom support as and when necessary came from the idea of making a barge sit on the bottom of Louisiana swamps with drilling platforms welded on top. Thus came the first submersible, which was a totally submerged, conventional sized barge with columns high enough to support a platform at a safe above-water distance. Pontoons on either side of the barge provided both stability and displacement control.

This type of unit is used in shallow waters such as rivers and bays, usually in waters up to 50 ft deep. One submersible, however, has been used in 175-ft water depths. The submersible has two hulls. The upper hull, sometimes referred to as the 'Texas' deck, is used to house the crew quarters and equipment, and the drilling is performed through a slot on the stern with a cantilevered structure. The lower hull is the ballast area and is also the foundation used while drilling.

The submersible is floated to location like a conventional barge and is then ballasted to rest on the river bottom. The lower hulls are designed to withstand the weight of the total unit and the drilling load.

Stability while ballasting these units is a critical factor although this problem was solved by attaching pontoons at each of the long ends. Initially one pontoon was ballasted until that end of the unit sat on the bottom and after stability was ensured, the other pontoon was ballasted until the unit rested on the bottom with topside-up every time. Today, however, submersibles are fading from the scene simply because current water depth requirements have surpassed their capabilities. Their water-depth restriction is because of the free-standing height of the units while these are towed from one place to the other.

3.3.1.2 Jack-Up

Jack-ups basically are made-up of four main components (Figure 3.33). Starting from the bottom those are

1. Footing
2. Leg
3. Jacking system
4. Hull

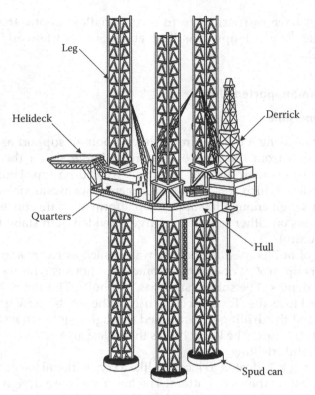

FIGURE 3.33
Jack-up drilling unit.

Footings are at the bottom of the legs and their purpose is to increase the legs' bearing area, thereby reducing the required capacity of the soil to provide a solid foundation to withstand the weight of the jack-up. There are two types of footings: (a) spud cans/tanks and (b) mats.

On the basis of the type of footings a jack-up uses, these are classified into two basic categories:

1. Independent leg type
2. Mat-supported type

In the case of the independent leg type, each leg has a separate footing whereas in the case of the mat-supported type, all the legs are connected to a common support at the bottom called a mat. The independent leg jack-up will operate anywhere currently available, but it is normally used in areas of firm soil, coral, or uneven sea bed. The independent leg unit depends on a platform (spud can) at the base of each leg for support. These spud cans are circular, square or polygonal, and are usually small. Most cans contain water jets and piping for the washing of material which may accumulate on top of the can and drain valves and vents to allow free-flooding capability. Figure 3.34 illustrates the footing configuration of various spud cans. The largest spud can being used to date is about 56 ft wide. Spud cans are subjected to bearing pressures of around 5000–6000 pounds per square foot (psf), although in the North Sea this can be as much as 10,000 psf. Allowable bearing pressures must be known before a jack-up can be put on location. Sometimes in softer soil, for stability, more weight may become necessary to be applied on the spud cans and for that independent leg drilling units are preloaded. The preload weight is usually about one-third greater than the weight of the platform plus the allowed variable load. This preload weight is obtained by pumping water into the tanks. While installing this load is given and once the maximum penetration is reached, the preload water is dumped.

The mat-supported jack-up is designed for areas of low soil shear value where bearing pressures must be kept low. The mat, which is connected to all of the legs, contains buoyancy chambers which are flooded when the mat is submerged. With such a large area in contact with the soil, bearing pressures of 500–600 psf usually exist. Figure 3.35 shows the line diagram of the mat footing.

An advantage of the mat-type jack-up is that minimum penetration of the sea bed takes place, perhaps 5–6 ft. This compares with a penetration of perhaps 40 ft on an independent leg jack-up. As a result, the mat-type unit requires less leg than the independent jack-up for the same water depth. One disadvantage of the mat-type unit is the need for a fairly level sea bed. A maximum sea bed slope of 1½° is considered to be the limit. Another problem with the mat-supported unit occurs in areas where there is coral or large rock formations. Since mats are designed for uniform bearing, the uneven bottom would probably cause a structural failure.

Legs of a jack-up are steel structures that support the hull when the unit is in the elevated mode and provide stability to resist lateral loads. Some

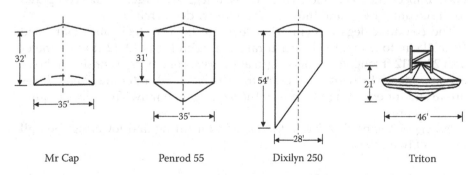

| Mr Cap | Penrod 55 | Dixilyn 250 | Triton |

FIGURE 3.34
Footing configurations of various spud cans.

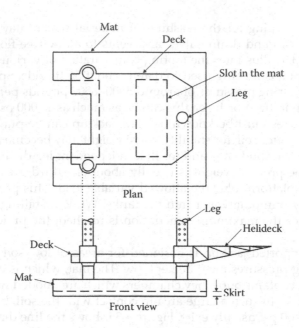

FIGURE 3.35
Mat footing.

independent leg units are able to cant (or slant) the legs for adding strength against overturning in deep water. The legs can be either (a) *trussed legs* or (b) *cylindrical legs/columnar legs*.

Trussed legs may be of triangular or square cross-section consisting of chords and braces. The braces provide the shear capacity of the leg while the chords provide the axial and flexural stiffness. At each corner of the trussed legs (Figure 3.36), the racks are provided for engaging with the gears or pinions in the jacking assembly. These racks resembling threads cut on a flat bar are fabricated and torch cut from special high-yield strength steel that has good impact characteristics. These trussed legs are larger in size compared to the columnar legs and thus require a larger deck area.

The columnar legs or cylindrical legs are hollow steel tubes fabricated from 1½ in. to 2¾ in. thick structural steel. Each leg is 10–12 ft in diameter and 225–312 ft long. Along its length at intervals of 6 ft, rings of six pin holes are provided which are spaced around the leg at 60° intervals. These holes are meant for engaging the pins in the jacking assembly. These legs require less deck area.

Jacking system or *elevating system* is used for lifting and lowering the hull and is of two types:

1. Rack-and-pinion type
2. Electro-hydraulic type

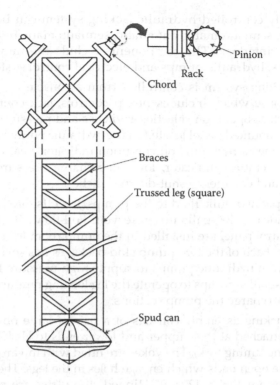

FIGURE 3.36
Components of a trussed leg of a jack-up.

1. Electrically operated rack-and-pinion jacking systems are designed for manual, individual leg or fully automatic operation. Features include pairs of pinions which engage a double rack to balance the spreading forces between the rack-and-pinion gear teeth. This jacking system provides a console in the control room of the rig with function switches for independent or simultaneous operation of the legs.

   Depending on the leg configuration, whether triangular or square cross-section, each leg has three or four leg-guide elevating assemblies, one at each corner of the leg. Each elevating assembly includes 3–6 motor gear units. Each motor gear unit consists of approximately 30 horsepower, 480- or 600-volt, AC, 600 rpm high-slip induction motor driving a pinion that engages the leg rack through a multi-gear train which provides 2000:1 speed reduction. Each electric motor is equipped with a spring-loaded brake which engages automatically when the motors are not energised. For smooth lifting or lowering operation, special attention is given to lubrication of the meshing surfaces in the gear train and the rack-and-pinion contacts.

2. Electrically controlled hydraulic jacking system can be operated manually, semiautomatic, automatic, semiautomatic interlock and automatic interlock. The components include the ram and yoke assemblies, hydraulic pumps and electrical control system.

The jacking system is controlled from a console located in the control house which includes pump controls, jack controls, function switches, operation selection switches and pressure gauges. A bulkhead mounted panel located in the hydraulic pump room below deck provides a hydraulic oil condition indicator, heat exchanger/water temperature indicator, low supercharge pressure indicator and reset and emergency shut-down switches.

To supply hydraulic fluid to the pumps, the offshore jacking system provides a 2400-gallon oil reservoir. Full-flow oil filters and a motor control panel are installed in the pump room for the hydraulic pumps. Each of the two pump skid packages has several electric motor-driven hydraulic pumps to supply high-pressure fluid to the leg jacks, smaller pumps to operate the leg locking pins and centrifugal units to charge the pump sections.

Each jacking assembly consists of a pair of large bore, hydraulic rams attached at their upper and lower ends to U-shaped yokes (resembling tuning fork). The yokes are fitted with inward-operating pins at the open ends which engage holes in the legs. The yoke pin's dimensions are $31\frac{1}{2}" \times 12" \times 9"$. Hinged stop plates are provided to hold the pins in attachment to the legs when desired; when the stop plates are lifted, the pins can be withdrawn from the leg holes. Six hydraulic cylinders and pistons provide the jacking force for the yokes; the heads are attached to the hull and the pins to the yokes.

The *hull* of a jack-up unit is a watertight structure that supports or houses the equipment, systems and personnel, thus enabling the unit to perform its tasks. When the jack-up unit is afloat, the hull provides buoyancy and supports the weight of the legs and footings equipment and variable load.

The hull is generally built of stiffened plate. The structure is configured to efficiently transfer loads acting on the various hull locations into the legs. Axial and horizontal loads are transferred into the legs through the hull leg interface connections and chords. Hence, a bulkhead terminates at each leg chord location.

3.3.1.2.1 *Installation of a Jack-Up*

During installation of a jack-up platform, the following steps are followed:

a. Transportation

The transportation of a jack-up unit while in transit mode from one location to another will depend on the method of its movement. Jack-up can be either self-propelled, propulsion assisted, or nonpropelled and hence needs some assistance to be moved from one place to the other.

Such nonpropelled jack-ups can be towed by either of the two methods, that is *wet tow* and *dry tow*.

Wet tow is the process of towing by tug boat when the jack-up unit is afloat on its hull and dry tow is that process when the whole jack-up unit mounts on the deck of another vessel as cargo. Wet tow can be further classified into three types: (i) *field tow*, (ii) *extended field tow* and (iii) *wet ocean tow*.

The field tow is such that the duration of towing is less than 12 h and naturally it corresponds to the condition where a jack-up unit is afloat on its hull with its legs raised and is moved to a very limited distance. However, even for such a short move, the prediction of weather and sea conditions is important.

The extended field tow is similar to the field tow except the duration of towing which is more than 12 h. The towing requires that the unit must always be within a 12-h tow of a safe haven should weather deteriorate.

Wet ocean tow is a float move lasting more than 12 h which does not satisfy the requirements of an extended field tow. In this condition, the jack-up unit is afloat on its own hull with its legs raised or removed and stowed on deck.

On average, towing speeds in calm seas are 4 knots using three 9000 hp tugs.

While being towed jack-up rigs are more vulnerable to accidents than the other mobiles. With the legs extended above deck the centre of gravity is raised and an unstable condition exists. So, for moving long distances, it is recommended to remove the legs and stow them on deck. Moving crews with special training have been developed by several contractors for in transit and going off or on location.

b. Arriving on Location

Upon completion of the transit mode, the jack-up unit is said to be in the arriving on location mode. Preparations include removing any wedges in the leg guides, energising the jacking system and removing any leg securing mechanisms installed for the transit thereby transferring the weight of the legs to the pinions.

c. Soft-Pinning the Legs

If an independent leg jack-up unit is going to be operated next to a fixed structure, or in a difficult area with bottom restrictions, the jack-up unit will often be temporarily positioned just away from its final working location. This is called 'soft pinning' the legs or 'standing off' location. This includes coordinating with the assisting tugs, running anchor lines to be able to 'winch in' to final location, powering up of positioning thrusters on the unit (if fitted), checking the weather forecast for the period of preloading and jacking up.

d. Preloading

Preloading of jack-up legs helps assure the suitability of the sea floor as a foundation. Preloading reduces the likelihood of a foundation shift or failure during a storm. The possibility does exist that a soil failure or leg shift may occur during preload operations and hence to alleviate the potentially catastrophic results of such an occurrence, the hull is kept as close to the waterline as possible without incurring wave impact.

e. Positioning the Platform

With the legs on bottom, jacking crews elevate the drilling floor to the height above wave action anticipated to be safe for the season and the area. Jacking speed at maximum load is approximately 1 ft/min for some jack-ups. Leg lengths normally permit from approximately 40–90 ft of air space between mean water level and the hull. Once the unit reaches its operational air gap, the jacking system is stopped, the brakes set and the leg locking systems engaged. The unit is now ready to begin operations.

The above operations have been sequentially presented in Figure 3.37. Also the final position of the jack-up is schematically shown in Figure 3.38.

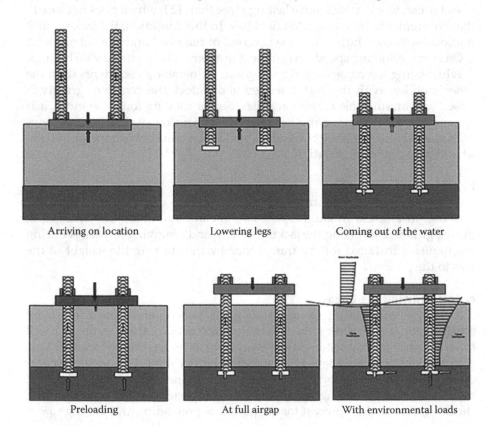

| Arriving on location | Lowering legs | Coming out of the water |

| Preloading | At full airgap | With environmental loads |

FIGURE 3.37
Installation of a jack-up.

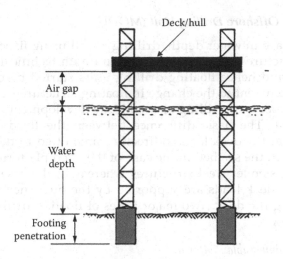

FIGURE 3.38
Final position of a jack-up.

3.3.2 Floating Units

The floating units are those that not only float while they are in transit mode but also remain floating during any oilfield operation like drilling or production. These are sometimes also called floaters. Such floating units or structures may be grouped as *neutrally buoyant* or *positively buoyant*. The neutrally buoyant means buoyant force is equal to the gravity force and the structures like semisubmersibles, ships or ship-shaped vessels, barges and spars fall under this category whereas positively buoyant means upward buoyant force is more than the downward gravity force and the structures like TLPs or mini-TLPs fall under this category.

Floating platforms may be grouped by their use for a particular operation like drilling or production.

Operation	Type	Structure/Unit
Drilling	Mobile offshore drilling units (MODU)	(i) Semisubmersible (ii) Drillship
Production	**Neutrally buoyant**	
	(a) Floating production system (FPS)	(i) Semisubmersible (ii) Spar
	(b) Floating storage and offloading (FSO)	Barge/Ship
	(c) Floating production, storage and offloading (FPSO)	Ship-shaped vessel
	Positively buoyant	
	(a) Conventional tension leg platform (TLP)	TLP
	(b) Mini-TLP	

3.3.2.1 Mobile Offshore Drilling Unit (MODU)

With the increase in water depth, drilling a well using fixed- or bottom-supported structures was becoming difficult both technically as well as economically and hence floating drilling units started becoming popular though concurrent to the change in floating structures, changes were made in the process of drilling also with the developments of new tools and equipment. The basic difference between the fixed and floating structure is that the deck load is directly transmitted to the foundation material beneath the sea bed in the case of the fixed platform through its typically long, slender steel structures whereas in the case of the floating structures, deck loads are supported by the buoyancy forces of the hull supporting the deck. Two major types of floating drilling units are described below.

3.3.2.1.1 Semisubmersible Platform

Semisubmersibles are multilegged floating structures with a large deck. These legs are interconnected at the bottom underwater with horizontal buoyant members called pontoons. Some of the earlier semisubmersibles resemble the ship form with twin pontoons having a bow and a stern. This configuration was considered desirable for relocating the unit from drilling one well to another either under its own power or being towed by tugs. Early semisubmersibles also included significant diagonal cross-bracing to resist the prying and racking loads induced by waves.

The introduction of heavy transport vessels that permit dry tow of MODUs, the need for much larger units to operate in deep water and the need to have permanently stationed units to produce from an oil and a gas field resulted in the further development of the semisubmersible concept. The next generation semisubmersibles typically appear to be a square with four columns and the box- or cylinder-shaped pontoons connecting the columns. The box-shaped pontoons are often streamlined eliminating sharp corners for better station-keeping. Diagonal bracing is often eliminated to simplify construction.

There are various designs of semisubmersibles such as the triangular design used on the Sedco series, four longitudinal pontoons used on the Odeco series and the French-designed Pentagone rig with five pontoons. Figure 3.39 shows sectional views of a semisubmersible indicating four components namely pontoons, stability columns, deck and space frame bracing.

Virtually all semisubmersibles have at least two floatation states: semisubmerged (afloat on the columns) and afloat on the pontoons, that is, raised at the surface of the water. It is to be noted that while the structure is raised or not semisubmerged, the pontoons are the principal elements of floatation and stability. The pontoons or hulls are used as ballast compartments to achieve the necessary drilling draft.

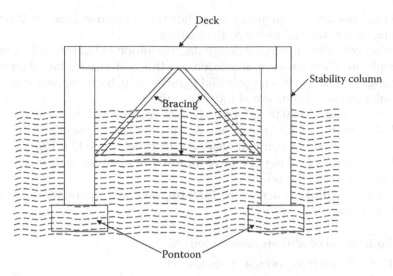

FIGURE 3.39
Semi-submersibles.

The deck provides the working surface for most of the drilling functions along with other auxiliary functions. Its weight is transferred to the columns through the structural connections, that is, the space frame bracing. Bracing systems are expensive to build and are a costly maintenance item regarding inspections and repairs.

Semisubmersibles permit drilling to be carried out in very deep water beyond 1000 ft of water depth and they are held on location either by a conventional mooring system or by dynamic positioning.

The motion that causes problems for the semisubmersible is heave or the vertical motion. Because of forces on the drill string when the vessel is heaving, these semisubmersibles with a low heave response are considered to be the most suitable. Heave is generated in response to exposed waterplane and is expressed as

$$T = \frac{2\pi}{\sqrt{\dfrac{gt}{d}}}$$

where T = time in seconds; t = tons per foot immersion; and d = displacement in tons.

Therefore, the smaller the waterplane area, or 't', the lower the heave response. This is achieved in the semisubmersible by submerging the lower pontoons and floating at the column or caisson level. With the loss of the waterplane area to reduce the heave response, a reduction in stability follows. Therefore, the designer must reach a compromise between acceptable heave response and adequate stability. There are, of course, other methods of reducing heave-induced forces on drill string.

Another consideration in the design and operation of the semisubmersible is propulsion. There are several opinions on this matter, each based on valid reasoning. Propulsion is a large initial expense which can be recovered in a reasonable period of time if mobility is required.

For mobilising although self-propelled semisubmersibles have shown a speed of more than 10 knots in calm sea, one of the newest semisubmersibles using tug boats with a nominal water depth rating of 2000 ft has a towing speed in calm sea of l0 knots. Some of the early semisubmersibles were rated at towing speeds of 3 knots.

Before the drilling rig arrives on location, several preparatory operations should be completed:

1. Bottom soil conditions determined
2. Prevailing winds and sea states known
3. Surveys made and buoy markers placed for exact anchor locations
4. Equipment conditioned and supplies inventoried
5. Lines of communication definitely established: operating company, drilling contractor, anchor handling boat and special crew

After its arrival, the anchor should be placed using a conventional mooring method following the spread mooring pattern. The details of anchoring will be covered later. Some of the semisubmersibles in deeper water can be dynamically positioned as well. After it is properly stationed, preparation for starting the drilling operation should be made.

3.3.2.1.2 Drillships

The last type of mobile drilling unit to be discussed is the drillship. As the name implies, it is simply a shipshape vessel used for drilling purposes. Earlier drillships were converted vessels, either barges, ore carriers, tankers, or supply vessels. However, although conversions are still being done, there are several new drillships being designed purely for drilling, such as the Glomar Challenger or the Offshore Discoverer. Drillships are the most mobile of all drilling units, but they are the least productive. The very configuration that permits mobility results in very bad drilling capabilities.

Drillships (Figure 3.40) are being used extensively to bridge the gap between the jack-up and the semisubmersible. However, it is the drillship that has drilled in the deepest water, over 1000 ft. As discussed earlier, heave is the major problem when using a floating vessel. The drillship, because of its surface contact with the sea, develops a very large heave response compared to the semisubmersible. It is possible, by means of stabilising tanks and other methods, to reduce roll on drillships but heave cannot be reduced.

FIGURE 3.40
Drillship. (Courtesy of Cliff/Wikimedia Commons/CC-BY-SA-2.0.)

A subsequent increase in 'rig downtime' or 'lost' time occurs. Because of this there is a bigger demand for the use of compensation devices.

Drillships similar to the semisubmersibles are held on location either by conventional mooring system or by dynamic positioning. However, there is one additional system that has been developed on a drillship and that is the 'turret' system.

Briefly then, drillships have many advantages like proven deepwater capability, capacity to transport much larger loadings of drilling supplies, faster travel time to remote locations, self-propelled and hence no need for tug, but the disadvantage is that it should only be considered for use in areas of small wave heights and low wind velocities.

3.3.2.2 Floating Production Unit

Once offshore operations extended beyond practical fixed platform limits, the production engineers borrowed concepts devised by the drilling engineers. They in turn had responded to the needs of the explorers with semisubmersibles and drillships as they moved out of shallow water. Thus, floating production units (plus, in many cases, the subsea completions) now provide the viable options in deepwater. Figure 3.32 shows different types of floating units used for oil and gas production.

Floating production units come in many sizes and shapes. Some provide more functions than others. In every case, they differ from fixed units by what holds them up – the buoyancy of displaced water, not steel understructure. Such units have four common elements:

Hull – the steel enclosure that provides water displacement. Floating unit hulls can be in ship shapes, pontoons and caissons, or a large tubular structure called a spar.

Topsides – the deck or decks have all the production equipment used to treat the incoming well streams plus pumps and compressors needed to transfer the oil and gas to their next destinations. Some have drilling and workover equipment for maintaining wells. Since almost all deepwater sites are somewhat remote, their topsides include living accommodations for the crew. In most cases, export lines connect at the deck as well.

Mooring – the connection to the seabed that keeps the floating units in place. Some combine steel wire or synthetic rope with chain, some use steel tendons. In some cases, they make a huge footprint on the seabed floor. This is discussed in detail in a subsequent chapter.

Risers – steel tubes that rise from the sea floor to the hull. A riser transports the well production from the sea floor up to the deck. The line that moves oil or gas in the other direction, from the deck down to pipeline on the sea floor, uses the oxymoron export risers.

3.3.2.2.1 *Tension Leg Platforms (TLP)*

The semisubmersible, used for years only for drilling, begat TLPs. By similar design, the buoyancy of a TLP comes from a combination of pontoons and columns. Vertical tendons from each corner of the platform to the sea floor foundation piling hold the TLP down in the water. Vertical risers connected to the subsea wells heads directly below the TLP bring oil and gas to dry trees on the deck.

Dry trees on the deck of the TLP control the flow of oil and gas production coming up through the conductor pipes. However, like other floating systems, it can receive production from risers connected to remote subsea wet tree completions. Most TLPs have subsea riser baskets, structural frames that can hold the top end of risers coming from subsea completions. Figure 3.41 shows a TLP used for production.

3.3.2.2.2 *Monocolumn TLP/Mini-TLP*

In shallower water or for smaller deposits in deepwater, and where no more drilling is planned, some companies use a smaller variation of the TLP called a mini-TLP, a monocolumn TLP, or sometimes a SeaStar (Figure 3.42).

FIGURE 3.41
Tension leg platform. (Courtesy of Derk Bergguist, South Carolina, Department of Natural Resources.)

The names monocolumn and SeaStar (a proprietary label) come from the underwater configuration of the floatation tanks, a large central cylinder with three star-like arms extending from the bottom. The cylinder measures about 60 ft in diameter and 130 ft in height. The arms reach out another 18 ft.

As with other TLPs, tendons secure the substructure to the sea floor, in this case two from each arm. The mooring system, risers and topsides are similar to any other TLP, except for the modest sizes. The absence of drilling equipment on board helps lower the weight of the topsides and allows for this scaled-down version.

3.3.2.2.3 Floating Production Storage and Offloading (FPSO)

This refers to a ship-shaped structure with several different mooring systems. From 400 yards away, most FPSOs are indistinguishable from oil tankers. In fact, while many FPSOs are built from scratch, the rest are oil tankers converted to receive, process and store production from subsea wells. FPSOs do not provide a platform for drilling wells or maintaining them. They do not store natural gas, but if gas comes along with the oil, facilities are onboard and the FPSO separates it. If there are substantial volumes, they are sent back down a riser for reinjection in the producing reservoir or some other nearby consumer.

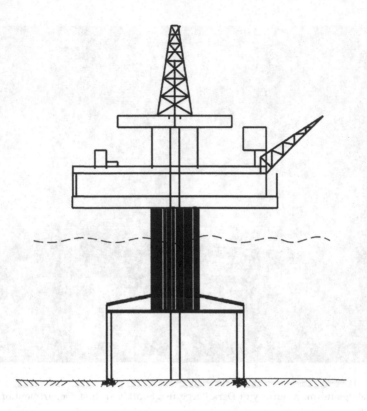

FIGURE 3.42
Mono column TLP.

The industry has found scores of remote or hostile environments that call for the FPSO design:

- At sea where no pipeline infrastructure exists
- Where weather is no friend, such as offshore Newfoundland or the northern part of the North Sea
- Close to shore locations that have inadequate infrastructure, market conditions, or local conditions that may occasionally not encourage intimate personal contact, such as some parts of West Africa

There are four principal requirements that drive the size of a typical FPSO:

1. Sufficient oil storage capacity to take care of the shuttle tanker turn-around time
2. Sufficient space on the topsides for the process plant, accommodation, utilities and so on

3. Sufficient ballast capacity to reduce the effects of motions of process plant and riser systems

4. Space for the production turret (bow, stern or internal)

As an FPSO sits on a station, wind and sea changes can make the hull want to *weathervane*, turn into the wind like ducks on a pond on a breezy day (Figure 3.43). As it does, the risers connected to the wellheads, plus the electrical and hydraulic conduits, could twist into a Gordian knot. Two approaches deal with this problem, that is, the cheaper way and the better way, depending on the ocean environment.

In areas of consistent mild weather, the FPSO moors, fore and aft, into the predominant wind. On occasions, the vessel experiences quartering or broadside waves, sometimes causing the crew to shut down operations.

In harsher environments, the more expensive FPSOs have a mooring system that can accommodate weathervaning. Mooring lines attach to a revolving turret fitted to the hull of the FPSO. As the wind shifts and the wave action follows, the FPSO turns into them.

The turret (Figure 3.44) might be built into the hull or cantilevered off the bow or stern. Either way, the turret remains at a permanent compass setting as the FPSO rotates about it.

The turret also serves as the connecting point between the subsea systems and the topsides production equipment. Everything between the seabed and the FPSO is attached to the turret-production risers, export risers, gas reinjection risers, hydraulic, pneumatic, chemical and electrical lines to the subsea wells, as well as the mooring lines.

Turrets contain a swivel stack, a series of fluid flow and electronic continuity paths that connect the seaside lines with the topsides. As the FPSO swings around the turret, the swivels redirect fluid flows to new paths, inbound

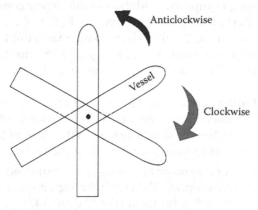

FIGURE 3.43
'Weathervane' concept.

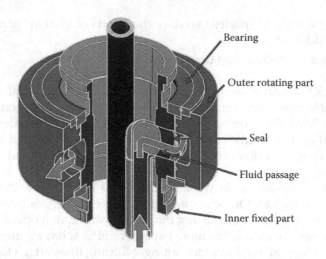

FIGURE 3.44
Turret of an FPSO.

or outbound. Other swivels in the stack handle pneumatics, hydraulics and electrical signals to and from the subsea systems.

In some designs, the FPSO can disengage from the seabed (after shutting in the production at the wellheads) to deal with inordinately rough seas, or other circumstances that might worry the ship's captain, like an approaching iceberg. A spider buoy, the disconnectable segment of the turret with the mooring, the riser and the other connections to the subsea apparatus, drops and submerges to a predesignated depth as the vessel exits the scene.

After the oil moves from the reservoir to the FPSO via the turret, it goes through the *processing* equipment and then to the *storage* compartments.

Shuttle tankers periodically must relieve the FPSO of its growing cargo. Some FPSOs can store up to 2 million barrels on board, but that still calls for a shuttle tanker visit every week or so. Mating an FPSO to the shuttle tanker to transfer crude oil calls for one of several positions.

- The shuttle tanker can connect to the aft of the FPSO via a mooring hawser and *offloading* hose. The hawser, a few hundred feet of ordinary marine rope, ties the shuttle tanker to the stern of the FPSO and the two-vessel weathervane together about the turret.

- The shuttle tanker can moor at a buoy a few hundred yards off the FPSO. Flexible lines connect the FPSO through its turret to the shuttle buoy and then to the shuttle tanker (Figure 3.45).

- Some shuttle tankers have dynamic positioning, allowing them to sidle up to the FPSO and use their thrusters on the fore, aft and

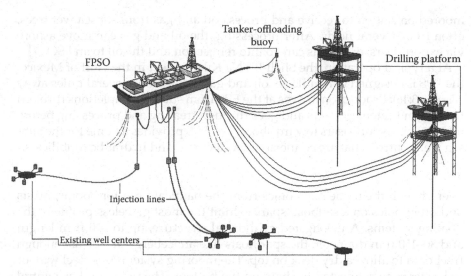

FIGURE 3.45
Offloading of an FPSO. (Courtesy of WikiDon/Wikimedia Commons/CC-BY-SA-3.0.)

>sides to stay safely on station, eliminating the need for an elaborate
>buoy system. The shuttle tanker drags flexible loading lines from the
>FPSO for the transfer.

Oil then flows down a 20-in. (or so) offloading hose at about 50,000 barrels per hour, giving a turnaround schedule for the shuttle tanker of about
one day.

3.3.2.2.4 Floating, Drilling, Production, Storage and Offloading (FDPSO)

Inevitably, the attributes of an FPSO plus a drillship or a semisubmersible
come together in the form of a floating drilling production storage and
offloading (FDPSO) to do it all in deepwater. The crucial technologies for this
union, still in the nascent stage, are motion and weathervane compensation
systems for the drilling rig.

3.3.2.2.5 Floating Storage and Offloading (FSO)

This specialty vessel stores crude from a production platform, fixed or floating, where no viable alternatives for pumping oil via pipeline exist. FSOs
almost always had a former life as an oil tanker and generally have little or
no treating facilities onboard. As with the FPSO, shuttle tankers visit periodically to haul the produced oil to market.

3.3.2.2.6 Floating Production System (FPS)

In theory, an FPS can have a ship shape or look like a semisubmersible or TLP,
with pontoons and columns providing buoyancy. Either way, the FPS stays

moored on station to receive and process oil and gas from subsea wet trees, often from several fields. After processing, the oil and gas can move ashore via export risers, or the gas can go into reinjection and the oil to an FSO.

At a typical operation, the Shell-BP Na Kika project in the Gulf of Mexico, the FPS is designed to handle six oil and gas fields, some several miles away (Figure 3.46). Production arrives at the FPS from subsea completions through flexible and catenary risers and goes through treating and processing before it leaves via export risers toward shore. The FPS provides a home for the subsea well controls that are connected via electrical and hydraulic umbilicals.

3.3.2.2.7 Spar

Even though the name *spar* comes from the nautical term for booms, masts and other poles on a sailboat, spars exhibit the most graceless profile of the floating systems. An elongated cylindrical structure, up to 700 ft in length and 80–150 ft in diameter, the spar floats like an iceberg – it has just enough freeboard to allow a dry deck on top. The mooring system uses steel wire or polyester rope connected to chain on the bottom. The polyester has neutral buoyancy in water and adds no weight to the spar, eliminating having to build an even bigger cylinder. Because of its large underwater profile, the huge mass provides a stable platform with very little vertical motion. To ensure that the centre of gravity remains well below the centre of buoyancy (the principle that keeps the spar from flipping), the bottom of the spar usually has ballast of some heavier-than-water material like magnetite iron ore.

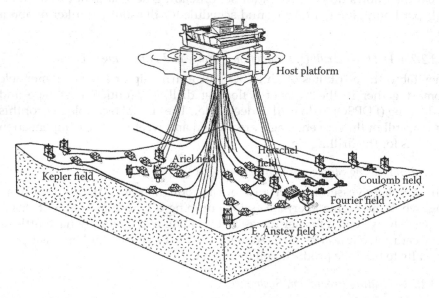

FIGURE 3.46
A typical FPS designed to handle six sub-sea fields (Na Kika Project). (Courtesy of Shell Oil Co.)

Because of the large underwater profile, spars are vulnerable not only to currents but also to the vortex eddies that can cause vibrations. The characteristic *strakes* (fins that spiral down the cylinder) shed eddies from these ocean currents, although the strakes add even more profile that calls for additional mooring capacity.

Drilling rigs operate from the deck through the centre of the cylinder. Wells connect to dry trees on the platform by risers, also coming through this core. Risers from subsea systems and export risers also pass through the centre.

Spars have evolved through several generations of design. The original concept had a single 600-ft steel cylinder below the surface.

Three types of production spars (Figure 3.47) have been built to date: the 'classic' spar, 'truss' spar and the third generation 'cell' spar. The basic parts of the classic and truss spar include:

1. Deck
2. Hard tank
3. Midsection (steel shell or truss structure)
4. Soft tank

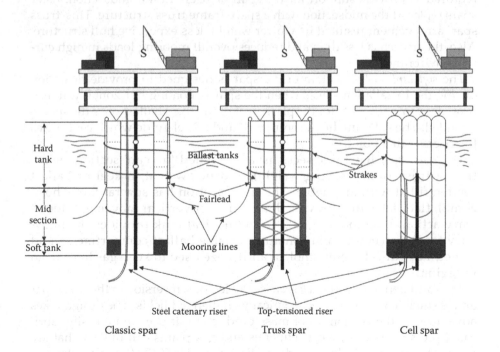

FIGURE 3.47
Types of production spar.

The topsides deck is typically a multilevel structure to minimise the cantilever requirement. For decks up to about 18,000 tons, the deck weight is supported on four columns, which join the hard tank at the intersection of a radial bulkhead with the outer shell. Additional columns are added for heavier decks.

The *hard tank* provides the buoyancy to support the deck, hull, ballast and vertical tensions (except the risers). The term 'hard tank' means that its compartments are designed to withstand the full hydrostatic pressure. The profile is shown in Figure 3.47. There are typically five to six tank levels between the spar deck and the bottom of the hard tank, each level separated by a watertight deck. Each level is further divided into four compartments by radial bulkheads emanating from the corner of the center well. The tank level at the waterline includes additional cofferdam tanks to reduce the flooded volume in the event of a penetration of the outer hull from a ship collision. Thus, there are up to 28 separate compartments in the hard tank. Typically, only the bottom level is used for variable ballast, the other levels being void spaces.

The *midsection* extends below the hard tank to give the spar its deep draft. In the early 'classic' spars, the midsection was simply an extension of the outer shell of the hard tanks. There was no internal structure, except as required to provide support for the span of risers in the midsection. Later spars replaced the midsection with a space frame truss structure. This 'truss spar' arrangement resulted in a lower weight, less expensive hull structure. Also, the truss has less drag and reduces overall mooring loads in high current environments.

The *soft tank* at the bottom of the spar is designed to provide floatation during the installation stages when the spar is floating horizontally. It also provides compartments for the placement of the fixed ballast once the spar is upended. The soft tank has a centre well and a keel guide which centralises the risers at that point.

The *truss spar* has three sections: a shortened 'tin can' section; below that, a truss frame (saving weight); and below that, a keel or ballast section filled with magnetite. The truss section has several large, horizontal, flat plates that provide dampening of vertical movement due to wave action. Like the original, the cylindrical tank provides the buoyancy for the structure and contains variable ballast compartments and sometimes tanks for methanol, or antifreeze used to keep gas lines from plugging.

The third generation, the *cell spar*, is a scaled down version of the truss spar and is suitable for smaller, economically challenged fields. The design takes advantage of the economies of mass production. It uses more easily fabricated pressure vessels, what refineries and gas plants call bullets, that are used to handle volatile hydrocarbons. Each vessel is 60–70 ft in diameter and 400–500 ft long. A cell, a bundle of tubes that looks like six giant hot dogs clustered around a seventh, makes up the flotation section extending below the decks. Structural steel holds the package together, extends down to the ballast section and can include heave plates. This design embodies a new construction technique using ring stiffened tubulars assembled in a hexagonal formation to form a spar. The first cell spar is designed for wet trees only. Because of the length of a spar, the spar hull cannot be towed upright. Therefore, it is towed offshore on its side, ballasted to a vertical attitude and then anchored in place. The topside is not taken with the hull and is mated offshore once the spar is in place at its site. The mooring cables are connected with predeployed moorings.

3.3.3 Buoyancy and Stability

Any seaworthy vessel must obey all the laws of hydrodynamics for the safe operation of drilling and production as well as during movement from one location to the other. Thus 'buoyancy' and 'gravity' become two very important factors which play a big role in determining the 'stability' of a floating unit.

3.3.3.1 Theory and Analysis

3.3.3.1.1 Buoyancy
Buoyancy is the apparent loss of weight of a body immersed in a fluid.

It was Greek scientist Archimedes (287–212 B.C.) who discovered that the weight of a body floating in a fluid equals the weight of the volume of the fluid displaced by the body. In precise terms, Archimedes' principle is stated as follows:

> A solid body wholly or partially submerged in a fluid is buoyed up by a force ($m_b g$) equal to the weight ($\rho_w V$) of the fluid displaced.

Referring to Figure 3.48, m_b is the mass of the fluid displaced by a ship of mass in air equal to m_o, V is the volume of water of constant density ρ_w displaced by the ship of weight in air $m_o g$.

Thus, the upward pressure of the displaced water, that is, buoyant force

$$m_b g = \rho_w V. \tag{3.1}$$

In Naval terminology, $V\rho_w$, that is, the weight of the volume of water below the waterline which is occupied by water is known as *'displacement.'* It is to be noted that dimensionally this is not the length but weight.

Now the above buoyant force ($m_b g$) acts at a point B (known as the 'centre of buoyancy') of the ship in a vertical direction opposite to that of gravitational force ($m_o g$), which acts at a point G (known as 'centre of gravity').

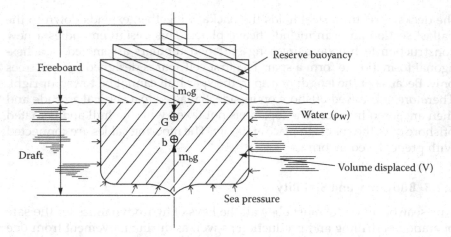

FIGURE 3.48
A partially submerged ship.

As the force of buoyancy is a vertical force and is equal to the weight of the fluid displaced by the body, the centre of buoyancy will be the centre of gravity of the displaced fluid of constant density.

Assuming a ship, of rectangular shape of length 'l', height 'h' and width (beam) 'b' and material density 'ρ' partially submerged in water of density 'ρ_w' (refer to Figure 3.49) up to a height 'd' (known as draft) from the bottom of the vessel, the height of centre of buoyancy can be found as follows:

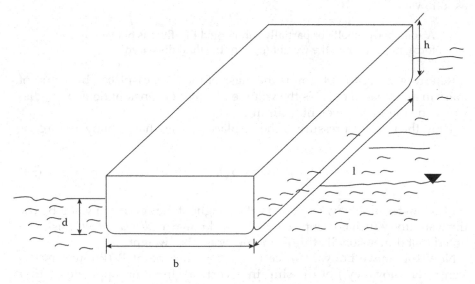

FIGURE 3.49
Rectangular shaped ship.

For equilibrium

$$\text{Buoyant Force } (m_b g) = \text{Gravity Force } (m_o g)$$

or,

the weight of water displaced = weight of the ship

or,

$$l.b.d.\rho_w = l.b.h.\rho$$

$$d = h \frac{\rho}{\rho_w}$$

Hence, height of centre of buoyancy

$$h_b = \frac{d}{2} = \frac{h}{2} \frac{\rho}{\rho_w} \qquad (3.2)$$

In the case of a two-liquid system for an arbitrarily shaped body, the location of B can be found out in a similar way as explained below:

Consider the gravity platform *partially submerged* in soft mud as shown in Figure 3.50.

Here we assume that the water-saturated mud layer behaves as a liquid whose density is ρ_m and the volume of displaced mud = V_m.

∴ Buoyant force at the centre of the displaced mud or sand vol. = $\rho_m V_m$

Portion of the caissons volume above the mud line = V_c

∴ Buoyant force for this portion of the caisson = $\rho_w V_c$

Submerged volume of each leg = V_1

Total number of identical legs = N

∴ Buoyant force on each leg = $\rho_w V_1$

∴ Total buoyant force on the legs = $N \rho_w V_1$

∴ Sum of these buoyant forces = $\rho_m V_m + \rho_w V_c + N \rho_w V_1 = m_b g$

This total force will be acting through the centre of buoyancy of the whole submerged structure.

So, if the height of the centre of buoyancy = h_b then equalising the total moment with the sum of individual moments, we get

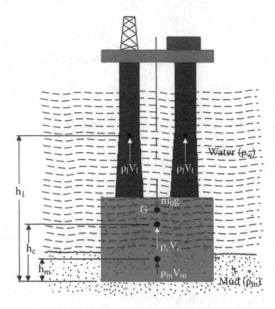

FIGURE 3.50
Gravity platform in upright position partially submerged in soft mud.

$$(\rho_m V_m + \rho_w V_c + N\rho_w V_l)h_b = h_m\rho_m V_m + h_c\rho_w V_c + h_l N\rho_w V_l$$

$$\therefore h_b = \frac{h_m\rho_m V_m + h_c\rho_w V_c + h_l N\rho_w V_l}{\rho_m V_m + \rho_w V_c + N\rho_w V_l} \tag{3.3}$$

Thus, the height of the centre of buoyancy for an upright position of any structure is calculated.

Once again referring to Figure 3.48, if a vessel is floating with some portion of its watertight hull still above the water surface, it has *reserve buoyancy*. That is, if additional weight was added to the vessel, it would still float. The total weight that would have to be added to cause it to sink is called '*reserve buoyancy*'. The word 'reserve' here means 'standby,' which is required for withstanding forces of wind, wave, current, accidental flooding, or shifting of weight aboard the unit.

Thus 'reserve buoyancy' is defined as the buoyancy above the waterline (i.e., the volume of the unit from the waterline up to the freeboard deck) that keeps a vessel upright or seaworthy, when subjected to wind, waves, currents, or other forces of nature, or to accidental flooding.

'*Freeboard*' is the vertical distance between the waterline and the uppermost continuous deck which is not watertight and this portion of the deck is known as the freeboard deck.

Opposite to freeboard is *'draft,'* which is the vertical distance between the waterline and the bottom of the vessel.

Depending on the weight being carried by the vessel, its draft, displacement, buoyancy, reserve buoyancy and freeboard can change. As the draft increases, the freeboard decreases. As displacement and buoyancy increase, the reserve buoyancy decreases.

Every vessel has a maximum draft to which it can be safely loaded. This draft, which provides the safe amount of reserve buoyancy, is called the *'load line'* and this load line is indicated on a vessel by a *'Plimsoll mark'* as illustrated in Figure 3.51. An act known as the British Merchant Shipping Act enacted in 1876 courtesy of Samuel Plimsoll, a member of British Parliament, calls for the placing of a mark at the maximum safe draft in order that all persons aboard a vessel could see that it was not overloaded.

3.3.3.1.2 Stability

Stability of a floating vessel is defined as its ability to remain upright or to return to an upright position when subjected to the forces of nature (i.e. because of wind, waves, etc.) or operation (i.e., by accidental flooding, etc.). In other words, to define simply, stability is nothing but the resistance of a vessel to capsizing.

As mentioned earlier, the criteria for stability revolves around two terms, that is, 'buoyancy' and 'gravity.' Also, a very important parameter that relates these two terms is the 'metacentre' and it is the 'metacentric height' which is considered to be an adequate criterion for stability.

FIGURE 3.51
Load line and plimsoll mark.

In Figure 3.52(a) and 3.52(b), the semisubmersible and the drillship are shown (1) in upright position where the centre of gravity (G) and the centre of buoyancy (B) lie in the same vertical plane and (2) in titled or rolling position being inclined by an angle of heel θ when the centre of buoyancy (B) moves to a new position B'. It is to be noted that when a vessel is rolling by some angle, this is known as the angle of HEEL and when a vessel is pitching by some angle, this is known as the angle of TRIM. For all practical purposes, stability against capsising refers to heeling.

Metacentre

As shown in Figure 3.52(a) and 3.52(b), when the floater is given a small angular displacement (heel) in the clockwise direction, the line of action of the force of buoyancy in this new position (B₁) will intersect the normal axis of the unit at some point M. This point M is called the metacentre. In other words, the metacentre may be defined as the point at which the line of action of the force of buoyancy will meet the normal axis of the floater when it is given a small angular displacement.

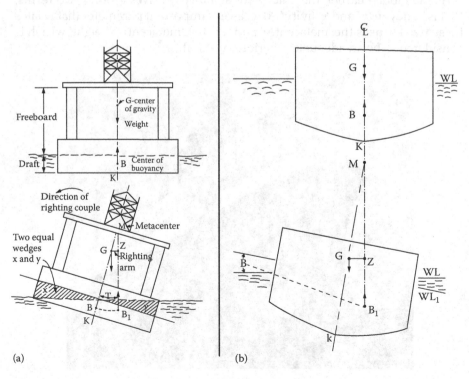

(a)   (b)

FIGURE 3.52
Upright and tilted floating units. (a) Column established drilling unit. (b) Drillship.

The distance GM, that is, the distance between the centre of gravity and the metacentre of a floating body, is called the 'metacentric height,' and the distance BM, that is, the distance between the centre of buoyancy and the metacentre of the floating body, is called the 'metacentric radius.' Referring to Figure 3.52(a) and 3.52(b), a relationship between metacentric height and metacentric radius can be established as follows:

$$GM = BM + KB - KG$$

The point K which is known as KEEL indicates the bottom of the unit.

3.3.3.1.3 Analytical Method of Calculating Metacentric Radius

In Figure 3.53, as the unit rolls through an angle θ, a wedge of the hull on one side becomes exposed and an equal wedge is immersed. The wedges are equal because the weight of the unit does not change. These two equal wedges form a moment. One wedge is immersed and picks up the force of buoyancy; the other wedge, no longer immersed, gains weight since it has lost buoyancy.

If the hull is heeled to a small angle θ, then the weight of one wedge is

$$F = \left(\frac{b}{2}\tan\theta\right)\left(\frac{b}{2}\right)(1)\left(\frac{1}{2}w\right) = \frac{b^2 1w}{8}\tan\theta$$

Taking moments about point B and using components of the weight F,

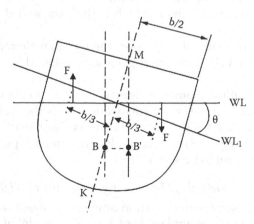

FIGURE 3.53
Metacentric radius determination (analytical method).

$$BB' \text{ (displacement)} = F(\cos\theta)\left(\frac{b}{3}\right)2$$

$$= \left(\frac{b^2 lw}{8}\tan\theta\right)(\cos\theta)\frac{2b}{3} = \frac{b^3 lw}{12}(\tan\theta)(\cos\theta)$$

$$= \frac{b^3 lw}{12}\left(\frac{\sin\theta}{\cos\theta}\right)\cos\theta = \frac{b^3 lw}{12}\sin\theta$$

But

$$BB' = BM(\sin\theta)$$

and

$$\text{displacement} = \text{vol. (w.)}$$

Therefore,

$$BM\,(\sin\theta)(vol.)(w) = \frac{b^3 lw}{12}\sin\theta$$

$$BM(vol.) = \frac{b^3 1}{12},\tag{3.4}$$

$$BM = \frac{b^3}{12\,vol} = \frac{b^3 l/12}{vol.} = \frac{I}{V}$$

For a box-shaped unit, the moment of inertia $I = b^3 l/12$ and the submerged volume, $V = $ l.b.d. where l is the *length*, b is the *beam* and d is the *draft* of a box-shaped vessel.

In general, for any shape of the vessel, the metacentric radius BM equals the moment of inertia (I) of the *water-plane area*, divided by the underwater volume (V).

It is to be noted that the moment of inertia is a function of the beam cubed and a slight reduction in the beam drastically affects the metacentric radius and thus the metacentric height. Naval architects check drilling units at various drafts to make sure the metacentric height remains positive and sufficiently great for the unit to be seaworthy.

3.3.3.1.4 *Experimental Method of Determining Metacentric Height*

This is for the purpose of demonstration only just to show how the metacentric height changes with the angle of heel. For this a model of a ship is floated on a water tank and the experiment is performed as explained below.

3.3.3.1.5 Theory of the Experiment

Consider a ship heeled by an angle θ with weight w, the centre of buoyancy B is changed to B′ (Figure 3.54). The line of action of gravity will usually diverge from the buoyancy's line of action. The separation of lines of action of two forces forms a couple. The righting moment C tends to restore the ship to an upright position and perpendicular distance GZ between the two lines of action is the righting arm.

$$C = W'. GZ$$
$$= W'. GM \, \mathrm{Sin} \, \theta \qquad (3.5)$$

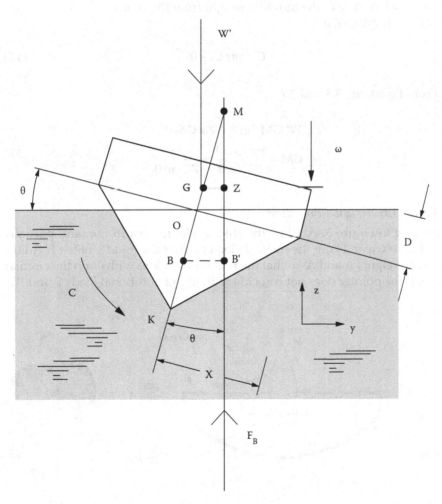

FIGURE 3.54
Metacentric radius determination (experimental method).

where W' = total weight of ship including w.

$$= W + w$$
$$W = \text{Weight of the ship model}$$
$$w = \text{Movable weight}$$

Summing the moments about the longitudinal axis at point O.

$$C = wx \, \text{Cos} \, \theta + w \, D \, \text{Sin} \, \theta \tag{3.6}$$

where x = Distance of the movable weight from the centre
For small value of θ

$$C = wx \, \text{Cos} \, \theta \tag{3.7}$$

From Equations 3.5 and 3.7

$$W' \, GM \, \text{Sin} \, \theta = wx \, \text{Cos} \, \theta$$
$$GM = \frac{wx \, \text{Cos} \, \theta}{W' \, \text{Sin} \, \theta} = \frac{wx}{W' \, \text{tan} \, \theta} \tag{3.8}$$

3.3.3.1.6 Experimental Procedure

Referring to Figure 3.55, allow the ship model to float in the tank partially filled with water. Hang the movable load w from the middle notch F. Adjust the side weights A and A' so that the pointer coincides with 0° on the circular scale. If the pointer does not coincide with 0° put additional loads P and P'.

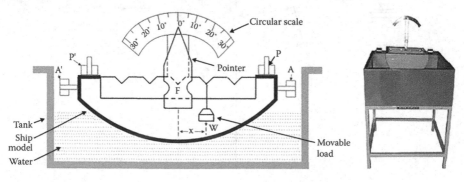

FIGURE 3.55
Metacentric radius determination apparatus.

After making the pointer read $0°$, move the load w to the next notch in the right side by a distance x and allow the ship to heel until it comes to an equilibrium position. Then take the reading of the heel angle θ on a circular scale. Repeat this process until the weight w reaches the last notch on the right side. Repeat this process by hanging the load w from the notches in the left side also from F. Note down values of the heel angle θ corresponding to the value of the distance x of the load w in tabular form.

Calculate the values of the metacentric height corresponding to the values of θ and x by using the following formula:

$$GM = \frac{wx}{W' \tan \theta}$$

Plot a curve of metacentric height GM versus heel angle θ.

(*Note*: Take x as positive right to F and corresponding angle of heel θ as positive. Take x as negative left to F and corresponding angle of heel θ as negative. Include weights A, A′, P and P′ in W′).

3.3.3.2 Conditions of Stability

Consider the stability of the gravity structure as shown in Figure 3.56 which 'floats' and rocks slowly in the mud layer. If the inertial forces are negligible, then $m_b g \approx m_o g$. If B and B′ remain always above G as it tips to a small angle θ, the structure will return to vertical equilibrium ($\theta = 0$). This configuration, shown in Figure 3.56a, is stable because the resultant buoyancy force and gravity force produce a couple opposite to the direction of rotation. The elastic restoring moment imposed by the soil beneath the caisson would be relatively insignificant in this case. If the design is such that B and B' remain below G, the structure may still be stable, but only if the metacentre M is above G, the 'floating' structure is dynamically stable because the restoring gravity-produced couple is in a direction to reduce rather than increase θ. If, however, M is below G, as shown in Figure 3.56c, the couple due to the buoyancy and gravity forces will increase θ, and the structure will topple in the absence of soil-restoring forces. This discussion is analogous to the elementary stability analysis of ships and other floating objects mentioned below.

3.3.3.2.1 Stability Analysis

As shown in Figure 3.52(a) and 3.52(b), there is a point Z such that the horizontal distance GZ is the distance between the force of buoyancy and the force of gravity. GZ is called the 'righting arm.'

This righting arm multiplied by the 'weight' or the 'displacement' gives the 'righting moment.' Graphs of the righting arm or moment similar to

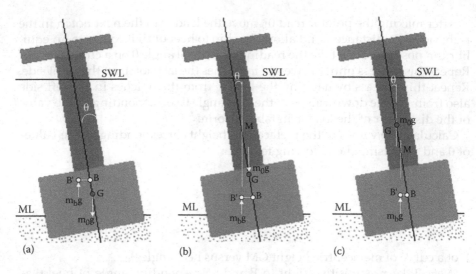

FIGURE 3.56
Stability criteria of a gravity platform (a) Stable. (b) Stable. (c) Unstable.

Figure 3.57 are prepared for drilling units. They show the righting arm for various angles of heel. A calculation for each draft has to be made. These graphs are called curves of static stability.

As shown in Figure 3.57, the righting arm increases with increasing angle of heel until some maximum value is achieved. This is sometimes associated with a deck going under so that the effective width decreases. Remember, the metacentric radius decreases drastically when the beam, or width, decreases. Eventually, the righting arm decreases to zero at an angle called the range of

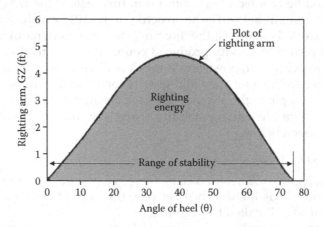

FIGURE 3.57
Static stability curve.

stability. At angles of heel greater than the range of stability, the unit continues to turn over.

3.3.3.2.2 Righting Energy

The area under the righting-arm curve is the righting energy. It is the righting energy that must combat the energy of the wind and seas and keep a unit upright. The righting energy is the inherent buoyant energy that is designed into the unit. Mathematically the area under the curve of Figure 3.57 when added up or integrated for each angle of heel will generate the curve as shown in Figure 3.58.

3.3.3.2.3 Drillship versus Drilling Unit

At one time, metacentric height was thought to be an adequate criterion for stability, that is, if the *GM* was positive, the vessel would be stable. This was true when all hull shapes were similar, but, especially with floating drilling units, this criterion was insufficient. Figure 3.59 shows righting-momentum curves for a typical drilling unit (or column-stabilised unit) and for a conventional ship. The ship has a much lower curve but a longer range of stability. Some ships have a range of stability greater than 90°, which means they can roll onto their sides and still return to an upright position. But the important factor is the righting energy, or the area under the curve. A drilling unit may not be able to survive a 90° roll as a ship can, but a drilling unit has much greater righting energy at lower angles of roll. One way to look at stability is to compare the overturning energy to the righting energy.

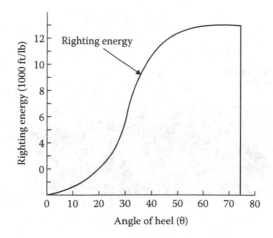

FIGURE 3.58
Righting energy curve.

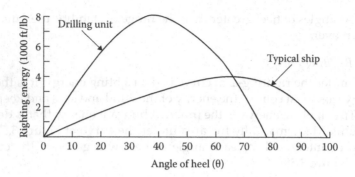

FIGURE 3.59
Righting energy curve of a ship and a drilling unit.

3.3.3.3 Static versus Dynamic Stability

Graphs of righting moment versus angles of heel are called righting moment curves which are also known as static stability curves (Figure 3.57). A floating vessel is known to be stable under *static condition* when there is steady wind force. But when a vessel is subjected to sudden gust blows along with a steady wind, that condition is known as *dynamic condition* and the stability of the vessel under such condition is known as dynamic stability. This is defined in terms of wind heeling moment as a function of the angle of heel (Figure 3.60).

The stability criterion that has been developed from years of experience and careful study is based on wind force. Because wind accounts for the

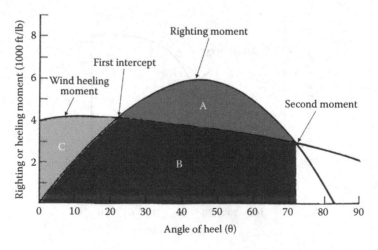

FIGURE 3.60
Dynamic stability curve.

largest overturning force, stability requirements for drilling units and ships are based on broadside wind force. For drilling units classified by the American Bureau of Shipping (ABS), a broadside wind force with a velocity of 100 knots is assumed. The force of the wind on the side of the drilling unit creates an overturning moment. This is calculated for several angles of heel. The result is plotted along with the righting moment. A typical curve known as dynamic stability curve is shown in Figure 3.60. The wind-heeling moment starts out much greater, but at angles of heel up to 72°, it is less than the righting moment. The two curves cross twice. The second crossing is called the second intercept. The areas under the two curves represent the righting energy and the wind-heeling energy. The ABS regulation in Rules for Building and Classing Offshore Mobile Drilling Units, 1973, states: 'In all cases, except column-stabilised units, the area under the righting moment curve to the second intercept or to the downflooding angle, whichever is less, must be 40% in excess of the area under the wind heeling moment curve to the same limiting angle. A figure of 30% is used for column-stabilised units.'

For ship-like vessels

Area (A + B) ≥ 1.4 Area (B + C)

For column-stabilised vessels

Area (A + B) ≥ 1.3 Area (B + C)

This excess area, or energy, is not all safety factor. If a drilling unit is rolling it can have considerable energy of motion when it is upright. If the unit has rolled to one side due to waves, as it passes a zero angle of heel it has some motion. This motion is energy, and this energy plus the wind-heeling moment must be resisted by the righting energy. This is a very complex dynamic situation, but some excess righting energy must be provided. The figures of 40% and 30% have been derived from years of experience with ships and drilling units and from studies of models and analytical investigations. It is based on these calculations that the freeboard, the maximum deck loads and the operating instructions for the unit are determined.

3.3.3.4 *Other Stability Considerations*

Governmental authorities of various countries require the following stability considerations.

Damaged stability – It requires that a vessel should be compartmentalised sufficiently to withstand flooding of any one major compartment. Further, in the damaged condition, the vessel should have sufficient stability to withstand a 50-knot wind (ABS MODU rules). The final waterline in the damaged condition is to be below the lower edge of any opening through which downflooding may occur.

Flooding of a compartment results in sinkage as well as trim. There are two methods of assessing stability in this condition:

Lost buoyancy method:

- Flooded volume treated as lost underwater volume
- Loss of water plane area calculated
- Sinkage and trim estimated
- Iterations carried out to get final position of vessel

Added weight method:

- Flooded water treated as added weight
- New displacement and KG evaluated
- Corrections for water plane area lost and displacement adjusted up to sinkage condition
- Repeat calculation to get convergent results

Both methods give equivalent results.

Free surface effect – Thus far stability considerations are based on a secured load whereas the movement of a liquid in a tank is not a secured load because the centre of gravity of the tank changes with vessel motion. This is known as free surface effect. Seagoing vessels are designed with partitioned tanks that will not present a stability problem unless they are cross-connected. Dangerous situations occur because of the free surface phenomenon and personnel responsible for ballasting the tanks or transfer liquid between tanks should have the knowledge of problems that can be caused by partially filled tanks.

The change of centre of gravity changes the moment and this is known as *moment of transference,* which is appreciably affected by the lateral extent of the tank and to a lesser extent by the height.

3.3.4 Station Keeping

Any floating vessel when used for carrying out an operation like drilling has to remain in its position at least for the duration of the operation because the vessel is always subjected to some motions under the influence of severe weather and sea states. Different forces that act on a floating drilling vessel are shown in Figure 3.61. There are six motions to which a floating vessel is always subjected. Out of six motions, three are translational and three are rotational (Figure 3.62). The translational motions are *surge, sway* and *heave* and the rotational motions are *roll, pitch* and *yaw.*

Surge: Translation fore and aft or bow and stern (along X-axis)

Sway: Translation port and starboard (along Y-axis)

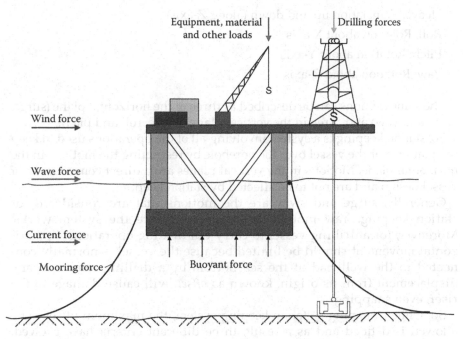

FIGURE 3.61
Forces acting on a drilling vessel.

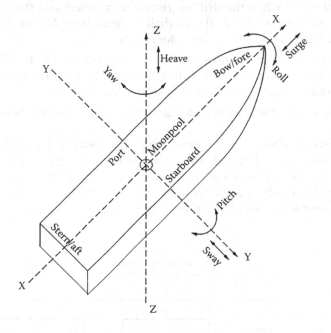

FIGURE 3.62
Six motions of a floating vessel.

Heave: Translation up and down (along Z-axis)

Roll: Rotation about X-axis

Pitch: Rotation about Y-axis

Yaw: Rotation about Z-axis

The same motions can be described as three in the horizontal plane (surge, sway and yaw) and three in the vertical plane (heave, roll and pitch).

So, station keeping is a system involving all of the operations used to keep the platform or the vessel over the borehole by restricting the motions in the horizontal plane. Motions in the vertical planes are a direct consequence of vessel design and are not to be affected by station keeping.

Generally, surge and sway are the motions that are considered for station keeping. Yaw motion gets decreased once the system works. Moreover, for a drilling vessel to carry out drilling operations, its horizontal movement should be limited because the vessel is normally connected to the wellhead at the sea bottom by a drilling riser and any displacement from its origin, known as *offset*, will cause damage to the riser, even snapping.

So, while designing a station keeping system, the maximum offset to be allowed is defined and as a result, three different criteria have evolved. Those are

1. *Operational* – when the drilling riser is connected and the vessel is close enough over the well for a drilling operation. At this stage, the offset is limited to 5%–6% of water depth.

2. *Non-operational but connected* – the riser is connected but no drilling operation is going on. At this stage, depending upon the operations, the limit of offset varies from 8%–10% of water depth.

3. *Disconnected* – the riser is no longer connected to the wellhead.

Other than the above criteria, designing or choosing a particular type of station keeping method, it is also necessary to know the percentage of time a vessel will be able to hold a station so that it can perform the operation. This requires the knowledge of environmental forces based on which the following station keeping methods are presently in vogue.

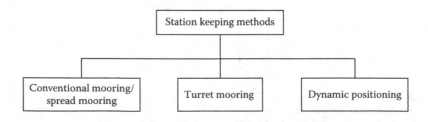

3.3.4.1 Conventional Mooring System

A conventional mooring system is made up of basic components like mooring line, anchors, connectors and handling equipment such as winches, windlasses, tension measuring devices and mooring buoys.

3.3.4.1.1 Mooring Components

Anchors are the most important component of a mooring system as they rely on the strength of the anchors. Figure 3.63 shows the most utilised anchor – a drag embedment anchor (DEA) or simply *drag anchor*, so-named because as it penetrates the sea bed, it is dragged along and uses soil resistance to hold the anchor in place until it reaches the required depth. It is mainly used for catenary shape mooring lines which arrive on the seabed horizontally. It does not perform well under vertical forces. Its main parts are *shank* to which the mooring line is attached and the *fluke* which actually digs into the soil. The angle between the shank and the fluke is known as the fluke angle and the maximum holding power of an anchor is sensitive to the fluke angle setting. Holding power of an anchor is often expressed in terms of a *'holding power ratio,'* which is defined as the mooring line tension on the anchor divided by the weight of the anchor in the air. An efficient anchor should have a high holding power ratio; that is, it should develop a maximum holding power for a minimum of weight. Ideally, the holding power ratio for a given anchor should exceed 10 for all types of bottom conditions ranging from hard sand and clay through soft mud. A more conservative rule of thumb is that holding power may be estimated as three times the anchor weight. In the case of soft mud, the setting of the fluke angle should be at 50° whereas for hard sand it is between 30° and 35° whereas for seabeds of unknown soil characteristics it varies from 35°–40°.

Where the vertical forces are predominant, the *suction anchors* or *suction piles* are used. Figure 3.64 depicts a deepwater floating production vessel moored with a taut station keeping system of a fibre rope using suction

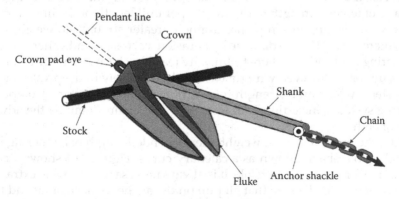

FIGURE 3.63
Drag anchor.

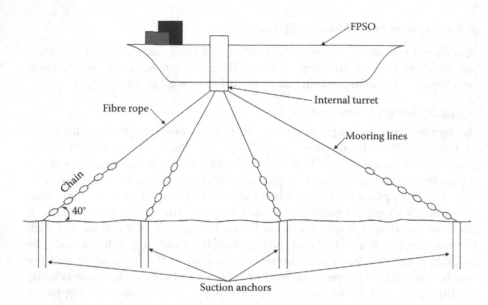

FIGURE 3.64
Deepwater FPSO using suction anchors.

anchors. The angle at the lower end is noted as being 40° to the horizontal. In this case, tubular piles are driven into the seabed and a pump sucks out the water from the top of the tubular, which pulls the pile further into the seabed. Figure 3.65 shows a typical suction anchor installation sequence.

Mooring lines are freely hanging lines connecting the floating platform to anchors or piles on the seabed positioned at some distance from the platform. The mooring lines are laid out, often symmetrically in plan view, around the vessel. Mooring lines can be either steel-linked chain or wire rope or a combination of rope and chain. In selecting the type of line one important parameter, that is, *strength-to-weight ratio* is very important. This is defined as the ratio of tensile strength to the weight per unit length of the line. In comparison with chain, wire rope has a much greater strength-to-weight ratio but is more susceptible to damage by abrasion, corrosion and general abuse. In mooring applications where technically feasible, chain is usually favoured over wire rope. However, wire rope is used extensively in deepwater mooring systems where high strength-to-weight ratios are important. Composite mooring systems, including both wire rope and chain, combine the advantages of both.

Under the force of its own weight, the suspended length of a mooring line will fall into a shape known as a catenary curve. Figure 3.66 shows a catenary curve for a semisubmersible. It is always necessary to provide extra line at the bottom to make sure that all pull on the anchor is horizontal and that it will not be pulled out vertically. The ratio of the total length of the mooring

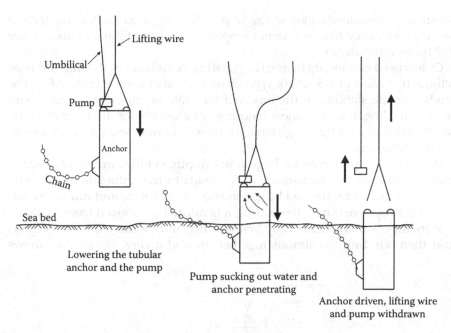

FIGURE 3.65
Suction anchor installation sequence.

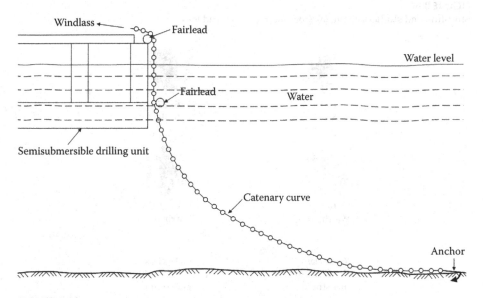

FIGURE 3.66
Catenary curve shape of a mooring line (chain).

line to the water depth is known as *'scope.'* Mooring lines for floating drilling vessels commonly have minimum scopes of 5.0–7.0 or total lengths that are 5–7 times water depth.

Chains used commonly in mooring floating vessels are of the *stud-link* type although *studless* or open-link type chains are also used (Figure 3.67). The studs provide stability to the link and facilitate laying down of the chain while handling. For permanent mooring, studless chains are recently being favoured because of their high strength-to-weight ratio and they increase the chain fatigue life.

Wire ropes are preferred for large water depth, yet they are more prone to kinks, abrasions and corrosions, which leads to the reduction of strength. Corrosion can be minimised by galvanising a wire rope and during its service if the rope parts near the middle, it is normally removed from service.

Wire rope is manufactured by first winding individual wires into strands and then winding the strands together around a core. Figure 3.68 shows

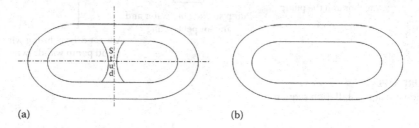

(a)   (b)

FIGURE 3.67
Stud-link and studless chain. (a) Stud link type. (b) Stud less type.

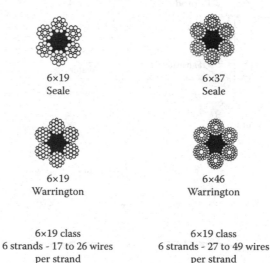

6×19
Seale

6×37
Seale

6×19
Warrington

6×46
Warrington

6×19 class
6 strands - 17 to 26 wires
per strand

6×19 class
6 strands - 27 to 49 wires
per strand

FIGURE 3.68
Cross-section of some wire rope.

cross-sections of some typical wire ropes. Almost all wire rope used in mooring systems is of either the 6×19 or the 6×37 class. The number 6 refers to the number of strands per rope and the number 19 or 37 refers to the number of individual wires per strand. The number of wires per strand can vary within each class in the illustration. The 6×19 class has relatively large sized individual wires; therefore, it has high resistance to corrosion and wear by abrasion, along with good flexibility and fatigue life. Smaller sized individual wires in the 6×37 class result in high flexibility and excellent fatigue life; however, corrosion and abrasion resistance is poorer. The two most common types of wire rope construction are *Warrington* and *Seale*. The Warrington construction has both large and small sized wires in the outer layer of each strand, offering good flexibility and fatigue life, but maintains poor resistance to abrasion and corrosion. Abrasion and corrosion resistance is improved in the Seale construction where an outer layer of large wires protects an inner layer of small wires. Usually, the Seale construction is specified for mooring applications.

Most of the wire rope cross-sections shown previously are constructed with fibre cores, which make them unacceptable for moorings where wire ropes must tolerate considerable abuse by bending over small radii, crushing on drums and rubbing on sheaves. Only ropes with metallic cores should be used. Two types are available, the independent wire rope core (IWRC) and the wire strand core (WSC). Generally, IWRC is preferred, inasmuch as WSC causes the rope to be much stiffer. IWRC increases rope weight about 10% and breaking strength about 7%, and increases resistance to crushing and abrasion.

The way in which the wires and strands are wound together is called the 'lay' of a rope. *Right lay* and *left lay* refer to the appearance of a rope to an observer as he looks along the rope axis from above. If the strands twist to the right, the rope is referred to as *right lay*, or standard lay. If they turn to the left, the rope is *left lay*.

Regular lay refers to ropes where the twist of the wires in the strands is opposite to the twist of the strands in the rope. Therefore, right regular lay, which is standard, indicates that the strands twist to the right and the individual wires in the strands twist to the left. In *lang lay* ropes, the wires and strands twist in the same direction. The outer wires of regular lay ropes run nearly parallel to the axis of the rope, whereas in lang lay ropes they run diagonally across the axis. Because the wires and strands of regular lay ropes are twisted in opposite directions, the tendency to kink and untwist is much less than for lang lay ropes. Also, lang lay ropes are more susceptible to crushing abuse.

Connectors or *connecting elements* are used to connect a section of chain to another known as chain fittings or a section of rope to another known as wire rope fittings. Some of these fittings can also connect chain to wire rope or to pad eyes on anchors or vessels. Mooring connectors are designed to take the full breaking strength of the chain or wire rope but their fatigue properties require special attention.

Chain fittings can be shackles, swivels (Figure 3.69), or detachable links. These fittings are fabricated from forged steel, not cast steel and should be inspected and tested frequently because their life expectancies may be about 30%–50% of the chain itself.

Although the proven strength of connector links is greater than the same size chain link, all commercial connectors have a lower fatigue life than chain. Consequently, they should be replaced more often than chain. Swivels are commercially available but are highly susceptible to fatigue failure and should not be used for mooring a drilling vessel.

To connect chain with anchor or with wire rope, an end shackle is used (Figure 3.70). *Wire rope fittings* must be manufactured properly in order to develop the full strength of a mooring system. Two common types of socket fittings are the swaged socket and the zinc-poured socket (see Figure 3.71). Swaged sockets are attached to the rope by inserting the rope into the shank hole and pressing or swaging the shank onto the end of the rope with a special tool. To attach zinc-poured sockets, the end of the rope is frayed, or 'broomed out,' to prevent the strands from untwisting. The broomed-out end is then inserted into the socket and the socket bowl is filled with molten zinc. All sockets should be made from forged steel rather than cast steel to provide extra strength in heavy-duty service.

Wire rope mooring lines are sometimes fitted with eyes and thimbles (Figure 3.72) rather than sockets. The best type of thimble has a plate insert to protect it from crushing when under a heavy load.

Handling equipment or the *shipboard equipment* depends on the type of line, that is, whether chain or wire rope is used for anchoring. Wire rope mooring line requires a *winch* and outboard *fairleads*. An arrangement is shown

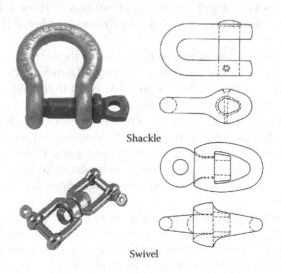

Shackle

Swivel

FIGURE 3.69
Chain fittings.

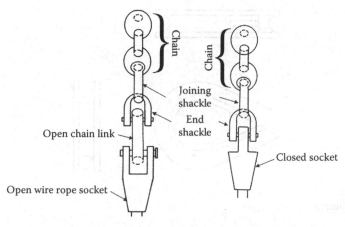

FIGURE 3.70
Connecting wire rope with chain through end shackle.

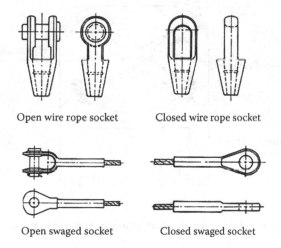

Open wire rope socket   Closed wire rope socket

Open swaged socket   Closed swaged socket

FIGURE 3.71
Wire rope fittings.

in Figure 3.73. A common wire rope tensiometer similar to the drilling line weight gauge is used.

In the case of chain mooring, *wildcat*, *chain stopper* and *fairleads* are used. An arrangement is shown in Figures 3.74 and 3.75.

Winches are used to handle and store wire rope. Dual drum winches (Figure 3.76) are most common; however, single drum and quadruple drum units have been used. The size of a winch depends on the amount of wire rope to be stored on the drum and the maximum line pull to be exerted. A winch should be able to pull half the breaking strength of a mooring line, and should be equipped with mechanical brakes which can hold full breaking strength.

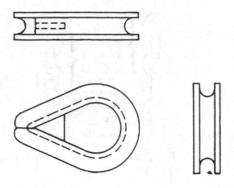

FIGURE 3.72
Wire rope thimble.

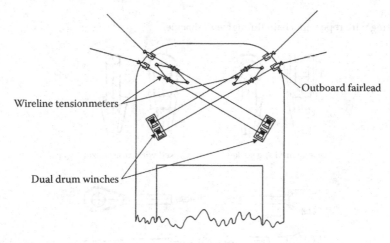

FIGURE 3.73
Arrangement of winch and fairlead on the deck of a vessel for wire rope mooring.

It is important to wind rope onto the winch drum properly. Correct winding depends on the direction of winding and the lay of the rope. Ideally, the winch drum should be grooved and the rope wound so that it does not cut across adjacent layers. This will minimise crushing and wear, and will extend the service life of the rope.

Windlasses are used to pull in and run out chain. The chain is not stored on the windlass itself, but in chain lockers below. To propel the chain, windlasses are fit with special gear-like sheaves, known as 'wildcats.' The wildcats are built with sprocket teeth, called 'whelps,' and pockets, which engage individual chain links and hold them securely. Figure 3.77 shows a windlass wildcat and part of the deck equipment used for handling chain. Chain may also be passed over a grooved sheave, also shown in Figure 3.77. Tension forces on the chain may bend the links lying flat against the sheave. The level

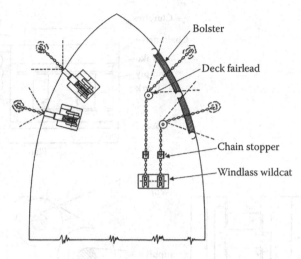

FIGURE 3.74
Arrangement of windlass and fairlead on the deck of a vessel for chain mooring.

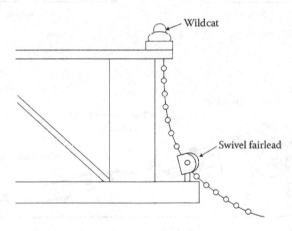

FIGURE 3.75
Location of wildcat fairlead on a semi-submergible.

of the bending stresses created will depend on the magnitude and direction of the tension forces in the chain.

Ahead of the windlass is a chain stopper which holds the entire tension of the mooring line when the windlass is not in use. A good fit between the chain and the whelps of the wildcat is important to avoid damaging the windlass. Size requirements for windlasses are similar to those for winches. A windlass should be able to pull approximately one-half the rated breaking strength of the chain passing over it.

Tension measuring devices differ for mooring lines made of wire rope as opposed to chain.

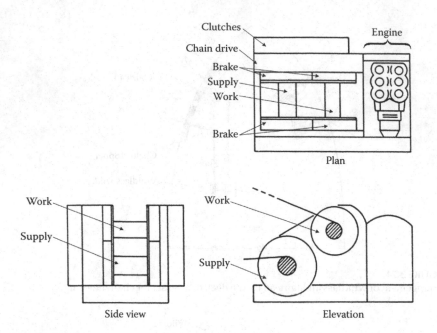

FIGURE 3.76
Dual drum winch.

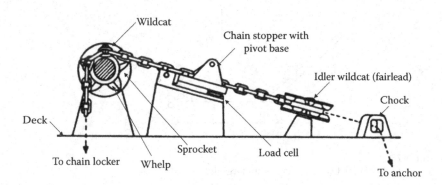

FIGURE 3.77
Windlass wildcat with other chain handling equipment.

Chain tension measurements are based on the physical principle of balancing forces and lever arms. An example is shown in Figure 3.78, where the chain stopper is mounted on a pivot. The chain tension is eccentric to the pivot, acting through a lever arm 'A'. An overturning moment on the chain stopper results, which is counterbalanced by the force in the load cell acting through the lever arm 'B.' The load cell force can be measured and multiplied by the ratio B/A to give the chain tension.

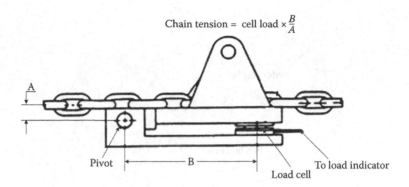

FIGURE 3.78
Chain line tension measurement.

Wire rope tension is measured by determining the force required to deflect the wire rope from a straight line. A three-sheave tension measuring device which incorporates this basic principle is shown in Figure 3.79. The deflection angle of the line and the force on the deflection sheave are uniquely related. The two outboard sheaves and the deflection sheave are designed to maintain a specified included angle, thereby fixing the geometry of the system. With the geometry fixed, the load cell reading can be calibrated to yield the wire line tension directly.

Mooring buoys are held at the water surface having been connected to one end of the pendant line whose other end is used to pull and lower the anchors. Considerable attention must be given to the design details of mooring buoys (Figure 3.80). They should be floated broadside down to minimise 'bobbing' behaviour, and should be designed to reduce movement in the propeller wash of the anchor handling boat. The shell of a mooring buoy is usually steel, and the interior is normally filled with polyurethane foam to prevent sinking in the event of puncture. Mooring buoys manufactured from synthetic materials such as rubber and plastics have been marketed; however, they are expensive and may not be sufficiently durable for extended service. In some

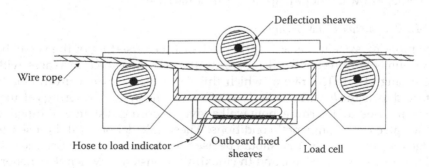

FIGURE 3.79
Wire line tension measurement method.

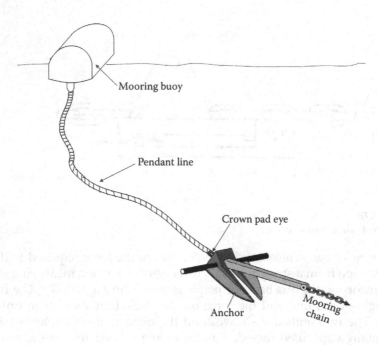

FIGURE 3.80
Mooring buoy with pendant line.

areas, mooring buoys can be hazards to navigation and legal regulations may require that they be equipped with lights, radar reflectors and other aids.

The *pendant lines* are wire ropes of 6×19 class whose diameter varies between 1¾ in. and 2¼ in. It is important that the pendant lines have the proper length. If a pendant line is too short, the mooring buoy may be pulled under and damaged in heavy seas. If too long, the excess line may pile up on the bottom and become tangled in the anchor, particularly in hard bottom. Common practice is to make the length of pendant lines 50–100 feet greater than the water depth. In shallow water, the extra length may be taken as a percentage of water depth, up to 25% in some cases.

3.3.4.1.2 *Principle of a Mooring System*

The mooring system acts as a spring to resist the offsetting of the vessel by environmental forces. Just like a spring the restoring force increases with increasing offset. The rate at which this force increases is conventionally referred to as the *hardness* or *stiffness* of a mooring system Mooring calculations are used to determine how well a given mooring system will function under given environmental conditions. These calculations will be used to judge the initial tension that should be used on the anchor line when the vessel is directly over the well, and the maximum tension to which the anchors should be set to avoid dragging an anchor under storm conditions. These calculations combined with environmental data will lead to an estimate of how much time the rig will be in operational mode, in the nonoperational mode and in the disconnected mode. The restoring force calculations are made using the catenary equation.

The catenary equation describes a line that is suspended at its two ends and allowed to sag under its own weight. By changing the boundary conditions from the fixed catenary, a mooring line can be represented mathematically. Figure 3.81 shows a diagram of a mooring line.

The general catenary equation for mooring use is

$$y = \frac{H}{w} \cosh\left(\frac{xw}{H}\right) \tag{3.9}$$

and the equations used for mooring calculations for one single weight line are

$$H = T - wd = T \cos \theta \text{ or } T = H + wd \tag{3.10}$$

$$\theta = \cos^{-1}(H/T)$$

$$V = \sqrt{T^2 - H^2} = H \tan \theta$$

$$s = \frac{V}{w} = \frac{H}{w} \tan \theta \tag{3.11}$$

$$x = \frac{H}{w} \ln\left(\frac{T+V}{H}\right) = \frac{H}{w} \ln (\sec \theta + \tan \theta) \tag{3.12}$$

$$L = x + A - s \tag{3.13}$$

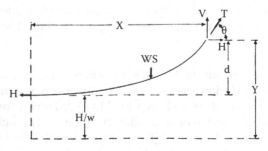

FIGURE 3.81
Parameters for calculating mooring line tension.

where:

 T = tension of the line, lb

 θ = angle of the line with respect to the horizontal, degrees

 H = horizontal restoring force, lb and is constant over the length of the
suspended line for any given value of T

 w = line weight per unit length, lb/ft

 s = suspended line length, ft

 d = water depth (should include height of outboard fairlead above water
line), ft

 y = ordinate = d + h/w, ft

 x = horizontal distance from the vessel to the point where the line touches
the seabed, ft

H/w = a translational boundary condition used to account for the force H, ft

 L = horizontal distance from the vessel to the anchor, ft

 A = total mooring line length, ft

The length of line, A is not critical to the calculation but it is critical to operations. Line lengths of five times water depth in water less than 600 ft and about three times water depth in deepwater are used. The minimum line length to avoid a vertical force on the anchor should be calculated at the maximum offset anticipated.

Calculation of tension and restoring force verses offset:

1. Calculate initial conditions using a calm water tension
   a. Choose a calm water tension
   b. Calculate H_o
   c. Calculate L_o
2. Choose a tension T_1 and repeat calculations for H_1 and L_1
3. Calculate the offset

$$\text{offset} = L_1 - L_o \text{ ft}$$

4. Choose tension T_i and calculate corresponding values of H_i and offset. Continue until the offset is greater than the anticipated maximum.

Thus, for different values of T_i different values of offset can be calculated. Hence from this calculation a sample table of offset versus tension and restoring force can be prepared, as is shown in Table 3.2. The figures shown in the table are for particular value of water depth, chain size and chain weight.

If a vessel is anchored by a single mooring line, the horizontal mooring line force H is equal to the horizontal environmental force imposed by wind,

TABLE 3.2

Example of Single Line Restoring Forces

Offset (% of Water Depth)	Tension (kip)	Line Angle (deg)	Single Line Restoring Force (kip)
10	121.2	39	94.3
8	106.1	42	79.3
6	94.1	44	67.3
4	84.4	47	57.6
2	76.6	50	49.7
0	70.0	52	43.2
−2	64.5	54	37.7
−4	59.9	56	33.0
−6	55.9	59	29.0
−8	52.5	61	25.6
−10	49.6	63	22.7

Source: R. Sheffield, *Floating Drilling: Equipment and its Use*, Gulf Publishing Company, 1980.

Note: Water depth = 400 ft; Chain size = 2½ in.; Chain weight = 66.8 lb/ft (in water).

waves and current. But in case the vessel is anchored by multiple lines, the environmental force is shared between them, as shown in Figure 3.82.

In order to maintain an equilibrium, the environmental force F must be balanced by a net restoring force from the mooring lines.

$$F = H_1 - H_2 \qquad (3.14)$$

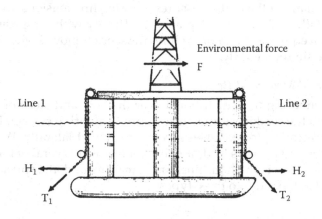

FIGURE 3.82

Multiple mooring lines used in a semi-submersible.

where F = total environmental force imposed on vessel by waves wind, and current

H_1 = horizontal component of tension in mooring line 1

H_2 = horizontal component of tension in mooring line 2

The important point to note in Equation 3.14 is that part of the tension H_1 is taken up in counterbalancing the tension H_2. When F is zero, the mooring line tensions simply balance each other.

In rough weather, the tension in windward–upwind lines, such as Line 1, can approach maximum allowable load. One means of lowering the tension in Line 1 is to slack Line 2, thereby reducing H_2.

A further result of Equation 3.10 is that the weight and breaking strength of a mooring line limit the water depth in which it can be used. The maximum horizontal restoring force H_{max} will be the difference between the maximum allowable tension in the line T_{max} and the slack line weight, wd.

$$H_{max} = T_{max} - wd \qquad (3.15)$$

Therefore, H_{max} decreases as d increases (provided T_{max} and w are fixed) for a given type of mooring line. Decreased H_{max} means that less environmental force can be resisted by the mooring system without excessive stresses.

Equation 3.15 typifies the limits of deepwater use of chain, which has a very low strength-to-weight ratio. Therefore, many deepwater moorings are designed with lighter, stronger wire rope. However, the disadvantage of lightness is that the catenary curve, length S, may be too long to be economical. Solving this dilemma requires the use of *composite mooring lines* such as those shown in Figure 3.83. Light, strong line (wire rope) at the top and heavy line (chain) at the bottom assures mooring line tensions are transmitted horizontally to the anchor. A practical difficulty with composite systems is that special equipment and procedures must be employed when using two types of line simultaneously.

3.3.4.1.3 Spread Mooring Patterns

A group of mooring lines distributed over the bow and stern of the vessel to anchors on the seafloor is known as a spread mooring system. It is meant to produce restoring forces when a vessel is moved laterally. With 8 or 10 mooring lines and anchors loads are distributed over several of the mooring lines. On the basis of which the lines are spread around, the spread mooring pattern can be divided into three types:

1. Symmetric pattern

2. Asymmetric pattern

3. Nonsymmetric pattern

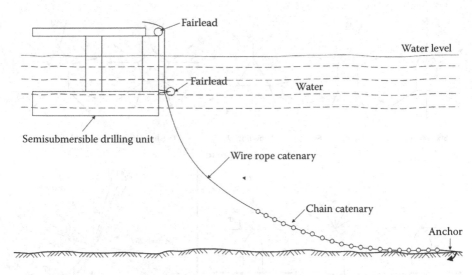

FIGURE 3.83
Composite mooring line.

Figure 3.84 depicts these three patterns. Although a symmetric spread of mooring lines is the simplest in terms of design where the included angle between two consecutive lines is the ratio of 360° to the total number of lines, it may not be the optimum in terms of performance. Criteria needing considerations are

1. Directionality of the weather, in particular if storms approach from a specific weather window; it may be advantageous to bias the mooring toward balancing these forces.
2. Subsea spatial layout; seabed equipment and pipelines may restrict the positioning of lines and anchors in this region.
3. Clashing of risers with mooring lines must be avoided and this may impose limitations on line positions.

In Figure 3.84, a nonsymmetric spread pattern is shown in a typical case where a large number of risers have been accommodated within a wide corridor by making close spacing between the number of lines.

3.3.4.1.4 Anchor Setting Procedure

Anchor handling is done by special crews and specially equipped anchor handling boats. A typical arrangement of the deck equipment on anchor handling boats is shown in Figure 3.85. The directional orientation of the vessel to prevailing or anticipated sea conditions, the type of anchors used, the compass direction of setting for each anchor and the proper length of the anchor lines are basic considerations. Semisubmersibles customarily use

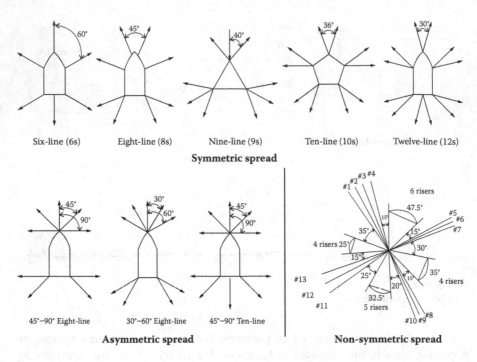

FIGURE 3.84
Spread mooring patterns.

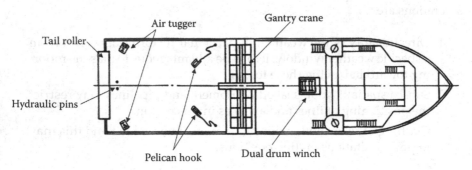

FIGURE 3.85
Deck of an anchor handling boat.

long lines. A typical case of order of setting and retrieving anchors for a particular pattern is shown in Figure 3.86.

Anchors are transferred by deck cranes from the rig to the anchor handling vessel. As the anchor handling boat moves away from the rig, windlasses or winches pay out the chain or wire line. The anchor handling vessels must have sufficient power to drag up to 3500 ft of 3-in. anchor chain. The individual anchors are transported on the deck of the anchor handling boat. Pendant

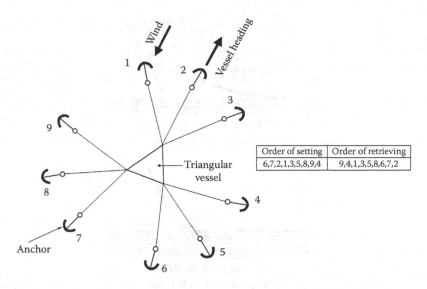

FIGURE 3.86
Order of setting and retrieving anchors.

lines support the anchor as it is being transferred, raised or lowered. Care is taken in paying out mooring lines to prevent chain or wire line from fouling.

When the anchor handling boat reaches the marker buoy, the anchor is lowered with the flukes down and set as shown in Figure 3.87. After all anchors have been run and set, anchors are *preloaded* by tensioning mooring lines to a maximum of one-half breaking strength. (Maximum allowable working load is one-third breaking strength.) This is above the design 'working load' of a mooring line and normally exceeds the holding power of the anchors. After preloading, mooring lines are *pretensioned*.

Pretensioning is adjusting the pull load of each line to the forces it is expected to experience on a continuing basis. These loads range from a minimum of 50,000 lb to as much as 200,000 lb in rough seas.

The anchor chain itself lies on the bottom at the anchor and contributes to the total holding ability of the anchor. Sometimes where soft bottom conditions require additional holding force, a '*piggyback*' or second anchor is attached. In piggybacking an anchor (Figure 3.88), the anchor handling boat attaches and sets the second anchor at the end of the mooring line. Actual installation procedures vary from one contractor to another. Many contractors simply detach the mooring buoy from the pendant line of the first anchor. An additional pendant line is attached to the second anchor, and this line is used to lower and set the second anchor while keeping the line to the first anchor tightly stretched.

Other contractors prefer to pull the first anchor on deck of the anchor handling boat, detach the pendant line and reattach a heavier line for the second anchor. This line is used to relower and set the first anchor. The heavy line

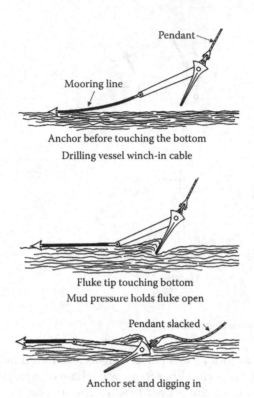

FIGURE 3.87
Anchor getting lowered, set and digging in.

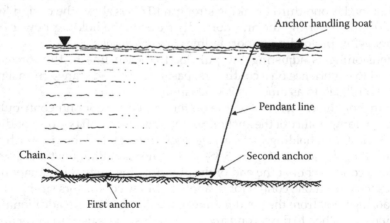

FIGURE 3.88
Piggybacking anchors.

from the first anchor and a pendant line are attached to the second anchor and it too is lowered and set as before.

Moving off location requires the use of anchor handling crews and equipment. In general, sequences of anchor setting are reversed. All anchors, chains and lines are returned to the rig and stowed for moving.

3.3.4.2 Turret Mooring

In conventional mooring, the symmetrical arrangement of anchors helps to keep the ship on its fixed heading location whereas in turret mooring, mooring turrets enable the ship's heading to be changed without moving the anchors or moving from the drilling position over the well. Turret mooring systems are the mooring systems that permit the vessel to freely 'weathervane' 360° thereby allowing normal operations in moderate to extreme sea conditions.

The turret mooring system got its name because of the use of a turret structure (turret means a small tower or tower-shaped projection having great height in proportion to its diameter and having an ability to revolve or rotate).

The turret mooring system consists of a turret assembly that is integrated into a vessel and permanently fixed to the seabed by means of a mooring system. The turret system contains a bearing system that allows the vessel to rotate around the fixed geostatic part of the turret, which is attached to the mooring system.

Depending on the location of the turret assembly mounting, this type of mooring system can be classified as 'moonpool turret mooring,' 'external turret mooring' and 'internal turret mooring.'

Moonpool turret mooring is used with drillships. From a roller-mounted turret in the ship's well beneath the derrick, eight anchor lines extend outward to the anchors (Figure 3.89). The marine conductor is the vertical axis around which the entire ship rotates. One bow thruster and two stern thrusters supply a total of 2250 hp to change headings. The vessel can be revolved 360° around the mooring plug by the bow and stern thrusters.

The other two types of mooring systems (Figure 3.90) are used with production vessels like FPSO or FSO where the system can also be combined with a fluid transfer system that enables connection of subsea pipelines to the vessel. The fluid transfer system includes risers between the pipeline end manifold (PEM) at the seabed and the geostatic part of the turret (Figure 3.44). In the turret, a swivel provides the fluid transfer path between the geostatic part and the free weathervaning vessel that rotates around the turret. The turret system is fully passive and does not require active vessel heading control or active rotation systems in the turret or swivels.

The *external turret mooring system* includes a turret system that is located at the extreme end of an outrigger structure attached to the bow of the vessel. This turret contains one large-diameter 3-raceroller main bearing that

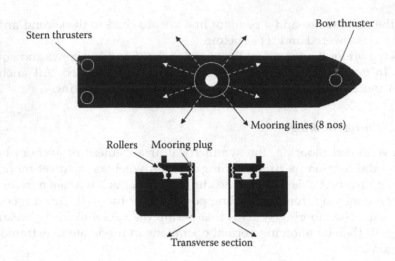

FIGURE 3.89
Moonpool turret mooring.

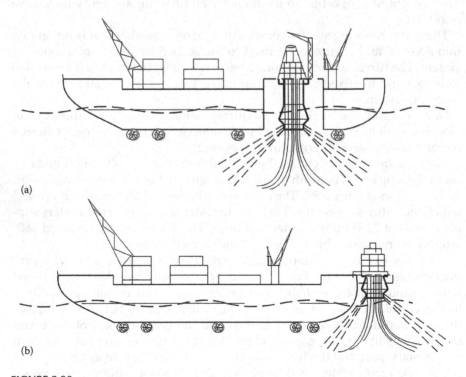

FIGURE 3.90
Two types of turret mooring for floating production unit. (a) Internal turret mooring and (b) external turret mooring.

transfers both the vertical and the horizontal loads from the riser and mooring system to the cantilever type outrigger structure.

In addition, it transfers the weight of the turret and the turntables above the main bearing to the outrigger. The turntables include the process facilities for commingling fluids from the risers using manifolds, as well as pigging facilities and/or chemical injection facilities to assure the flow in the risers. The swivel stack is located at the highest deck of these turntables and transfers fluid and electrical power to the turret access structure located on the outrigger. This access structure also provides support access to the different turntable decks.

The external turret including (a part of) the outrigger can be installed as one integrated structure on the vessel bow, minimising the scope of work for the integration of the turret into the vessel. Furthermore, this integration work can be executed while the vessel is afloat. Typically, the cantilever type outrigger structure is integrated above the main deck of the vessel. However, depending on the departure angle of the mooring lines and on the size of the outrigger structure, part of the bow might need to be removed to avoid clashes between the vessel and the mooring lines.

The *internal turret mooring system* includes a turret system that is integrated into the hull structure at the bow of the vessel. The internal turret is a slender structure that is connected to the vessel structure via a large-diameter 3-race roller main bearing at the top via a large bearing at the bottom.

The slender turret shaft ensures that the horizontal loads from the risers and mooring lines are transferred via the sliding bearing to the vessel, while the remaining forces from the turret and the vertical loads from the risers and mooring lines are transferred via the 3-race roller bearing. The supports of the bearings are integrated into the turret casing, which is part of the modified vessel hull structure. Typically, the turret casing is a cylindrical structure that is integrated between the bottom and the deck of the vessel and transfers the turret loads via the vessel web frames to the vessel hull structures.

The turntable decks are positioned on top of the 3-race roller bearing. The turntables include the process facilities for commingling fluids from the risers using manifolds, as well as pigging facilities and/or chemical injection facilities to assure the flow in the risers. In addition, the turntables provide space for subsea control facilities, to enable safe operation of the subsea manifolds and well heads. The highest deck level of the turntables provides space for the swivel stack, which transfers fluid and electrical power to the turret access structure that is located on the vessel main deck. This access structure also provides support access to the different turntable decks.

3.3.4.3 Dynamic Positioning System

Dynamic positioning (DP) is a computer-controlled system to automatically maintain the horizontal position of the moonpool over the bore hole and also the heading by using its own propellers and thrusters. Dynamic positioning is advantageous at places where mooring or anchoring is not feasible due to deepwater, congestion on the sea bottom (pipelines, templates, etc.), or other problems. DP makes weathervaning possible and extends water depth capability beyond the realm of conventional mooring.

3.3.4.3.1 *Principle of the DP System*

As was discussed earlier, any floater will experience surge and sway motion due to different environmental forces in the horizontal plane. As a result, there will always be some offset. The objective of the station keeping method is to control this offset within a given limitation. In the DP system to control the offset, the first thing that is necessary is to measure or determine this offset and then to send command to the thrusters. But for sending command that is the exact value of the thrust to the thrusters it is necessary to know the amount of force that is causing the measured offset and the direction of the environment forces.

So, it is found that there are three basic elements that comprise the dynamic positioning system. Those are

1. *Referencing/sensing system,* which continuously measures the position with respect to the ocean bottom borehole. The position referencing sensors along with wind sensors, motion sensors and gyrocompasses provide information to the computer.
2. *Control system* or controller, which is a major part of the on-board computer which after receiving the information from the earlier one determines the magnitude and direction of the thrust to be provided and also sends command to the thrusters.
3. *Propulsion system or thruster,* which is instrumental in providing the thrust as per command received from the controller.

It is to be noted here that the function of these three elements has to be so accurate, fast and synchronised that from the time of occurrence of any offset to ultimately providing thrust, the on-board computer should not take more than fraction of a second and thus it has to be a highly sophisticated system. Figure 3.91 depicts the basic two elements and their components.

3.3.4.3.2 *Components and their Functions*

3.3.4.3.2.1 *Reference Systems*   This comprises (1) position reference system, (2) heading reference system, and (3) sensors.

Position Reference System
All commercial drilling vessels employ acoustics as the primary source for determining the position of the vessel relative to the wellhead. This is called the *acoustic system.* There is also a backup position reference system which is known as the *taut wire system.*

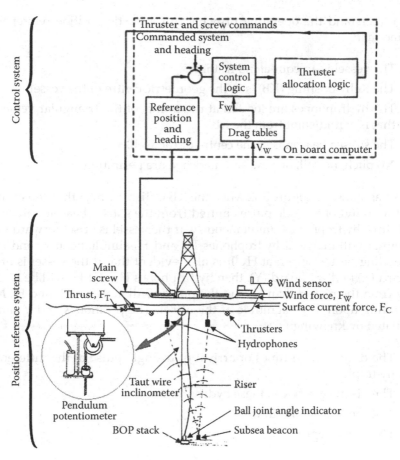

FIGURE 3.91
Components of dynamic positioning (DP) system.

In an *acoustic system*, two things are necessary, that is, one to produce the sound signal and the other to receive the signal. As shown in Figure 3.91 there are *hydrophones* mounted at the bottom of the vessel to receive the signal and there are *subsea beacons* located on or near the wellhead.

Two types of subsea beacons are employed: a *free-running pinger* that continuously transmits signals at a specified rep rate, or a *transponder* that responds only to a command from the surface. The transponder is more accurate because it measures the water depth and requires no additional measurement for tides. In dynamic positioning, a transponder will allow the computer to select the most advantageous rep rate for the particular location.

The receiving hydrophones on the vessel must be carefully installed in dry dock with the vessel perfectly level. The tolerance allowed for this installation varies between manufacturers and recommendations by the manufacturer should be followed carefully.

To understand the operating principles of acoustic position referencing, assume that:

1. The vessel is an equilateral triangle.
2. The Kelly bushing (KB) is in the geometric centre of the vessel.
3. The hydrophones are located at the points of the triangular vessel, that is, equidistant from the KB.
4. The subsea beacon is in the centre of the well.
5. No pitch, no roll, no yaw and no heave are permitted.

Now, as shown in Figure 3.92 when the KB is directly over the acoustic beacon, the sound of a single pulse emitted from the subsea beacon will arrive at all three hydrophones simultaneously. If the vessel is offset forward only, the signal will arrive at hydrophones H_1 and H_3 simultaneously and later (depending on the offset) at H_2. It is also evident that if the vessel is offset forward (X) and starboard (Y), then hydrophones H_2 and H_3 will be farther away from the pinger than H_1 and the pulses will arrive at H_2 and H_3 later depending upon the magnitude of the offset. The position X and Y can be calculated by knowing:

1. The difference in times of arrival of a single pulse at the different hydrophones
2. The distance between these hydrophones
3. The speed of sound in water
4. The water depth

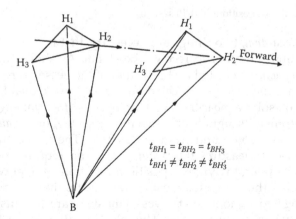

FIGURE 3.92
Principle of acoustic position reference system.

In fact, for the case with the beacon in the centre of the well, yaw can be permitted because all calculations would be in *ship's* coordinates (fore/aft, port/starboard). Since beacons will not survive in the well bore and must be offset, the position reference system must be aware of *Earth's coordinates* (north/south, east/west). This and other aspects of the ideal situation explained above are altered by information put into the device by hand or by external sensing equipment.

Corrections are required for pitch and roll, that is, the inclination of the hydrophone plane. The inclination of the vessel is measured by a vertical reference unit consisting of a pendulum or gyroscope. This is often a weak link in the system, and some of the 'electronic jitter' in various commercial systems is caused by the vertical reference unit.

The *taut wire system* uses a small diameter line attached to a weight on the sea bottom. The weight is of such size that its dead weight will stress the line 10%–20% of its breaking strength. Similarly, on top a counterweight is rigged to effect a constant tension of 5%–19% of the breaking strength of the line. For example, a counterweight of 500 lb might be used to tension a 1/8 in. line against a 1000 lb weight on the bottom.

A two-axis inclinometer is fit with a guide and gimbaled to pick off the angle of the wire line with respect to the vertical. The vertical angle is referenced to a shipboard coordinate system of bow-stern axis inclination and port-starboard axis inclination. Knowing water depth and the angle of inclination, the offset can be easily calculated (Figure 3.93).

Heading Reference System

Gyrocompasses are normally used to determine heading. A sine-cosine potentiometer is also used which is driven by a repeater from the main gyrocompass. The more advanced methods are ring-laser gyroscopes and fibre optic gyroscopes.

Sensors

Besides position and heading, other variables are fed into the DP system through sensors:

1. *Motion reference units (MRU)* or *vertical reference units (VRU)* determine the ship's roll, pitch and heave.

2. *Wind sensors* are fed into the DP system feed forward, so that the system can anticipate wind gusts before the ship is blown off position.

3. *Draught sensors* are used because any change in draught influences the effect of wind and current on the hull.

3.3.4.3.2.2 Control Systems The basic control in DP system has three degrees of freedom, that is, surge, sway and yaw. No attempt will be made to control roll, pitch and heave by the DP system.

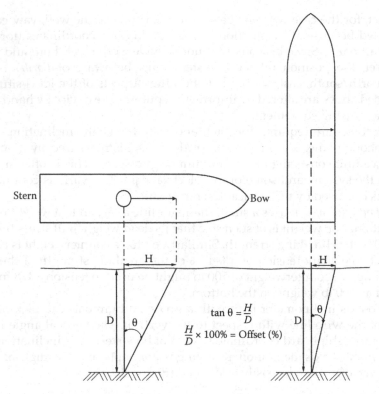

FIGURE 3.93
Principle of a taut wire system.

For vessel control in dynamic positioning, each degree of freedom is controlled separately Thus, the overall control is divided into three separate channels. Basically, these channels consist of a fore-aft position controller, a port-starboard position controller and a heading position controller. The force or thrust demand output of these three controllers is input to the thruster logic section which determines the actual thruster response to the output commands.

The final output of the three controllers includes a command signal in the surge axis for longitudinal thrust in that axis, a command signal in the sway axis for lateral thrust in that axis and a command signal in the yaw axis for a turning moment in that axis.

To understand the functioning of controller operation let us consider that the DP controller is a computer that takes input data from various sensors and estimates the thrust required to correct the vessel's position. A diagram of the logical sequence of events is shown in Figure 3.94 and is described below:

1. The computer receives the vessel position from the position referencing system and the heading from the gyrocompass.

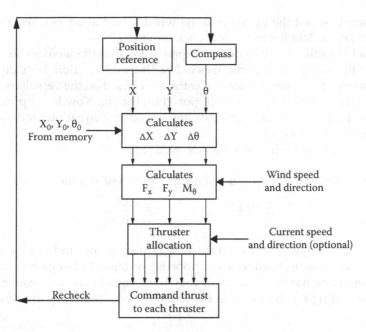

FIGURE 3.94
Logical sequence of events in a control system.

2. The computer compares the position and heading of the vessel with X_o, Y_o and θ_o that would place the vessel directly over the well and maintain the proper heading.

3. The forces required to bring the vessel over the well are estimated. Vessel position, heading and wind data are used.

4. The controller allocates thrust to the individual thrusters.

5. Then the commands are sent to the thrusters and it is checked to see if they responded properly.

6. And, thus, the cycle is repeated.

The total lapsed time for the whole sequence of events is about ½ to 3 sec.

Force estimations are based on two types of information: feedback and feed-forward data.

Feedforward data are used to estimate the force required to resist wind gusts. Before the vessel is built for dynamic positioning, wind tunnel tests are run to determine the forces and moment that wind will impose on the vessel. Tables of wind effects at different angles and velocities are developed and stored in the computer memory. In operation, the wind speed and direction are measured, and the resulting forces and moment are estimated from the tables so that the thruster forces can be adjusted for wind gusts. Steady-state wind forces are not the purpose of the wind information, because they are taken into consideration in the feedback equation.

Feedback information does not anticipate a force as the feedforward information, but merely reacts to changes in heading and position. For example, if a wave moves the vessel, the force is estimated based on the vessel's response to that wave and the other forces imposed on the rig. Now let us pick one of the forces F_X, F_Y, moment M_θ, and look at the feedback equation for estimating these forces and moment.

The offset in the X direction $= X - X_o = \Delta X$ and

Force in X direction	Proportional term	Differential term	Integral term
$F_X =$	$A\,\Delta X +$	$B, \dfrac{d\Delta X}{dt} +$	$C \int \Delta X\, dt$

A, B and C are dynamic response factors called gains, and can be considered analogous to the hardness of a mooring system. Higher gains will cause the system to be harder. Gains will vary with vessel mass and water depth. To give a feeling for the terms in the feedback equation, they are discussed below:

   $A\,\Delta X$ is the proportional term and increases the restoring thrust as the
      offset increases.
   $B\dfrac{d\,\Delta X}{dt}$ is the differential term that increases the thrust relative to the
      speed at which the vessel is moving away from the well. This term
      may reverse the thrust as the well is approached depending on the
      vessel speed.

$C \int \Delta X\, dt$ is the integral term. Without this term, the vessel would wander back and forth over the well and would not settle down. This term gains experience and determines the steady state forces required to hold the vessel over the well. Time is required to build up this integral when the system is first put into use.

3.3.4.3.2.3 Propulsion System/Thrusters   The element of a dynamic positioning system which produces the forces and moment to counter the disturbance forces and moments acting on the vessel is the thrusters. Thrusters include any commendable device that produces thrust and in this context all propellers are thrusters. Basically, in a propulsion system or thrusters system, two main components are there: (1) Thrust producing mechanism or device, that is, thrusters and (2) thrust control system. The purpose of the thrust control system is to implement the command variable in the thrust producing mechanism. For example, for a variable pitch propeller the control system will set the pitch of the propeller to be equal to the commanded pitch. The thrust control system can be electrical, hydraulic, mechanical or a combination. In many thrusters systems, the control system controls the prime mover which furnishes power to thrusters, and in such cases electrical controls are more suitable for electric motors being used as prime movers.

Two types of thrusters are used for dynamic positioning. *Constant pitch, variable rpm* units with DC motors are the old standby of the shipping industry. They are very efficient and economical when the rpm is kept nearly constant, as on a cruise. In dynamic positioning, however, the thrust must be varied rapidly at times which tends to decrease the efficiency. DC motors are limited in their ability to reverse thrust rapidly.

The second type of thruster is the *variable pitch, constant rpm* thruster. Variable pitch thrusters use the same principle as airplane propellers. They use AC motors, and the pitch is controlled hydraulically. These thrusters can reverse thrust more rapidly than can the DC drive motors, but some of the large variable pitch units have experienced hydraulic problems. The common example of both these types is the *open propeller type thruster* as shown in Figure 3.95. It can be either fixed-pitch with controllable speed (the first type) or fixed-speed with controllable pitch (the second type). It acts like a screw in the water and, therefore, is generally called a *screw*. It can have a minimum of three blades and as many as six blades. The screws are capable of a few hundred pounds of thrust and as much as hundreds of thousands of pounds of thrust. The example of variable or controllable pitch and constant speed thrusters are *ducted propeller* (Figure 3.96), ducted propeller in tunnel type installation or *tunnel thruster* (Figure 3.97), *azimuthing ducted propeller* (Figure 3.98) and *cycloidal thrusters* (Figure 3.99).

In a ducted propeller, the open propeller is placed in a specially shaped housing called a duct. This thruster is illustrated in Figure 3.96. The primary reason for adding the ducting to the open propeller is to increase its efficiency or thrust output per unit horsepower. The increase in efficiency can be as much as 20% at the low speeds used during dynamic positioning.

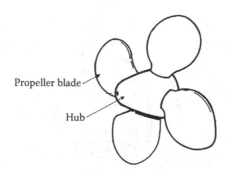

Propeller blade

Hub

FIGURE 3.95
Open propeller-type thruster.

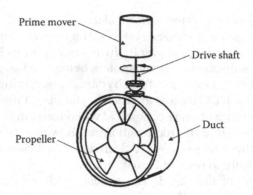

FIGURE 3.96
Ducted propeller.

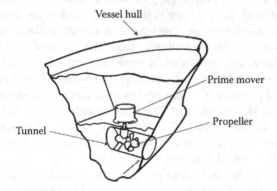

FIGURE 3.97
Tunnel type thruster.

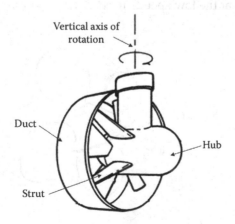

FIGURE 3.98
Azimuthing thruster.

The ducting can either be as shown in Figure 3.96 or as shown in Figure 3.97 where the open propeller is placed in a tunnel. Commonly, in the latter case, the thruster is called a tunnel thruster as opposed to a ducted thruster.

Ducted thrusters are primarily used on dynamically positioned vessels as controllable-pitch or variable-pitch propellers with a fixed axis of thrust output at right angles to the centreline of the vessel, that is, the lateral direction. These two propellers are fixed-axis propellers. Because the fixed-axis propeller can only produce thrust along a single azimuth with relation to the centerline of the vessel, the propeller is less able to counter disturbance forces which are not collinear with its line of thrust.

For this reason, another type of thruster which can change the azimuth of its thrust output with respect to the vessel coordinate system has been produced. The azimuthing thruster can be an open or ducted propeller and can have controllable pitch or controllable speed. However, the duct must be free to rotate and, therefore, cannot be a tunnel in the hull of the vessel.

As Figure 3.98 shows, a typical azimuthing thruster not only has a prime mover to rotate the propeller to produce thrust, but also has a drive mechanism to rotate the azimuth of the thruster. Thus, the rate at which the direction of the thrust can be changed depends on the speed of the azimuthing drive system. If the azimuthing drive system is too slow, the advantages gained by being able to steer the thruster in the direction of the forces acting on the vessel and to maximise thruster efficiency will be lost.

Another type of thruster which can steer its thrust direction is the cycloidal propeller shown in Figure 3.99. Not only does the cycloidal propeller steer its thrust output differently than the azimuthing thruster shown in Figure 3.98, but also it produces its thrust in a different manner.

The rotational axis of the cycloidal propeller is vertical. Through a complicated mechanical coupling, the pitch of each vertical blade is changed as the blade makes one revolution. The name of the propeller is derived from the way the pitch of the blade traces out a cycloidal pattern as the blade completes a revolution. By properly positioning the pitch control mechanism, the thrust magnitude and direction can be controlled.

Because of their highly flexible thrust azimuthing capability, cycloidal thrusters have been used on vessels requiring high manoeuvrability, such as tug boats as well as on the dynamically positioned vessels. The variable pitch thrusters are very efficient for DP, and are used even for the main propulsion screws.

The configuration of thrusters in a vessel is also shown in Figure 3.100 where these are designated as main screws and lateral thrusters. Main screws, which control the surge motion, are positioned on the stern side and are normally two in number with very high HP capacity of the order of 3000 to 8000 hp each whereas lateral thrusters, which are positioned along the longitudinal axis to control sway and yaw motion of the vessel and are generally 4 to 11 in number, have the capacity varying from 750 to 2500 hp each.

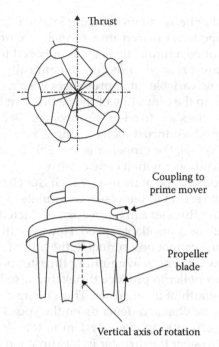

FIGURE 3.99
Cycloidal thruster.

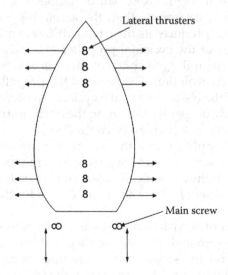

FIGURE 3.100
Typical thruster configuration.

Obviously, a DP system is quite complicated. Many things require consideration, but the most important things are reliability and good experience of the manufacturers of the central DP controller, the thruster control mechanism and the thrusters themselves.

Bibliography

Chakrabarti, S. K. *Handbook of Offshore Engineering*, Vol. 1, Elsevier Ltd., 2008.

Chakrabarti, S. K. *Handbook of Offshore Engineering*, Vol. 2, Elsevier Ltd., 2008.

Dynamic Positioning 2015 (July 2, 2015). http.//en.wikipedia.org/wiki/Dynamic Positioning.

ETA Offshore Seminars Inc., *Compilation, the Technology of Offshore Drilling, Completion and Production*, Pennwell Publishing Company, 1976.

Graff, W. J. *Introduction to Offshore Structures*, Gulf Publishing Company, 1981.

Hall, R. S. *Drilling and Producing Offshore*, Penn Well Publishing Company, 1983.

IMD, ONGC, India, Offshore Production Technology, PD-08, 1997.

Leffler, W. L., R. Pattarozzi, and G. Sterling, *Deepwater Petroleum Exploration & Production – A Nontechnical Guide*, Pennwell Publishing Company, 2003.

Lessons in Rotary Drilling, Petroleum Extension Service, 1976.

Mitra, N. K. *Fundamentals of Floating Production Systems*, Allied Publishers Pvt. Ltd., 2009.

Morgan, M. J. *Dynamic Positioning of Offshore Vessels*, The Petroleum Publishing Company, 1978.

Sheffield, R. *Floating Drilling: Equipment and its Use*, Gulf Publishing Company, 1980.

Wilson, J. F. *Dynamics of Offshore Structures*, John Wiley & Sons, Inc., 1984.

4

Offshore Drilling

4.1 Outline of Normal Drilling Operation

Regardless of the type of drilling platform and wherever it operates, the drilling rig it supports will always be comprised of certain fundamental items of equipment that are common to every one of the thousands of rotary rigs in existence in the world, both onshore and offshore.

Hence, a brief discussion on the basic components of rotary drilling operation follows.

4.1.1 Rotary Drilling Rig Operation and Its Components

Drilling of a well by rotary method is done by rotating the bits connected at the bottom of the drill string through which drilling fluid is circulated and as the drilling progresses, drill pipes are added to the string. Thus, to achieve the purpose of making a hole by this method, three main operations are involved, which are

1. Hoisting
2. Rotation
3. Circulation

Of course, to provide power for the above operations, a prime mover is one of the very important components.

4.1.1.1 Hoisting

The hoisting system is used primarily to raise and lower the drill-pipe in the hole and to maintain the desired weight on the bottom.

The principal components of the hoisting system are the (1) block-and-tackle system, (2) derrick, (3) draw works and (4) miscellaneous hoisting equipment such as hooks, elevators and weight indicators.

4.1.1.1.1 Block-and-Tackle System

The block-and-tackle system is comprised of the (1) crown block, (2) traveling block and (3) drilling line. The drilling line passes from the draw works to the top of the derrick. From here it is sheaved between the crown block and traveling block to give an 8-, 10- or 12-line suspension. The line is then clamped to the derrick floor by the deadline anchor. The principal function of this block-and-tackle system is to provide the means for removing equipment from or lowering equipment into the hole. The block-and-tackle system develops a mechanical advantage, which permits easier handling of large loads. The positioning of the drilling equipment in the hole is also a function of the block-and-tackle system, as well as providing a means of gradually lowering the drilling string as the hole is deepened by the drill bit. A schematic diagram of the hoisting elements of a rotary rig are shown in Figure 4.1.

The block-and-tackle system has two distinct advantages in hoisting operations:

1. Actual horsepower requirements can be less because the rate of doing work can be reduced.

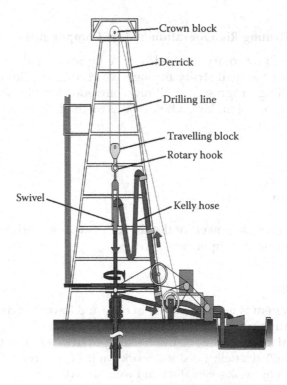

FIGURE 4.1
Hoisting elements of a rotary rig. (Courtesy of Mudgineer/Wikimedia Commons/CC-BY-SA-3.0.)

2. Engine torque requirements will be considerably less, depending upon the number and arrangements of lines in the block-and-tackle system.

The number of sheaves and the arrangement of the drilling line through these sheaves is important when derrick loads are considered. One phenomenon of the block-and-tackle system is that the actual load on the supporting structure, in this case the derrick, may be considerably larger than the weight of the load actually lifted. Figure 4.2 shows three possible block-and tackle combinations. View A shows a single sheave at the top of the derrick. View B shows three sheaves in the crown block and two sheaves in the traveling block, with the end of the line (called the dead line) attached to the derrick floor. View C shows the same sheave arrangement except that the dead line is now attached to the traveling block. Summation of the total pull on the derrick for the three arrangements is shown below.

			LOADS ON DERRICK
View A	–	–	2.0W
View B	–	–	1.5W
View C	–	–	1.2W

Important conclusions developed from an analysis of the various possible loading combinations are

1. Actual loads imposed on the derrick may be much larger than the weight to be lifted.
2. As the number of sheaves in the block-and-tackle system is increased, the actual load on the derrick is decreased.
3. Attaching the dead line to the traveling block rather than to the derrick floor reduces the actual load on the derrick.

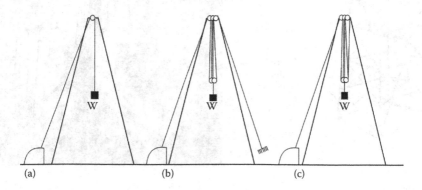

(a) (b) (c)

FIGURE 4.2
Three possible block-and-tackle combinations.

From the preceding discussion it is obvious that when the block-and-tackle system is used, the force required to move a given weight is reduced. This ratio of the weight lifted to the amount of force required to lift the weight is called the mechanical advantage (MA). Thus, in view 'A', MA = W/W = 1, in view 'B', MA = W/(W/4) = 4 and in view 'C', MA = W/(W/5) and hence it is evident that the mechanical advantage is exactly equal to the number of lines which are actually helping lift the load.

4.1.1.1.2 Derrick

The conventional derrick is a four-sided truncated pyramid (Figure 4.3) ordinarily constructed of structural steel, although tubular steel is used infrequently for certain parts of the derrick. Derricks may be either portable or nonportable, the portable derrick being commonly referred to as a mast. The nonportable derrick is usually erected by bolting the structural members

Gin pole
Water table
Monkey board
Safety platform
Derrick floor

FIGURE 4.3
Derrick. (Courtesy of Starzycka/Wikimedia Commons/Public domain.)

together. After the well has been drilled, the derrick can be disassembled, by unbolting, and erected again at the next location.

The principal considerations involved in the design of a derrick are

1. The derrick must be designed to carry safely all loads which will ever be used in wells over which it is placed. This is the collapse resistance caused by vertical loading, or the dead-load capacity of the derrick. The largest dead load which will be imposed on a derrick will normally be the heaviest string of casing run in a well. However, this heaviest string of casing will not be the greatest strain placed on the derrick. The maximum vertical load which will ever be imposed on the derrick will probably be the result of pulling on equipment, such as a drill pipe or casing, that has become stuck in the hole.

2. The derrick must also be designed to withstand the maximum wind loads to which it will be subjected. Not only must it be designed to withstand wind forces that will act on two sides at the same time (the outer surface of one side of the derrick, and the inner surface of the opposite side), but cognizance must also be taken of the fact that the drill pipe maybe out of the hole and stacked in the derrick during periods of high winds.

4.1.1.1.3 Derrick Substructures

In drilling operations, a working space must be provided below the derrick floor (Figure 4.4). The actual space required will depend upon the type of rig being used and the formation pressures that will be encountered in the drilling of the well. To control excessive formation pressures and prevent the wells from getting out of control, blowout preventers (BOP) as shown in Figure 4.18 must be attached to the innermost string of casing in order that they can be closed around the drill pipe, or, if the drill pipe is out of the hole, they can be used as a valve to close the casing completely. Some blowout-preventer designs will close the opening at the top of the casing regardless of whether the drill pipe is in or out of the hole. The closing element is comprised of a hydraulically expanded rubber sleeve designed to withstand any formation pressures which might be encountered. The rubber element in this type of blowout preventer is sufficiently flexible to permit the rotation or even the complete withdrawal of the drill pipe while maintaining a positive seal. Other types of blowout preventers are fitted with rams, which are made of steel faced with hard rubber. The closing elements in this type of blowout preventer are not flexible and therefore will close around only one geometrical configuration. If these blowout preventers are equipped to close around the drill pipe and it becomes necessary to close the blowout preventers while the drill pipe is out of the hole, then at least one joint of drill pipe must be

FIGURE 4.4
Derrick substructure. (Courtesy of www.featurepics.com.)

placed in the hole before an effective seal can be maintained: The rams, or closing elements of these blowout preventers can be changed to fit the size of drill pipe being used, or they can be fitted with blank rams, which give complete closure with the drill pipe removed from the hole. In many areas, at least two blowout preventers are used in series, and in this case several feet of space beneath the derrick floor is required.

To provide ample work room, the foundations are not used. Spreading the load of each derrick leg over a relatively large area can be accomplished by placing each corner of the substructure on a base which provides the proper bearing area. These bases may be made of structural steel members laid on a wooden timber matting. The steel bases may be welded and the timber mats bolted to reduce the set-up and tear-down time when the rig is moved. To determine the size of each mat required, it is necessary to know the load-bearing capacity of the soil and the maximum derrick loads anticipated.

4.1.1.1.4 Draw Works

The draw works (Figure 4.5) have often been referred to as the power control centre of the drilling rig because on it are located most of the

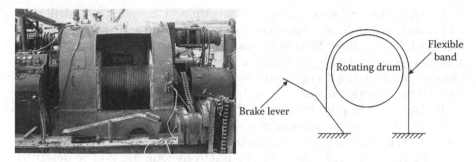

FIGURE 4.5
Draw-works with friction type mechanical brake. (Image courtesy of Schlumberger.)

controls required to run the rig. The principal parts of the draw works are (1) hoisting drum, (2) catheads, (3) brakes and (4) clutches. The *hoisting drum* is probably the most important single item of equipment on the draw works because it is through this drum that power is transmitted to remove the drill string, and on this drum the drilling line is wound and unwound as equipment is manipulated in the hole. Power for the drum comes from the power plant through suitable transmission devices, either mechanical, hydraulic or electric. The proper design of a hoisting drum required careful balancing of several objectives, some of which conflict with each other. From the standpoint of supplying the motive power for hoisting, the ideal drum would have a diameter as small as possible and a width as great as possible. This is true because the input power required to rotate the drum will be a function not only of the line pull but also of the diameter of the drum.

Solely from the standpoint of the drilling line, the drum should be as large in diameter as possible and also as wide as possible. The bending stresses in the drilling line are decreased as the diameter of the drum around which they are wrapped is increased. Also, line cross-over reduces the effective life of the drilling line.

There are maximum safe speeds at which the line can be wrapped around a drum, which vary directly with the drum diameter, large drum diameters permitting greater line speeds.

Another major factor to be considered in drum design is the overall size of the draw works. Strictly from a size standpoint, the drum should be as small as practical. In consideration of economy, the drum should be large enough to permit fast line speeds, yet small enough to present no insurmountable moving problems.

In the final analysis, in drum design all the aforementioned factors must be considered, with the end result a compromise compatible with all the various considerations.

Catheads are small rotating drums located on both sides of the draw works. They are used as a power source for many routine operations around the rig, such as screwing and unscrewing ('making up' and 'breaking out') drill pipe and casing, and pulling single joints of drill pipe and casing from the pipe rack up to the rig floor. Most of the earlier 'catheading' was done manually and was accomplished by wrapping a Manila line several turns around the cathead and applying the necessary power by the friction hold of the Manila rope on the rotating cathead.

The *brakes* are important units of the draw works assembly as they are called upon to stop the movement of large weights being lowered into the hole. When a round trip is being made, the brakes are in almost continuous use. Therefore, the brakes must have a relatively long life, and be designed so that the heat generated by braking can be dissipated rapidly. The principal brake on a hoisting drum is usually a friction type mechanical brake. A diagram of this brake is shown in Figure 4.5. It is essentially comprised of a flexible band (made of asbestos or other heat- and wear-resistant material) wrapped around a drum. One end of this flexible band is permanently anchored, while the other end is movable and attached to a lever by means of which the band can be either loosened or tightened against the rotating drum.

Two types of auxiliary brakes have been developed, a hydraulic brake and an electro-magnetic brake. The hydraulic brake utilises fluid friction to absorb some of the work done in lowering equipment into the hole. The hydraulic brake is designed to impel fluid in a direction opposite to the rotation of the drum, thus tending to slow the drum rotation. The electro-magnetic brake is essentially two opposed magnetic fields, the magnitude of which are determined by the speed of rotation of the drum and the amount of external current supplied. For either the hydraulic or electromagnetic brake to become effective, some movement of the drum is required. Therefore, neither of these types of brakes will bring the hoisting drum to a dead stop. It remains for the friction brake to effect complete stoppage. Some type of auxiliary brakes are standard equipment on most medium- and deep-drilling rigs.

The draw-works *clutch* is used to engage mechanically the hoisting drum with the transmitted power. With the advent of internal combustion engine power plants, the earlier type jaw clutch was not found satisfactory, since it caused shock loads to the system. Pneumatic clutches were designed to overcome the disadvantages of the old mechanical clutch. The pneumatic or air-operated clutch is actually a friction clutch. It consists basically of an expanding element on the inside of which is attached many friction shoes which cause the clutch to become engaged as the pneumatic element is expanded.

4.1.1.1.5 Miscellaneous Hoisting Equipment

Other items of equipment necessary for proper handling of the drill string and other equipment are hooks, links swivel, and elevators (Figure 4.6). The *hook* provides a connection between the traveling block and the swivel. The *swivel* is a device that permits the drill pipe to rotate without causing rotation of the traveling block and lines. *Links* and *elevators* are lifting tools used

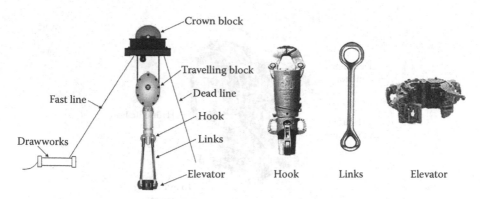

FIGURE 4.6
Miscellaneous hoisting equipment – hook, links and elevator.

in removing or inserting the drill pipe and casing in the hole. The links or bails fit onto the lugs of the hook and elevators. Once in position the flaps are closed and locked using a bolt. There are many types of elevators which can be categorised as:

Centre latch: mostly used for drill pipes and have protruding handles.

Side door: mainly used for casing and drill collars. The opening handle is more flush with the body.

The conventional hook may be replaced by a rotary connector.

4.1.1.1.6 Weight Indicators

A weight indicator is a device for measuring the weight of equipment hanging from the traveling block. This device, as originally designed, was intended principally to show the load on the derrick at all times, thus preventing derrick overload.

Three basic types of weight indicators have been developed and used to some extent: (1) an instrument attached to the traveling block or hook which measures the load directly; (2) an instrument placed on the dead line which causes a bend in the line and, by measuring the tendency of the line to straighten when placed under a load, converts this measurement into an approximate load on the line itself; and (3) an instrument composed of two elements, a pressure transformer and an indicating gauge, connected by a hydraulic hose. The pressure transformer may be one of several different designs (Figure 4.7) and it may be attached to either the dead line anchor or the dead line itself. In either case, the indicating gauge is actuated by the forces on the pressure transformer. This latter type of instrument is generally accepted as the most reliable and rugged weight indicator available at the present time, principally because of certain limitations in the other two types.

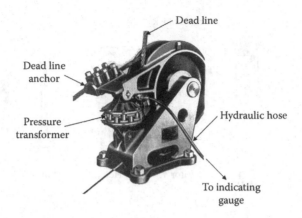

FIGURE 4.7
Weight indicator – pressure transformer fitted in deadline.

4.1.1.2 Rotation

A new hole drilled by rotating the drill string and bit is commonly referred to as rotary operation. The system of rotating equipment includes the swivel, kelly, rotary table, drill pipe, drill collars and the drill bit as shown in Figure 4.8.

The *swivel* performs three functions: (1) it suspends the kelly and drill string; (2) it permits free rotation of the kelly and drill string; and (3) it provides a connection for the rotary hose and a passageway for the flow of drilling fluid into the top of the kelly and drill string. Accordingly, the chief operating parts of the swivel (Figure 4.9) are a high-capacity-thrust bearing, which is often of tapered roller bearing design, and a rotating fluid seal consisting of rubber or fibre and metal rings which form a seal against the rotating member inside the housing. The fluid seal is arranged so that the abrasive and corrosive drilling mud will not come in contact with the bearings.

The swivel is suspended by its bail from the hook of the traveling block. The fluid entrance at the top of the swivel is a gently curving tube, which is referred to as the *gooseneck*, which provides a downward pointing connection for the rotary hose. In this manner the rotary hose is suspended between the upper nonrotating housing of the swivel and the standpipe, which extends part way up the derrick and conveys mud from the mud pump. The fluid passageway inside the swivel is commonly about 3 or 4 in. in diameter so that there is no restriction to mud flow. The lower end of the rotating member of the swivel is furnished with left-hand threads of API tool-joint design.

Kelly, which is also known as a *grief stem*, is used mainly to transmit the torque from the rotary table to the drill string. It has a typical cross-section (Figure 4.10). Internally it is hollow, cylindrical in shape whereas externally it is usually square although sometimes a hexagonal shaped kelly is used. It is

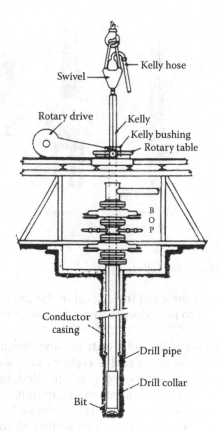

FIGURE 4.8
Rotating equipment in a drilling rig.

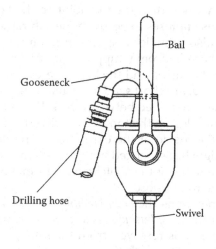

FIGURE 4.9
Swivel and its operating parts.

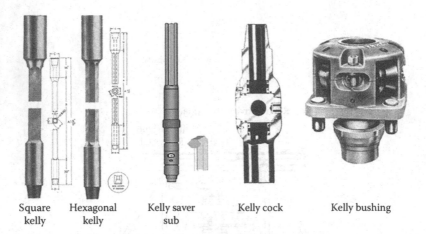

| Square kelly | Hexagonal kelly | Kelly saver sub | Kelly cock | Kelly bushing |

FIGURE 4.10
Kelly, kelly saver sub, kelly cock and kelly bushing.

connected between the bottom of the swivel to the top of the drill pipe and allows the drilling fluid to flow downward. The size of its internal diameter is normally 3 in.

Another notable feature of the kelly is its top and bottom end connections. The tool joint thread on the lower end is right-handed similar to that of the drill pipe whereas the thread on the top is left-handed, so that normal right-hand rotation will tend to keep all joints made up tight.

A *kelly saver* sub (Figure 4.10) is used between the kelly and the upper joint of the drill-pipe string. The use of this short section eliminates the necessity of unscrewing the lower end of the kelly itself during drilling operations and thereby prevents thread wear on the kelly joint. The tool joint on the kelly saver sub should be weaker than the tool joint on the kelly. This protective arrangement will cause failures to occur in the sub rather than in the kelly. The kelly saver sub also provides a space for mounting a rubber protector which will prevent the kelly from whipping against the inside of the casing, so that wear on both the casing and the kelly is prevented.

Kelly cocks (Figure 4.10) are short sections containing a valve which may be manually closed. They are placed between the kelly and the swivel. The use of a kelly cock permits closing the top of the drill string so that flow cannot take place through the inside of the drill string. The flows of concern are usually those that might be associated with drill-stem testing or other operations wherein the pressures of the subterranean formations might otherwise be applied against the rotary hose.

During drilling operations the *kelly bushing* (Figure 4.10) remains on the kelly. The outside of the *kelly bushing* is commonly square, about 16 in. on each side, so that when it is removed from the rotary table, the largest bits used in the well will pass through the rotary table. Torque is applied from the rotary table, through the kelly bushing, and thence to the kelly stem itself. The kelly is free to slide through the kelly bushing (or rotary bushing) so that the drill string can be rotated and simultaneously lowered or raised during drilling operations.

Nowadays the kelly is slowly getting replaced by power swivels or top drives (Figure 4.11) but still the kelly maintains its importance as a back-up unit.

A *rotary table* is the driving force behind any rotary drilling operation. Until recently the rotary table was fixed at the rig floor and it is the component that drives or rotates the drillstring. Nowadays modern rigs have a power sub known as a 'top drive' that has incorporated rotary into it (Figure 4.11). But still there are more rigs having rotary tables than top power subs, as many of today's holes are drilled with down hole motors where the rotary table still plays an important role.

The rotary table (Figure 4.12) itself is a very simple design and has changed little over the years. The rotating table is usually cast of alloy steel and fitted underneath with a ring gear which is shrunk onto the table proper. The table is supported either by ball or tapered roller bearings capable of supporting the dead load of the string of the drill pipe or casing which may be run into the well. Provisions in the form of suitable bearings must also be made for holding the table in place against any possible tendency to upward movement imparted during rig operations. Suitable guards are arranged so that mud or water cannot get into the oil bath provided for the gears and bearings. The ring gear and its driving pinion gear are customarily of spiral bevel construction, which provides for smoother operation than straight bevel gears. The speed reduction from the pinion shaft to the table is of the order of 3 or

FIGURE 4.11
Top drive/power swivel. (Courtesy of Shutterstock.)

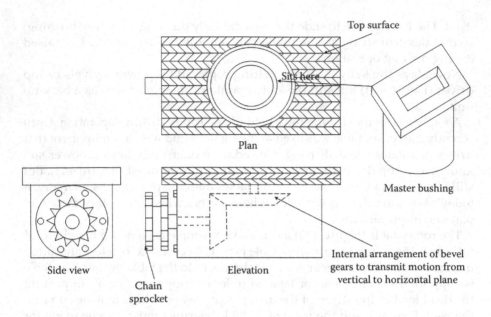

FIGURE 4.12
Rotary table and master bushing.

4 to 1. The pinion shaft is customarily furnished with antifriction bearings, often of tapered roller design. It is also equipped with oil seals and guards to prevent entry of mud or water into the interior of the equipment.

Power for driving the rotary table is most often taken from the draw works and transmitted to the rotary table by a chain and sprocket drive which rotates the horizontal pinion shaft. Thus, the spiral bevel pinion gear rotates the ring gear, which in turn rotates the vertical drill string.

The rotary table is designed in such a way that it can hold different sizes of pipe and accordingly it is rated ranging from 48 in. on a big rig to 17½ in. on a small rig. This size indicates the maximum pipe size that can go through the rotary; however, any pipe below the rated size can be held in the same rotary opening by changing the combination of 'master bushing' and 'slips' that hold the pipe.

As the main drive in the rotary table is round, for turning a pipe through this table without spinning, a cog is needed and that is done by using a 'master bushing' (Figure 4.12) which is rectangular in shape externally and fits in the opening of the table. To hold the pipe inside this bushing while drilling is stopped or halted, 'slips' are used to prevent it from dropping and during drilling for rotating the drill string. Kelly bushing either fits in the master bushing using four 'drive pins' or directly gets 'square fitted' depending on the type of master bushing.

Rotary 'slips' are conventional tools for handling drill pipes and casing and necessary tools for coupling and uncoupling operations. Drill pipe slips come in a range of sizes and in many case the dies that grip the pipe can be interchanged or a set of slips are so designed that different sized slip dies fit the same body. Its handles are designed in such a way that the people handling them can hold the handle from below.

The *drill pipe* is the major component of the drill string. Its upper end is supported by the kelly stem during drilling operations. The drill pipe rotates with the kelly, and the drilling fluid is simultaneously conducted down through the inside of the drill pipe and subsequently returned to the surface in the annulus outside of the drill pipe. In a deep well, the top portion of the drill pipe is under considerable tension while drilling, since most of the weight of the drill string is supported from the derrick.

The drill pipe (Figure 4.13) in common use is a hot-rolled, pierced, seamless tubing. Drill pipes are of many grades as per API and of many sizes. The API grades are D, E, X 95, G-105 and S-135. API Grade D drill pipe has a minimum yield strength of 55,000 psi, and Grade E drill pipe has a minimum yield strength of 75,000 psi. The X, G and S have 95,000, 105,000 and 135,000 psi minimum yield strengths, respectively.

The drill pipe most commonly used is a Range 2 pipe, which has an average length of 30 ft per individual length (joint) of pipe. The addition of tool joints produces an average made-up length of about 31 ft per individual length of pipe.

The individual lengths of pipe are fastened together by means of tool joints, which means that there are complete tool joints spaced at 31-ft intervals throughout the length of the drill pipe. The male half of a tool joint is fastened to one end of an individual piece of pipe and the female half is fastened to the other end.

Drill collars form the lowermost section of the drill string. The name derives originally from the short sub which was used to adapt the threaded joint of the bit to the drill string. However, modern drill collars are each about 30 ft in length, and the total length of the string of drill collars may be from about 100–700 or more feet. The purpose of the drill collars is to furnish weight and stiffness in the bottom portion of the drill string. During drilling, all the drill pipe should be in tension, since the drill pipe is essentially a tube of medium

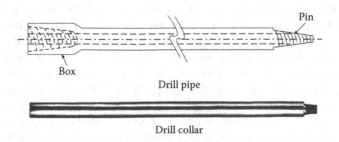

Pin

Box

Drill pipe

Drill collar

FIGURE 4.13
Drill pipe and drill collar.

wall thickness and has but little resistance to bending by column action. This means that the total weight of the string of drill collars should be determined by the weight carried on the drill bit.

Drill collars (Figure 4.13) are usually made with an essentially uniform OD and ID. A tool-joint pin is cut on the lower end and a tool-joint box is cut into the upper end; however, when the individual lengths are coupled together, they present a smooth exterior. The internal bore of the drill collars is usually 2½ in. or 2⅞ in., but the smaller bore tends to give excessive fluid-flow pressure losses, and very little difference in weight results from using the larger bore. The OD of the drill collars is limited by the size of the hole being drilled.

A *drill bit* which actually cuts the rock is attached to the lower end of the drill collar. In the rotary system of drilling, a hole is made by lowering the string of the drill pipe and drill collars until the bit touches or approaches the bottom of the hole. Circulation of drilling fluid is established down through the drill pipe, and the fluid is discharged through ports or nozzles in the bit so that the bit and bottom of the hole will be kept clean. Rotation of the drill string is established by means of the rotary table. The top of the drill string is then gently lowered by means of the draw works or hoist until suitable weight for drilling is applied to the bit.

The type of bit which should be used in any particular instance is governed primarily by the characteristics of the rock to be drilled and the conditions under which the rock must be drilled. Different types and variations of bits are in common use in drilling wells based on various cutting actions like chipping, grinding, crushing and scraping action.

Beside these technical considerations, economic considerations involved generally require that the hole should be drilled at the lowest possible cost per foot. The total cost per foot depends upon, among other considerations, the average rate of drilling and the total feet drilled per bit.

Based on the above techno-economic considerations, the different types of bits can be drag bit, disc bit, rolling cutter bit, diamond bit, PDC bit and so on.

Drag bits – Drag bits (Figure 4.14) have no independently moving parts, and the term is applied especially to bits of the blade type. The simplest of the drag bits are the *fishtail bits* with two blades spaced 180 degrees from each other. Bits having three or four blades are also used, but such bits are usually fingered so that the total length of the cutting edges does not exceed the length of the hole diameter by more than about 20%. The blades on modern bits are furnished with tungsten carbide inserts or are otherwise hard surfaced to reduce wear. Short blades are preferred, for their use permits the mud-discharge nozzles to be positioned a short distance above the bottom of the hole so that maximum jet energy can be utilised in the drilling. The mud streams flowing out of the discharge nozzles are directed to the bottom a short distance ahead of each cutting edge. Drag bits are used in drilling soft formations, and under ideal conditions the drilling action probably

Drag bit   Tricone roller bit   Two cone roller bit

FIGURE 4.14
Drag bit and rolling cutter bit.

resembles the turning of earth by a plow or better known as scraping action. The mud streams directed to the bottom of the hole break up material loosened by the bit and carry it upward to the surface. In many soft formations, a hole can be made by the jetting action of the drilling fluid.

Disk bits – These bits are interesting from a historical standpoint. They are a form of drag bit in which the cutting edges are mounted on disks. The disks are mounted off-centre with respect to the axis of the drill string, so that as the drill string is rotated, the scraping action on the bottom causes the disks to rotate slowly. In this manner the total cutting edges available for drilling are increased by comparison with the stationary blades of the drag bit. Two or four disks are mounted in the bits. The bottom of the drilled hole is rounded, a form adapted to the flushing of cuttings by the mud stream. The disk bit inherently does not have the weight-bearing capacity of the drag bit and does not provide as much clearance at the bottom of the hole for removal cuttings.

Rolling cutter bits (Figure 4.14) are the most widely used bits. Structurally the bits are classified as *cone-type bit* and *cross-roller bits*. Rows of teeth are cut into the rolling members, so that these bits are also referred to as *toothed-wheel bit*. The teeth are hard surfaced with such material as tungsten carbide to obtain longer bit life. The toothed wheels rotate independently; as the drill string is rotated, the rolling cutters turn by their contact with the bottom of the hole. For drilling of very hard rock, rounded tungsten carbide inserts have been substituted in place of the teeth ordinarily cut on the rollers, and these latter bits are referred to as *button bits*. With respect to the size of the mud fluid-discharge ports and their arrangement, the bits are generally classified as conventional or *jet bits*.

Rolling cutter bits are designed for soft, medium and hard formations. For drilling in softer formations such as the younger shales, it is desirable to have widely spaced, long, slim teeth so that any tendency for cuttings to pack between the teeth will be minimised.

Rolling cutter bits for use in drilling hard formations have closely spaced and shorter teeth. This feature provides for the maximum strength and tooth surface that can be utilised in drilling the formation and therefore increases the footage drilled per bit. The rolling elements are designed so that they have a true rolling action or maximum crushing on the bottom of the hole, since sliding on the bottom tends to wear the teeth of the bit. For very hard formations where ordinary bits get quickly worn out, tungsten carbide bits are used whose tips are made of sintered tungsten carbide.

For drilling formations that are intermediate between the very soft and the very hard, the several bit manufacturers have a sufficient choice of bits of intermediate characteristics to drill economically the formations encountered. As per API classification, rolling cutter bits can be from A-type to K-type suitable for different formations.

The mud-discharge ports in the conventional type of bit are often referred to as *watercourses*. The theory of positioning the conventional watercourses so that the mud fluid is directed down onto the top of the cutting elements is obviously to maintain the cutters in a clean condition so that they will be free to drill into the bottom of the hole.

Diamond bits – Diamond bits (Figure 4.15), similar to drag bits, have no independently moving parts. They drill by direct abrasion or scraping against the bottom of the hole. Diamonds are much harder than the common rock minerals and are the only material that has been used economically for such severe service. These diamond bits can be either core bits or plug type bits.

In manufacturing a diamond bit, a steel bit blank is machined with threads suitable to attaching it to a core barrel or the drill collars. Diamonds are placed in a bit mould and covered with powdered metal. With the bit blank held in position, the assembly is heated until the powdered metal fuses, and then it is slowly cooled. In this manner, the diamonds are held in a matrix that is attached to the steel bit blank. The matrix is composed partly of hard particles such as tungsten carbide, together with a softer alloy which acts as the binder material. The binder material must melt at a sufficiently low temperature so that the diamonds will not be injured in the process. The hard particles in the matrix are necessary in order that the matrix will not erode

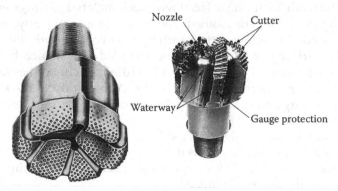

FIGURE 4.15
Diamond bit and PDC bit.

or wear away, thus causing loss of the diamonds and failure of the bit. Used in the hardest formations such as chert, it can drill faster than the hardest conventional bit. The spiral waterways on the bit (plug type) serve to channel the drilling fluid around to pick up the cuttings. A bit such as this may contain as much as 900 carats of industrial diamonds.

PDC bits – Polycrystalline diamond compact (PDC) bits (Figure 4.15) use polycrystalline diamond materials which are one of the most advanced materials for oil drilling tools. Fixed-head bits rotate as one piece and contain no separately moving parts. When fixed-head bits use PDC cutters, they are commonly called PDC bits. Since their first production in 1976, the popularity of bits using PDC cutters has grown steadily, and they are nearly as common as roller-cone bits in many drilling applications. PDC bits are designed and manufactured in two structurally dissimilar styles.

1. Matrix-body bit
2. Steel-body bits

The two provide significantly different capabilities and because both types have certain advantages, a choice between them would be decided by the needs of the application.

'*Matrix*' is a very hard, rather brittle composite material comprising tungsten carbide grains metallurgically bonded with a softer, tougher, metallic binder. Matrix is desirable as a bit material because its hardness is resistant to abrasion and erosion. It is capable of withstanding relatively high compressive loads, but compared with steel, has low resistance to impact loading.

Steel is metallurgically opposite of matrix. It is capable of withstanding high impact loads, but is relatively soft and, without protective features, would quickly fail by abrasion and erosion. Quality steels are essentially homogeneous with structural limits that rarely surprise their users.

PDC is extremely important to drilling because it aggregates tiny, inexpensive, manufactured diamonds into relatively large, intergrown masses of randomly oriented crystals that can be formed into useful shapes called diamond tables. Diamond tables are the part of a cutter that contacts a formation. Besides their hardness, PDC diamond tables have an essential characteristic for drill-bit cutters: They efficiently bond with tungsten carbide materials that can be brazed (attached) to bit bodies. Diamonds, by themselves, will not bond together, nor can they be attached by brazing.

Diamond grit is commonly used to describe tiny grains ($\approx$ 0.00004 in.) of synthetic diamond used as the key raw material for PDC cutters.

To manufacture a diamond table, diamond grit is sintered with tungsten carbide and metallic binder to form a diamond-rich layer. They are wafer-like in shape, and they should be made as thick as structurally possible because diamond volume increases wear life. Highest-quality diamond tables are $\approx$ 2–4 mm, and technology advances will increase diamond table thickness. Tungsten carbide substrates are normally $\approx$ 0.5 in height and have the same cross-sectional shape and dimensions as the diamond table. The two parts, diamond table and substrate, make up a cutter.

Forming PDC into useful shapes for cutters involves placing diamond grit, together with its substrate, in a pressure vessel and then sintering at high heat and pressure.

4.1.1.3 Circulation

In rotary drilling it is essential to continuously remove the rock cuttings from the bottom of the hole so that the bit gets a fresh surface to cut and for this purpose some fluid, be it air, water or any other fluid must be pumped or sent through the hollow drill pipe past the nozzles in the bit so that cuttings may be brought back through the annulus between the outside of the drill pipe and the wall of the hole. In oil well drilling, simple air or water is not suitable because of the larger depth and hence a drilling fluid, commonly known as mud which is made heavier than water by mixing bentonite and other additives in certain proportion with water, is used. There are different types of drilling fluids used for different conditions and different situations. The most common is a water-base mud. This drilling fluid serves many other purposes other than simply cleaning the hole and carrying the cuttings. Those other purposes are to maintain a hydrostatic pressure (in overbalanced drilling) more than the formation pressure, to cool and lubricate the bit, to maintain and preserve the hole by preventing caving and many more.

In a rotary rig fluid circulation system, mud typically travels in the following sequence, as shown in Figure 4.16.

1. From mud pit or mud tank to mud pump
2. From mud pump to standpipe/kelly hose/swivel
3. From swivel to kelly
4. From kelly to drill string/bit
5. From bit up the annular space
6. Up the annular space through the BOP stack
7. Through the BOP stack to the return line
8. From return line to the shale shaker
9. From the shale shaker to the setting pit
10. From the setting pit to sump/storage/reserve pit

The major components of the circulating system are briefly described below.

Mud pits and tanks – These are required for holding an excess volume of mud at the surface. The volume so held in the circulating system is usually between 300 and 700 barrels. Mud pits are scraped out in the earth with a bulldozer and are usually two in number. Steel pits or tanks are readily

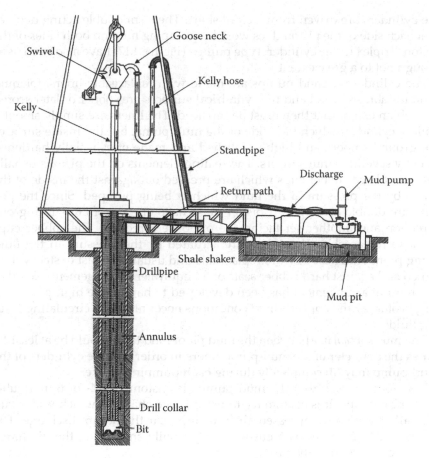

FIGURE 4.16
Fluid circulation system.

transported on the highways by trucks. Three such steel pits are often provided, and they are coupled together with one or more 12–18 in. diameter conduits. Four-inch-diameter steel pipe is often placed around the rims of such steel mud pits and furnished with connections where mud guns (jets) may be mounted for jetting down onto the surface of the mud. Jets are usually operated from an extra or auxiliary mud-mixing pump similar to but smaller than the circulating-mud pump. Steel mud pits and tanks have the advantage over pits dug in the earth that there is less chance of the mud's becoming contaminated with blowing sand or other materials. The drill cuttings and cavings and any mud discarded from the circulating system are generally run or jetted into the *reserve pit*. This is an area, adjacent to the mud pits, about 100 ft square, surrounded by earthen banks about 3 ft high.

Mud pumps – The pump normally used for this service is reciprocating piston, two-cylinder (duplex), double-acting type in which the pistons inside the cylinders are driven from a crankshaft. The term double acting denotes that each side of the piston does work by pumping mud on both sides of the piston. Triplex (three cylinder) type pumps (Figure 4.17) have also been used though not to a great extent.

The cylinders in mud pumps are made with replaceable liners, for mud contains abrasive sand and the cylindrical surfaces are sooner or later scored to such an extent that they must be replaced. The liners are simply smooth, hollow cylinders which fit inside of the mud pump, but the inside surfaces are ground smooth and highly polished and made of especially hardened steel by several manufacturers. The sealing elements on the pistons usually consist of hard rubber cups which are pressed out against the inside of the liners by the pressure in the fluid which is being pumped. Since the pistons are double acting, each piston must have one set of cups facing one direction and another set facing the opposite direction. The rubber cups are also replaceable. The valves are actuated by the pressure in the fluid being pumped. These also are replaceable and usually consist of steel valves which seal against hard rubber seats or an equivalent arrangement. No other mechanical arrangement has been developed to handle the high pressures, large volumes and abrasive fluid conditions encountered in circulating drilling fluids.

The mud *suction line* between the mud pit or mud tank should be at least 1½ times the diameter of the mud-pump liners in order that the cylinders of the mud pump may fill completely during each pumping stroke.

The *discharge line* from the mud pumps is customarily 4- or 6-in. double extra-heavy pipe. It is customary to use a surge chamber which will retain a small volume of compressed air in its top near the pump discharge. The compressed air serves as a cushion for partially smoothing the discharge pressure surges from the pump.

The *rotary hose* or the *kelly hose* conducts the mud from the upper end of the stand-pipe, the vertical steel pipe which extends about halfway up the

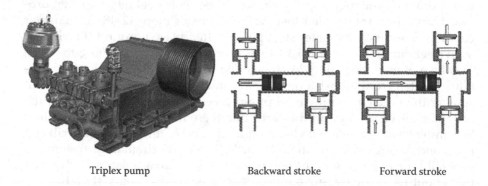

Triplex pump   Backward stroke   Forward stroke

FIGURE 4.17
Triplex double-acting pump. Triplex pump (courtesy of www.sdrljx.made-in-china.com).

derrick, to the swivel which supports the top of the drill string. Rotary hoses are customarily 3 in. or more in internal diameter in order that no appreciable fluid-flow pressure losses will occur in the hose. The flexible feature permits raising and lowering the drill string during drilling operations while mud is being pumped down through the drill string.

The *swivel,* which was described earlier, permits rotation of the drill string and conducts mud through it. The internal diameters of the drill pipe, tool joints and drill collars, and the total cross-sectional areas of the discharge jets in the bit, determine the major portion of the fluid-flow pressure losses, which in turn determine the pressure which must be supplied by the mud pump.

The flow of drilling fluid returns to the surface in the annular space outside of the drill pipe. The greater diameter of the annular space in the lower section of the hole is generally determined by the drill bit size, and in the upper section by the diameter of the smallest size of casing which has been set and cemented in the well. The upward flow of mud passes through the blow-out preventer and then into the nearly horizontal mud flow line under the derrick floor, which leads the mud back to the mud pits.

The *blowout preventer (BOP)* (Figure 4.18), which remains under the rig floor on the casing head, consists of one preventer, which can seal the annulus around the kelly or the drill pipe, and one or more ram type preventers. Also included here are a connection to fill the hole with mud when making a trip known as a kill line, which permits mud to be pumped down the annulus to restore the pressure balance, and another connection to release the annular pressure known as relief lines or choke lines.

FIGURE 4.18
Blow out preventer. (Courtesy of Robert Nunnally/Wikimedia Commons/CC-BY-SA-2.0.)

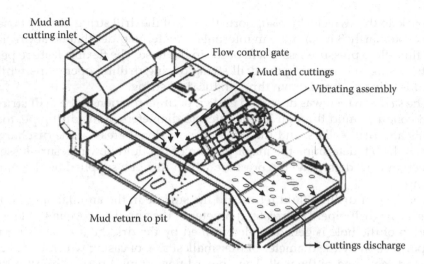

FIGURE 4.19
Shale shaker.

A *shale shaker*, which customarily sits on top of the first mud pit, receives the mud returning from the flow line. The shale shaker (Figure 4.19) is essentially a screen that is used to separate drill cuttings and cavings from the mud. Two types are in common use. In one type the screen is in the form of a tapered cylinder which is rotated by the flow of drilling fluid. The other is a rubber-mounted sloping flat screen which is vibrated by the rapid rotation of an eccentric mass (cam) driven either by electric or hydraulic motor. In both, the drilling fluid falls by gravity through the screen and the drill cuttings and any cavings present pass over the end of the screen. The screen openings are generally rectangular.

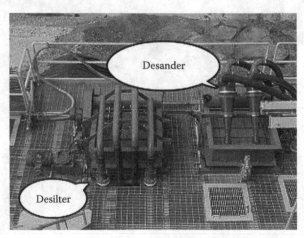

FIGURE 4.20
Desander/centrifuge. (Courtesy of Wikimedia Commons/Public domain.)

Other pieces of equipment which are sometimes used include *vacuum chambers* or *degasser,* which removes gas from the mud, and *centrifuges* or *desander* or *cyclone.* Centrifuges (Figure 4.20) are used to recover high specific gravity mud weighting material for maintaining high density muds; they are also used for rejecting sand and similar material from the mud, for the purpose of maintaining minimum density drilling muds.

4.1.2 Types of Configuration of Wells and Drilling Methods

There are various types of wells which are designated differently depending on the mode of operation, type of technology and the trajectories used. The most common types of such wells whose configurations are shown in the Figure 4.21 are defined below.

Conventional wells – These wells are vertical or directional (i.e., deviated from the vertical) and are drilled using standard drilling technology and equipment.

Horizontal wells – These wells lie at an inclination that is greater than 80–85 degrees. Generally, wells are at a 90-degree angle in the reservoir if they are drilled in a direction parallel to the structural contours. Wells of this type require more advanced drilling technology and sometimes the use of rotary steerable assemblies (RSA) guidance technology, including measurement while drilling (MWD). Using special equipment, quite long horizontal wells can be drilled.

Horizontal wells can be further divided into sub-types such as ultra-short radius (USR), short radius, medium radius and long radius, depending on the degree of curvature between their position on the surface and the entrance to the reservoir. Ultra-short radius and short radius wells become horizontal a short distance from the vertical location on the surface.

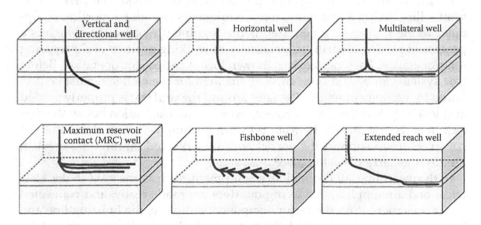

FIGURE 4.21
Configuration of wells. (From Jorge S. Gomes and Fernando B. Alves, *The Universe of the Oil & Gas Industry – From Exploration to Refining,* Partex Oil and Gas, 2013.)

Multilateral wells – These are basically two or more horizontal wells (laterals), which branch out from the original well (mother borehole). These wells are generally designed to produce from multiple areas of the reservoir. The diagrams in Figure. 4.21 (fishbone, multilateral and MRC wells) exemplify this type of well.

MRC wells – MRC means *maximum-reservoir contact*, referring either to wells with many branches (they are in fact a kind of multilateral wells) or to wells that are very long. These wells may have different configurations, such as the fork or even fishbone wells, mentioned above. They are used not only in advantageous situations for reservoir development but also in situations where there are constraints in the surface location.

Extended reach wells – These are very long horizontal wells that enter the reservoir at a distance far from the vertical position of their surface location.

USR wells – These are ultra-short radius wells which in terms of configuration are the opposite of extended reach wells. USR wells have very sharp bends from their vertical surface location. The distance between their surface location and the point of entry into the reservoir ranges from 10 to 30 m. These wells have very small diameters (more or less 3⅞ and 4½ in.) and, in most cases, require the drilling mast to be tilted at the surface or the use of coiled tubing drilling (CTD) techniques.

Slim holes – These are wells where the initial casing is reduced. It may start at 7 in. and end in a 3-in. production casing. CTD techniques may be used.

4.1.2.1 Vertical and Directional Drilling

All vertical wells or those with certain deviations use the standard drilling technology. When certain wells are intended to be deviated, then certain special tools are used which will be described later.

As far as the standard drilling procedure is concerned, the first step is to mobilise the rig along with all its components and all supporting equipment to the rig site. The next step is the 'rigging-up' operation, which is a process of placing all drilling machinery in working position and preparing it for operation. Each rig requires its own layout pattern for operation. Before this system can be laid out, however, the area for the well site must be prepared for equipment set up. The land around the well site is properly graded and levelled. A 'cellar' or 'basement,' which is an excavation below the derrick floor, is first dug for blowout prevention equipment and above this the substructure to support the derrick, draw works and engines are erected. Finally, the derrick is set in place.

In the meantime, all the auxiliary equipment, such as the mud tanks, electrical and air supplies, and living quarters are made ready, and connected to rig operation. With the major components in place, the hoisting line and traveling block are hung and the primary drilling tools are prepared for operation.

The rigging-up phase of a full-phase derrick might take as long as a week. After the derrick is erected, drilling commences with a large hole (usually about 1 ft in diameter and more than 100 ft deep) into which a string of protection, or 'conductor'/'surface' pipe, will be lowered and cemented. The purpose of this pipe, known as a 'casing,' is to prevent formation fluids encountered at depth from mingling with and damaging freshwater-bearing sands on which the landowners or local towns may be dependent. Other functions of casings are described later.

The main drilling operation is then continued beneath the surface pipe, usually with a 7- to 9-in.-diameter bit which is screwed onto a 30-ft joint of drill pipe and lowered into the hole. The drill pipe is then attached to a 40-ft, hollow stem called the kelly, which is turned by the rotary table. The kelly, along with the string of drill pipe and the bit, is suspended at the draw works by means of a steel line running over the crown blocks atop the derrick down to the traveling block and swivel in which the kelly turns.

As drilling proceeds, the bit is lowered in the hole by releasing line from the draw works until a new length of drill pipe is needed. Additional joints are added until the driller reasons that the bit is worn out.

To replace the bit, it is necessary to do the 'pulling out' operation that is bringing up all the drill pipe. This pipe is racked by the derrickman from a platform to one side of the derrick in double or treble combined joints, called stands. Two or three drilling personnel on the floor handle the massive wrenches (or *power tongs*) for breaking and making up the pipe connections, as well as the *slips* for holding pipe in the rotary table bushing when it is not supported by the draw works.

To do the 'running-in' operation, that is, to go back into the hole, the drill pipe is made up, stand by stand, and lowered into the hole until the new bit rests on the bottom. The mud pump is started and a circulating stream of fluid, sucked out of the pits, is forced down the pipe through jet holes in the rotating bit, cutting soft rock and carrying the drill cuttings back to the surface. The drill cuttings are passed over a screened 'shale shaker' to provide a sample for the geologist and to be eliminated from the mud circulating system. Chemicals are mixed and added to maintain certain essential features of the drilling mud: (1) caking qualities, to plaster the walls of the hole and prevent caving of soft formations; (2) viscosity, or thickness, to help float the rock cuttings out of the hole; (3) weight, to keep the mud from invading porous reservoirs and to help prevent high-pressure reservoir blowouts; and (4) high water retention to circulate bonds within the mud.

The problems, the cost and the hazards of drilling increase rapidly with depth. Drilling mud becomes a very expensive chemical preparation. The deeper the hole, the more mud is needed. Such large quantities of mud may be lost in formation cavities that none returns to the pits (this is known as 'lost circulation'). When there is no heavy column of mud in the hole, gas, under high pressure, may blow out: blowing gas may ignite and do untold damage.

Steeply dipping rock strata divert the drilling string away from the verti-cal. To avoid crooked holes, periodic 'straight hole' surveys are made of the deviation from vertical, and most drilling operations restrict this to a few degrees.

Surveying Instruments – There are two basic types of surveying instru-ments to reveal the course of the bit. One reveals the angle of deviation only whereas the other reveals both the angle and the direction (azimuth) of the deviation. The angle of vertical deviation can be measured by a plumb-bob or pendulum which maintains its vertical position under all conditions. When an instrument containing this pendulum makes an impression over a rotating chart which is mounted on a clock, the position of the pendulum or its angle with vertical at any time can be read from the chart. If a time versus depth log (using a synchronised surface watch) is kept, then at any depth the angle of deviation can be read.

The direction of deviation or the horizontal direction can be obtained from a compass reading which always points to magnetic north. By making a reading of the compass the survey chart reveals the direction of the hole as compared with north. The records of both vertical deviation and horizontal direction can be made in a single instrument.

Instruments utilising a magnetic compass require shielding from the mag-netic disturbance caused by the drill string. So, a special nonmagnetic drill collar made of K-monel metal is run just above the bit to house the instru-ments. One or more than one reading can be taken in one trip based on which there are single shot instrument or multiple shot instrument and these can be lowered along with the drill pipe or by wire line or by free drop or go-devil operation.

There is a unique surveying device known as a gyroscopic multiple shot surveying instrument (Surwel) shown in Figure 4.22 which utilises the abil-ity of a gyroscope to maintain the same directional orientation over a con-siderable time period. Rotation at 10,000–15,000 rpm is induced by power from batteries contained in the instrument. The timing device is set to take pictures at the desired intervals with a record of time, vertical deviation (indicated by X in Figure 4.22) and direction (indicated by arrow in Figure 4.22). A separate log of depth versus time is again kept at the surface with a synchronised watch. Surveys are taken both while going in and coming out of the hole. This device may be run on a drill pipe, tubing, or other standard rig equipment and is the only instrument that can be run inside the casing on a wire line.

Deviating a Well – As such maintaining a truly vertical hole is very critical because the bit might penetrate rock bedding planes at an angle other than 90° and this will make the bit walk away from vertical. So, the best preven-tive measure is to drill lightly by applying less and controlled weight on the bit.

However, there are situations when deviating a well becomes essential and some such situations are depicted in Figure 4.23 where directional drilling is

Timer Bubble shell Camera   Angle unit   Gyro

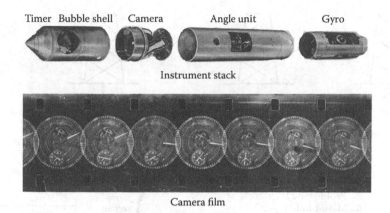

Instrument stack

Camera film

FIGURE 4.22
Gyroscopic surveying instrument (camera film showing direction by arrow, vertical deviation by cross and time (depth) at which picture was taken).

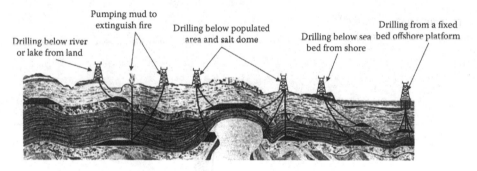

FIGURE 4.23
Situations under which wells are deviated.

being done. To reach the targeted point in the reservoir, different deflection patterns or well profiles are followed which give different shapes to the wells. Basically, three different shapes or profiles are common, namely, (1) build-and-hold shape or profile I, (2) S-shape or profile II and (3) continuous build or profile III. These profiles or shapes are shown in Figure 4.24 where different terminologies used for various sections have been mentioned.

For deviating a well, certain specialised tools, known as deflecting tools, are used. Whipstock and knuckle joint are two common types of such tools used in hard rocks. A *whipstock* (Figure 4.25) is a wedge that is set in the hole. The flat edge of the wedge is on the bottom of the hole. The drill string is rotated and lowered. Forced against the side of the hole, it starts to cut out of the original hole. After drilling below the whipstock, the bit is pulled out of the hole again. The whipstock is pulled out with the assembly. At the surface, the whipstock can be removed. The drilling assembly is run back in. The ledge at the bottom of the hole will be worn and cut away as operations continue.

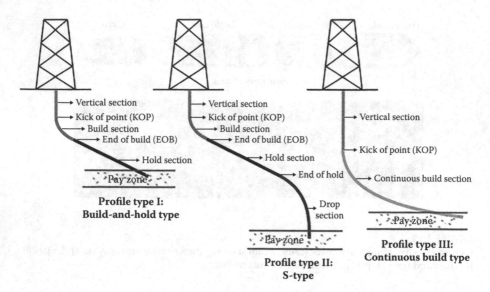

FIGURE 4.24
Directional well profiles.

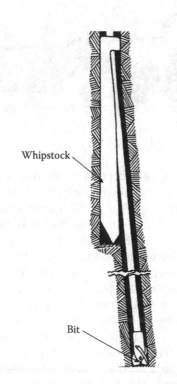

FIGURE 4.25
Whipstock.

Whipstocks are also used to deviate the well out of casing. Instead of a drill bit, a mill is used to cut the metal of the casing. In this application, the whipstock is left in place to guide tools through the cut left in place in the casing. Thus, the whipstocks can be either removable or fixed depending on its use.

The *knuckle joint* (Figure 4.26) is essentially an extension of the drill string incorporating a universal joint at the junction. Thus, rotation at different angles is possible. Before lowering this knuckle joint in the hole, it is oriented and set for a particular degree of deviation. The initial deviation will be in the order of 3 to 5 degrees. The direction (azimuth) of course will be determined by the direction of the target.

After lowering on its new course, the directional control is made by proper variation of rotary speed, weight on bit, pump pressure and so on. Much weight and slow rotary speed make the bit build angle or increase deviation.

For allowing accurate hole orientation of these deflection tools, several instruments and techniques are used. Nowadays continuous monitoring of the profile is also possible using measurement while drilling (MWD) tools.

Recent developments in advanced drilling are the 'steerable motor' and 'rotary steerables.'

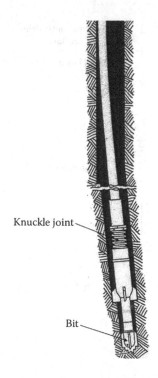

FIGURE 4.26
Knuckle joint.

A *steerable motor* is a downhole motor, powered by pumping mud through it. The lower part of the motor has an adjustable bend (Figure 4.27). Before running the motor in the hole, several things are done to set it up:

1. The bend is adjusted for the directional performance required of the motor. The bend will vary from 0° to something less than 2°.

2. The tools above the motor that transmit navigational information back to the surface are connected to the motor and calibrated, so that the driller can see which direction the bend points during drilling. These tools are called MWD tool.

3. Other components in the system will be adjusted or tailored for the required directional performance.

Referring to Figure 4.27, the main components of the system from bottom upward are described below.

Drill bit – When mud is pumped down the drill string, the motor turns. As the bit sits on a bent housing, it does not point straight ahead. A side force at the bit results from this, which causes the bit to drill a curved hole.

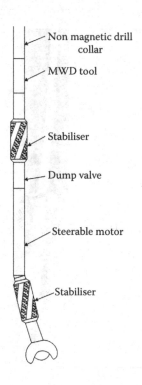

FIGURE 4.27
Steerable motor.

Undergauge stabiliser – Behind the bit is usually an undergauge (smaller diameter than the bit) stabiliser. This forms a fulcrum, with the motor behind acting as a lever, to allow the side force to be generated at the bit.

Motor – Above the stabiliser is the motor itself, with the bottom part having the adjustable bend.

Dump valve – At the top of the motor is a dump valve. This allows mud to be diverted at the top of the motor if need be.

Stabiliser – A stabiliser above the motor acts as the far end of the lever, exerting an opposing force at the drill bit.

MWD – Measurement while drilling tool measures and transmits directional information to the surface. It transmits three pieces of information – the wellbore inclination and azimuth and also the tool face azimuth (TFA). This information allows the driller to guide and correct the direction that the well drills.

NMDC – It is the nonmagnetic drill collar placed above the MWD tool to avoid any interference.

While drilling, the motor can be orientated in a desired direction, and the bit drills a curved hole. This can be used to turn the hole, drop angle, build angle, or some combination of build/drop and turn. The whole drill string can also be rotated so that the bit will drill straight ahead. Therefore, it is called a steerable motor – it has the capability to really be steered to cut the required path, which can be quite complex.

Steerable motors are very commonly used for initially kicking off the well from vertical and for continuing to drill and control the well path after the kick-off. As the well inclination increases by more than 60°, it starts becoming difficult to make such an assembly slide (drill without rotating the drill string) while drilling. Another issue is keeping the hole clean; rotating the drill string greatly improves the transport of cuttings out of the hole, and at high inclinations, sliding the string can lead to cutting beds building up and sticking the drill string. What is needed is a *steerable tool* that works with the drill string rotating, and they do now exist. The drawback is that they are expensive.

The *rotary steerable* is run immediately above the bit, a kind of replacement for the near bit stabiliser. Modern tools will have three blades positioned close to the drill bit, much like stabiliser blades. These blades move in and out. The control system works so that as the tool turns, the blade turning opposite the desired direction of deviation pushes against the side of the hole, imposing a side force on the bit to make it drill a curved hole.

A computer within the tool controls the movement of the pads. Surface commands can be sent down to the tool by creating pressure pulses in the drilling fluid.

With a steerable motor, adjusting the well path in a series of slide drilling and rotary drilling does not create a nice smooth curve. The hole deviates with many small but sharp doglegs, with straight intervals in between. One of the big advantages of the rotary steerable tool is that it creates a nice smooth curved hole. This then leads to a more stable wellbore and less resistance when tripping in and out of the hole. At high inclinations, it would be easier to run casing or logging tools through a smooth curve.

When drilling long horizontal hole sections, it becomes very difficult to drill while sliding. Rotating the string breaks the friction between the pipe and the hole. For long horizontal or extended reach wells, rotary steerables have allowed holes to be drilled that are not possible with sliding motor assemblies.

Historically, a major cause of lost horizontal holes (which could not be completed to reach their objective) was inadequate hole cleaning. Cuttings build up in the hole next to the sliding drill pipe. Rotating the pipe stirs up the cuttings and gets them into the moving fluid stream. Since the cost of this system is very high, use of such tools is justifiable in the case of complex well paths, long horizontal holes, or when geosteering is required.

Casing a Well – During the course of drilling, it is necessary to run casing at various depth intervals, that is, to lower the desired length of the casing or pipe into the hole and cement it in place. The number and size of the casing strings used vary with the area, depth, anticipated producing characteristics of the well and the choice of the operator.

Casing serves as a structural retainer in the well, excludes undesirable fluids, and acts to confine and conduct oil or gas from subsurface strata to the surface. Casing must be capable of with standing external collapse pressure from fluid surrounding the casing, the internal pressures encountered in conducting oil or gas from the producing formation and the tension resulting from its own suspended weight. It must also be equipped with threaded joints that can be made up easily and provide leakproof connections.

The American Petroleum Institute has developed specifications for casing meeting the major needs of the oil and gas industry and has published these as API standard 5A.

The casing strings are lowered and cemented in a well to ensure smooth drilling and production of hydrocarbons from the producing intervals. The cement slurry is pumped down the pipe and up the annulus between the casing and the open hole. The amount of slurry will depend on the annular volume and fill-up height of cement. After the pumping is stopped, the cement is allowed to set for several hours before further drilling starts. It is very important to ensure a good cement bond between casings/liners and the formation for which 'cement bond logging' is performed.

In general, more than one casing string is lowered in a well for different reasons at different stages of drilling (Figure 4.28). The diameter of casing pipe lowered can vary from as low as 3" ID to as high as 30" ID. Starting from the surface, these casings with sequentially reduced diameter are described below with their major functions:

Conductor casing – This is the largest diameter of casing used in a well and is lowered only for a few meters and is cemented to its entire length to the ground level. The size of conductor casing generally varies from 16 to 24" depending upon size of the spudding bit. It has the following functions:

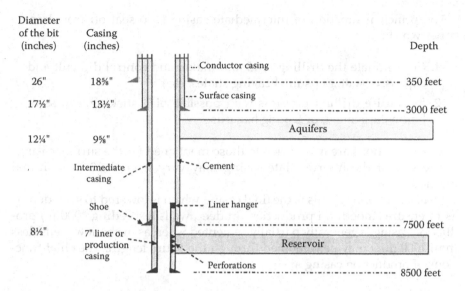

FIGURE 4.28
Typical arrangement of casing strings in an onshore well. (From Jorge S. Gomes and Fernando B. Alves, *The Universe of the Oil & Gas Industry – From Exploration to Refining*, Partex Oil and Gas, 2013.)

1. It guides the spudding bit.
2. It avoids washing and eroding of surface soil and prevents the forming of cavity around the spudded well.

Surface casing – This casing is landed to few hundred metres and is cemented to the top. The functions of this type of casing are

1. It curtails caving and washing out of poorly consolidated surface beds.
2. It furnishes a means of handling the return flow of drilling fluid.
3. It protects fresh water sands from possible contamination by drilling mud or oil, gas and/or salt water from lower zones.
4. It allows attachment of blow out preventers.

The size of casing of this type varies between 12 and 16 in.

Intermediate string – This casing is lowered to varying depths of a few hundred metres to thousands of metres depending upon geological conditions and drilling problems in the area. Depending upon the geological conditions of a specific area, one or more than one intermediate casing may be required to be lowered. These casings may or may not be cemented up to the surface.

The principal function of intermediate casing is to seal off troublesome zones which:

1. Contaminate the drilling fluid and make mud control difficult and expensive (salt, gypsum, heaving shales, etc.).
2. Jeopardise drilling progress with possible pipe sticking, excessive hole enlargement, or fishing hazards.

These functions are in addition to those mentioned for the surface casing.

The size of the intermediate casing may vary from 9–12 in. nominal diameter.

Production Casing – This is the final casing which is lowered to such depths as to ensure smooth oil production. In deep wells (exceeding 1000 m) production casing is generally partially cemented whereas in shallow wells (less than 1000 m) it may be cemented through the entire length. The chief functions of production casing are

1. It provides a support for installation of well head assembly and Christmas trees for production and BOP for workover operations.
2. It provides a smooth passage for production of oil and gas and performing workover operations which may be required from time to time during the life of a well.
3. It isolates the undesirable pay zones.
4. It enables selective production from the zone or a part of particular zones.

Whereas the landing of final casing through the producing zone provides a number of advantages from an operational viewpoint, it reduces the production from the particular producing zone. It also increases the cost of the completion of a well. In such cases, in place of production casing, *production liners* are used which are suspended from the intermediate casing via a liner hanger. The bottom end of each casing/liner segment is called the *shoe*.

4.1.2.2 Horizontal, Multilateral and Extended Reach Well

Horizontal drilling is a drilling process in which the well is turned horizontally at depth. It is normally used to extract energy from a source that itself runs horizontally, such as a layer of shale rock.

Since the horizontal section of a well is at great depth, it must include a vertical part as well. Thus, a horizontal well resembles an exaggerated letter 'J.' When examining the differences between vertical wells and horizontal wells, it is easy to see that a horizontal well can reach a much wider area of formation and extract the oil and gas that is trapped within it.

4.1.2.2.1 Drilling Methodology

Most horizontal wells begin at the surface as a vertical well. Drilling progresses until the drill bit is a few hundred feet above the target rock unit. At that point, the pipe is pulled from the well and a hydraulic motor is attached between the drill bit and the drill collar.

The hydraulic motor is powered by a flow of drilling mud down the drill pipe. It can rotate the drill bit without rotating the entire length of the drill pipe between the bit and the surface. This allows the bit to drill a path that deviates from the orientation of the drill pipe.

After the motor is installed, the bit and pipe are lowered back down the well and the bit drills a path that steers the well bore from vertical to horizontal over a distance of a few hundred feet. Once the well has been steered to the proper angle, straight-ahead drilling resumes and the well follows the target rock unit. Keeping the well in a thin rock unit requires careful navigation. Downhole instruments are used to determine the azimuth and orientation of the drilling. This information is used to steer the drill bit.

The first horizontal wells, designed in the mid-1980, were drilled with two main objectives: (1) to intersect and connect natural networks of fractures present in fractured reservoirs and (2) to prevent water coning and gas cusping in reservoirs with small oil columns. Over the years, the technology of horizontal wells has improved and today the applications of these wells are much more varied. Companies use horizontal well technology to develop very thick reservoirs, reservoirs with low permeability, reservoirs with heavy oil, as well as to produce the remaining reserves (attic oil, bypassed oil, cellar oil, etc.) and to extract rich liquid at the bottom of gas-condensate reservoirs. Another very common application in the oil industry is the drilling of horizontal wells as water injectors. The fact that the wells are horizontal may help increase significantly the areal sweep efficiency in the reservoir, thereby improving the recovery factors.

World experience in horizontal wells indicates that the production of a horizontal well in homogeneous reservoirs is on average two to four times greater than a vertical well in the same reservoir. For this reason, the fact that the operational costs of a horizontal well are higher cannot be considered a limiting factor in its implementation. In economic terms, the profitability of the well as a source of hydrocarbons is usually higher. It is mainly for economic reasons that many oil companies are now embarking on a drilling philosophy that necessarily includes the drilling of horizontal wells.

From a technical and economic point of view, the feasibility of horizontal wells is the rule both in their simplest (unilateral) and more complex form (multilateral). From a general perspective, horizontal wells offer economic advantages and make it possible to develop thinner reservoirs (neglected in the past).

Depending on the degree of curvature, the horizontal well can be described as ultra-short radius (USR), short radius, medium radius and long radius with curvatures of 10–30 m, 140 m, 250 m and 300 m, respectively.

Using special technology (titanium tubing with a diameter of 3⅞ or 4½ inches) and advanced control systems (rotary steerable assemblies), the *USR well* can be horizontalised (moving from a vertical to a horizontal position) a few metres away, ranging typically from 10–30 m. The horizontal extension of the well depends on its degree of curvature. The higher the degree of curvature, the smaller the range of the horizontal well. Generally, these wells do not exceed 200 m in length. USR wells are mainly used in old fields to drain hydrocarbons near existing wells, either by crossing remaining oil pockets or avoiding nearby water coning. USR wells are also used to reactivate existing wells, if they are damaged or blocked mechanically.

Multilateral wells can be defined as two or more lateral wells (laterals), which are hydraulically connected to the mother borehole by a junction through which fluids are produced or injected (Figure 4.29). Multilateral wells are drilled to greatly increase the space in the well exposed to the reservoir. It allows more oil or gas to be produced with lower pressures. Further, it improves the recovery factor by speeding up the increased rates. It reduces capital expenditure (CAPEX), due to the decrease in the total number of producers and injectors in the field and reduces the geological risk (shorter wells in different azimuths). This is in fact a kind of MRC (maximum reservoir contact) well.

Its disadvantage is that only one main wellbore, or mother borehole, is used as a junction between the various multilaterals and the surface. The use of only one main borehole increases the risk of mechanical failure in completion (should it be complicated), with disastrous results in the production mechanism. For example, if there is a malfunction in the completion of the main wellbore, which has access to three bifurcations (three multilaterals), all three of the wells stop producing instead of just one. To minimise this problem, the solution is to install

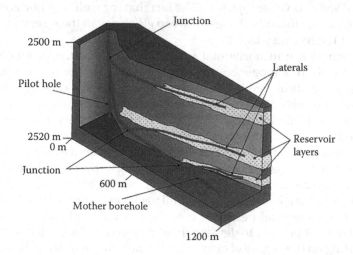

FIGURE 4.29
Multilateral wells. (From Jorge S. Gomes and Fernando B. Alves, *The Universe of the Oil & Gas Industry – From Exploration to Refining*, Partex Oil and Gas, 2013.)

the most reliable completion technologies (smart completions) – controlled from the surface – in the mother borehole. Another disadvantage is the difficulty of abandoning a multilateral, especially when it comes to older wells with corrosion problems in the mother borehole or when it starts to hinder production (producing water, for example). The consequences are reflected in increased workover costs, that is, high operating expenditure (OPEX).

The applications of multilateral wells have broadly followed that of horizontal wells, namely:

1. Intersection of fractured reservoirs.
2. Development of continuous reservoirs with well-defined microstratigraphy (layered systems).
3. Development of discontinuous and anisotropic reservoirs with alternate good and poor areas (non-productive).
4. Drainage of unswept oil pockets near oil wells.
5. Application of EOR techniques to recover additional reserves near the wells.

However, before this technology can be applied, a detailed analysis of the benefits/costs/risks must be performed and the life cycle of the well/field must be taken into account.

Extended reach wells in terms of configuration are opposite USR wells in that these enter the reservoir at a distance far from the vertical position of their surface location. Normally an extended reach well is the one where the horizontal displacement of the well is at least twice the vertical depth. An example of this type of well is in Wytch Farm oil field in the South of England which being an environmentally sensitive area was developed economically by drilling under the sea from the land (Figure 4.30). At that time, the wells expanded the boundaries of extended reach wells (i.e. twice the vertical depth).

While drilling extended reach wells, the rotary steerable tools are used along with logging while drilling (LWD) tools for sending commands to them so that the bottom hole assembly (BHA) steers the bit in the reservoir without having to communicate with the surface.

4.1.2.3 Underbalanced, Managed Pressure, and Dual Gradient Drilling

Underbalanced drilling (UBD) is a procedure used to drill oil and gas wells where the pressure in the wellbore is kept lower than fluid pressure in the formation being drilled. So, fluids in the reservoir flow toward the well while drilling, which is the opposite of what happens with conventional overbalanced drilling methods.

By definition, an 'underbalanced' condition is generated any time the effective circulating downhole pressure of a drilling fluid (the pressure exerted by the hydrostatic weight of the fluid column, plus whatever pump pressure is applied

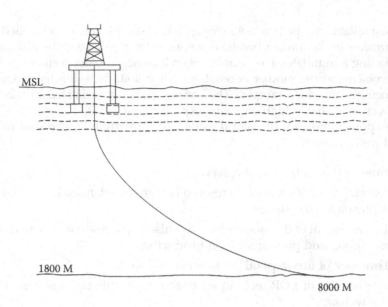

FIGURE 4.30
Extended reach well.

to the fluid system to circulate it, and the associated frictional pressure drop) is less than the effective pore pressure in the formation adjacent to the sand face. Most formations, unless abnormally pressured, are naturally placed in an overbalanced state when water-based fluids of normal density are used in typical oilfield operations. In some abnormally high-pressured formations (and in some normally pressured formations), a naturally underbalanced condition can be generated using conventional oil- or water-based drilling fluids. This condition, if it occurs during a drilling operation, is normally termed 'flow drilling.'

The flow of fluids toward the well during the drilling process has many advantages, such as:

1. Reducing the effects of damage (skin) of the reservoir near the well.
2. Increasing the rate of penetration (ROP).
3. Minimising circulation losses in depleted or fractured reservoirs.
4. Reducing the possibility of blocking the drill string.
5. Reservoir characterisation – since fluids are produced during drilling, it is possible to know which areas are the most productive. This information is essential to assess fractured intervals or highly productive flow units.

The industry's biggest interest in this technology is that of minimising the effects of reservoir deterioration, substantially reducing the skin effect. In this context, the main objective is to maintain the high production rates of the wells. High rates mean an increased production of the field, which can result in significant improvements in the recovery factor.

The UBD concept is not new in the industry. Drilling fluids, where a normally inert gas is injected to reduce the density and hence the hydrostatic pressure, have been used over the years, especially in impermeable formations, to increase the ROP and the lifetime of the wells. However, the application of UBD in permeable formations with simultaneous production was never developed for safety reasons.

Today in both vertical and horizontal wells, there is a greater ability to control UBD. Since the reservoir pressure does not vary with the lateral extension of the well, circulation systems can be designed for a given downhole pressure, thereby minimising operational risks. This UBD technology was tested in fractured reservoirs and compared with conventional methods of overbalanced drilling. With conventional methods, there was always the need to resort to sealing materials (LCM; lost circulation materials) to meet the heavy loss of circulation that occurred along the fracture system. The loss of fluids during drilling caused various problems not only in terms of costs but also in terms of the reservoir (contamination of the reservoir with mud and LCM). With the introduction of UBD, the use of LCM was avoided, operating costs decreased and damage to the reservoir (clogging of fractures) with the injection of bad fluids was prevented.

Figure 4.31 provides a schematic illustration of a typical UBD process using a closed surface control system. The vast majority of UBD operations, which have been executed to date, particularly in Western Canada, have utilised a jointed drill pipe and the direct injection of the base drilling fluid and some

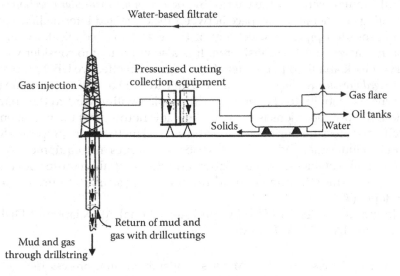

FIGURE 4.31
Schematic illustration of UBD process.

type of non-condensable gas directly down the centre of the drill string. This is not necessarily because this is the best technology to use, but because it is the least expensive and easiest to execute with the number of rotary rigs and conventional drilling equipment available.

The majority of these through string injection UBD projects have been conducted using water-based fluids (often produced water) and nitrogen. The average gas injection rates have typically been in the range of 30–40 m³/min (1.5–2.0 MMscf/day), although this is highly dependent on the reservoir pressure and the amount of reservoir gas which can be produced to assist in the UBD operation once the producing zone has been penetrated. In UBD operations executed in gas reservoirs, once penetration of the zone occurs, the UBD operation may be partially self-supporting with the produced gas, and very little supplemental gas may be required with sufficient fluid injection required only for adequate downhole motor and drilling operations.

Liquid circulation rates vary widely depending on the bottom hole pressure required and the amount of fluid being produced from the formation. The liquid circulation rate is a dominant factor in the control of the effective bottom hole pressure. Typical liquid flow rates in many recent UBD operations have been in the 0.3–0.8 m³/min (2400–6500 bbl/day) range.

Real time data recording and gathering systems, both down hole and at the surface, are essential to a properly executed UBD operation. This allows for the monitoring of the performance of the entire operation to maintain an optimum underbalanced condition.

So far as the actual operational system is concerned, it is mainly the injection of gas either using *parasite injection string*, ported to the casing side, along with the intermediate casing string allowing the gas to be injected directly in the vertical annular section of the well, or by the use of *microannular injection string* by creating another annular space between the cemented intermediate casing and the inserted pipe. To select any of these systems, technical, safety and economic factors must be considered. It is also important to consider various reservoir rock and fluid parameters for design of an effective UBD process.

Managed pressure drilling (MPD) in contrary to OBD or UBD is a unique process where flexibility to change the drilling plan with regard to the changes in the well bore conditions is the key issue. It is nothing but the management of well-bore-pressure during drilling and several methods are proposed for it.

MPD includes many ideas that describe techniques and equipment developed to limit well kicks, lost circulation and differential pressure sticking, in an effort to reduce the number of additional casing strings required to reach total depth (TD).

A formal definition of MPD by the International Association of Drilling Contractors (IADC) is as follows:

> Managed pressure drilling (MPD) is an adaptive drilling process used to more precisely control the annular pressure profile throughout the well bore. The objectives are to ascertain the downhole pressure environment

limits and to manage the annular hydraulic pressure profile accordingly. This may include the control of back pressure by using a closed and pressurised mud returns system, downhole annular pump or other such mechanical devices. Managed pressure drilling generally will avoid flow into the well bore.

In this definition, the word 'adaptive' is the key. MPD prepares the operation to change for fulfilling pressure profile objectives while drilling.

It is implied that this drilling method uses a single-phase drilling fluid treated to produce minimal flowing friction losses.

MPD's ability to dramatically reduce non-productive time (NPT) as such lost in conventional drilling methods in mitigating problems like lost circulation, stuck pipe, well kicks, fractured formations and so on in today's high rig rate market makes it a technology that demands consideration in any drilling or development program.

The basic techniques covered under MPD are as follows.

Constant bottom-hole pressure (CBHP) is the term generally used to describe actions taken to correct or reduce the effect of circulating friction loss or equivalent circulating density (ECD) to stay within the limits imposed by the pore pressure and fracture pressure.

Pressurised mud-cap drilling (PMCD) refers to drilling without returns to the surface and with a full annular fluid column maintained above a formation that is taking injected fluid and drilled cuttings. The annular fluid column requires an impressed and observable surface pressure to balance the downhole pressure. It is a technique to safely drill with total lost returns.

Dual gradient (DG) is the general term for many different approaches to control the up-hole annular pressure by managing ECD in deepwater marine drilling.

To understand the basics of MPD, first it is necessary to understand the pore pressure and fracture pressure and how these two determine the drilling window.

Pore pressure is the pressure of the fluid in the bore spaces that increases from zero at the surface at a rate that is equal to a column of water extending from the point of interest to the surface, or at about 0.43 psi/ft in freshwater basins and 0.47 psi/ft in saline or marine environments. These are considered the 'normal' pressure gradients:

$$\text{Pore pressure} = \text{Formation water gradient} \times \text{TVD}^* \qquad (4.1)$$

However, the straight-line increase may be offset because of transition zones, faults, or geologic discontinuities; and this leads to problems in avoiding well kicks and setting casing depths.

---

* TVD = True vertical depth.

So, in a simplistic sense,

Pore pressure = Formation water gradient × TVD × Lateral stress   (4.2)

Subnormal pressured formations have pressure gradients less than normally pressured formations. Subnormal pressures can either occur naturally in formations that have undergone a pressure regression because of deeper burial from tectonic movement or, more often, because of depletion of a formation because of production of the formation fluids in an old field.

Abnormally pressured formations have pressure gradients greater than normally pressured formations. In such formations, the fluids in the pore spaces are pressurised and exert pressure greater than the pressure gradient of the contained formation fluid.

Models and correlations are developed to estimate pore pressure – Eaton's correlation is one of those.

Fracture pressure is the amount of pressure a formation can withstand before it fails or splits. It can be also be defined as the pressure at which the formation fractures and the circulating fluid is lost. Fracture pressure is usually expressed as a gradient, with the common units being psi/ft or ppg. Deep formations can be highly compacted because of the high overburden pressures and have high fracture gradients. In shallow offshore fields, because of the lower overburden pressure resulting from the seawater gradient, lower fracture gradients are encountered. Many of the formations drilled offshore are young and not as compacted as those onshore, which results in a weaker rock matrix.

As per Eaton, the fracture pressure can be calculated from the following formula:

$$P_f = \left(\frac{v}{1-v} \right) \sigma + P_p \qquad (4.3)$$

where v = Poisson's ratio
$\quad P_p$ = Pore pressure
and σ = effective stress

The effective stress is defined as the difference between pore pressure and total stress:

$$\sigma = S - P_p \qquad (4.4)$$

where
$\quad \sigma$ = effective stress
$\quad S$ = total stress or overburden pressure

The overburden pressure is not a fluid-dependent pressure. Hence, it is expressed mathematically as follows:

$$S = \rho_b \times D \tag{4.5}$$

where
   ρ_b = average formation bulk density
   D = vertical thickness of the overlying sediments

The bulk density of the sediment is a function of rock matrix density, porosity within the confines of the pore spaces and pore-fluid density. This is expressed as

$$\rho_b = \phi\rho_f + (1-\phi)\rho_m \tag{4.6}$$

where
   ϕ = rock porosity
   ρ_f = formation fluid density
   ρ_m = rock matrix density

Now all the techniques covered under MPD share a common problem and that is the effect of equivalent circulating density (ECD) or annular pressure drop. When circulating a drilling fluid, friction increases the well-bore pressure over the static condition. The equivalent circulating density at any point accounts for the sum of hydrostatic pressure of a column of fluid and frictional pressure loss above that point. Thus, at any point of interest, the dynamic equivalent density, ECD, is higher than the static equivalent mud density, EMD. The ECD is calculated as

$$ECD = EMD + \frac{\Delta P}{0.052 \times TVD} \tag{4.7}$$

where
   EMD = static equivalent density of a column of fluid that is open to the atmosphere
   ΔP = annular frictional pressure loss (APL)
   TVD = true vertical depth

In the above equation, it is assumed that the fluid properties are constant but in HPHT wells this assumption may not be true and there the calculation of ECD becomes more complicated.

ΔP or APL is calculated by following simple formulae where a simple calculator and Fann V-G meter can predict it just at the drilling site.

For Bingham flow,

$$\Delta P = \frac{YPL}{225(d_b - d_p)} + \frac{\mu_{app} \times \upsilon \times L}{1500(d_b - d_p)^2} \tag{4.8}$$

Since the second part of the term is small, often the first part is used for the estimate of annular pressure loss. (Annular fluid was assumed to be in laminar flow.)

For turbulent flow,

$$\Delta P = \frac{\rho \upsilon^2 L}{5000(d_b - d_p)} \tag{4.9}$$

where
  ΔP = pressure drop, in psi, annular pressure loss
  L = length, ft
  d_b = hole diameter, in.
  d_p = pipe diameter, in.
  ρ = mud density, ppg
  YP = yield point, lb/100 ft^2
  μ_{app} = plastic viscosity, cp ($R_{600}/2$)
  υ = annular velocity, ft/sec
  R_{600} and R_{300} = VG meter readings at 600 and 300 rpm

These equations give only estimated answers that generally tend to be high – greater than the actual value. This is particularly true when using the Bingham term to estimate ΔP (annular pressure loss) caused by pipe movement. Annular pressure loss is a major challenge when using constant bottom-hole pressure.

Managing the well-bore pressure in a small operating window between pore pressure and fracture pressure involves manipulating the circulating density while using a minimum mud density (Figure 4.32). The static pressure of mud is very close to formation pore pressure. Pore pressure is not always the lower critical pressure, especially in directional wells. Well-bore stability that is a function of stress and well direction may form the base for the lowest practical mud density as well as modifying the fracture pressure.

Controlling the ECD within the upper and lower limits of the window is often referred to as constant bottom-hole pressure (CBHP) management.

For details on this technique, the readers may refer to Reference 9.

In the second technique, that is the *pressurised mud cap drilling* (PMCD), mud and water are pumped down the well bore and drill pipe to prevent kicks and control loss of circulation while drilling in fractured formation or

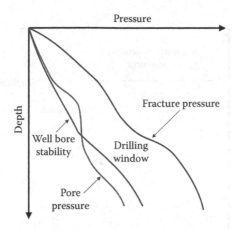

FIGURE 4.32
Drilling window – area between fracture and pore pressure/wellbore stability.

in a layered formation with different pressure regimes. Further, it is called pressured mud cap because it places a column of mud in the annulus that is lighter than required to balance the formation pressure. Here drilling is conducted through a rotating head with the well shut in at the surface for which a special surface equipment set-up is required. Details of this technique are available in the reference already cited.

Dual-gradient drilling (DGD), the third technique under MPD, is defined as 'two or more pressure gradients within selected well sections to manage the well pressure profile.' It refers to offshore drilling operations (covered in detail later) where the mud returns do not travel through a conventional, large-diameter drilling riser. The returns are either dumped at the seafloor (pump and dump) or returned to the rig, from the seafloor, through one or more small-diameter return lines. 'Pump and dump' and 'riserless muds return' are methods of DGD that can be presently used to drill the top-hole section of offshore wells.

Beyond the surface casing, the present DGD techniques are to return the mud to the rig through small-diameter return lines. A seafloor or mud-lift pump takes returns from the well annulus at the seafloor and pumps it back to the surface. By adjusting the inlet pressure of the seafloor pump to near seawater hydrostatic pressure (HSP), a dual-pressure gradient is imposed on the well-bore annulus, much the same way riserless drilling imposes the seawater hydrostatic pressure in the annulus of the well.

As can be seen in Figure 4.33, the seafloor pump reduces the pressure imposed on the shallow portion of the well, while the higher-density mud below the seafloor achieves the required bottom-hole pressure to control the formation pore pressure. The high mud density is imposed on a shorter vertical distance, while above the seafloor, seawater hydrostatic pressure is imposed.

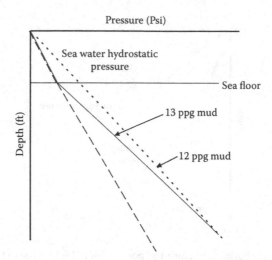

FIGURE 4.33
Conventional single gradient versus dual gradient concept.

The major problems associated with conventional riser drilling in ultra-deep water is deck space and tremendous deck loads imposed by the extremely long risers required for drilling. These extremely long risers require large volumes of drilling fluid just to fill them, as much as 3700 bbl for a 10,000 ft long, 19.5 in. inside diameter riser, costing well over $400,000 for synthetic-based drilling fluid. Not only is the cost high, but the volume of mud to fill the riser may be much greater than the storage capacity of the rig itself.

In addition to logistical challenges encountered when handling a drilling riser, reaching the geological objectives becomes more difficult as water depths become greater with a conventional mud-filled riser system. Geologic targets tend to be deeper below the mud line in deep waters, resulting in additional casing strings. Not only do the deep targets increase the number of casing strings required, but the effective window between the pore pressure gradient and fracture pressure gradient narrows with increasing water depth. The narrow window also increases the frequency of casing points. With the current marine risers, an operator can quickly run out of usable hole size before the geologic objectives are met.

Figure 4.34 shows the effective narrowing of the pore/fracture gradient window from A to B as water depth increases from 1000 ft to 5000 ft. This will require more casing strings to reach the total depth.

The widening of the pore/fracture gradient window can be achieved by DGD which will allow the operator to reach the total depth with fewer casing strings and a larger final well-bore size. This is critical, not only from the time and cost standpoints of the additional casing, but also because it allows larger production casing to be run, which prevents the production rate from

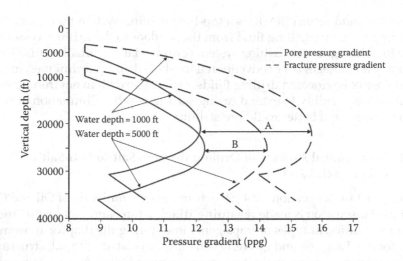

FIGURE 4.34
Narrowing of pore/fracture gradient window with increase in water depth from 1000 ft to 5000 ft.

being choked back by small production tubing. The greater window also allows the operator to plan the well with sufficient trip and kick margins.

Most of the problems associated with the conventional marine riser can be either minimised or eliminated with the dual gradient achieved through the use of the mud-lift principle. Thus, the advantages to using DGD are less deck space required for the small-diameter (6-in. outside diameter) return line, smaller deck loads and less drilling mud required to drill a well. DGD also allows for smaller second- and third-generation floating rigs to be upgraded to drill in deeper water. This would increase the rig availability for deepwater drilling. Additionally, the ability to meet geologic objectives with fewer casing strings, allowing a larger, optimised-diameter production tubing, allows the well to produce at high rates, which in turn can make the wells more economically attractive. This can also reduce drilling costs by reducing plateau times while drilling a well.

Although DGD can minimise or eliminate many of the problems associated with conventional riser drilling, there is still some bottlenecks in the implementation of this unconventional system. At the recent Offshore Technology Conference, 2014 in Houston, a panel of 10 key players opined that increased use of dual gradient is inevitable based on drilling challenges presented by many of the best available prospects but adoption could be gradual. To drill the first deepwater well in the U.S. Gulf of Mexico using DGD, companies like Chevron and Statoil are seeking to win approval from the U.S. Bureau of Safety and Environmental Enforcement (BSEE) because the well control is very critical.

Much of the research and most of the publications on DGD concern the use of riserless mud return technology only after the surface casing is set.

Riserless mud return (RMR) is a top-hole drilling system that uses a sub-sea pump to return drilling fluid from the seafloor to the drilling vessel and is the first dual-gradient drilling system commercially available. This system has many advantages over conventional top-hole drilling techniques, including the use of engineered drilling fluids, capacity to drill in environmentally sensitive areas, ability to extend casing setting depths, elimination of intermediate liners and better well-bore stability.

4.1.3 Geochemical Aspects of Drilling through Salt to Sub-Salt and Pre-Salt Layers

In Chapter 1 under Section 1.2.4 (Structure for Accumulation of Oil and Gas) a detailed discussion is made regarding 'diapiric trap' and 'salt diapir' and in Figure 1.18 different types of traps generated during the diapiric movement were shown. Large oil and gas reservoirs are associated with salt structures. These structures are normally tectonic structures known as salt domes, salt ridges, salt tongues and so on, but there can be undeformed bedded sedimentary salt as well and at the same time mixed domains like in the Gulf of Mexico, South Atlantic margin basins (Brazil, Angola), Precaspian basins and so on. Because of viscous behaviour at modest stresses and temperatures, salt can be tectonically mobilised solely because of density differences between salt (2.16 gm/cm^3 for pure NaCl) and other sediments (2.3–2.6 gm/cm^3).

In the course of geological depositions, a substantial amount of sediments remain under the salt deposits, that is, autochthonous salt layer which contains a huge reserve of oil and gas. These are known as *pre-salt* layers. It is the geological layers that were laid down before a salt layer accumulated above them and the petroleum that was formed in the pre-salt layer could not move upward because the salt layer acted as an impervious rock. As a result, a large accumulation of oil and gas took place. Though the actual reserve is not known, the local report available from Petrobras and other companies indicates that the oil and natural gas lie below an approximately 2000 m deep layer of salt, itself being an approximately 2000 m deep layer of rock under 2000–3000 m of the Atlantic containing 50 billion barrels of oil which is four times greater than the current reserve in the Brazilian continental shelf. But the drilling through the salt to extract the pre-salt oil and gas is very expensive.

Exploring these pre-salt deposits is also a big challenge. These autochthonous salt layers are regionally extensive which obscure very deep (pre-salt) sedimentary targets for hydrocarbon exploration and require alternative seismic acquisition and processing solutions. Reverse time migration (RTM) is a significant advancement in seismic imaging below salt layers which provides the most accurate view of pre-salt prospects, discoveries and fields.

The extensive accumulation of the autochthonous salt is formed through the evaporation of seawater and may get covered with deposited sediment,

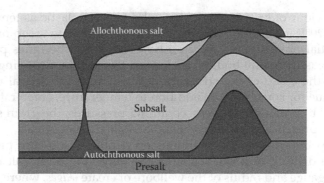

FIGURE 4.35
Concept of subsalt and presalt. (From Dwight 'Clint' Moore, *Pioneering the Global Subsalt/ Presalt Play: The World beyond Mahogany*, Field, AAPG Annual Conference and Exhibition, New Orleans, LA, April 11–14, 2010.)

becoming buried under an increasingly large overburden. As mentioned earlier because of the density difference with other sediments and due to some tectonic and compaction forces, salt lying below may intrude within the overburden and then migrate up. As a result, it becomes stratigraphically above the autochthonous source layer and is emplaced as subhorizontal or moderately dipping sheets like salt diapirs. This is known as allochthonous salt. Figure 4.35 shows both allochthonous (meaning a deposit or formation that originated at a distance from its present position) salt and autochthonous (meaning a deposit or formation formed in its present position) salt.

Thus, pre-salt is beneath an autochthonous salt layer and is stratigraphically older rock and lies at a much greater distance, whereas that which lies beneath an allochthonous salt layer is known as *sub-salt*, which is stratigraphically younger rock or more commonly known as younger sediments. Worldwide there are more than 100 salt tectonic basins and sub-salt plays a very important role. In the U.S. Gulf of Mexico, over 40 apparent subsalt fields have been discovered to date and projected potential recoverable reserve is from 7 to 18 billion of oil equivalent.

Both sub-salt and pre-salt are considered to be huge resources of oil and gas but the very nature of salt makes the job of exploration, drilling and completion quite difficult. Different types of problems that are faced during drilling operations because of the mobility and certain specific physical properties of salt bodies are discussed hereafter.

4.1.3.1 Drilling Challenges

It is to be noted that while discussing different problems or challenges in drilling, not only is the salt problematic but also the sub-salt layer which is an unconsolidated formation that poses altogether a different challenge. The reason being their opposing physical and chemical properties. So, first let us concentrate on the important property of salt, that is, *salt creep*.

The behaviour of salt changes from solid to semi-plastic as temperatures approach 220°F, a condition that allows the substances to creep in response to differential pressure and forces. Above 400°F, salt becomes plastic and flows very easily. As salt bodies rise toward the surface, dropping temperatures slow the creep of the structure upward, although lateral movement may continue for some time. In addition to temperature, creep rate of salt is determined by differential stress, confining pressures, the grain size of the salt and the presence of inclusions of other solids, water or gas.

Salt creep imposes many conditions on drilling, requiring precautions. During the drilling of a salt section, especially long ones, salt creep will reduce the gauge and radius of the wellbore or create *ledges*, where downhole tools and pipe can hang up. Sufficient mud weight must be maintained to resist salt creep. Also, the salt section must be drilled as quickly as possible to reduce the impact of creep on open-hole gauge.

High temperatures and pressures associated with depth make mud weight control imperative. Freshwater pills in the drilling fluid will dissolve the troublesome sections, but they can pose problems in wellbore enlargement. Numerous operators have reported hang-up problems in drilling long salt sections.

In drilling salt sections, drillers confront horizontal stresses that are usually equal to vertical stress, which usually correlates to overburden depth. This is because salt will not compact with greater temperature and pressure, and in a plastic or creep state, will flow horizontally as well as vertically when the hydrostatic pressure of the mud is less than formation stress.

Hence, drillers need to pay careful attention to drilling fluid weight and hydrostatic pressure of the fluid as a function of depth and higher horizontal stresses in salt sections.

In drilling long salt sections, salt creep can reduce borehole gauge before the section is drilled, so drillers must plan for salt creep in advance. If wells have been drilled previously in the salt section, sufficient information may be available on the creep rate. However, there is enough variation in creep rates from adjacent wells and from the top of the salt structure to the bottom that operators prefer to conduct *characterisation of the salt* section being drilled.

To characterise the salt, drillers typically core a section of salt after entering the salt body, and sometimes again in the lower sections. Salt specialists can analyse salt cores to estimate the creep rate and provide a full characterisation of the salt body. The analytical procedure involves slicing the core into thin sections and checking the microstructure with magnification. From this, the analyst obtains grain boundaries, fluid inclusions and other factors.

Using information provided by the core analysis, drilling engineers can then calculate creep rates and closure rates for boreholes drilled in the salt structure from which the core was obtained.

Salt bodies rarely exhibit uniform creep rates, and when wellbore temperatures rise, as they frequently do in long salt sections, the salt can exhibit different characteristics. If the operator suspects that the creep rate or inclusion volume may change with depth, then additional cores should be taken and adjustments made accordingly.

Now that drillers have worked out the drilling fluid weight needed to counter the *borehole closure* by creeping salt, they can turn to another salt drilling problem that is quite the opposite of closure – *wellbore enlargement*.

Enlargement or washout is another common problem in drilling salt sections. This occurs when undersaturated drilling fluid dissolves the borehole walls, but can also take place when salt inclusions are leached out by saturated fluids, weakening and collapsing the wellbore walls.

A characterisation of the salt core will indicate whether the inclusions are present in sufficient quantities and what types of fluid components will leach them from the salt. Drillers then can take precautions in drilling fluid formulation, fluid circulation rates, or in rate of penetration.

An enlarged borehole presents serious problems in drilling fluid volume, movement, setting and cementing in casing strings and circulating cement above the salt top. Unlike conventional wells, remedying an enlarged borehole in a salt section is extremely difficult because drilling fluid chemistry and physics requirements eliminate many measures that might be taken.

The initial problem in the enlarged section is the accumulation of drill cuttings as the drilling fluid slows through the expanded annulus. The cuttings can collapse around bottom hole assemblies being run or complicate the running of the initial casing string.

Freshwater pills are commonly used to free stuck drill strings or casings, but frequent sticking problems or overuse of the pills can create wellbore dissolution elsewhere.

Later, after the initial casing string is run, wellbore washouts do not allow even placement of cement around the string. Formation loading on the casing is not uniform and ovaling or bending can begin before curing of the cement. Figure 4.36 shows both closure and enlargement of the borehole.

Knowing the nature of salt and the type of problems encountered during drilling through it, it is now necessary to know the nature of unconsolidated sediment just below the salt structure.

The *unconsolidated field* beneath the salt section is generally porous and has a tendency to quickly absorb large volumes of drilling fluid when an overbalanced condition exists. However, too great an underbalance would allow formation flow if the contents of the unconsolidated zone are highly pressured.

The driller has to approach the lower boundary of the salt section slowly. The unconsolidated interval may be as thick as 1000 ft but generally averages 100–300 ft in the U.S. Gulf of Mexico.

The subsalt interval has been disturbed by the movement of the salt body, which has left it much weaker than the salt above it or the shale formations below it. This impacts on the choice of drilling fluid weight when the drill bit approaches the subsalt interval.

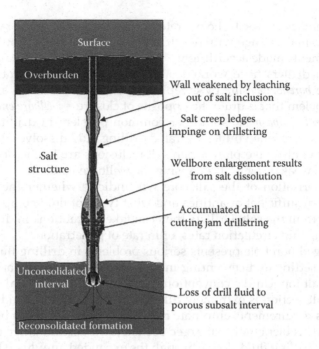

FIGURE 4.36
Closure and enlargement of bore hole in a salt section.

4.1.3.2 Mitigating the Challenges Posed by Salt and Sub-Salt

At the time of drilling, several measures can be taken by the driller to deal with known creep rates and loss of wellbore radius.

Controlling the fluid pressure – The conventional method of keeping the creep rate constant is to increase the mud weight by a 1:1 proportion for every lb/gal horizontal stress increase in the salt sections.

Reaming – In addition to other measures, back or side reamers are incorporated in the bottom hole assemblies to deal with the loss of the wellbore gauge while the well is being drilled.

Undersaturation of drilling fluid – By undersaturating the drilling fluid slightly, the rate of wellbore wall dissolution can match the creep rate of the salt, although achieving this balance is difficult. Few drillers will take risks in undersaturation since the penalty for overestimating the creep rate is wellbore enlargement.

To maintain near-gauge boreholes, *drilling fluids* must minimise hole closure and washouts or enlargement. So, the choice is among three types of drilling fluids, that is, water-base mud (WBM), oil-base mud (OBM) or synthetic drilling fluid.

WBM with low salt concentrations try to balance salt erosion and dissolution with creep rate to maintain hole size. However, because salt creep and dissolution change across thick salt sections, this can be problematic and hole size may vary with depth. High-salt concentration, WBMs dissolve enough salt to offset creep, but can become undersaturated at high temperatures and enlarge the hole.

Dissolution of encroaching walls and ledges can be halted with the use of oil-based fluids, if salt creep is minimal. However, some types of oil-based fluids can leach out inclusions without touching the salt, weakening the wellbore walls. With higher creep rates, some salt dissolution is necessary where higher fluid weight cannot be maintained. This occurs when the drill bit nears the unconsolidated subsalt section and the driller wants to avoid heavy loss of drilling fluid to a permeable interval.

Synthetic drilling fluids can solve problems in salt and subsalt formations, but if these fluids are lost to the subsalt formation, the well could become very expensive. Synthetic muds cost $200–$800/bbl, and it would not be uncommon to lose $100–$200 bbl before adjustments could be made.

For reasons of cost, flexibility and familiarity, many drilling engineers prefer to use supersaturated water-based fluids through the salt section and below. Generally, variations depend upon the temperature gradient through the salt section.

The next important step is to find out a suitable *bit* to drill through salt and subsalt by maintaining proper rate of penetration (ROP).

Salt is weak and soft, so polycrystalline diamond (PCD) and other mill-tooth insert cutters, which make holes by scraping, are used. Stronger inserts may be needed to penetrate caprock formed on the top of some salt layers by groundwater leaching of minerals. Side-cutting, eccentric or bicentred reamers above bits have been proposed to open up hole diameters that are larger than the bit and allow for some salt creep before the borehole becomes undergauge.

After drilling into salt, heavier than expected mud weights may be needed to control salt flow. Drilling speeds vary among operators, but reasonably fast penetration rates – 60–150 ft/h (18–46 m/h) – are required, so wells can be cased quickly. Good hole cleaning and periodic back-reaming, however, should not be sacrificed just to make a hole faster. Circulating a small volume of freshwater can remove salt restrictions and free stuck pipe, but care must be used to prevent washouts. Enlarged or undergauge holes make directional control difficult.

Thick salt bodies can affect temperature and pressure in surrounding formations. Salt thermal conductivity is high compared to other sediments, so overlying formations are heated and underlying formations are cooled. Because salt is a barrier to basin fluids, if outward flow is insufficient to achieve normal compaction, high pressure may develop below salt. As disrupted sediments below salt are penetrated, fluid losses or flow can occur, depending on mud weight and formation pressures, unless drillers proceed slowly and carefully.

Once drilling through the salt and the unconsolidated sediment is done, *casing* has to be done early as possible. The normal casing procedure if followed in this case leads to the following two situations because of salt creep.

Wellbore shift – The entire cased section is shifted laterally or bowed out at 60–90% of the salt flow creep rate in the salt section (Figure 4.37). Creep rates are rarely uniform through the entire section.

Localised loading – Varying salt creep rates, poor cement placement, or weak areas along the casing string result in bending, deformation, ovaling and loss of gauge.

Wellbore shift can become a severe problem during the later stages of well production or workover when the casing string experiences stretching, severe bends and shearing.

Shear zones (shear planes) surrounding the salt structure are where most of the stretching, collapse and shearing of the casing is likely to occur. The shear zones range from 100 ft to 400 ft in thickness.

In the shear zones, salt gradually changes to the surrounding formation, usually cap rock at the top and unconsolidated zone at the base of the salt.

To take care of these problems, drillers install either a single string of heavy wall casing or concentric strings of regular wall casing through the salt section (Figure 4.38).

The concentric casing program consists of the primary casing plus a liner hung off above the salt section, providing two casing strings and two layers of cement to resist loading. Today, the choice is usually concentric strings because the tests indicate that casing is more sensitive to wall thickness (concentric strings) than to yield strength (heavy wall) because of non-uniform loading, which is the case with salt section loading. A uniform load, typical of that imparted by fluid pressure, can usually be resisted by yield strength alone.

Casing strength in the salt section is designed to suit the depth at which the salt occurs. Generally, for each linear foot of depth, casing should be able to withstand one additional pound for every square inch. Thus, a casing string at 10,000 ft should be able to take 10,000 psi of load.

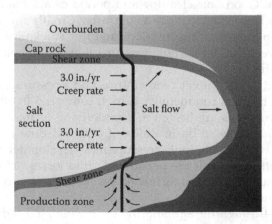

FIGURE 4.37
Wellbore shift in a salt section.

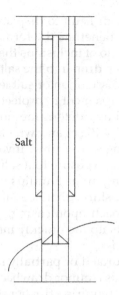

FIGURE 4.38
Concentric casing strings used in a rapidly moving salt.

On the issue of where to case, drilling engineers are split in their opinion. Most engineers will drill to the base of the salt section and case immediately, with the liner run concentrically and hung off at the top of the salt section. This approach is conservative in that the time allowed for salt creep between drilling and casing is minimised.

However, some drillers will try to drill through the unconsolidated zone beneath the salt section and into shale or firmer formations below before setting casing. This provides support for the casing and cement layer, and allows more flexibility in the casing program, but it exposes the open borehole to a longer period of salt creep.

Either of these approaches will likely depend on the rate of salt creep and accommodation of the unconsolidated formation below the salt to the weight and chemistry of the drilling fluid. If the driller expects problems upon entering the unconsolidated formation, or the salt creep rate is high, then it might be safer to case at the base of the salt.

Casing cement plays a more important role in salt sections than anywhere else in the well. Not only does it prevent annulus flow between formations and supports the casing, as in conventional wells, but it prevents salt ledges from encroaching on the casing in salt sections.

One difficulty in cementing casing in salt sections is that the borehole is rarely uniform. The process of removing salt ledges, and certainly the running of freshwater pills, creates washouts where the borehole becomes enlarged.

These enlarged areas or pockets tend to gather drill cuttings that are not removed by the drilling fluid. The pockets will also contain a higher volume of drilling cake that becomes difficult to displace with the cement. Further, the cement does not properly penetrate the formation in pockets.

If the salt contains a high ratio of inclusions that do not dissolve uniformly with the salt, the inclusions can drop into the salt pockets also. Alternatively, the inclusions may slow the dissolution of salt ledges, resulting in pinching of the casing in spots and little cement to protect the casing.

Additives to the cement slurry determine how the mixture reacts with the salt, how quick the slurry will set and whether chemical changes in the salt will induce physical changes over time. Invariably, the chemistry of the cement will change because of exposure to a saline environment.

The timing of cement setting in the annulus is critically important in salt section casing installation. A slurry that sets up too slowly (over-retarded) can allow salt zones to encroach upon casing, resulting in thin cement or none at all. A cement that sets up too quickly may not allow for proper displacement of the drilling fluid.

The cement can be salt-saturated or partially saturated to properly retard the setting process, but salt has a minor drawback. The saline mixture tends to over-retard cement setting, requiring further additives to control fluid loss and mixing.

However, undersaturation of the cement can bring about problems later in the life of the well, especially wells where the salt formation contains modest amounts of magnesium ions. Over time, the cement experiences an ion exchange between the positively charged calcium in calcium chloride and magnesium ions. The result is a weakening of the cement.

Since drilling through the salt section is not an easy matter, some operators have attempted to avoid the salt section to reach the subsalt layer by adopting directional drilling around ledges and overhangs. But another problem arises with this option. The wellbore often must turn horizontal to avoid the salt and intersect the thin subsalt interval which makes this well very costly, almost equal to the cost of drilling a thick salt section. It is hoped that operators would soon develop a reliable procedure for drilling salt sections and avoid the wellbore contortions associated with directional drilling. In future years, operators will have a better idea of the forces within the salt sections, loading on casing strings and how to resist these forces.

4.2 Difference in Spudding from Land Drilling

The word 'spudding' in the context of drilling means the very start of drilling operations in a new well. In other words, it is that moment when the drilling bit penetrates the surface utilising a drilling rig capable of drilling the well to a targeted depth. But to start the drilling operation some preparations are required prior to that and those involve the mobilisation of all the machinery and equipment to the site, erection and placement of these at their designated places and interconnecting them simultaneously with the source of electricity, air and water. In the case of drilling from the land, all the equipment and so on are erected on the ground surface and after connecting with each other, the spudding occurs with the drilling bit touching the ground surface only.

But in the case of offshore drilling, all the drilling equipment and its ancillaries have to be transported over the sea/ocean surface by means of a barge or ship and then needs to be placed and erected on the deck of a platform, be it fixed, compliant or mobile type. The place atop the platform where the drilling and other auxiliary equipment including crew quarters are placed is commonly known as *topside*. The next step before spudding is to make the drilling bit touch the ground surface, in this case the sea bed or mud line, for which a conduit between the topside and the sea bottom is essential which is commonly known as *conductor* or *drilling riser* depending on the type of platform being used. But to connect this conductor or riser at the sea bottom, some provision has to be made and that is known as *subsea facility* or *subsea unit*. Among the subsea units, common ones are subsea template, subsea wellhead and subsea BOP.

Thus, it is observed that though the actual procedure of drilling remains the same as in the case of onshore or land drilling, the sea water in between the topside and the subsea makes the difference. Hence, before discussing the actual process of offshore drilling, it is essential to know about the topsides and the subsea units for different types of platforms.

4.2.1 Topsides

Topsides come in many sizes and shapes but can be loosely categorised into three main types:

1. Skid-mounted equipment
2. Integrated decks
3. Modularised decks

1. Skid-mounted equipment

In the early days, the most effective way to build topsides was to have a structural deck fabricated first. The drilling equipment was then brought to the deck fabrication site and installed on and in the deck. Each item like pumps, engines, quarters buildings, helicopter landings and so on had to have enough structural support steel to provide the strength and rigidity needed to transport and install it. That is, each piece was mounted on its own *skid* so that the skid steel for each equipment item did not share in the load-carrying needs of the entire deck.

Many of these equipment skids did require considerable structural steel to allow them to be lifted into place, either offshore or at the deck yard site. Much of the necessary piping and electrical, pneumatic and hydraulic lines could already be installed on each component skid when it arrived. The rest would be added after the skids were set; for example, the skids were 'hooked up' to each other, which often occurred offshore. One advantage of this design was that the structural deck could first be used to accommodate the drilling and completion rigs and then later receive the production skids. A disadvantage was the reality that doing offshore hook-ups cost 5 to 15 times as much as those done onshore.

2. Integrated deck

For integrated deck and topsides, the individual pieces of drilling/production equipment are placed in the structural deck as it is being fabricated, with the steel structural deck members directly integrated into the support system for each piece of equipment. The electrical, pneumatic and pressure piping systems are fabricated and installed for the entire set of equipment, not just a hook-up of individual skids. That saves structural steel and overall topsides weight. However, the equipment and structural work scheduling is much more intense, given the limited deck space.

An integrated deck may be divided into many levels and areas depending on the functions they support. Typical levels are as follows.

Main (upper) deck, which supports the drilling/production systems and several modules (drilling, process, utilities, living quarters, compression, etc.)

Cellar deck, which supports systems that need to be placed at a lower elevation and installed with the deck structures, such as pumps, some utilities, pig launchers/receivers, Christmas trees, wellhead manifolds, piping and so on.

Additional deck levels, if needed. For example, if simultaneous drilling and production operations are planned, some process equipment may be located in a *mezzanine deck*.

3. Modular deck

As the deepwater decks became larger and larger, it was a necessity to split the deck into smaller, liftable pieces and to allow them to be built more cost-effectively.

A modular deck may be divided into many pieces and modules depending on the functions they support and the installation equipment available. Typical modular deck components are as follows.

Module Support Frame (MSF) – It provides a space frame for supporting the modules and transferring their load to the jacket/tower structure. The MSF may also be designed to include many platform facilities, such as the storage tanks, pig launching and receiving systems, metering/proving devices and the associated piping systems.

Modules – These provide a number of drilling/production and life support systems. For example:

Living quarters module generally supports a heliport, communication systems, canteen, office and recreational facilities.

Utilities module generally supports power generation and electrical and production control systems including a control room.

Wellhead module generally supports the wellheads, well test and control equipment.

Drill rig module contains the derrick, draw-works, drillers and control rooms, drill pipe and casing storage racks and pipe handling systems. Drill rig module is located over and supported by the wellhead module.

Production module contains the oil/gas/water separation and treatment systems, other piping, control systems and valves for safe production, metering and transfer of the produced liquids and gas to the offloading system.

Power module contains either gas turbines or diesel engines for generation of electricity to drive rotating equipment such as compressors and pumps.

Pumps and compressors module contains various pumps for different purposes in drilling and production as well as compressors for pneumatic lines and also for either transportation of gas to the shore or injecting gas into the formation during the production stage.

As an example, the Mars TLP topsides (Figure 4.39) was built in five major modules and then installed module by module at the shore quayside. The module setting was then followed by several weeks of intra-module hook-up, testing and commissioning.

Hook-up and commissioning field work is initiated when all the topsides modules or skids are placed on the host structure. Successful projects plan these activities from the early stages of concept development. Even though

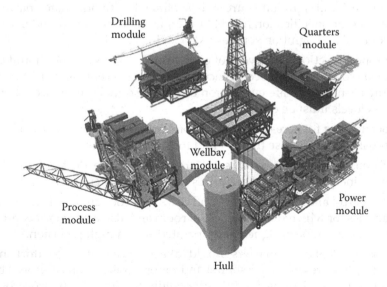

FIGURE 4.39
Mars TLP topsides. (Courtesy of Shell E&P.)

the modules shown in Figure 4.46 may well be individually complete and functionally tested, there will remain a lot of intra-module connections and testing to do before the entire system is integrated and fully functionally confirmed which is done under the overall guidance of engineers, technicians and specialists.

The best place to do this work is onshore or at a quayside, before the facility is moved to its offshore location. TLPs and concrete gravity-based platforms may all have their topsides decks and facilities fully integrated before they leave the quayside on their way to the installation site. Spars and fixed platforms always have their topsides modules set after they are in place. This requires that some of the hook-up and commissioning be done offshore, at a considerable cost premium.

Preparing the layout of topsides

Nowadays with the help of computer aided design (CAD) model, the topside facilities can be properly laid out keeping in mind the following necessary and important parameters.

1. To ensure safe movement of the operators inside the deck structure around any equipment without any potential clashes.

2. To ensure correct orientation of valves so that ample access to equipment exists and its removal for maintenance purpose or replacement becomes easy.

3. To prevent occurrence of any hydrocarbon ignition, the deck layout has to be such that the equipment under two broad categories – 'fuel sources' and 'ignition sources' is separated with adequate spacing. As per the specification API 14 J the equipment, classified under 'fuel sources' and 'ignition sources' is given in Table 4.1.

4. Proper selection and layout of control and safety systems should be made to ensure low occurrence of accidental events and in case of any eventuality, safe evacuation of the personnel within acceptable risk levels must be ensured.

5. Scaffolding for accessing any equipment should be avoided as far as possible to minimise the cost and for safety reasons.

6. To ensure easy access to such areas within the deck where the machinery or equipment requires frequent maintenance.

7. To control and manage safety and drilling/production critical emergencies for which the main control room and also emergency control room have to be designed and operated with a high precision.

8. The layout of crew quarters should take into account the comfort and safety features of the personnel and for obvious reasons it should be far from the drilling module, the wellheads and the hydrocarbon handling equipment.

TABLE 4.1

Equipment – Fuel Sources versus Ignition Sources

Fuel Sources	Ignition Sources
Wellheads	Fired vessels
Manifolds	Combustion engines (including gas turbines)
Separators and scrubbers	Electrical equipment (including offices and buildings)
Coalescers	Flares
Oil treaters	Welding machines
Gas compressors	Grinding machinery
Liquid hydrocarbon pumps	Cutting machinery or torches
Heat exchangers	Waste heat recovery equipment
Hydrocarbon storage tanks	Static electricity
Process piping	Lightning
Gas-metering equipment	Spark producing hand tools
Risers and pipelines	Portable computers
Vents	Cameras
Pig launchers and receivers	Cellular phones
Drains	Non-intrinsically safe flashlights
Portable engine-driven equipment	
Portable fuel tanks	
Chemical storage	
Laboratory gas cylinders	
Sample pots	

Source: API 14J.

Besides the above parameters, the general layout of topsides should pay special attention to the following items:

1. *Control and safety systems*, which include either local or central operational control system, data acquisition systems, well control and shut down systems, fire detection systems, combustible gas monitoring systems and so on.

2. *Firefighting equipment*, which should be accessible in any location on the deck.

3. *Safe work areas* for carrying out minor fabrication or repair works, welding and gas-cutting and so on.

4. *Storage* of diesel fuel, chemicals, spare parts and other consumables.

5. *Ventilation* to be provided for dispersing any hazardous vapours and accumulation of gases.

6. *Escape routes* – minimum two in number to be provided from each location and *evacuation* can be provided through use of boat landing, survival craft and helideck.

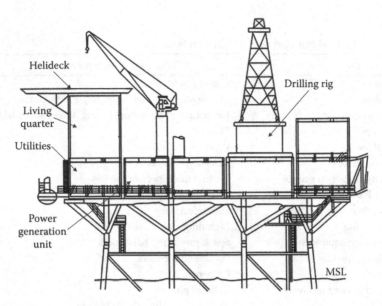

FIGURE 4.40
Topsides layout of a fixed platform.

Thus, it is found that for the main function, that is, drilling in this case, standard equipment similar to land drilling like derrick, draw works, pumps, engines, circulating system and so on will be the major part of the topside but besides these items auxiliary systems to support the main function will equally occupy the space in this. A general layout of a fixed platform is shown in Figure 4.40.

Further, a detailed schematic of the topside layout in the case of a jack-up platform is shown in Figures 4.41 and 4.42.

Regarding topside layout in the case of a floater it has been discussed in detail in Chapter 6.

4.2.2 Linkage between Topside and Subsea

Between topside and sea bed there needs to be a device that can connect them and also guide the drill string and other equipment and tools to the bottom for carrying out drilling from the top. Conductors or risers are such devices which act as conduits from the topside to the ocean floor and primarily act as guides for the drill string. The conductors are used in the case of fixed platforms whereas risers are used with the compliant platforms and mobile units. There is also a guidance system that serves to run and position different equipment and tools down to the sea bottom. Following are the descriptions of each of these devices.

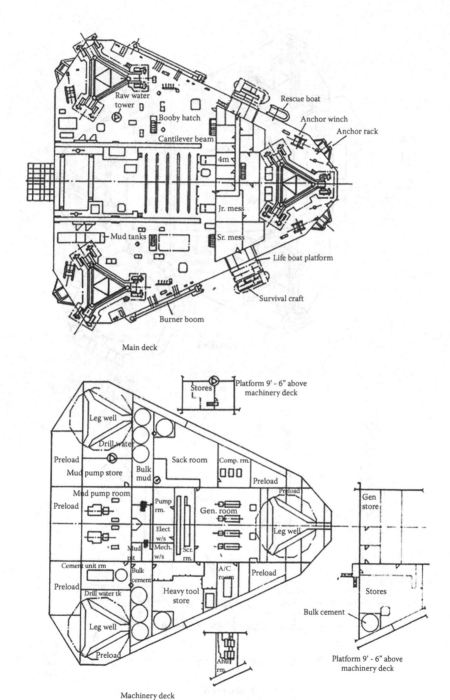

FIGURE 4.41
Topsides layout of a jack-up (main deck and machinery deck). (Courtesy of ONGC.)

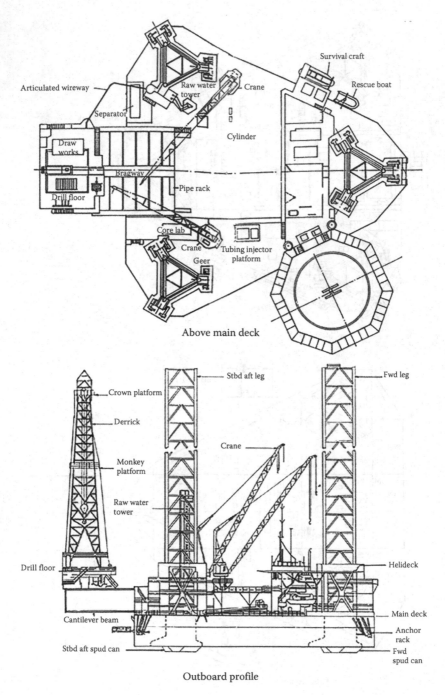

Above main deck

Outboard profile

FIGURE 4.42
Layout of a jack-up (plan and elevation). (Courtesy of ONGC.)

4.2.2.1 Conductor

In a conventional fixed platform, individual conductor pipes must be provided for each well to be drilled through them for protecting the surface casing from the natural forces of the sea as well as geomechanical conditions such as mudslides. These conductors are large diameter steel pipes usually 26–36 in. in size which are driven into the soil at the sea bed through *guides* connected to a jacket or within the legs of gravity platforms and the deck structure of the platforms.

Conductor guides for positioning and guiding the conductors are framed at various elevations within the jacket and decks to provide support for the conductors such that the usual effects of environment such as waves, winds, current and the like can be safely withstood by the conductors and to maintain conductor alignment.

There are two kinds of conductor guide systems that have been utilised in the offshore industry. The first system, which is more common and older, consists of guides that are rigidly connected to the jacket and deck framework. The conductors are placed through these guides. The conductor guides and structural framework provide support for the conductors at various levels throughout the jacket and deck or decks. This type of system generally includes three kinds of conductor guide assemblies. The first kind are those within the horizontal framework levels of the jacket and typically consist of vertical guides made from steel tubes welded to the horizontal jacket tubular members. The other types are located in the upper and lower deck levels. The lower deck level guides are similar to those of the jacket except that they are rigidly connected to the deck floor beams. These guides are located in line with the jacket guides. The upper deck level assembly consists of a grid of beams bolted to the permanent upper deck beams supporting removable hatches which line up with the conductor guides in the lower deck. Access is provided to the lower deck level, which is typically the conductor termination level, by removing the hatches. While advantages of this type of system include the fact that conductor guides and framing are normally built within the jacket and deck during land fabrication, when jackets are set over existing wells, offshore construction thereof is required but the problem is that this type of arrangement may not sufficiently withstand extreme environmental effects such as mud slides or ice movement.

The second type of arrangement, which has recently been utilised in areas of extreme environmental loading, such as mud slide zones, consists of jacket conductor guides positioned inside a large diameter pile which has previously been driven through a jacket sleeve. The pile protects the conductors from environmental loading. The typical jacket conductor assembly consists of a series of horizontal guide frames connected to a central post and supported by the pile at its top. Additional guide assemblies similar to those detailed with respect to the first type of arrangement are provided in the decks of the platform. Because the jacket conductor guide assembly must be erected offshore after the jacket and piling are installed, it is required that the conductor guides for the deck sections be built offshore to conform to the orientation of the conductor guides in the pile.

While the conductor is lowered through the guides located at different levels in the jacket and the decks, different types of centralisers are used to hold them centrally inside those guides. Figures 4.43 and 4.44 show the arrangement of the guides and different centralisers.

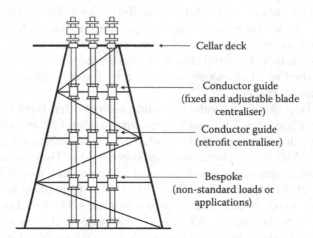

FIGURE 4.43
Conductor guides.

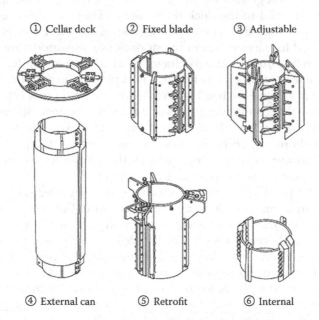

FIGURE 4.44
Conductor guide centralisers.

These centralisers are as follows:

1. *Cellar deck centralisers*: Uppermost centraliser, in air, helps to stabilise Christmas tree placed above the wellhead.
2. *Conductor guide centralisers (fixed)*: Used at various levels on the conductor from the splash zone to the seabed to stabilise the conductor.
3. *Conductor guide centralisers (adjustable)*: Similar to the fixed version but with adjustable blades to set the gap.
4. *Can type guide centralisers*: Designed for deepwater applications where there is significant vertical conductor movement.
5. *Retrofit guide centralisers*: Used to replace lost or missing guide centralisers.
6. *Internal or cementing centralisers*: To space out internal strings – allows cement flow.

4.2.2.2 Drilling Riser

The term 'riser' is used in the context of both drilling as well as production and so when it is used for drilling operations, it is known as a drilling riser and for production operations it is known as a production riser.

A *drilling riser* is a conduit that provides a temporary extension of a subsea oil well to a surface drilling facility. Drilling risers are categorised into two types – (1) *marine drilling riser* and (2) *tie-back drilling riser*. Marine drilling risers are used by a floating vessel when the blowout preventer (BOP) is placed at the sea bottom and tie-back drilling risers are used with bottom supported units like jack-up platforms or compliant types like TLPs when the BOP is placed at the platform deck.

In exploratory drilling operations carried out with a jack-up rig, the 'tie-back riser' runs from the drilling deck to the wellhead at the sea bottom (Figure 4.45). The drilling fluid is pumped down the drill string and the mud along with cuttings returns through the annulus between the drill string and the casing up to the subsea wellhead and further up between the drill string and the riser up to the top surface. Ordinarily, when the targeted depth is reached, the well is tested and capped at the bottom and the rig moves out to another drilling site. So, the riser string gets detached and reused at the new site. After the rig has moved out, these capped wells are connected to a protector platform by tie-backing them using tie-back strings. A tie-back riser can be either a single large diameter high pressure pipe, or a set of concentric pipes extending the casing strings in the well up to a surface BOP. Because the BOP is at the surface, the tie-back riser must contain full well pressure.

For floating drilling operations using a drillship or semisubmersible, a 'marine riser' connects the drilling deck to the BOP at the sea bottom but this riser requires more hardware to accommodate the vessel's heave, sway and surge and this extra hardware like telescopic joint, tensioners and so on

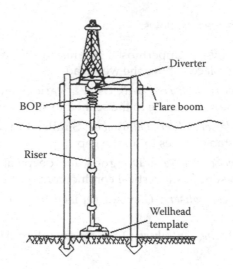

FIGURE 4.45
Tie-back riser.

makes it different from a tie-back riser. In this case, since the BOP is below it, marine risers do not need to contain full well pressure. BOP control lines and choke/kill lines are strapped to the riser for well control. For deepwater applications, it includes buoyancy modules to reduce the hanging weight or the amount of tension required to maintain stability of the riser. These buoyancy modules make them close to neutrally buoyant when submerged. Since the marine riser is a more complex system with many component parts, a detailed description of this henceforth follows.

Marine Riser
The main purposes of using a marine riser are to provide a return flow path between the wellbore and the drilling vessel and to guide the drill string or casing to the BOP stack on the ocean floor.

Components of a riser system must be strong enough to withstand high tension and bending moments, and have enough flexibility to resist fatigue, yet be as light as practicable to minimise tensioning and flotation requirements. These considerations should be given to all components when selecting a riser system.

The components of a marine riser system (Figure 4.46) from bottom to top are hydraulic connector, lower flexible joint (ball joint), flexible piping for choke and kill lines, riser pipe and connectors, choke and kill lines and connections, telescopic (slip) joint, diverter system or bell nipple and riser tensioning equipment.

In some instances, an annular preventer is included between the hydraulic connector and lower ball joint. This practice permits replacement of worn annular preventer elements without removing the BOP stack. Usually a

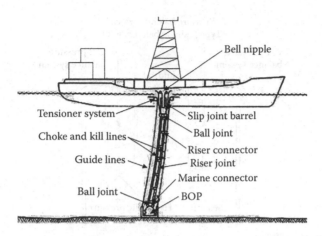

FIGURE 4.46
Components of a marine riser system.

marine riser guide frame encloses a hydraulic connector, an annular pre-venter, a flexible joint and choke-and-kill flex lines as a single unit which is sometimes called a *lower marine riser package.*

The *hydraulic connector* joins the bottom section of the marine riser to the top of the BOP stack. Clamps and release mechanisms are activated by hydraulic pressure from the BOP system. This type of connector is known as a Mandrel-type connector which secures locking dogs around flanges at the top of the subsea BOP. By design speedy mating of the connector with the BOP stack is possible and disconnect can be accomplished in even less time. By construction the connector can withstand the great bending and tensile stresses to which it is subjected by the drilling platform's horizontal move-ment and the tensioning of the marine riser system.

Flexible or ball joints at the base of the marine riser package accommodate up to 10° of deviation from the vertical to allow for any horizontal movement to which the drilling vessel might be subjected. When using a ball joint, the weight of the riser and the drilling fluids within it creates an unbalanced force on the socket section, tending to force it down onto the ball section. This compression force is balanced by pressurised lubricating oil between the top socket section and the ball section. The hydrostatic head of drill-ing fluid in the marine riser and the overpull of the riser tensioners cre-ates an upward (or tension) force on the socket section, acting on the lower face of the ball section. These upward forces are variable and are again com-pensated by pressurising the lubricating oil between the ball section and the lower socket section. A single ball flex joint pressure-balance system is shown in Figure 4.47. Hydraulic fluid from the BOP stack control system is applied to the base of the floating piston in another fluid-oil separator. This transfers the pressure into the lubricating oil, maintaining the required bal-ancing force to compensate for the overpull and mud weight. The required

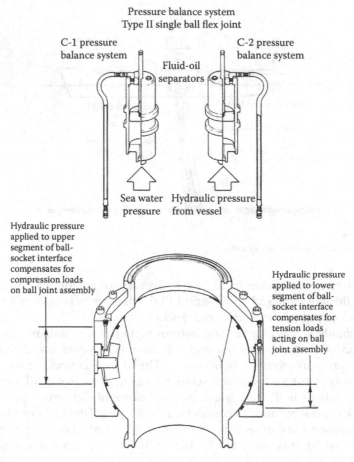

FIGURE 4.47
Flex joint pressure balance system. (Courtesy of GE Oil & Gas.)

hydraulic pressure is dependent upon the tension load on the riser, water depth and mud weight.

To avoid twisting the flexible choke-and-kill lines around the flex joint, an antirotation pin is incorporated to prevent the socket from rotating relative to the ball section.

Because of the pressure requirements anticipated for a ball joint in deep water (3000–6000 ft), a nonpressurised flex joint was developed with the high tensile capability to handle the deep-water subsea equipment. Vetco's Uniflex (Figure 4.48) is an example. Since the Uniflex joint requires no hydraulic balance pressure, its operation is simplified and service and maintenance requirements are substantially reduced. The inner surfaces, subject to drill pipe wear, carry removable bushings. Primary flexing takes place at each of the two bearing rings in the upper and lower sections (Figure 4.49). The two

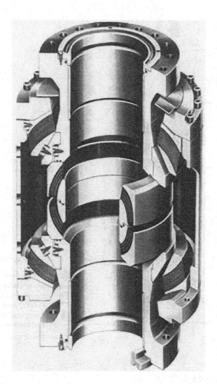

FIGURE 4.48
Uniflex joint cross-section. (Courtesy of GE Oil & Gas.)

pieces in the middle, the seal assembly, are composed of the same flexing material and mainly seal between the internal mud pressure and the external ambient pressure. The flex material is laminated layers of steel and rubber. The action is more like a sliding, compressive loading than the pivotal loading on the spherical surface of the single ball joint.

Flexible piping is required for the choke and kill lines around the ball joint. Usually these are referred as choke-and-kill flex lines. These vary from 5000 psi working pressure hoses to coiled or spiral pipe loops.

Riser pipe is a seamless pipe with mechanical connectors welded on the ends. Its size is determined by the bore of the BOP stack and wellhead with allowance for clearance in running drilling assemblies, casing and casing hangers.

An example illustrating riser pipe equipment is

Zapata Ugland – semisubmersible, 1974:

(Subsea two stack system)

685′ of 24″ OD pipe with 45′ stroke telescoping joint

685′ of 16″ OD pipe with 45′ stroke telescoping joint

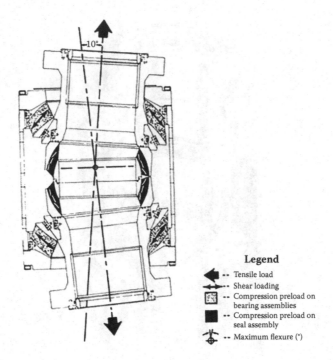

FIGURE 4.49
Flexing uniflex. (Courtesy of GE Oil & Gas.)

Connectors and riser pipes are designed for fast positive latching, and choke and kill lines are included as integral parts (Figure 4.50). Sometimes lines for BOP power fluid are integrated, as are the kill/choke lines. The riser can be run in a manner similar to a drill pipe, by stabbing one stalk at a time into the string and tightening the connector. Connectors used in making up integrated choke and kill lines and riser pipe connections are designed to require a minimum number of operations. Connectors can be actuated hydraulically, making diver assistance unnecessary. TV cameras run on guide lines have been used to assist in alignment, stabbing and attaching connectors. Special clamps are used instead of some of the flange-type connectors used on land. If divers are used, fast operating air powered tools are used to reduce diver time on the bottom.

The *telescopic, or slip joint,* is used at the top of a marine riser and performs these functions:

1. Compensates for vertical movement of the vessel while drilling and for added dimension required for any horizontal displacement of the platform.
2. Provides fitting for choke and kill line hoses.
3. Provides for connecting bell nipple or diverter assembly.
4. Provides for attachment of riser tensioner system.

FIGURE 4.50
Integral marine riser pipe with choke and kill lines. (Courtesy of GE Oil & Gas.)

It consists of a hollow inner barrel which telescopes into an outer barrel. The outer barrel of the telescopic joint is attached to the marine riser assembly and the inner barrel is attached to the drilling vessel. The bell nipple or diverter assembly is attached to the inner barrel which is suspended from the rotary support beams of the rig. The strength of the telescopic joint in tension is sufficient to support the weight of the BOP stack and marine riser. Resilient seal or packing elements between the inner and outer barrels provide a pressure seal. Oil or water lubrication reduces wear.

A *diverter system* controls flow of gas or other fluids which may enter the wellbore under pressure before the BOP stack and the casing which supports it have been run. The diverter is an integral part of the bell nipple. It may be a bag-type unit or a modified rotating BOP. In other words, a diverter is a low pressure annular preventer that seals off the riser bore

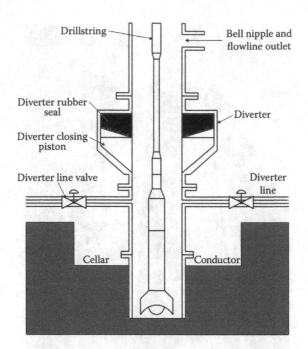

FIGURE 4.51
Diverter system and its components.

(Figure 4.51). The diverter redirects the flow of mud and cuttings during a kick when the BOPs are not used and the diverted fluids flow overboard. Control is from a hydraulic system. It is valuable at the time of drilling through shallow gas zones or for diverting gas kicks in deep high-pressure zones.

Riser tensioners support a major portion of the weight of the marine riser system plus the drilling mud it contains. These tensioners also provide a means of compensating for heave motions of the platform. Without the riser tensioning system, only a very short riser could be supported by the BOP stack. Drilling at greater water depths has shown the need for flotation equipment or buoyancy module to support part of the weight of the marine riser.

These modules may be thin-walled air cans or fabricated syntactic foam modules that are strapped to the riser. *Air cans* have a predictable buoyancy, and this buoyancy can be controlled from the surface. However, they slow down the running operation more than syntactic foam modules do. *Syntactic foam* modules began to appear in the early 1970s. They are convenient because they become part of the joint, they require only a little additional care in handling, and they do not have lines to be attached. Various compositions of foam have been tried with varying degrees of success.

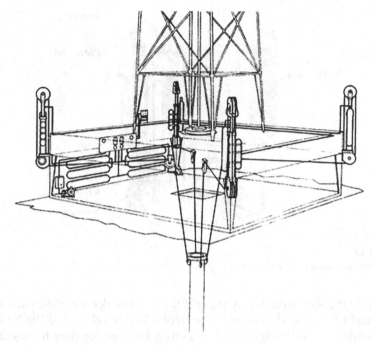

FIGURE 4.52
Riser tensioner system.

The major components of a pneumatic riser tensioning system (Figures 4.52 and 4.53) are

1. Tensioner cylinders and sheave assemblies
2. Hydropneumatic accumulators and air pressure vessels
3. Control panel and piping manifold
4. High pressure air compressors
5. Standby air pressure vessels

In Figure 4.53, the lower sheaves are attached to the cylinder, the upper sheaves are attached to the rod, and the piston rod applies a force tending to separate the upper and lower set of sheaves. This separating force determines the tension in the line reeved on the sheaves. Tension is maintained by pressure transmitted to the piston face by oil that is pressurised by the air in the hydropneumatic accumulator. To increase the tension, air is added to the reservoir through a line from a compressor and to decrease the tension, air is vented from the accumulator.

Oil is used on both sides of the piston for lubrication and corrosion inhibition. All systems use hydraulic dampening as a safety device to keep the

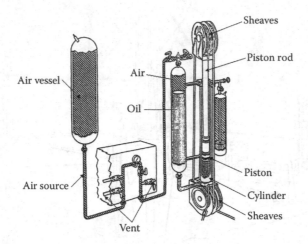

FIGURE 4.53
Components of a riser tensioner system.

rod from being shot from the cylinder if the line breaks. Reeving varies from four to eight turns, and decreases the stroke length relative to the heave. In field operations, four to eight tensioners may be used for riser tensioning.

Tie-Back Riser

Drilling with a surface BOP and high-pressure tie-back riser is a reversal of the standard techniques used in drilling from a floating drilling vessel. With this arrangement, a smaller, less complex BOP at the surface can be maintained and operated more easily. The high-pressure drilling riser assembly is essentially an extension of the BOP, designed to withstand the maximum pressure and stresses anticipated during drilling of the well under all weather conditions. The tie-back riser assembly consists of the lower riser package, the intermediate drilling riser joints and the top section. Figure 4.54 shows this assembly which is being used with a TLP. In the case of a jack-up platform, the tie-back riser may be simpler than the one described here although there also it has to withstand the forces caused by waves and currents.

The lower riser package is run at the bottom of the string and locks the riser to the subsea wellhead. It permits angular deflection of the riser with respect to the subsea wellhead while used with a TLP. The higher working pressure connector with metal-to-metal seals provides pressure integrity to the subsea wellhead. It will withstand riser tension and transmit bending moments without overstress or fatigue. The riser joints are stronger in both tension and bending than the pipe body. Each joint contains three integral hydraulic lines for control of the connector in the lower riser package (lock, primary and secondary release). The riser pipes are 50 ft long, and pup joints space out as required.

The top section of the drilling riser system includes the upper flex joint, drilling riser tensioner, tensioner spool and BOP stack. The upper flex joint

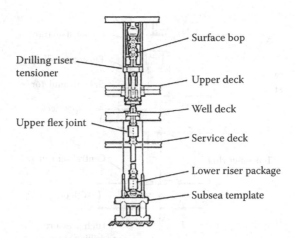

FIGURE 4.54
Tie-back riser assembly. (Courtesy of GE Oil & Gas.)

is located just below the TLP deck levels and allows larger riser angles than could otherwise be accommodated within the TLP deck arrangements. Bearing points on the lower structural level of the TLP decks are required to maintain the upper portion of the drilling riser in a vertical position.

Maintaining proper tension in the drilling riser is important to prevent buckling and to ensure a long fatigue life. Riser deflections due to winds, waves, currents and platform offset require that the tensioning system allow changes in the apparent length of the riser relative to the TLP decks. The tensioner system supports the surface BOP from a point below while providing the 1,00,000-lb working tension anticipated for the high-pressure drilling riser. The tensioner module shown in Figure 4.55 is a hydropneumatic system consisting of support beam structure, hydraulic cylinder, accumulators, air bottles, interconnecting piping, control console, air supply and compressors with dryers.

The drilling riser tensioner is designed to allow passage of the lower wellhead connector and high-pressure flex joints so that the riser may be run through the drilling riser tensioner, permitting a straightforward, simple running procedure.

4.2.2.3 Guidance Systems

At the start of drilling, it so happens that before lowering the riser other subsea units need to be lowered for which some sort of guidance is necessary without which positioning or landing heavy equipment exactly above the well becomes almost an impossible task. It is like threading a needle. So, for this, two types of methods are used.

The first method uses wire ropes as *guidelines* to maintain mechanical communication between the vessel and the wellhead, and to guide the equipment into position for landing or entry into the well.

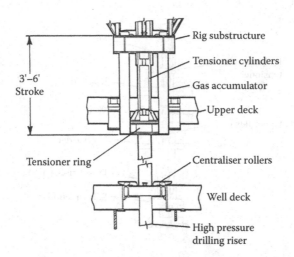

FIGURE 4.55
Tensioner module in a tie-back riser system. (Courtesy of GE Oil & Gas.)

Re-entry is the term for the second method. It commences without mechanical communications between the wellhead and vessel, using mechanical guidance and alignment only in the final stages of landing the equipment. The re-entry technique replaces guidelines on some of the dynamically positioned vessels.

Guidelines are used to lower all equipment from the rig deck to the ocean floor prior to installation of the marine riser. Usually four lines are employed which at their bottom ends are connected to the four corners of the temporary guide base set at the sea bed. These are affected by exactly the same disturbing forces which act on the marine riser. Consequently, guideline tensioners are necessary. Makers of tensioning systems use the same engineering principles for both, the guideline tensioners having less dynamic load capacity. Both are operated from the same control panel. After the marine riser is in place, guidelines can be pretensioned and attached to the outer barrel of the telescopic joint. Television cameras for use in subsea inspections are run on these lines.

Re-entry is used for remotely re-establishing guidelines and on dynamically positioned vessels that do not deploy guidelines. In the latter case, re-entry is used for running drilling equipment into the well before the riser is run, for landing the BOPs and for reconnecting the riser when necessary.

Thus, re-entry means re-establishing broken communications with a subsea well even when nothing has been entered.

Three tools used for re-entry are

1. Television
2. Acoustic devices for active targets
3. Acoustic devices for passive targets

Television is very important and should be used with the sonic devices. Subsea vision is limited, and acoustic devices are needed for position determination until the string is brought within visual range of the well.

Acoustic devices have the advantage of being able to 'see' 500 ft or more through even muddy water; however, the signature (picture) seen on acoustic signatures depends on sound reflection or absorption and may or may not exhibit the same picture as a visual observation. Resolution, being able to distinguish objects that are close together, is a minor problem, but is inherent in sonics.

Acoustic devices employing active targets are usually included with dynamic positioning equipment. The targets are acoustic beacons on the well and on the re-entry string. Position of the wellhead and the re-entry string is determined by the position referencing equipment on the vessel. These positions are displayed relative to each other on a scope.

The acoustic devices commonly used to detect inactive targets are scanning sonar systems that operate like radar. Acoustic energy is emitted from a rotating scanning head. Objects in the energy path reflect the sound back to the scanner and relative positions of the objects are displayed on a scope. Since this system sees the well along with other objects below, a specific signature for the well should be designed using reflectors in specific configurations. For final re-entry, a TV camera is still recommended.

4.2.3 Subsea Drilling Unit

The subsea drilling unit consists of a base that sits at the sea bed and further guides the other subsea units to be lowered accurately on it. It is known as a *guide base*. When more than one well is to be drilled from the same platform and to be placed adjacent to other ones, then a template is lowered which is known as a *drilling template*. The other common subsea drilling units are *subsea wellhead* and *subsea BOP*. Description of all these units has been made below.

4.2.3.1 Guide Base and Drilling Template

There are two types of guide bases that are in use: (1) temporary guide base (TGB) and (2) permanent guide base (PGB).

1. Temporary Guide Base (TGB)/Mud Mat

To get the well spudded, a heavy steel template to guide the bit to the right spot on the ocean floor is necessary. A temporary guide base (TGB) serves as a foundation for all other sub-sea equipment and as an anchorage for the guidelines on which the equipment will be run down to the sea bed. The TGB although is synonymous with mud mat but in their construction, they are a little different which is obvious from Figures 4.64 and 4.65 and especially when the mud mat is the expandable type.

FIGURE 4.56
Temporary guide base (TGB). (Courtesy of GE Oil & Gas.)

The TGB (Figure 4.56) or mud mat is a circular, octagonal, or square flat steel frame of about 400 sq. ft. in area that has compartments in which ballast materials can be placed. It is designed to be installed through the moonpool of a drilling vessel.

The expandable mud mat (Figure 4.57) has four expandable wings to be opened by remotely operated vehicle (ROV) once it enters the water. With the wings folded, the mud mat has dimensions of 20 ft × 20 ft. Each of the wings is 18 ft × 4.2 ft giving a total soil bearing area of approximately 625 sq. ft when opened.

The unit alone weighs about 4–8 tons, but is heavily weighted with bags of cement, barite, or other heavy materials before being lowered to the seabed on the end of a string of drill pipe. A special running tool for lowering and releasing the TGB is fitted to the drill pipe string, and this connects with a slot in the steel guide base frame. On the underside of the frame four spikes or stabbing stakes project to dig into the seabed and firmly anchor the unit.

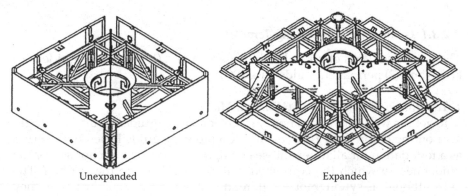

Unexpanded   Expanded

FIGURE 4.57
Mudmat – expandable. (Courtesy of GE Oil & Gas.)

Four wires or guidelines are attached with the edges of the TGB and when it has been landed these are tensioned up and used for guiding other items of equipment down to their locations above the TGB. There are also two smaller lines for running TV cameras down for monitoring operations from the doghouse on the drill floor.

In the centre of the TGB frame is a wide circular aperture with a funnel shape projecting above it into which the bottom of another frame – the permanent guide base – will fit. All subsequent down hole operations will be conducted through this aperture.

To drill the first hole for running the conductor casing of 30 in. size, a 36-in. bit is guided on the guideline by retrievable guide arms (Figure 4.58) that are released by the parting of shear pins or shear screws after the bit has entered the circular aperture in the TGB. The guide arm is later retrieved to the surface by attached tugger lines.

2. Permanent Guide Base (PGB)

The PGB (Figure 4.59) is another heavy, steel frame, about 3 tons in weight and square in shape, that has a wide central aperture and a tall post in each corner through which the four guidelines run.

The PGB serves as a landing seat for the wellhead and as a guide for drilling tools and the BOP stack, which is eventually located above the wellhead. The posts are used to locate the BOP stack, which has arrangements to accept them within its own frame.

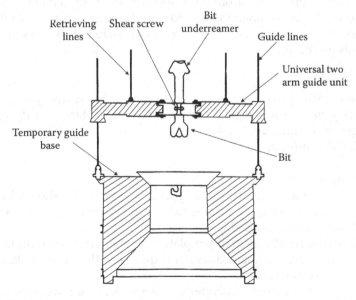

FIGURE 4.58
Retrievable guide arm above TGB.

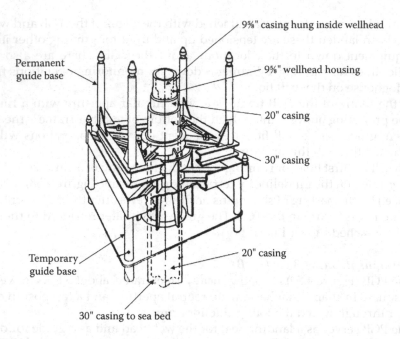

FIGURE 4.59
Permanent guide base (PGB). (Courtesy of GE Oil & Gas.)

The PGB is run down the guidelines to connect with the TGB and there is a funnel-shaped projection around the aperture on its underside that inserts into the TGB's funnel-shaped top aperture and ensures an accurate fit.

The conductor casing (30 in. diameter) is connected at the bottom of PGB which rests on the TGB.

Drilling Template

The subsea drilling templates are of three types. These are *spacer template, unitised template* and *modular template*. Each of these template designs has advantages in certain applications and limitations as well under certain other drilling conditions.

Spacer Template

The spacer template is the simplest type of template. As shown in Figure 4.60, each well slot on the template is topped by a funnel on which a retrievable guide structure is landed.

Since this is normally a small template, and the wellhead receptacles are gimballed, the spacer template does not require levelling if the bottom slope is less than three degrees.

Spacer templates are recommended for use with six or fewer wells and are designed to accept standard 6-ft radius guideline drilling equipment and

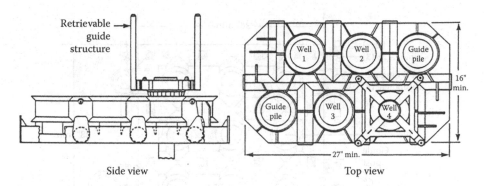

Side view   Top view

FIGURE 4.60
Four-well spacer template. (Courtesy of GE Oil & Gas.)

BOP stacks. Spacer templates are also small enough to be lowered through most moonpools or spider deck openings, without the necessity of keelhauling the template.

Spacer templates can also be used from a jack-up with mudline suspension equipment.

Unitised Template

The unitised template is generally recommended for use with six or more wells. It is used when an operator has a firm idea of the number of wells that will be drilled in a certain location.

Unitised templates are fabricated from large tubular members and incorporate a receptacle for each well and a 3- or 4-point levelling system. Drilling equipment guidance is achieved using integral guide posts or retrievable guide structures.

The basic components of unitised templates are

1. The *basic template structure*, which is made of welded tubular members
2. The *pile leveling receptacles*, which receive the pile guide housings with slips inside for template levelling
3. The *wellhead receptacles*, which receive the 30-in. wellhead housings
4. *Cantilever bumper pile modules*, used for locating and drilling the jacket bumper or guide piles
5. *Replaceable guide posts*, mounted on the template in guide post receptacles. A four-post retrievable guide structure, mounted on the 30″ housing may also be used to guide the drilling equipment.

Figure 4.61 shows a nine-well unitised template with a 3-point levelling system and integral guide posts. The unitised template offers extreme water depth capability and currently up to 32-well capacity.

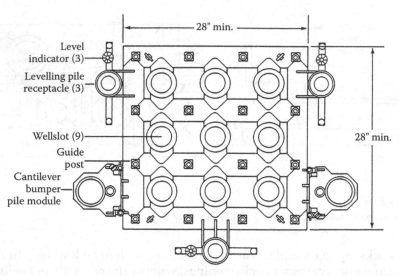

FIGURE 4.61
Nine-well unitised template. (Courtesy of GE Oil & Gas.)

Modular Template

A modular template system is recommended for guideline drilling systems and for use with drilling programs where flexibility is a requirement. The modular template system employs a template structure that is smaller than the unitised template and is made up of several interlocking modules. Figure 4.62 shows a 2-well modular template.

Modular templates are generally chosen for use when the number of wells to be drilled has not been firmly established when drilling begins. Use of modular templates requires a lower capital investment to determine reservoir characteristics while providing the capability to expand the system, as desired.

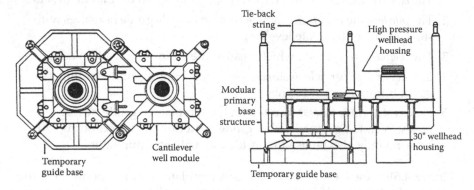

FIGURE 4.62
Two-well modular template. (Courtesy of GE Oil & Gas.)

Referring to Figure 4.62, it becomes obvious that the bottom part of the modular template is much like a TGB above which the modular primary base structure is placed and the cantilever well modules are attached to this primary base structure.

A modular primary base structure can also take the place of a standard PGB allowing the operator to index additional well slots, enabling production from a subsea production system. This is possible because the modular primary base structure has been designed with enough flexibility to accept a retrievable plumbing structure for subsea production or other peripheral structures required for a subsea system.

There can be two types of base structures which function as the foundation for additional cantilever modules. One type, the modular primary base structure, is a single-well structure designed with hang-off preparations for the addition of cantilever well modules.

The second type of base structure has provisions for three wells. This structure is installed on a gimbaled base and is then used as the foundation for additional cantilever modules.

Both types of base structures can also accommodate cantilever pile modules, in addition to cantilever well modules.

The modular template system gives the operator the option of selecting either a tie-back or subsea production development program after the exploratory well has been drilled. In addition to well modules and cantilever pile modules used in a tie-back system, the template base structure can accommodate modules for subsea production. A flowline module is cantilevered from the base structure for a single-well subsea production system or a combination of well and plumbing modules is used for a multiwell system.

The modules are designed to be run through the moonpool of a drilling vessel. After fabrication and nondestructive and functional testing, the modules are loaded onto a work-boat for transportation to the drilling location.

The components of a modular template system are installed through the moonpool of the drilling vessel, without assistance from cranes or barges. Modules are run individually, beginning with the temporary guide base.

The temporary guide base is run on a drill pipe, using a J-type running tool. Figure 4.63 shows the temporary guide base and J-type running tool. The temporary guide base establishes guidelines to assist in running the modules, and acts as a receptacle for the primary base structure and 30" conductor casing. Four 18" stabbing stakes and a 100-sq.-ft load-bearing area stabilise the temporary guide base on the ocean floor.

When the temporary guide base has landed, the running tool is released by slacking off the weight of the drill pipe and turning the J-type running tool one-eighth turn to the right.

The modular primary base structure is run next, assembled with the 30" conductor housing. The primary base structure is similar to the PGB. It has alignment and latching devices to attach the modules that will be cantilevered from it.

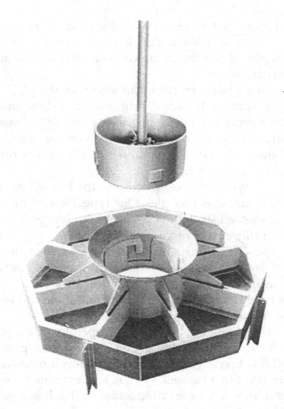

FIGURE 4.63
J-type running tool for lowering TGB. (Courtesy of GE Oil & Gas.)

After the first well has been drilled, up to eight cantilever well modules can be added, using the primary base structure as the foundation. Each well conductor hole is drilled before the next module is added. Figure 4.64 illustrates one possible configuration in which the cantilever well modules can be attached to the primary base structure. Modules accommodate additional wells and jacket location piles for platform alignment.

4.2.3.2 Subsea Wellhead

Wellhead is an equipment attached to the top of the tubular materials used in a well to support the strings (i.e. casings and tubing), provide seals between the strings, and control production from the well.

In the case of a surface wellhead used either on land or at the deck of a platform, it comprises casing heads, casing hangers and tubing head, and tubing hanger along with sealing elements. Figure 4.65 shows a typical surface wellhead assembly consisting of different casing heads joined one above the other and the casing hangers seat in the bowl of the casing head to hang or suspend the next smaller casing string securely and provide a seal between the suspended casing

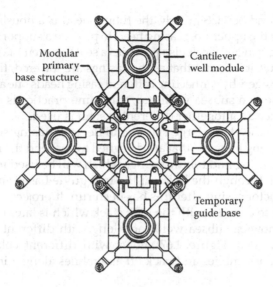

FIGURE 4.64
Five-well modular template system. (Courtesy of GE Oil & Gas.)

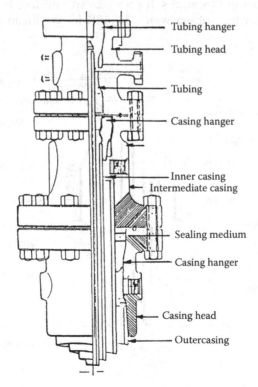

FIGURE 4.65
Surface wellhead assembly.

and the casing head bowl. Similarly, the tubing head is a housing, attached to the top flange on the uppermost casing head to provide a support for the tubing through a tubing hanger which also provides a seal between the tubing and the tubing head. Thus, it is found that as the casings are lowered, they are landed from different hangers by connecting different casing heads one above the other.

But in the case of a subsea wellhead, this same practice is not practically feasible and hence an altogether different system is used.

The subsea wellhead is a large cylindrical device housing several internal fittings called 'casing hangers' that are designed to suspend the required number and sizes of casing and tubing strings which will be used in the well. The wellhead locates through the holes in the two guide bases and fits into the top of the conductor casing after this has been run. It projects above the PGB and is designed to connect with the BOP stack which is later run above it.

Figure 4.66 shows a subsea wellhead along with different casings set at various depths. Also, Figure 4.67 shows with different colours different housing, hanger assemblies and pack off assemblies along with the PGB.

Wellhead Housing

The subsea wellhead housing (typically 18¾ in.) is, effectively, a unitised wellhead with no annulus access. It provides an interface between the subsea BOP stack and the subsea well. The 18¾ in. wellhead will house and

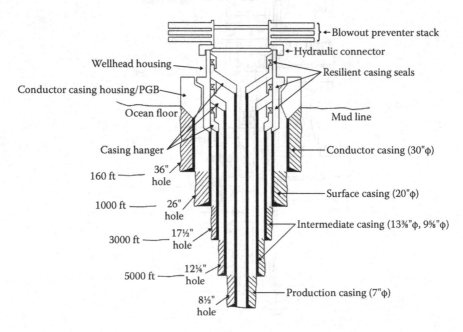

FIGURE 4.66
Subsea wellhead.

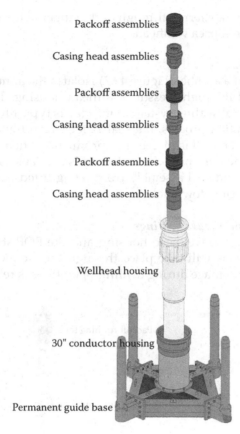

Packoff assemblies

Casing head assemblies

Packoff assemblies

Casing head assemblies

Packoff assemblies

Casing head assemblies

Wellhead housing

30" conductor housing

Permanent guide base

FIGURE 4.67
Subsea wellhead housing with hanger assemblies, pack off assemblies and PGB. (Courtesy of FMC.)

support each casing string by way of a mandrel-type casing hanger. The ID of the 18¾ in. wellhead provides a metal-to-metal sealing surface for the seal assembly, when it is energised around the casing hanger. The wellhead provides a primary landing shoulder in the bottom ID area to support the combined casing loads, and will typically accommodate two or three casing hangers and a tubing hanger. The minimum ID of the wellhead is designed to let a 17½ in. drilling bit pass through.

Casing Hangers
All subsea casing hangers are mandrel type, as shown in Figure 4.67. The casing hanger provides a metal-to-metal sealing area for a seal assembly to seal off the annulus between the casing hanger and the wellhead. The casing weight is transferred into the wellhead by means of the casing hanger/ wellhead landing shoulder. Each casing hanger stacks on top of another, and

all casing loads are transferred through each hanger to the landing shoulder at the bottom of the subsea wellhead.

Pack Off Assembly

The seal or pack off assembly (Figure 4.67) isolates the annulus between the casing hanger and the high-pressure wellhead housing. The seal incorporates a metal-to-metal sealing system that today is typically weight-set type. During the installation process, the seal is locked to the casing hanger to keep it in place. If the well is placed into production, then an option to lock down the seal to the high-pressure wellhead is available. This is to prevent the casing hanger and seal assembly from being lifted because of thermal expansion of the casing down hole.

Bore Protectors and Wear Bushings

Once the high-pressure wellhead housing and the BOP stack are installed, all drilling operations will take place through the wellhead housing. The risk of mechanical damage during drilling operations is relatively high, and

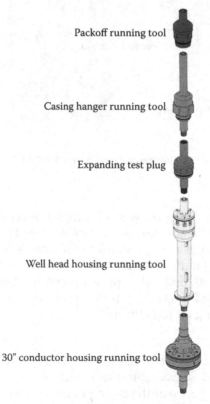

Packoff running tool

Casing hanger running tool

Expanding test plug

Well head housing running tool

30" conductor housing running tool

FIGURE 4.68
Subsea wellhead running, retrieving and testing tool. (Courtesy of FMC.)

the critical landing and sealing areas in the wellhead system need to be protected with a removable bore protector and wear bushings.

Running and Test Tools

The standard subsea wellhead system will include typical running, retrieving, testing and reinstallation tools which have been shown in Figure 4.68.

4.2.3.3 Subsea Blowout Preventer (BOP)

The blowout preventer used in the case of onshore drilling has been discussed in Sections 4.1.1.1 and 4.1.1.3. Here the main discussion is on subsea BOP, that is, the BOP that rests at the sea bottom with seawater all around it.

Figure 4.69 shows a typical BOP stack with the major components normally used in a typical BOP stack. Though the size and pressure ratings may vary, all major BOP stacks in use for subsea drilling today contain the major components shown in Figure 4.70. These major components are: (1) hydraulic wellhead connectors; (2) ram type BOPs; (3) annular type BOPs; (4) hydraulic or electrohydraulic control system for actuation of all the components on the BOP stack; and (5) a four-post guide frame used to guide the BOP stack down the guidelines to the wellhead landing base and line up the BOP stack with the wellhead for proper seating.

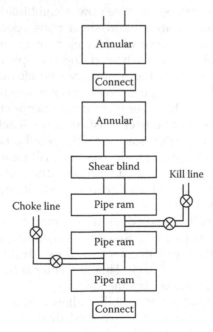

FIGURE 4.69
Subsea blowout preventer stack.

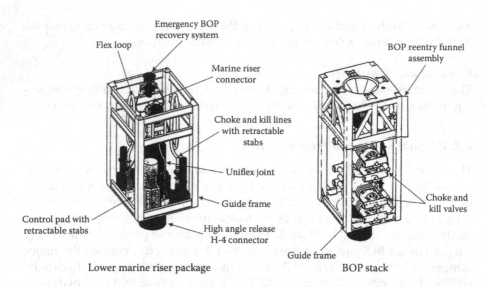

FIGURE 4.70
Subsea BOP stack with lower marine riser package.

Ram BOPs contain pipe rams sized to fit the drill pipe being used. In some cases, rams are installed to fit the various size casing strings being landed through the BOP stack. At least one chamber in the ram BOPs will contain blind rams or, as in common practice now, a combination blind and shear ram assembly. The combination blind/shear rams operate under standard conditions as blind rams to close off the well bore at any time that the drill pipe or tools are removed from the hole giving a complete closure of the hole.

However, in an emergency these rams can be closed effectively on the drill pipe severing the drill pipe in an emergency when it is not feasible to remove the pipe from the hole. For instance, if anchor chains or lines break while drilling and the marine riser must be immediately removed to keep from breaking off from the BOP stack, or if the possibility of a blowout going uncontrolled through the drill pipe exists, the drill pipe must be severed to stop the blowout at the BOP stack rather than at the surface.

The annular BOP, however, is designed to close off any tools that may be in the hole. In most cases, it is a drill pipe, but may include casing, tool joints on a drill pipe, drill collars and so on. In an emergency, the element of the annular BOP can be completely closed on the open hole in the same manner as the blind rams in the ram preventers. In subsea drilling, the annular BOP is really the heart of the BOP stack. This preventer is the most versatile and most used preventer in subsea drilling.

Unlike drilling on land where visibility allows the location of the kelly or the drill pipe and tool joints to be determined, drilling through a subsea BOP stack does not allow the location of the tool joints to be determined in relation to the BOP stack. It is not feasible to close the pipe rams and ram preventer without knowing the location of the tool joints. If an attempt is made to close the pipe rams and the rams close on a tool joint, the BOP could possibly be damaged. Damage could involve the rams and/or the tool joint to the extent that it may part, dropping the drill string to an emergency situation.

This leaves only one alternative. The annular preventer should be closed first in all cases when necessary to close the BOP on the pipe. Since the annular BOP can close on both tool joints and the drill pipe, there is no danger of damaging the BOP or the tool joint itself. Therefore, the annular preventer is used both to close off the hold and to locate tool joints. The forces required to strip a tool joint through the annular BOP are much greater than the forces required to strip the drill pipe up or down through the element of the annular BOP.

The drill pipe can then be raised and, with a noticeable change in the weight of the drill string, will indicate a tool joint passing through the element of the annular BOP. As the tool joint is pulled up through the element of the annular BOP, it is positively located. Then it is possible to safely close the ram BOP on the drill pipe, not on the tool joint.

Since a floating vessel is being used, the drill pipe will be in continual action at all times with the heave or movement up and down the drilling vessel caused by the sea. When the BOP is closed, the pipe must still be free to move up and down to follow the heave of the vessel. In the past, the stripping life of the BOP has been very minimal. Due to its inherent design features, the annular BOP has been able to strip a greater footage of pipe than the ram preventers.

When feasible, a set of pipe rams should be closed and the drill string lowered until a tool joint is supported on top of the pipe rams. This is possible only when the drill pipe is being handled in elevators on the rig – not when the kelly is attached.

Hydraulic Control Manifold

This is truly the 'heart of the system.' It is a recommended and accepted practice now to place this item in a safe, yet accessible, location on the vessel. Contained in the hydraulic manifold are pilot control valves (normally one per hydraulically actuated stack component) complete with air operators and hydraulic pressure gauges for the various pressures required on the BOP stack, complete with pressure transmitters to convert these hydraulic pressures to either air master panel or the electric driller's panel.

Also located on this item are the pilot regulators that remotely control the subsea hydraulic regulators located in the pod. These pilot regulators are remotely controlled from the driller's panel (air or electric). There are normally three regulators on each hydraulic control manifold – one for the ball joint, one for the annular preventer(s) and one for the remaining stack functions. Explosion-proof boxes containing pressure switches are included for remote panel indicating light operation, and air solenoid valves enable remote operation of the pilot control valves and pilot regulators.

Hydraulic hose bundles are used to transmit pilot signals and power fluid from the hydraulic control manifold to subsea pods.

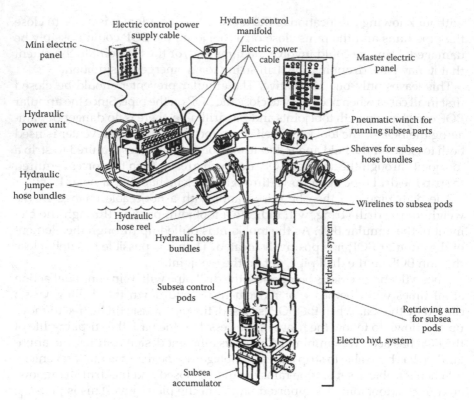

FIGURE 4.71
Subsea BOP control system.

The *subsea pod* contains the pilot-operated control valves and pilot operated regulators required to direct the hydraulic fluids to the various stack functions. Figure 4.71 shows the BOP control system in which the subsea control pod is shown with both hydraulic as well as electro-hydraulic systems.

Electro-Hydraulic Control System

The electro-hydraulic control system is similar to the hydraulic system, except that an electric signal is sent subsea to a solenoid valve which supplies hydraulic pilot pressure to the subsea control valves. One of the main advantages in the electrohydraulic system is the reduction in signal time to almost zero for any water depth. The electro-hydraulic control system costs are higher than the costs of comparable all-hydraulic systems for shallow water application, but the reverse begins to be true for water depths between 1500 and 2000 ft. Another advantage of the electro-hydraulic system is that it has more readback capabilities.

Running the BOP Stack

Running the subsea BOP stack with its choke and kill lines, hydraulic hose bundles and control pods and the top and bottom connectors of the preventer assembly is done on the marine riser.

The process of landing a BOP stack is described sequentially below:

1. A riser running tool is made up on the drill pipe, lowered through the rotary table and connected to the riser stab assembly.

2. The preventer stack is positioned on the spider beams and guidelines strung through guide funnels.

3. The riser stab assembly is made up on top of the preventer stack and the assembly is pressure tested.

4. Next the riser running tool is released and the next joint of riser is made up on top of the riser stab assembly.

5. The assembly is lifted to clear the spider beams; the spider beams are removed, and the assembly is lowered to hang off on the first joint of the riser.

6. The remaining joints of the riser pipe are added as the assembly is lowered.

7. The telescoping joint is added to the top of the riser; the riser tensioning system is connected and the BOP stack landed on the wellhead.

8. The connector is latched.

9. The diverter, bell nipple and choke and kill lines are connected at the top of the riser.

10. The preventer stack and choke and kill lines are pressure checked.

In considering the effectiveness of blowout preventers it should be noted that the protection afforded by the entire system is limited by the weakest component wellhead, fittings, valves, clamps, connectors and casing.

The preventer system must be maintained in peak operating condition at all times through periodic tests and inspections. Drillers must be trained and skilled in operating BOP equipment to utilise the full capabilities of the equipment at their disposal.

4.2.4 Heave Compensator and Bumper Sub

It has been observed in the components of the marine riser that the 'riser tensioners' are used to nullify the effect of heaving of the vessel so that the riser string always remains tensioned without any slackness which is important while the riser is under 'operational' mode. But during the same operational mode the derrick and in turn the whole drill string comes under the influence of the heaving motion and as a result the drill string and the bit which are hung from the hook of the travelling block move up and down as the vessel heaves. Naturally at one time the bit comes under a heavy compressional load and at the other time it simply hangs and rotates freely without any cutting action. Hence this up-and-down motion of the drill string needs to be compensated and this is done by the drill string motion compensators which can be (1) heave compensators and/or (2) bumper sub.

4.2.4.1 Heave Compensator

The heave compensators work on the same principle as that of riser tensioners but the difference between them lies in their tolerances to the load fluctuation. The riser tensioners tolerate a larger force fluctuation of up to 15% of the mean load whereas heave compensators strive for as small a load fluctuation as practicable. For example, if a BOP stack is to be landed with a perfect heave compensator, lowering the travelling block 1 ft relative to the rig floor should lower the stack 1 ft relative to the wellhead, regardless of the heave of the vessel. This negligible tolerance becomes necessary for operations like drilling where maintaining a constant bit weight is essential or during landing the BOP stack where a soft landing following utmost safety in rough sea conditions is very much essential.

Heave compensators have a typical stroke of 18 ft. The weight of the string pulls down, compensators push up. Weight on the bit remains constant as long as vertical motions are within the limits of the motion compensators capability. An additional benefit of heave compensation is better directional control.

Heave compensators can be either 'traveling block compensators' or 'crown block compensators' depending on their positioning.

Traveling block compensators depending on the working of the piston inside the cylinder can be either *tension type* or *compression type*. These compensators locate the operating cylinder on the traveling block, between the block and the hook. Large air reservoirs are located on the deck or below, and pressurised fluid is piped to the cylinder by flexible hoses. Dual or single pistons may be used to support the load.

The Rucker heave compensator (Figure 4.72) uses *compression type cylinder* where high-pressure air is applied on the piston face or on the blind side of a cylinder between the travelling block and hook. The low-pressure oil accumulator is used on the rod side of the cylinder for lubrication and dampening. This type uses a chain being reeved around the cylinder for absorbing lateral motion of the hook. A chain reeved in two parts of lines delivers a compensation stroke twice that of the cylinder stroke, that is, an 8-ft cylinder stroke will deliver 16 ft of compensation motion.

The Vetco heave compensator (Figure 4.73) can be a either a dual piston or single piston type but Western Gear makes a single piston type only (Figure 4.74). Both these types use a *tension type cylinder* where high pressure air is applied on the rod side of the cylinder between the traveling block and the

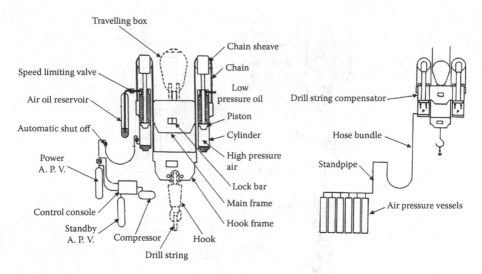

FIGURE 4.72
Rucker heave compensator – compression type.

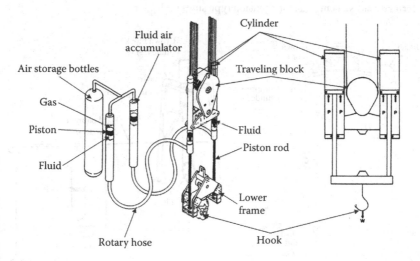

FIGURE 4.73
Vetco heave compensator – tension type double cylinder.

hook. To provide lubrication to the cylinder, an air/oil accumulator is used on the rig floor or on the derrick (mast) and oil flows through the hose loop to the cylinder.

Crown block compensators are pneumatic springs but the mechanical compensation is accomplished by taking in and paying out on the fast and the dead drilling lines. This is a *compression type* compensator where the motion is achieved by moving sheaves attached to the piston rods and idler sheaves are positioned on arms as shown in Figure 4.75.

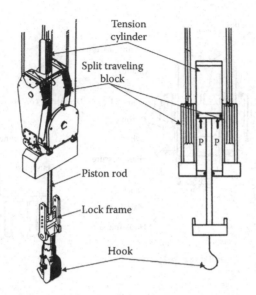

FIGURE 4.74
Western gear heave compensator – tension type single cylinder.

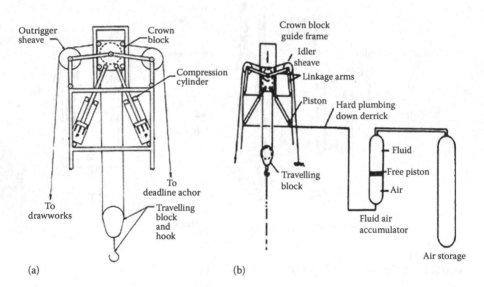

FIGURE 4.75
Crown block heave compensator – compression type.

The compression cylinders support the crown block. The crown block is mounted on a framework supported dolly atop the derrick, which allows vertical motion of the crown block. Here the use of rigid hydraulic lines instead of flexible ones would decrease the probability of failure of hydraulic lines. Since this whole equipment is massive and placed at the top of the derrick (mast), the CG of the vessel tends to rise slightly and, as a result, wind heeling moment of the vessel increases. The spreading of the idler sheaves causes additional geometric effect which needs a careful design.

4.2.4.2 Bumper Sub

A 'bumper sub' is a slip joint or telescopic joint placed in the drill string to decrease the transfer of vessel heave to the bit and transmit torque at every position in its stroke. Its 5-ft stroke is adequate for the 2- or 3-ft averages of vertical motion in 7- to 8-ft seas. Normally bumper subs fit into the drill collar string although they do not have the same ruggedness of drill collars. The reason is that these are two concentric cylinders telescoping one into the other and at the same time these are rotating and also allowing the drilling fluid to pass through them. Hence, while designing extra precaution is required as compared to the slip joint used in the marine riser. Thus, two general types of subs are available: *balanced* and *unbalanced*.

Unbalanced bumper subs are typical telescopic joints where internal pressure operates on the end of the inner cylinder and the annular pressure operates on the outside area of the outer cylinder. The result is an opening force on the sub.

Balanced bumper subs have special internal chambers and porting that equalises the areas acted on by the internal and annular pressures.

Bumper subs should be maintained and operated as per the manufacturer's specifications. Replacing seals, worn parts and lubricating grease is important. A good workmanship and maintenance are the key words in using any bumper sub.

4.3 Procedure of Offshore Drilling

It has been already mentioned that the drilling technique used in offshore is the same as that used onshore. The only difference in techniques lies in the type of platform from which the drilling is being done.

In the case of 'fixed platforms,' many wells are drilled through conductors (described before) and the wellheads and BOPs of all the individual wells are placed at the cellar deck of the platform. The 'conductor' which is normally piled into the sea bed acts as the first casing, that is, a conductor casing. After drilling through this up to a required depth, a surface casing is run and cemented. At the top of the surface casing, the casing head is fixed and then the subsequent casings are run and cemented by hanging them from different casing hangers as is the practice in onshore drilling. All the intervening operations and the drilling equipment and tools to finally reach the target depth by drilling from a larger diameter to a smaller diameter depending on the casing programme remains the same as that of onshore drilling. Hence, the difference that arises is mainly in the case of compliant platforms and mobile platforms. Here, the drilling technique used in the case of two types of mobile units, that is, jack-up and floating unit, has been separately dealt with in detail. The drilling technique from a compliant platform like TLP is more or less the same as the jack-up with little modifications like using a subsea wellhead and tie-back riser, for which the reader may look into Reference 27.

4.3.1 Drilling from a Jack-Up

As is known when a jack-up is installed at a location, it acts like a fixed platform and the drilling procedure will be the same with a BOP placed at the deck but since after completing the well it has to move out, the wells are temporarily abandoned and for this a *mud line suspension system* is required which is different from a subsea wellhead.

A *mud line suspension system* (Figure 4.76) provides a means of hanging off the various casing strings at or below the mud line while providing a method of disconnecting all casing strings. In addition, this method of supporting the casing strings reduces the load carried by the drilling platform since the weight of the casing strings is supported at the mud line. The selected concentric casing strings extend from the mud line to the drilling platform where conventional BOPs are installed.

In most cases, a mud line suspension system will utilise the standard $30'' \times 20'' \times 13\frac{3}{8}'' \times 9\frac{5}{8}'' \times 7''$ diameter offshore exploration casing program. As seen in Figure 4.76, there is a $30''$ landing ring at the top of the $30''$ casing and this conductor casing extends to the drilling structure where diverter equipment is installed for protection during drilling for surface casing. This $20''$ surface casing gets suspended on the $30''$ landing ring by the $20''$ mud line casing hanger and is extended back to the deck of the platform. All subsequent casing strings are run and suspended in a similar way, each string supported at or near the mud line by a mud line casing hanger of the appropriate size with threaded preparation for both running tool and tie-back tool.

After completion of drilling, for leaving the site, wells are temporarily suspended by stripping off each string from the casing hangers by the same running tools which were used to run each casing string. When the portion of casing strings from mud line to drilling platform are retrieved, a corrosion cap (Figure 4.77) is installed to reduce marine growth and corrosion in critical areas of the casing hangers.

Later on, depending on the type of completion, these wells can be tied-back to a fixed jacket platform by means of tie-back assemblies or alternately these can be completed at subsea.

A typical well (Figure 4.78) drilled with a jack-up is described sequentially below.

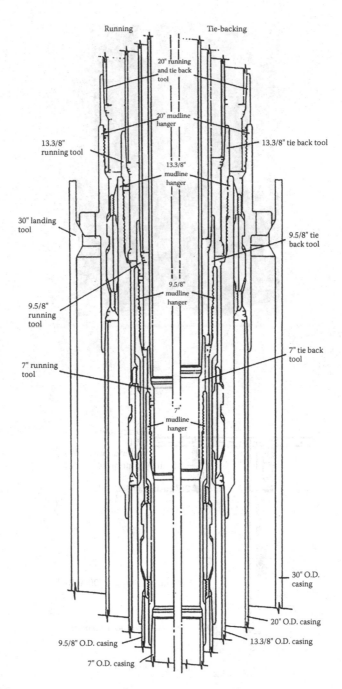

FIGURE 4.76

Mud line suspension system. (Courtesy of GE Oil & Gas.)

FIGURE 4.77
Corrosion cap.

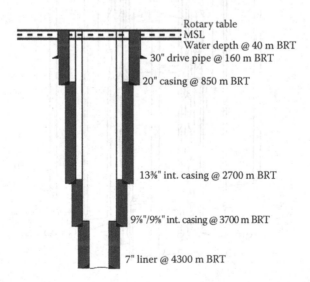

Rotary table
MSL
Water depth @ 40 m BRT
30" drive pipe @ 160 m BRT

20" casing @ 850 m BRT

13⅜" int. casing @ 2700 m BRT

9⅞"/9⅝" int. casing @ 3700 m BRT

7" liner @ 4300 m BRT

FIGURE 4.78
A typical well drilled from a jack-up.

Moving and Installing the Jack-Up

The jack-up rig is being towed from a previous location and installed at the current location – a detailed description is given in Chapter 3. The air gap for the operation is 41 ft, the distance from RT (rotary table) to MSL (mean sea level) is 30 m. All the surface equipment is erected and tested. Checking of all necessary tools, supplies and so on is done.

Driving 30" Conductor

The conductor pipe is normally driven into place with the use of a pile hammer, if the sea bed is soft. In the case of a hard bed, either jetting or drilling is to be done. In this case D-62/Modified D-55 diesel hammer is used to pile the conductor. The 30" joint containing the 30" landing ring is the third joint from the bottom of the conductor. The pipe is driven around 80 m below the mud line, leaving the mud line landing ring at 2 m above the mud line. This provides the seat for the mud line suspension (MLS) system and works as a support for the 30" MSP (medium service pressure) diverter at the deck.

Drilling 24" Hole

Picking up a 24" bit and recommended BHA (bottom hole assemblies) drilling is done up to 850 m. The hole is circulated with prehydrated bentonite mud.

Running 20" Casing and Cementing

A 20" casing is made up and run. MLS assembly is pinned up with a 20" joint and connected to an MLS running tool. From the MLS assembly to rotary table again a 20" casing is made up and run. Then landing of 20" string from casing shoe to MLS on a 30" landing ring is done. Then tension is provided at the surface (i.e., deck) equivalent to 20" casing buoyed weight from MLS to surface and the casing is set on slips in the rotary.

After this cementing of the annulus is done using conventional cementing unit so that 15 m of cement remains on the top of the float collar.

Installation of 20¾" Wellhead and 21¼" BOP

Removing the diverter from the top of the 30" casing head and adjusting the length of the 20" casing the 20¾" wellhead is installed which is comprised of two sections. The bottom section is 20¾" pin top and 20" VG lock bottom and the upper section is 20¾" box bottom and 20¾" studded top.

Then at the studded top, the 21¼" BOP with double ram, one annular preventer, bell nipple and choke and kill lines is installed and tested.

Drilling 16" Hole

A hole is drilled with a 16" bit and recommended BHA up to 2700 M. Wire line logs and MDT (modular formation dynamics tester) tools are also run.

Running and Cementing 13⅜" Casing

Similar to running a 20" casing, this is also run, in between connected to 13⅜ mud line hanger and extended to the top with the remaining string after which it is landed at the MLS. Next the cementing head is installed at the deck and the cementation of the annulus is done.

Installing 20¾" × 13⅝" Casing Spool and ⅝" BOP Stack

The same procedure as followed in onshore drilling is followed here also.

Drilling 12¼" Hole and Running and Cementing 9⅞" Casing

A 12¼" hole is drilled through primary targets, pressure transition zone and charged sands up to 3700 m and a 9⅞ casing is run and cemented by following the same procedure as mentioned before to cover two prospective sands and facilitate the penetration of the potential productive high-pressure reservoir sections. It is to be noted that during drilling LWD and mud logging trends are continuously monitored for any anticipated increase in formation pressure.

Installing 13⅝" Wellhead Casing Spool and 13⅝" BOP Stack

After the setting of cement the 'C' section (13⅝" box bottom × 13⅝" flange top) wellhead is installed and tested. Then at the top of the wellhead flange a 13⅝" double ram BOP is installed. This BOP has 3½"–5" variable rams in the lower cavity and a 5" pipe ram in the upper cavity. Besides 13⅝" annular BOP, bell nipple, choke and kill lines are also installed and tested as per the recommended standard.

Drilling 8½" Hole to Total Depth

A 8½" bit along with LWD and recommended BHA are picked up and drilling carefully the cement, float collar and shoe track, it proceeds further to reach the target depth.

Running and Cementing 7" Liner

The liner, though resemble casing, does not extend up to the MSL and is suspended from the bottom of 9⅞" casing by means of liner hangers. Then it is cemented. Logging tools are lowered to do cased hole logging.

Actual drilling up to the TD finishes at this stage and further well program will be based on the testing requirement or plugging and temporary abandonment.

4.3.2 Drilling from a Floating Unit

The real challenge in offshore drilling is faced when a floater semi or ship is used for drilling and both the wellhead and BOP stack are installed on the seabed. Hence, for better understanding and grasping the whole process of drilling, the main operations in drilling a typical offshore well are illustrated with the aid of a hypothetical exploration well programme (Figure 4.66) using a semisubmersible. Sequentially these operations are as follows:

Moving Rig onto Location and Running Anchors
Rig-moves and anchoring operations are discussed in Chapter 3.

Rigging up
This basically involves preparing the rig for drilling.

Running the Temporary Guide Base (TGB)
The temporary guide base, the heavy steel frame that will serve as the foundation for other sub-sea equipment, is first weighted up with bags of cement or barite and lowered to the sea bed on the end of a string of drill pipe. The end of the drill string is fit with a special running tool that releases the guide base when it is in position. When this has been done, the four guidelines and two TV camera lines are tensioned up from the rig and the underwater TV camera is deployed to monitor further operations from the doghouse.

Spudding in and Drilling 36" Hole to 160' BSB
After the TGB has been positioned, a 36" diameter bit called a 'hole opener' is lowered to the sea bed inside a 'utility guide frame' that runs down the four guidelines (Figure 4.58). This large bit enters the central aperture in the TGB and the guide frame automatically detaches itself. A relatively short section of hole is now drilled with the hole opener; in this particular well, drilling stops at 160' below the sea bed (BSB). Sea water is initially used as the circulation fluid through the drill string, and this returns with the drilling bit's cuttings to the sea bed where the cuttings are deposited. Every 30 ft an 'inclination survey' is made to ensure that the hole is vertical.

After the hole has been drilled it is normally filled with a mixture of water and bentonite to make a thick gel substance to prevent 'sloughing' or caving-in, and filling. The hole opener is then pulled out of the hole and back up to the rig.

Running 30" Casing and Landing the Permanent Guide Base (PGB)

A 30" conductor casing, sometimes called the 'outer conductor,' is next run into the 36" hole to prevent sloughing. With the permanent guide base installed on its top, it also serves as the landing base for the BOP which will be run later.

Before the last joint of 30" casing is run, the PGB is attached to its top, leaving about 5 ft of casing protruding above it. This will provide an anchorage for the next string of casing, which will be 20" wide. The PGB, with the casing suspended from its aperture, is lowered on a special running tool to the sea bed. The guide lines running through the four posts on the PGB guide it into position and it slots into the TGB's aperture with a funnel-shaped bottom projection that guarantees an accurate fit.

Cementing the 30" Casing

Now the casing has to be anchored to the wall of the hole. This is the first of several 'cement jobs' that will be performed during the well programme. Since this string is short compared with the 'surface', 'intermediate' and 'oil' strings, its cementing is less complex, although a good bond must still be achieved. A drill string is run into the 30" casing, with a special 'stinger string' through which cement will be pumped extending it to the 'shoe' at the bottom of the casing. Cement is pumped through this at high pressure until returns are seen by the TV camera coming out on the sea bed. The cement is then left a few hours to set (called 'waiting on cement'), after which the permanent base for the main drilling operations should have been established on the sea bed.

Drilling 26" Hole to 1000' TVD

A hole now has to be drilled for the 'surface' casing, which is 20" in diameter. This will require a hole 26" wide, and a 26" bit of a type appropriate for the nature of the formation is selected. The bit is again guided down to the top of the PGB by the utility guide frame, and once again the circulating fluid is discharged onto the sea bed, since no riser has yet been run to enable it to return to the rig. For about a day and a half the 26" bit drills to a true vertical depth (TVD) of 1000 ft, which might not equate with the measured depth (MD) if the hole deviates from the vertical at all. Later sections of the well will be made to deviate, but this particular section is not. To make sure it stays vertical, inclination surveys are made every 200 ft.

To prevent 'sloughing' (caving-in), the hole is filled with a gel-water fluid which is pumped down through the drill string. Following this, the bit is pulled back to the drilling unit.

Running and Cementing 20" Casing, and Running the 18¾" Wellhead

The shoe at the bottom of the 20" inner conductor casing is guided into the aperture in the PGB by automatically detaching arms on the utility guide frame, as with the 30" outer conductor. There are 1000' of 20" casing run, to the top end of which is attached the wellhead, a long, cylindrical device with internal fittings called casing hangers that suspend the various sizes of casing and tubing strings that will be run during the remainder of the well programme. The upper end of the wellhead, which has an internal hole diameter of 18¾" in this case, is designed to closely latch onto the 18¾" BOP stack when this is run, making a gas-tight connection. The wellhead is run with the last joint of the 20" surface casing, after which the casing is cemented into the hole through the stinger string, with cement returns again being discharged to the sea bed.

Running the 18¾" BOP Stack and the Marine Riser

When the surface casing string and wellhead have been set and cemented, the BOP stack, in this case a 15,000 psi stack with an 18¾" aperture, is run attached to the lower end of the 21" bore marine riser. This will act as a conduit for tools and for drilling fluid and cuttings returning from the well.

As the BOP stack nears the wellhead it becomes liable to damage by landing heavily or bumping, since the rig will be heaving up and down to some degree if there is the slightest sea running. As the final joints of riser are added on the drill floor, therefore, the riser tensioner wires are connected just below the slip joint, and part of the load is transferred to them. A proportion of the total load is thus taken by the rig's surface motion compensator system while the remainder is held by the riser tensioning wires, enabling the stack to be landed with a minimum of jarring.

The BOP stack is latched onto a special connector on the wellhead, which both supports it and provides a gas-tight seal. After it has been landed, all subsequent strings of tubulars and casing will run through the 18¾" apertures in both the stack and the wellhead. Once the stack is in position on the wellhead, it is hydraulically pressure-tested to ensure a good seal.

At the top of the riser, above its telescopic slip joint and underneath the drill floor a diverter is fitted. This is a large housing in which an emergency sealing device can shut off the vertical 21" wide access to the drill floor and divert well fluids to either a narrow flow line outlet or a vent line outlet. The vertical access can be sealed around the kelly, drill pipe or casing that happens to be running through it when it is closed, so it is really a form of annular blow-out preventer.

Drilling fluid returning from the well normally passes out through the flowline to the shale shaker, and this is the same line through which mud is delivered from the fill-up line when the drill string is being stripped out of the hole.

Drilling 17½" Hole to 3000' TVD (3219' MD)

The cement-filled shoe of the 20" casing is next drilled out with a 17½" bit, after which a 17½" hole is drilled to a vertical depth, in this well's case, of 3000'. However, not the entire distance is drilled vertically. At 1800 ft TVD there is a 'kick-off point' where the drill bit will be deviated by directional drilling techniques. By the time the bit arrives at a level 3000' directly below the sea bed it will actually have travelled 3219' along the hole. This is termed the 'measured depth' (MD).

From the kick-off point drilling is done by means of a bit rotated by a downhole drilling motor or 'mud motor' and a device called an MWD tool is incorporated in the drill string.

Logging

When 3000' TVD has been reached, the well is logged with electric and sonic wireline logging devices to determine conditions in the hole before running casing. Before logs can be run, however, mud must be circulated through the well to remove any cuttings, and to allow the mud to be 'conditioned.'

Running and Cementing 13⅜" Casing

About 3000 ft of 13⅜" 'intermediate' casing is now run and cemented to seal the wall of the hole thus far drilled. A powerful cement mixing and pumping plant onboard, called the 'cement unit,' plant consists basically of a large mixing hopper and water pipes leading to a powerful pump which delivers the mixed cement slurry to a cementing head on the drill floor where it enters the casing. The amount of slurry required is very carefully calculated from a knowledge of the width and depth of the hole and the size of the casing, and various additives are mixed with it to a precise formulation that is monitored by sophisticated equipment.

Drilling 12¼" Hole

The cement shoe of the 13⅜" casing is now drilled out with a 12¼" bit and directional drilling continues with the mud motor for approximately 10 days to a depth of 5000'.

Logging

When 5000 feet has been reached, mud is circulated down the hole to clean out any cuttings, and the mud is conditioned to restore its quality. Electric logs are then run, as in the 17½" hole.

Running and Cementing 9⅝" Casing

This is done as with the 13⅜" casing, with a pressure test following the cementing. This casing string is the second intermediate string.

Drilling 8½" Hole to Total Depth

An 8½" bit is now used to drill to total depth (TD). MWD devices are used every 500' during this final phase of drilling to determine the accuracy of the bit's travel as it nears its target. This involves only a brief pause in drilling, when sensors in the instrument cause a transmitter to send pulses up through the mud to the rig, giving a digital readout of angle and azimuth in the doghouse. When TD is reached, the 'pay zone' or zones should have been penetrated, but there is not a 'gusher' like there would have been in the old cable tool wells. The well is still controlled by the weight of the drilling mud, and any formation fluids present in the pay zones will not be able to flow until special casing called a 'liner' has been set through the zones and has been 'perforated' to allow access for the fluids to the rig. But this liner will not necessarily be run unless the results of a 'coring' programme yield encouraging results.

Coring

A 'core barrel' with a 'coring bit' at its lower end is now fit to the bottom of the drill string and approximately 600 ft of core samples are removed from the formation at a rate of 6–8 ft an hour. These will be sent ashore for laboratory analysis and will largely determine if the test programme will be run.

Logging

Following coring the well is again cleaned out by circulating and there is a 3-day period of well logging of various kinds to determine hole and formation conditions.

Running and Cementing the 7" Liner

Core analysis having yielded 'shows' of hydrocarbons in the pay zone formations, the operator has decided to run a 7-in. liner in which test equipment can be installed to allow well fluids to flow to surface under control. This liner looks much like any other type of casing, but it does not run all the way from the well bottom to the sea bed like ordinary casing. Instead it is suspended from the bottom of the deepest string of casing run (the 9⅝" in this case) by means of a liner 'hanger.' The liner is cemented at the same time as it is run, and a packer or plug is set at its top to isolate the test zones inside it from the cased hole above. If a string of narrow tubing is now run down through this packer with special valves that allow the controlled entry of well fluids, it will be possible to channel pressurised well fluids to the surface under control. But to allow the fluids access to the inside of the liner, it has to be perforated. This is one of the operations in the period of 'well testing' and later 'well completion.'

Installing 'Corrosion Cap'

After installing the liner or the production casing, the 'corrosion cap' (Figure 4.77) is placed over the BOP and later on the special crew for 'well testing' and 'well completion' comes and puts the well on production.

Bibliography

A Primer of Oil Well Drilling, Petroleum Extension Service and American Association of Oil Well Drilling Contractors, 1958.

Austin, E. H. *Drilling Engineering Handbook*, International Human Resources Development Corporation, 1983.

Bais, P. S. *Drilling Operations Manual*, Inst. of Drilling Technology, ONGC, India, 1984

Bennion, D. B., B. Lunan, and J. Saponja, Underbalanced drilling and completion operations to minimize formation damage – reservoir screening criteria for optimum application, *The Journal of Canadian Petroleum Technology*, Sep., 1998, 37(9), 36–50.

Bern, P. A., D. Hosie, and R. K. Bansal, *A New Downhole Tool for ECD Reduction*, IADC/ SPE 81642, 2003.

Bern, P. A., W. K. Armagost, and R. K. Bansal, *Managed Pressure Drilling with the ECD Reduction Tool*, SPE 89737, 2004.

Conductor guide centralizers, 2015, http://claxtonengineering.com/products-services /asset-life-extension/conductor-guide.

Conductor guide system for offshore drilling platform, 2015, http://www.google.co .in/patents/US4561803.

Devereux, S. *Drilling Technology in Nontechnical Language*, Penn Well, 1999.

Drilling riser, 2015, http://en.wikipedia.org/wiki/Drilling_riser.

Drilling through salt, Technical Newsletter, SCOR Global P & C, May 2013.

Moore, D. *Pioneering the Global Subsalt/Presalt Play: The World beyond Mahogany*, Field, AAPG Annual Conference and Exhibition, New Orleans, LA, April 11–14, 2010.

Encyclopedia of Hydrocarbons, Vol 1/Exploration, Production and Transport.

ETA Offshore Seminars Inc, the Technology of Offshore Drilling, Completion and Production, Pennwell Publishing Company, 1976.

Farmer, P., D. Miller, A. Pieprzak, J. Rutledge, and R. Woods, *Exploring the Subsalt*, *Oilfield Review*, Spring 1996, 50–55.

Frick, T. C. and R. William Taylor, *Petroleum Production Handbook*, Vol. I, SPE of AIME, 1962.

Gatlin, C. *Petroleum Engineering – Drilling and Well Completions*, Prentice-Hall Inc., 1960.

Gomes, J. S. and F. B. Alves, *The Universe of the Oil & Gas Industry – From Exploration to Refining*, Partex Oil and Gas, 2013.

Horizontal drilling & directional drilling: Natural gas wells, 2015. http://geology .com/articles/horizontal-drilling.

Leffler, W. L., R. Pattarozzi, and G. Sterling, *Deepwater Petroleum Exploration & Production – A Nontechnical Guide*, Pennwell, 2011.

Le Blanc, L. *Drilling, Completion, Workover Challenges in Subsalt Formations*, *Offshore*, June 1994 (Part I), July 1994 (Part II) and August 1994 (Part III).

Lynch, P. *A Primer in Drilling & Production Equipment*, Vol. 2 and 3, Gulf Publishing Company, 1981.

Maclachlan, M. *Marine Drilling*, Houlder Offshore Limited. Dayton's Publishing.

Maurice, B. D., V. Maury, F. Sanfilipo, and F. J. Santarelli, *Drilling through Salt: Constitutive Behaviour and Drilling Strategies*, Univ. of Waterloo, Ontario, Canada, 2005.

McCray, A. W. and F. W. Cole, *Oil Well Drilling Technology*, Univ. of Oklahoma Publishing Division, 1960.

PDC drill bits, 2015, http://petrowiki.org/PDC_drill_bits.

Rehm, B., J. Schubert, A. Haghshenas, A. S. Paknejad, and J. Hughes, *Managed Pressure Drilling*, Gulf Publishing Company, 2008.

Sereda, N. G. and E. M. Solovyov, *Drilling of Oil and Gas Wells*, Mir Publishers, 1977.

Sheffield, R. *Floating Drilling – Equipment and its Use*, Gulf Publishing Company, 1982.

Stewart, H. *Drilling and Producing Offshore*, Pennwell Publishing Company, 1983.

VETCO Templates and Tie-Back Systems, 1984.

Weijermars, R., M. P. A. Jackson, and A. Van Harmelen, *Closure of open wellbores in creeping salt sheets*, Geophysical Journal International, 2014, 196, 279–290.

What is horizontal drilling and how does it differ from vertical drilling? 2015, http://energy.wilkes.edu/pages/158.asp.

5

Offshore Well Completion

Well completion is a process of making a well ready for production. In other words, once after drilling a well a decision is taken to move ahead with developing it, the completion operation is undertaken. It incorporates the steps taken to transform a drilled well into a producing one.

5.1 Basics of Well Completion

Completion of a well is the most important phase of a well's life and irrespective of whether it is an offshore or onshore well, its basics remain the same. The production technique future workover possibilities, well productivity, downhole problems and so on all depend on how the well is completed. Generally speaking, immediately after the cementation of the production casing and liner, the completion process starts. First, the well is emptied of drilling mud by displacing it with a completion fluid composed of brine (i.e. water plus sodium chloride or calcium chloride or zinc bromide). This brine performs the same pressure control function as drilling mud while allowing completion work to be done in the wellbore. The displacement of drilling mud with brine takes place through a tubing workstring being run through the wellbore. After the complete displacement, an operation called perforation of the casing/liner is carried out. The next operation is gravel packing to stop the production of sand. Then the workstring is replaced by the completion string comprising mainly tubing along with packer, safety devices, circulation and communication devices and accessories. After setting the packer, the completion fluid is displaced with a permanent, corrosion-resistant fluid. Once the job is done up to this, the surface equipment, that is, the BOP is removed and in its place the Christmas tree is installed over the wellhead and the flowlines are connected to it.

Thus, broadly, in any well completion technique either for production or injection well, three major aspects are involved: (1) well completion equipment, both surface as well as subsurface, (2) completion-specific operations and (3) well completion design.

5.1.1 Well Completion Equipment

Well completion equipment is broadly divided under two headings: (1) subsurface completion equipment and (2) surface completion equipment. The equipment connected from below the wellhead is known as subsurface equipment and that above it is known as surface equipment.

5.1.1.1 Subsurface Completion Equipment

Figure 5.1 represents a typical single zone completion showing subsurface completion equipment which includes starting from the bottom of the well upward wireline entry guide (WEG), perforated joint, packer, landing nipple, polished nipple, flow coupling/blast joint, sliding sleeve, side pocket mandrel and downhole safety valve.

Wireline entry guide is a small fitting on the end of a tubing string which is intended to make pulling out wireline tools easier by offering a guiding

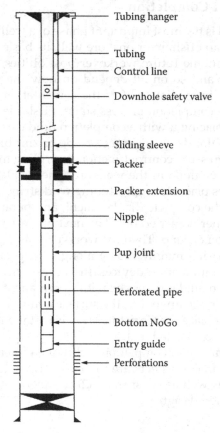

Tubing hanger

Tubing

Control line

Downhole safety valve

Sliding sleeve

Packer

Packer extension

Nipple

Pup joint

Perforated pipe

Bottom NoGo

Entry guide

Perforations

FIGURE 5.1
Subsurface completion equipment.

surface for the toolstring to re-enter the tubing without getting caught on the side of the shoe. Its bottom end is a beveled shape for which it is called a *mule shoe*. Sometime this entry guide comes with a 'No-Go' port fitted above it which is called a *pump-out plug*.

A *perforated joint* is a length of tubing with holes punched into it. If used, it will normally be positioned below the packer and will offer an alternative entry path for reservoir fluids into the tubing in case the shoe becomes blocked.

A *production packer* is a down-hole tool designed to assist in the efficient production of oil and gas from a well with one or more productive horizons. It is used to provide a seal between the outside of the tubing and the inside of the casing to prevent the movement of fluids due to pressure differential above and below the sealing point.

A packer assembly is made up of the sealing element, the circulating valve, slips, friction springs and blocks, safety joints and hydraulic hold down. Some of these features may not be found in certain packers because the particular application may not require them. Figure 5.2 shows the common elements in a packer.

In conventional packers, the *seal* is provided by a hollow rubber cylinder that is compressed longitudinally. This causes the rubber to expand laterally to come in contact with the casing. The application of sufficient force will seat the rubber tightly to prevent movement of fluids around it.

The *slips* are used to support the packer against the casing while the force is applied to expand the rubber element. These are similar to the slips used

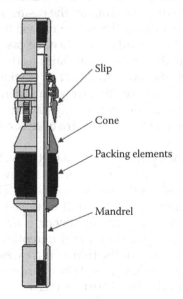

FIGURE 5.2
Common elements of a packer.

to support the drill pipe when making a trip in the hole, except that they are designed to work on the inside of the casing. There are usually three or more segments having serrated surfaces that are pressed against the casing by a slip-and-cone arrangement. As the weight of the tubing or drill pipe is applied, the slips move outward on a cone-shaped device and dig into the casing, and downward movement is prevented. The pipe weight is then picked up by the packer assembly, and the sealing element expands. Some packers have two sets of slips, working in opposite directions so that the packer will not move up or down, regardless of the direction of applied pressure differential. Once the sealing element is expanded, the hole below the packer is isolated from the casing annulus. This may be a disadvantage if (1) it is desired to circulate drilling mud out of the annulus; (2) it is desired to equalise pressure before unseating the packer; and (3) it is desired to circulate fluids while cementing, fracturing, treating or testing a well. For these reasons, most packers are equipped with a *circulating valve* (sometimes called an equalising valve) located above the sealing element. This valve can be opened by a drill pipe or tubing manipulation (rotation, pulling up, pushing down, or a combination of these operations). With the valve open, communication is established between the inside of the tubing and the casing annulus.

Friction springs are similar to the springs on casing centralisers. They serve a different purpose, however. On packers, the slips are held in a closed position until the packer has been lowered to the proper position in the well. The slips are then released, usually by rotation of the tubing in a right-hand direction. Normally, the packer body would rotate at the same time, but the friction springs provide sufficient drag on the casing so that the packer does not turn as the tubing is rotated. Thus, the release mechanism is actuated, and the slips are allowed to spread out against the casing. Application of tubing weight will complete the setting of the slips as described above.

Friction blocks perform the same function as friction springs but are thin blocks having cylindrical surfaces that drag on the inside of the casing and prevent rotation of the packer during the slip-release operation.

Safety joints are used with packers that are used for well testing or treating called retrievable packers.

One of the functions of a packer is to isolate the well's pressure inside the tubing and relieve the casing of any internal strain. This imposes a large pressure differential across the packer in most cases. This differential is large enough, in some wells, that the packer will be forced upward in the casing. To prevent this, a *hydraulic hold-down*, or anchor, is used below the packer.

The hydraulic anchor uses the pressure differential across the tubing as a working force. As the pressure in the tubing increases, the holding action of the anchor increases. The anchor consists of a series of slips or serrated buttons on the outside of a cylindrical surface (Figure 5.3). These slips are free to move in and out, radially, as the pressure varies. A pressure increase will cause the slips to move out and bear against the inside of the casing. Further

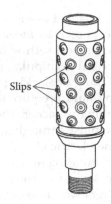

Slips

FIGURE 5.3
Hydraulic hold-down.

pressure increase causes the slips to be forced more tightly against the casing, preventing upward movement of the anchor and the packer.

Opening the circulating valve on the packer will equalise the pressure
across the tubing and the anchor and the slips on the anchor will retract,
permitting the packer to be moved.

Packers may be classified according to a number of criteria, such as their
retrievability, setting mechanism, or application. Commonly they are classified by the retrievability such as 'permanent packers' or 'retrievable packers'.

Permanent packers are those packers that cannot be entirely retrieved and
reinstalled in the well. This type of packer is normally run and set separately
on electric cable or slickline, a workstring or tubing and the production tubing is then either stabbed into or over the packer. A permanent packer may
also be run integrally with the tubing string, if there is a means of disconnecting the tubing above the packer. Permanent packers must be milled out
to remove them from the wellbore.

Retrievable packers are those packers that are designed to be retrieved
and reinstalled in the wellbore. Retrievable packers are normally run integrally with the tubing string and are set with either mechanical manipulation or hydraulic pressure. Certain retrievable packers may also be set on
electric cable. They are unset either by a straight pull or by a combination
of rotation and a straight pull or with a special retrieving tool on a drill-
pipe. Once unset, the compressible packing elements and slips or hold-
down buttons relax and retract, allowing the packer to be removed from
the wellbore.

All packers are set by applying a compressive force to the slips and rubber
packing elements. This force may be created in many ways, including tubing
rotation, slacking off weight onto the packer, pulling tension through the
tubing, pressuring the tubing against a plug, or sending an electric impulse
to an explosive setting tool. The techniques for creating the setting force are
referred to as packer setting methods.

Packer setting methods are classified as mechanical, hydraulic, or electric based on which they are also known as *mechanical packers, hydraulic packers* or *electric set packers*. A mechanical setting method refers to those techniques that require some physical manipulation of the completion string, such as rotation, picking up tubing or slacking off weight. A hydraulic setting method refers to applying fluid pressure to the tubing, which is then translated to a piston force within the packer. The third method, electric line, involves sending an electric impulse through an electric cable to a wireline pressure setting assembly. The electric charge ignites a powder charge in the setting assembly, gradually building up gas pressure. This pressure provides the controlled force necessary to set the packer.

There are many manufacturers who make different types of packers for which the given references may be cited by the readers.

A *landing nipple* is a short tubular nipple with tubing threads that is run in the well on the tubing string to a predetermined depth. Landing nipples are internally machined to receive a locking device which has a precision-machined profile that locks a production flow control device in the tubing string. The landing nipple is honed to receive high-pressure and high-temperature packing for sealing purposes. The packing is contained on the removable locking device. Landing nipples are furnished in all nominal tubing sizes, weights and threads, with or without ports, and are available in two basic types, selective and nonselective. Landing nipples are normally constructed of special alloy steels, stainless steels or monel, with strength ratings equal to or greater than the tubing string.

A *nonselective landing nipple* (Figure 5.4) is receiver for a locking device. As illustrated in this figure, it utilises a no-go principle (reduced I.D.) to locate the locking device in the landing nipple. This requires that the outside

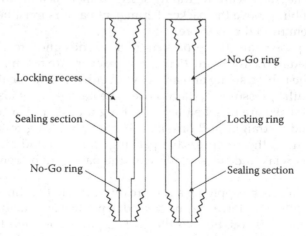

FIGURE 5.4
Nonselective landing nipple.

diameter of the locking device be slightly larger than the smallest internal diameter of the nipple.

A *selective landing nipple* is essentially full-opening. More than one can be run in a tubing string if all have the same internal dimensions (Figure 5.5). All selective landing nipples utilise a mechanical principle for locating removable equipment.

Some advantages of using a landing nipple when completing a well are

1. Plug well from above, below, or both directions.
2. Test tubing string.
3. Set tubing safety valve.
4. Set bottom-hole regulator.
5. Set bottom-hole choke.
6. Land slim-hole packer.
7. Hang bottom-hole pressure gauge with or without packoff.
8. Hang sand screen.
9. Locate and land pump with or without hold-down.
10. Set standing valve.
11. Hang extension pipe.
12. Set nipple stop.
13. Reference point for checking measurements.
14. Set hydraulic packers.
15. Set injection safety valve.

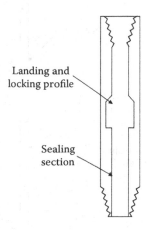

Landing and
locking profile

Sealing
section

FIGURE 5.5
Selective landing nipple.

A *polished nipple* is a short tubular nipple with tubing threads. It is constructed of the same materials as the landing nipple. A polished nipple does not contain locking recesses, but is machined and internally honed to receive a sealing section. Polished nipples may be used in conjunction with landing nipples, sliding sleeves, blast joints and other completion equipment. For example, in Figure 5.6 a landing nipple is attached to the top of a blast joint and a polished nipple is attached to the bottom of the same blast joint. The landing nipple receives the removable locking and sealing device for an attached spacer pipe. The lower sealing section is positioned in the polished nipple. The removable assembly permits the isolation of this blast joint in the event of communication caused by erosion.

Flow coupling is tubular in construction, normally 2–4 ft long, and usually made of high-grade alloy steel. The flow coupling is machined with coupling-size outside dimensions and full tubing inside dimensions which furnish a greater wall thickness as protection against possible internal erosion and corrosion. Flow couplings are positioned immediately above and, on some occasions, below a landing nipple designed to receive a production control

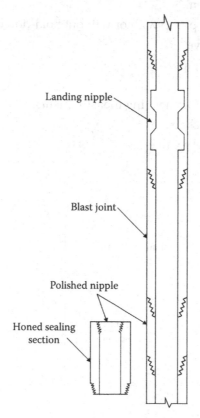

Landing nipple

Blast joint

Polished nipple

Honed sealing section

FIGURE 5.6
Polished nipple.

such as a tubing safety valve, bottom-hole regulator, bottom-hole choke and so on. Figure 5.7 depicts the use of a flow coupling around a landing nipple and a subsurface safety valve.

Blast joints play an important role in a planned completion utilising other wireline completion equipment. They are constructed of various types of materials, with external and internal dimensions similar to those of flow couplings. Fluids entering perforations may display a jetting behavior. This fluid-jetting phenomenon may abrade the tubing string at the point of fluid entry, ultimately causing tubing failure. So, to delay the erosional failure at the point where fluids enter the wellbore and impinge on the tubing string, blast joints are used. These are joints of pipe with a wall thickness greater than the tubing. These joints are run in the completion opposite the casing perforations, as shown in Figure 5.8.

To provide communication between the inside of the tubing string and the tubing-casing annulus, either above a packer or between two packers, flow control equipment known as *circulating devices* are used. Commonly used devices are ported nipples, sliding sleeves and side pocket mandrels.

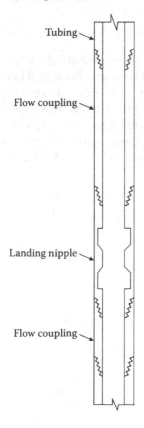

FIGURE 5.7
Flow coupling.

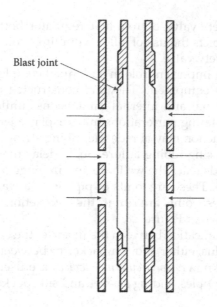

Blast joint

FIGURE 5.8
Blast joints.

A *ported nipple* (Figure 5.9) is essentially a landing nipple that contains ports and internally honed sections above and below the ports which receive packing sections of a subsurface control. Ported nipples have coupled outside diameters and have the advantage that the subsurface commingling device may be removed from the tubing string to repair or alter flow courses. In multizone completions utilising side-port equipment, the zones are commingled during service operations and offer a restriction with a flow-control in place

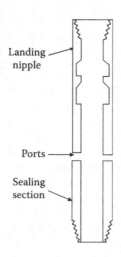

Landing
nipple

Ports

Sealing
section

FIGURE 5.9
Ported nipple.

in order to prevent commingling. Therefore, if more than one ported-nipple assembly is installed in the same tubing string, it is necessary to remove the upper control to retrieve the lower. Ported nipples generally receive different types of production controls, such as side-door choke, separation tool, cross-over assembly and so on.

A *sliding sleeve* is a cylindrical device with an internal sleeve mechanism. Both the inner sleeve and outer body are bored to provide matching openings. The inner sleeve is designed to move upward and downward, using a wireline shifting tool. When the sleeve is shifted to the open position, the sleeve openings mate with the openings in the outer body, thereby establishing tubing/annulus communication. When the sleeve is shifted to the closed position, the sleeve openings are displaced from the outer body openings, which are then isolated by the inner sleeve wall. An example of this device is shown in Figure 5.10.

Sliding sleeves are versatile and flexible components. They can be used at virtually any point in the completion string where circulation, injection or selective production is required. The device can be opened and shut by wire line and therefore does not require the completion to be retrieved after circulation is established. Depending on the design of the communication ports, the device can offer a circulation area even greater than the internal diameter of the tubing.

Side pocket mandrels or retrievable valve mandrels (Figure 5.11) were initially designed to receive retrievable gas-lift equipment. However, since this mandrel receives retrievable locks and sealing devices, it has been used in

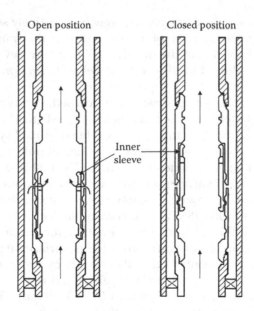

FIGURE 5.10
Sliding sleeve.

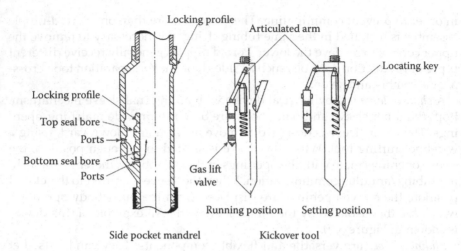

FIGURE 5.11
Side pocket mandrel and kickover tool.

the same manner as ported nipples and sliding sleeves. This mandrel offers advantages similar to side-port equipment, in that the retrievable flow-control devices may be retrieved by wire line, and advantages similar to a sliding sleeve in that it has full tubing internal dimensions with a flow device positioned in the side pocket. Because of the full internal bore provided, it is necessary that the outside diameter be larger than the tubing coupling at the side-pocket section.

Gas-lift valves and subsurface control may be selectively set and retrieved from the mandrel through use of a kick-over tool that positions the flow device into the side pocket of the mandrel. With a flow device in this offset position, the possibility of sand or tubing debris falling on top of the flow device is reduced.

Thus far as *downhole safety systems* are concerned, two types are possible: (1) Tubing safety system and (2) annulus safety system. These two systems may be wire-line retrievable, tubing retrievable or mixed types. These systems may be controlled by an independent control fluid or by differential pressure variation on either side of the valve. The first one is referred to as hydraulically controlled valve or subsurface safety valve and the second one is referred to as velocity downhole safety valve or storm chokes.

A *subsurface safety valve (SSSV)* is a control device used to shut off production from the well in an emergency situation, for example, if the surface control system is damaged or destroyed. The subsurface safety valve's opening and closing may be controlled at the surface by pressure supplied in a hydraulic line or directly by subsurface well conditions.

Surface-controlled subsurface safety valves (SCSSV) (Figure 5.12) incorporate a piston on which hydraulic pressure acts to open the closure mechanism. A spring acts in the opposite direction on the piston to close the closure

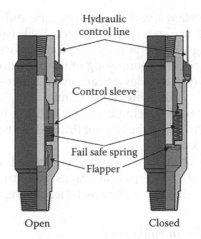

Open   Closed

FIGURE 5.12
Surface controlled subsurface safety valve.

mechanism as hydraulic pressure is lost. In most SCSSV designs, well pressure acts in conjunction with the spring to oppose the hydraulic pressure and close the valve.

Subsurface-controlled subsurface safety valves (SSCSV) are operated directly by well pressures and require no hydraulic line for their operation. They are normally open while installed in the well and typically require flowing the well outside its normal production regime to close. The inability to control SSCSVs from the surface limits their use to applications that are outside the capabilities of SCSSVs.

Subsurface safety valves are classified according to their method of retrieving and their internal closure mechanisms. Tubing retrievable safety valves (TRSV) are valves that are an integral part of the tubing string, requiring the tubing to be removed for retrieval of the safety valve. Wireline retrievable safety valves (WLRSV) are installed inside the tubing with a locking device to secure them inside a safety valve landing nipple. WLRSV's can be installed and retrieved without removal of the tubing.

The most common types of closure mechanisms for subsurface safety valves are the ball and the flapper. Either mechanism can be used with a tubing-retrievable or wireline-retrievable safety valve. These SSV's are placed below the surface (tubing head) as is deemed safe from any possible surface disturbance including cratering caused by the wipeout of the platform. Where hydrates are likely to form, the depth of the SCSSV below the seabed may be as much as 1 km which allows the geothermal temperature to be high enough to prevent hydrates from blocking the valve.

A *storm choke* also known as a 'velocity safety valve' is not operated by hydraulic fluid, but by a pressure differential of the produced or injected fluid. It operates by fluid velocity and closes when the fluid flow from the well exceeds preset limits.

This type of safety system is very hard to regulate and calls for good knowledge of the well (GOR, flow-rate, pressure, etc.). Each valve of this type must be regulated according to the characteristics of each well, which will mean having logistics and a regular monitoring of the pressure changes on a field.

Therefore, this type of safety device can only be wire-line retrievable with the drawbacks of seal and diameter constraints.

This type of safety device is being used less and less in completions nowadays, but can be used to help if the control line should become blocked (postpone the date of a work-over).

The forerunners to modern subsurface controlled safety valves, storm chokes were used in offshore applications as a contingency device in the event of a catastrophic failure of surface facilities during a storm or hurricane.

5.1.1.2 Surface Completion Equipment

The surface completion equipment includes wellhead, tubing hanger and Christmas tree.

The wellhead has been dealt with in detail in Chapter 4 and on the top of the wellhead the tubing hanger sits to act as the main support for the production tubing.

A *tubing hanger* is a device that provides a seal between the tubing and the tubing head which is attached to the top flange on the uppermost casing head or wellhead. There are different types of tubing hangers available which basically are latched around the tubing and placed inside the tubing head bowl for permanently supporting the weight of the tubing and simultaneously providing a complete sealing or packing.

The next important surface equipment is the *Christmas tree*, which is nothing but an assembly of valves and fittings used to control production and provide access to the production tubing string. It includes all equipment above the tubing head top flange. Figure 5.13 shows a typical Christmas tree assembly. Just above the tubing head flange there is a tubing head adapter flange for connecting the *master valve* and providing a support for the tubing.

Valves used on wellheads are basically of two types: gate valves and plug valves. Both are available with flanged end connections. Gate valves can be divided into lubricated and nonlubricated, wedging and nonwedging types.

Full-opening valves must be used in the vertical run of the Christmas-tree assembly (i.e. master valve) to provide access to the tubing. Full-opening valves must also be used on tubing-head outlets and casing-head outlets equipped for valve-removal service (i.e. casing valve).

Restricted-opening valves may often be used as wing valves, without loss of efficiency or utility, to effect an economic saving of 20–30% of full-opening valve cost.

Other Christmas-tree fittings include tees, crosses and other connections necessary to provide the most desirable arrangement for the particular application.

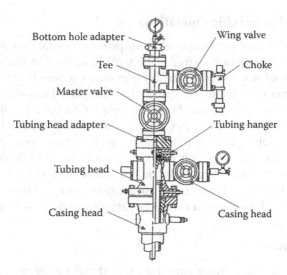

Bottom hole adapter

Wing valve

Tee

Choke

Master valve

Tubing head adapter

Tubing hanger

Tubing head

Casing head

Casing head

FIGURE 5.13
A typical Christmas tree assembly.

The size of the vertical run may vary from 2–4 in. but must be consistent with the master-valve and tubing-head adapter-flange size to give full-opening access to the tubing for wire-line tools and instruments.

The outlet on the tee or cross and wing assembly must be of sufficient size to handle the production requirements without undue restriction. Outlets vary in size from 2–4 in., although the 2-in. size is normally adequate and is most commonly used.

All Christmas tree assemblies should be assembled, pressure-tested to hydrostatic test pressure and checked with a drift mandrel to assure full opening before installation.

A *bottom-hole test adapter* is a device attached to the top of a Christmas tree assembly to provide fast and safe adaptation of a lubricator for swabbing or testing. It may also include an internal thread to act as a lift thread for setting or raising the Christmas tree and tubing.

It is available in sizes from 2–4 in. and in working pressures from 960–10,000 psi.

A *choke* is a device attached downstream from the wing valve to restrict, control and regulate flow. It may be of the positive or adjustable type.

A positive choke is composed of a body with internal provisions for receiving removable orifices of various sizes which are installed manually.

An adjustable choke is similar to a positive choke except that a stem with visible graduations indicating the effective orifice size is used to adjust the flow opening.

Both positive and adjustable chokes are available with threaded or flanged end connections.

5.1.2 Completion Specific Operations

Perforation and gravel packing are the important operations which are carried out during completion to make the well fluid flow to the wellbore continuously without any problem. In certain cases where there is damage to the formation near the wellbore, stimulation operation (mostly acid stimulation) is carried out though not being a completion specific operation. Out of all these, perforation operation is quite common specially in the case of cased-hole completion and in all offshore well completions, perforation of the production casing/liner is a must. In case of sand exclusion problems, gravel packing operation is carried out or alternately slotted/screen liners are lowered below the oil string (i.e. production casing or liner) opposite the producing zone. The details of these operations are dealt with below.

5.1.2.1 Perforation

Perforation of casing or liner may be carried out either in underbalanced conditions using tools such as tubing conveyed perforation (TCP) or in overbalanced conditions, using perforation tools that are deployed down the wellbore on wireline.

This process involves running a perforation gun and a reservoir locating device into the wellbore, via a wireline/slickline or coiled tubing. Once the reservoir level has been reached, the gun then shoots holes in the sides of the well to allow the hydrocarbons to enter the wellbore. The perforations can either be accomplished by firing bullets into the sides of the casing or by discharging jets or shaped charges into the casing. The degree of penetration varies between 10 and 20 in., depending on the type of perforating tool being used. When the guns are fired at a speed of 8000 m/sec and with an impact pressure of around 3×10^6 psi, the casing, cement and rock which comes in its path get completely pulverised. The technique of bullet perforating and jet perforating is briefly described below.

Bullet perforating – The bullet perforator is essentially a multibarreled firearm designed for being lowered into a well, positioned at the desired interval, and electrically fired at will from surface controls. Penetration of the casing, cement and formation is accomplished by high velocity projectiles or bullets. Current equipment permits the selective firing of one bullet at a time, selective firing of independent groups of bullets, or simultaneous firing of all bullets, depending on the operator's need.

Many bullet types are available, each being tailor-made for a particular purpose. Bullet guns designed for use in virtually all sizes of casing are readily available from many service companies.

An interesting application of projectile equipment is the formation fracturing tool or bear gun shown in Figure 5.14. This device fires a large (1½ in. diameter) projectile vertically downward. The missile passes through a vertical barrel and is deflected 90° as it leaves the muzzle. Horizontal penetrations

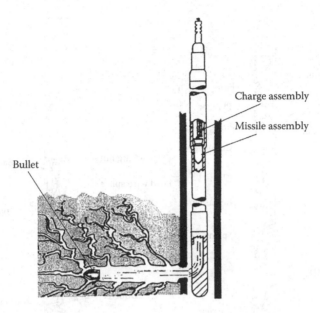

FIGURE 5.14
Formation fracturing tool/bear gun.

of 2–5 ft as well as considerable fracturing of the formation are estimated to be obtainable with this device.

Jet perforation – Two basic types of jet perforating equipment, retrievable and expendable guns, are available. As implied, a retrievable gun is composed of a cylindrical steel carrier with the charges opposite ports facing radially from the vertical axis of the carrier. Expendable guns are composed of materials that disintegrate into small particles when the gun is fired. A small diameter jet gun which can be run inside tubing and fired in the casing below is shown in Figure 5.15. The materials commonly used for the carrier are aluminum or cast iron while the cases housing the charge are constructed from glass, aluminum, plastic, cast iron or ceramic material.

The sequence of events in Figure 5.16 shows a generally accepted pattern of the jet's development during succeeding stages of the explosion. Penetration of the target is obtained from the jet stream's high velocity impact developed by the liner's inward collapse and partial disintegration. The velocity of the jet is on the order of 10,000 m/sec, which causes it to exert an impact pressure of some 4 million psi on the target. The crumpled portion of the liner, called the carrot, is an undesirable feature of jet perforating as it may sometimes lodge in the perforation and obstruct flow. Improved charge designs have, however, been developed which eliminate this problem and produce an essentially carrot-free jet.

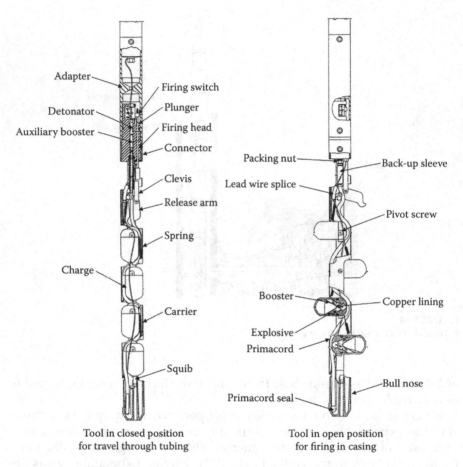

Tool in closed position
for travel through tubing

Tool in open position
for firing in casing

FIGURE 5.15
Tubing type jet perforating gun.

5.1.2.2 Gravel Packing

The completion of a well in an unconsolidated sand is not as simple a job because the additional problem of excluding any sand produced with the oil must be solved. Sand production, if unchecked, can cause erosion of equipment, and wellbore and flow string plugging to the extent that well operation becomes uneconomical.

The most common methods of excluding sand employ some means of *screening.* These techniques include:

1. Use of slotted or screen liners
2. Gravel packing

The basic requirement of these methods is that the openings through which the produced fluids flow must be of the proper size to cause the

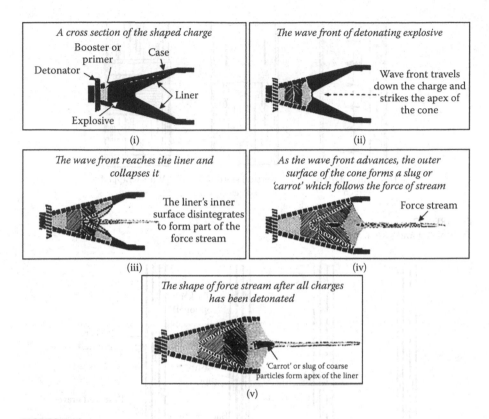

FIGURE 5.16
Various stages in jet development.

formation sand to form a stable bridge and thereby be excluded. This, of course, requires some knowledge of the sand size. The appropriate liner slot or screen width is then taken as twice the 10-percentile size indicated from the screen analysis. For gravel packs, the gravel size is generally selected as 4–6 times the 10-percentile sand size.

Typical slot and *screen* arrangements are shown in Figure 5.17. It is extremely important that the sand face be free of mud cake before the liner is set to prevent plugging. This is commonly accomplished by either using clay-free completion fluids or washing the well with salt water before hanging the liner.

Gravel packing may be performed in several ways in either perforated or open hole intervals. Figure 5.18 shows sequential operation of the gravel placement method commonly being practiced. Since the screen is used to exclude gravel only, the slots may be larger than in the previous case and are usually only slightly smaller than the gravel. The required thickness of the gravel pack is only four or five gravel diameters. The formation sand then bridges within the pores of the gravel pack while gravel entry is prevented by the screened liner. Uniform placement of gravel is facilitated by using a

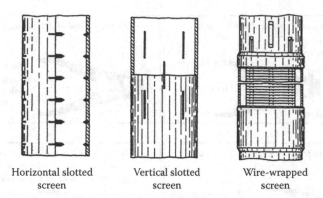

| Horizontal slotted screen | Vertical slotted screen | Wire-wrapped screen |

FIGURE 5.17
Typical slot and screen arrangements.

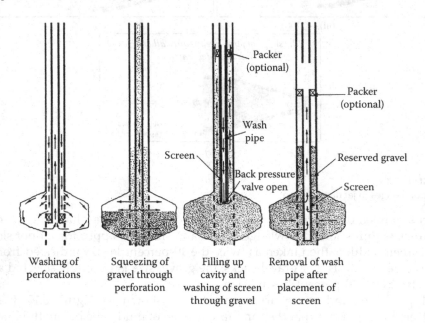

| Washing of perforations | Squeezing of gravel through perforation | Filling up cavity and washing of screen through gravel | Removal of wash pipe after placement of screen |

FIGURE 5.18
Sequential operation of gravel packing method.

penetrating (high fluid loss) fluid such as oil or salt water. Perforation washing is also necessary to ensure that all perforations are open and to allow proper gravel placement.

5.1.3 Well Completion Design

Wells represent the major expenditure in reservoir development. Oil wells, gas wells and injection wells present unique problems depending on the specific operating conditions. The individual well completion must be designed to yield maximum overall profitability on a field basis.

The ideal completion is the lowest cost completion (considering initial and operating costs) that meets or nearly meets the demands placed upon it for most of its life. To intelligently design a well completion, a reasonable estimate of the producing characteristics during the life of the well must be made. Both reservoir and mechanical considerations must be evaluated.

Basic decisions to be reached in designing the well completion are: (a) the method of completion, (b) the number of completions within the wellbore, (c) the casing-tubing configuration, (d) the diameter of the production conduit and (e) the completion interval.

(a) Method of Completion

Basically, there are three methods to complete a well.

1. Open hole completion
2. Perforated completion
3. Liner completion

In the case of open hole completion, the production casing is set on the top of the pay zone.

In the case of perforated completion, the pay zone is covered by the production casing and later it is perforated.

In the case of liner completion, the casing is set on the top of the pay zone and a liner is hung. The liner can be cemented in which case the liner will be perforated afterward for producing. The other one is a perforated liner, which is used in case of open hole completion normally.

In the case of offshore, only the perforated completion technique is utilised. The advantages are

1. Various producing intervals can be isolated effectively, which helps in selective treatment and controlled production.
2. Multilayer completion.

(b) Number of Completions

Based on the number of horizons to be produced through a single well, completion methods are categorised as either 'single-zone completion' or 'multiple-zone completion'.

Single-zone completion

Factors leading to selection of single-zone 'conventional' completions, as opposed to miniaturised completions, multiple inside-casing completions, or multiple tubingless completions are high producing rates, corrosive well fluids, high pressures, governmental policies and operator tradition.

Various hookups are possible depending on objectives. Basic questions concern uses of tubing and packers. Many wells are produced without tubing.

Valid reasons for tubing may include:

1. Better flow efficiency.
2. Permit circulation of kill fluids, corrosion inhibitors, or paraffin solvents.
3. Provide multiple flow paths for artificial lift system.
4. Protect casing from corrosion, abrasion, or pressure.
5. Provide means of monitoring bottom-hole flowing pressure.

Tubing should be run open-ended, and set above the highest alternate completion interval to permit through tubing wireline surveys and remedial work.

A packer should be run only where it accomplishes a valid objective such as:

1. To improve or stabilise flow.
2. To protect casing from well fluids or pressure; however, it should be recognised that use of a packer may increase pressure on casing in the event of a tubing leak.
3. To contain pressure in conjunction with an artificial lift system or safety shut-in system.
4. To hold an annular well-killing fluid.

Where packers are used, landing nipples to permit installation of bottom-hole chokes or safety valves are sometimes desirable. Also, a circulating device is sometimes desirable to assist in bringing in or killing the well.

Multiple Zone Completion

Factors leading to selection of multiple completions are: higher producing rate, faster payout and multireservoir control requirements. Numerous configurations are possible utilising single or multiple strings of tubing.

Single tubing with single packer – There is both tubing and annulus flow (Figure 5.19). This is the lowest cost conventional dual.

Single tubing with dual packer – Again, there is both tubing and annulus flow (Figure 5.20). The advantage is that cross-over choke permits the upper zone to be flowed through tubing.

Parallel tubing with multiple packer – This is shown in Figure 5.21. It can lift several zones simultaneously. Concentric tubing and wireline workovers are possible in all zones.

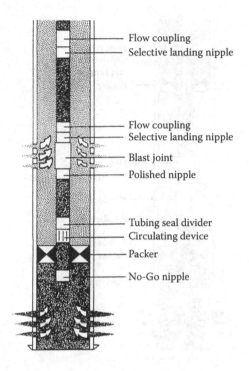

FIGURE 5.19
Single tubing with single packer. (From Thomas O. Allen and Alan P. Roberts, *Production Operations*, Vol. 2, Oil & Gas Consultants International Inc. Copyright © 1984. Reprinted with permission of PetroSkills, LLC.)

Single tubing with multiple packer – selective zone – This is shown in Figure 5.22. Here producing sections can be opened or closed by use of wireline.

(c) Casing – Tubing Configurations

Multiple 'Tubingless' Completion

The multiple tubingless completion is an outgrowth of permanent well completions (PWC) and concentric tubing workover technology. The overall concepts of PWC had the objective of eliminating the necessity of pulling tubing during the life of a well. Figure 5.23 illustrates the basic arrangement for permanent well completions with and without a packer. An essential feature of PWC is setting the bottom of the tubing open-ended and above the highest anticipated future completion zone.

Primary developments needed to make the system feasible form the basis of current well completion technology. These developments include:

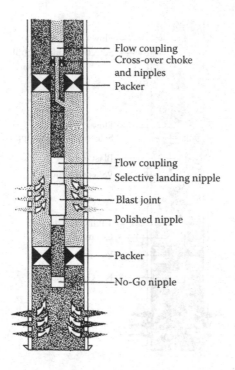

FIGURE 5.20
Single tubing with dual packer. (From Thomas O. Allen and Alan P. Roberts, *Production Operations*, Vol. 2, Oil & Gas Consultants International Inc. Copyright © 1984. Reprinted with permission of PetroSkills, LLC.)

1. The through tubing perforator, and along with it the concept of underbalanced perforating (differential pressure into the wellbore) to provide debris-free perforations.

2. A concentric tubing extension run and set on a wireline to permit circulating to the desired point in the well. Later the wireline tubing extension was replaced using a full string of small diameter tubing which could be run through the normal producing tubing using a small conventional workover rig. Use of this small tubing is termed concentric tubing workover.

3. Low fluid loss, below frac pressure, squeeze cementing, to provide slurry properties so that cement can be placed at the desired point (in the perforations or in a channel behind the casing) and the excess cement subsequently reverse circulated out of the well.

4. Logging devices, gas lift valves, bridge plugs and other necessary tools designed to be run through tubing on a wireline or electric line.

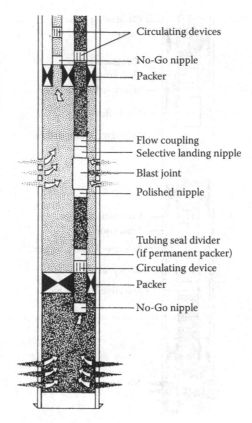

Circulating devices

No-Go nipple

Packer

Flow coupling
Selective landing nipple

Blast joint

Polished nipple

Tubing seal divider
(if permanent packer)

Circulating device

Packer

No-Go nipple

FIGURE 5.21
Parallel tubing with multiple packers. (From Thomas O. Allen and Alan P. Roberts, *Production Operations*, Vol. 2, Oil & Gas Consultants International Inc. Copyright © 1984. Reprinted with permission of PetroSkills, LLC.)

The tubingless completion involves the cementing of one or more strings of 2½ in. or 3½ in., as production casing in a single borehole. Figure 5.24 shows a comparison between conventional and tubingless completion in a multipay field.

The original effort was aimed at reducing the initial investment. However, the major economic benefits have been in reducing well servicing and workover costs, with particular application to triple completions in lenticular multireservoir fields, and to dual offshore wells.

This type of completion is not necessarily restricted to low-return, low-volume, short-lived wells. Single or multipay gas fields are excellent candidates for tubingless completions. Recently for optimising production without costly well intervention, intelligent completion technology has been practiced.

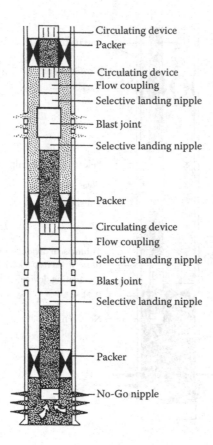

FIGURE 5.22
Single tubing with multiple packer. (From Thomas O. Allen and Alan P. Roberts, *Production Operations*, Vol. 2, Oil & Gas Consultants International Inc. Copyright © 1984. Reprinted with permission of PetroSkills, LLC.)

This *smart* or *intelligent completion* consists of some combination of zonal isolation devices, interval control devices, downhole control systems, permanent monitoring systems, surface control and monitoring systems, distributed temperature sensing systems, data acquisition and management software and system accessories. It enables operators to collect, transmit and analyse downhole data; remotely control selected reservoir zones and maximise reservoir efficiency by:

1. Increasing production – Commingling of production from different reservoir zones increases and accelerates production and shortens field life.

2. Increasing ultimate recovery – Selective zonal control enables effective management of water injection, gas and water breakthrough and individual zone productivity.

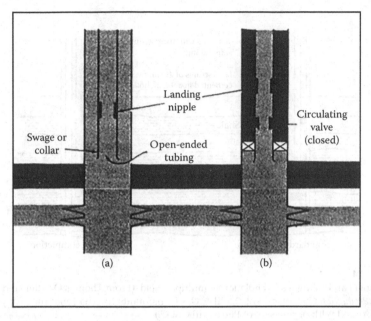

FIGURE 5.23
Permanent well completion (a) without and (b) with a packer. (From Thomas O. Allen and Alan P. Roberts, *Production Operations*, Vol. 2, Oil & Gas Consultants International Inc. Copyright © 1984. Reprinted with permission of PetroSkills, LLC.)

3. Reducing capital expenditure – The ability to produce from multiple reservoirs through a single wellbore reduces the number of wells required for field development, thereby lowering drilling and completion costs. Size and complexity of surface handling facilities are reduced by managing water through remote zonal control.

4. Reducing operating expenditure – Remote configuration of wells optimises production without costly well intervention. In addition, commingling of production from different reservoir zones shortens field life, thereby reducing operating expenditures.

As an example, in the case of horizontal wells, which are usually completed with uncemented liners in the horizontal section, there is sometimes the need to apply water production control valves in case the well intersects a fractured zone or fault which communicates with an aquifer or another water-saturated zone (Figure 5.25). In these cases, it is common to use inflatable packers that swell when in contact with formation fluids, known as swelling packers. There are packers that swell in the presence of oil (oil swelling packers) and others in the presence of water

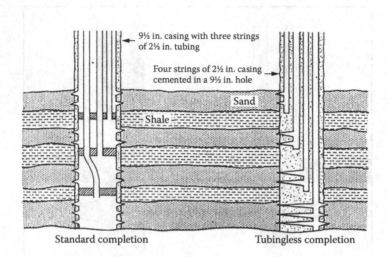

FIGURE 5.24
Conventional and tubingless completion in multipay field. (From Thomas O. Allen and Alan P. Roberts, *Production Operations*, Vol. 2, Oil & Gas Consultants International Inc. Copyright © 1984. Reprinted with permission of PetroSkills, LLC.)

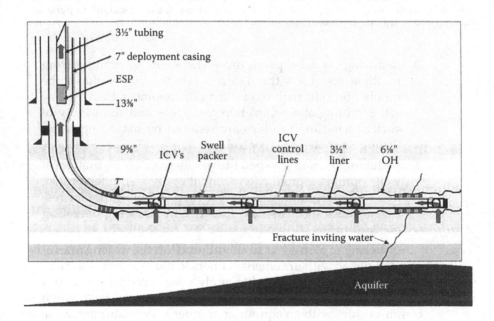

FIGURE 5.25
Intelligent completion in a horizontal well. (From Jorge S. Gomes and Fernando B. Alves, *The Universe of the Oil & Gas Industry – From Exploration to Refining*, Partex Oil and Gas, 2013.)

(water swelling packers). Swelling is irreversible. When inflating, the packers isolate the horizontal section of the well into very defined areas which, with the aid of inflow control valves (ICV), can control the part of the horizontal well that should or should not flow.

(d) Sizing Production Conduit

The size of the production string/casing depends upon the diameter of flow conduit (single or multiple) needed to produce the desired flow stream, the method of artificial lift, if required, or specialised completion problems such as sand control.

Sizing of the production tubing depends primarily on the desired production rate. Maximum production rate in a given well depends upon:

1. Static reservoir pressure.
2. Inflow performance relation (IPR), which represents the ability of a well to produce fluids against varying bottom hole pressures.
3. Pressure drop in tubing.
4. Pressure drop through the wellhead constrictions.
5. Pressure drop through the flow line.
6. Pressure level in the surface separating facilities.

Where maximum flow rate is an objective of well completion design, all of these factors must be considered.

(e) Completion Interval

Selection of the completion interval is dictated by many interrelated factors. In thin multireservoir fields, the choice is usually obvious. In thicker multilense zones comprising single reservoirs considerable study of geologic and reservoir fluid flow mechanisms may be required for intelligent interval selection. A well in a water drive reservoir should be completed near the top of the zone, and a well in a gas drive reservoir should be completed near the bottom.

Similarly, in a massive reservoir with good permeability and vertical communication, opening a limited interval will provide high capacity and, at the same time, facilitate future workovers. In a multizone field having marginal stringers, commingling separate reservoirs is a possibility.

In case a stimulation job is required to be carried out, it is simplified by limiting the number of perforations and the intervals open. For future workover possibilities, it is often sound practice to open only a limited interval initially, and complete in other zones later, if needed, by means of low-cost workovers.

5.2 Offshore Well Completion Methods

Although the basics of completion methods remain the same in offshore wells, the main differences between the onland and offshore completion are reduction in weight and to minimise space requirements. These can be done by using composite valves and closer spacing between the wells, with only just sufficient room for safe and efficient operation of the valves. Normally the distance between two wells varies from 6–10 ft. While doing so, some changes in the construction of wellheads and Christmas trees are also made although functionally they remain the same as that of onland wells. For added safety, a SSSV is installed in the tubing string below the mud line (sea floor) in order to close the well automatically in case of any eventuality.

The most preferable technique that is used in the case of offshore is the multiple completion techniques which permits two or more layers to be produced simultaneously.

Additionally, in the case of offshore, the well can either be completed at the deck level or subsea based on which they are classified as 'deck level completion' or 'platform completion' and 'subsea completion'.

5.2.1 Deck Level Completion

In the deck level completion, all the casings are brought to the deck level, and all casing hangers, tubing hangers, well heads and Christmas trees are located at the deck level. Therefore, all the components of well heads and Christmas trees are of dry type standard systems.

The salient features of the system are (1) most preferred completion, (2) well-established technology, (3) field proven equipment, (4) know how is available for drilling, production operations and remedial functions, (5) workover, artificial lift, water injection/enhanced recovery are possible with standard and established tools and technology, (6) constant monitoring by personnel is possible and (7) higher platform cost.

The deck level completions are possible with fixed bottom or compliant type platforms like steel jacket platforms, steel gravity platforms, concrete gravity platforms, tension leg platforms, guyed towers and so on.

While drilling from mobile drilling structures such as jack-up or semisubmersible, the strings of downhole casing (typically 20″, 13⅜″, 9⅝″ and 7″) are run and suspended at the mud line (refer to Chapter 4). After the drilling is completed, the well is plugged and capped. The mud line suspension equipment includes tieback devices to which the abandonment caps can be attached.

To connect the casings at the deck level, first a production platform which is normally a jacket structure is to be positioned and installed over the subsea template. Once the jacket is in place, the wells must be tied back to it with casing extensions, called *risers*. The number of strings tied back varies according to preference and conditions. The largest size casing string is tied back first and the smallest size last. Following are the method of jacket installation over the template and then the tie-backing procedure.

5.2.1.1 Installation of Jacket over Template

Prior to the installation of the jacket, in one of the methods, the bumper piles are set to establish the alignment between the jacket and the template. The bumper piles must also absorb the impact loads that occur as the jacket is lowered into position.

The cantilever bumper pile guide module is run into position over two guide posts on the modular or unitised template and engages the bracket on the template corner. The cantilever bumper pile guide module landed on a unitised template is shown in Figure 5.26. The module is run with a J-type running tool using a drill pipe. The hole for the pile is drilled through the funnel of the module.

After the hole is drilled for the pile, a bumper pile spacer insert is installed with a J-type running tool. There are 64″ funnels with 54″ pile spacer inserts used for 54″ piles. Other sizes can be made available. Guidance for installing the insert is provided by two guidelines attached to the module on a 6-ft radius, thereby allowing a standard utility guide frame to be used.

After the bumper pile is run and cemented, the guide pile module is retrieved over the flush O.D. of the pile. The guide pile module can now be used to locate subsequent bumper piles. To protect the template from jacket impact loads, the guide pile module is removed from the last bumper pile prior to installation of the jacket. In this way, the jacket bumper piles are not connected to the template in any way during jacket installation.

The jacket is then installed over the template any time after the bumper piles are in place. This method of installation technique known as 'bumper pile jacket installation' is shown in Figure 5.27. The jacket is first delivered to the template area and ballasted to the vertical position with the legs approximately 30 ft above the sea floor. Winch lines are then rigged to fixed mooring

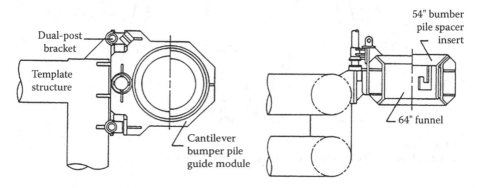

FIGURE 5.26
Cantilever bumper pile guide module on a unitised template. (Courtesy of GE Oil & Gas.)

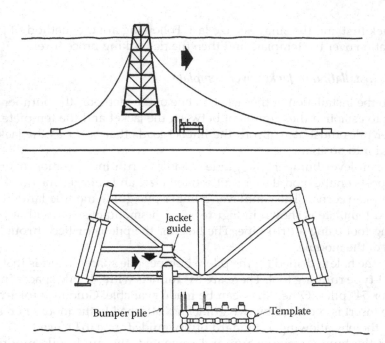

FIGURE 5.27
First technique of jacket installation. (Courtesy of GE Oil & Gas.)

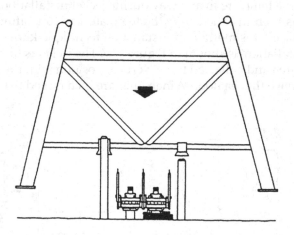

FIGURE 5.28
Second technique of jacket installation. (Courtesy of GE Oil & Gas.)

points. After prealigning the jacket, it is slowly winched toward the template until the jacket guides contact the bumper piles. The jacket is then ballasted and fixed in position.

Another method of locating the platform jacket over the template known as 'guide pile jacket installation' is shown in Figure 5.28, where the jacket is lowered over guide piles of uneven height. These piles are installed either just prior to, or just after, the wells are drilled. The uneven pile height allows the jacket to be landed on the taller pile first. The jacket is turned on the taller pile until properly positioned over the shorter pile. It is then lowered to the bottom and fixed in place. This approach is used primarily for lightweight platforms.

5.2.1.2 Tie-Back Methods

Once the platform has been properly positioned and installed over the subsea template, the outer, or largest, casing tie-back string can be connected to the subsea wellhead. This first casing tie-back string is generally 20″ casing. Casing tie-back strings are selected for each well program and normally include 20″, 13⅜″ and 9⅝″ sizes.

The first tie-back string must be accurately guided into position on the subsea wellhead using the platform conductor guides (refer to Section 4.2.2.1) for guidance. It is recommended that the platform conductor guides, the 20″ tie-back string with its attached centralisers and the tie-back adapter, used to physically align and connect the 20″ tie-back string to the subsea wellhead, be designed as an inter-related system.

There are two basic tie-back methods, both of which are compatible with template wells. These methods employ two different tie-back adapters, the *standard tie-back adapter* and the *external lock tie-back adapter.*

The *standard tie-back adapter* incorporates a downward-looking funnel that serves to align the tie-back string with the wellhead. It has a threaded internal lock-down sub. The funnel is stronger than the casing string in terms of resistance to bending.

The weight of the 20″ casing is used, sometimes with additional weight added, to seat the tie-back adapter over the wellhead housing. Once the adapter is landed, an internal lockdown sub is made up into the left-hand running threads of the wellhead housing by a torque tool that is run on a drill pipe inside the 20″ casing. In this way, the 20″ casing and the tie-back adapter are not rotated during make-up of the internal lockdown sub. The lockdown sub in the funnel-shaped tie-back adapter is shown in Figure 5.29.

Since the 20″ tie-back casing is not rotated during makeup of the tie-back adapter, guide ribs or centralisers, shown in Figure 5.30, may be used on the outer tie-back string. These centralisers can be either concentric or eccentric, depending on the severity of offset or angular misalignment encountered when tying-back the wells. Concentric centralisers that impose no offset to the tie-back conductor can be used if wellhead-to-platform offset is minimal (<18″).

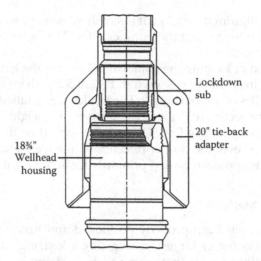

FIGURE 5.29
Lockdown sub in the tie-back adapter. (Courtesy of GE Oil & Gas.)

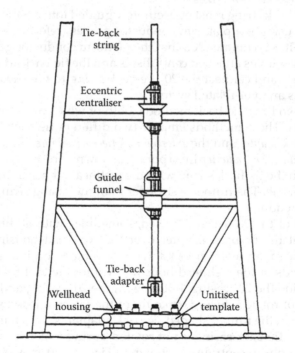

FIGURE 5.30
Eccentric centraliser on the outer tie-back string. (Courtesy of GE Oil & Gas.)

However, if the offset and/or angular misalignment between the well-heads and the platform jacket are significant, the tie-back string may require eccentric centralisers.

The following is a suggested tie-back system installation procedure for use with an 18¾" × 13⅜" × 9⅝" wellhead system. Due to the variety of conditions under which this system may be used, certain steps may have to be modified as conditions dictate.

1. The corrosion cap is retrieved from the 18¾" wellhead housing the nominal seat protector running and retrieving tool.

2. Then the distance from the cellar deck to the top of the wellhead housing is correctly determined.

3. A jet sub is run in the hole to clean thoroughly the wellhead housing and running threads.

4. The wear bushing is retrieved and again run in the hole with a jet sub. The wellhead housing is cleaned.

5. The 20" tie-back adapter is made up on the bottom of the 20" tie-back casing.

6. Then the 20" tie-back casing is run through the jacket funnels with the centralisers located at the elevation of the jacket guide funnels.

7. The 20" tie-back adapter is carefully landed on the wellhead housing; the measurements are checked to ensure that the 20" adapter is properly seated.

8. The 20" torque tool is run on a drill pipe, inside the 20" casing tie-back string. As shown in Figure 5.31, the 20" torque tool is landed in the 20" tie-back adapter lock-down sub. The 20" lock-down sub is rotated to the left, slowly at first to allow the drive keys to engage the slots in the lock nut. Continued rotation engages the running threads on the inside of the wellhead housing. This should require only a nominal amount of torque, about 1000 ft-lb, to make up the coarse left-hand threads.

9. The 20" tie-back torque tool is recovered and a cup tester is run if a low pressure test is required at this stage.

10. The 13⅜" tie-back casing, with a 13⅜" tie-back adapter is run on the bottom, as shown in Figure 5.32, through the 20" casing and carefully stabbed into the wicker threads located in the 20" lock-down sub; then slacked off slightly on the hook. Pressure testing of the 13⅜" tie-back to the casing pressure rating is done using a cementing head and testing against the bridge plug, or using a cup tester.

11. The 9⅝" tie-back casing is run through the 13⅜" casing and landed the 9⅝" tie-back tool on the 18¾" × 9⅝" casing hanger.

12. The 9⅝" torque tool is run on a drill pipe through the 9⅝" tie-back casing, landed the tool and rotated, slowly at first to engage the

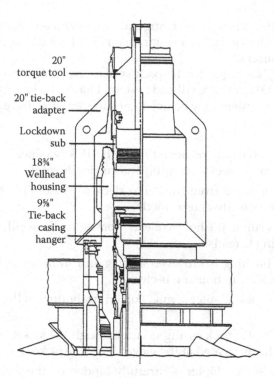

20"
torque tool

20" tie-back
adapter

Lockdown
sub

18¾"
Wellhead
housing

9⅝"
Tie-back
casing
hanger

FIGURE 5.31
Tie-back adapter made up with torque tool. (Courtesy of GE Oil & Gas.)

drive keys. The lock-down sub is rotated to the right until a minimum torque of approximately 10,000 ft-lb is reached. The torque tool
is then removed and the 9⅝" casing is pressure tested to the desired
pressure rating, using a 9⅝" cup tester.

The wellhead is completely tied back after Step 12. The well may now be
completed from the surface using conventional methods.

The *external tie-back adapter* is an external lock tie-back casing connector
that locks onto the external profile of the 20" wellhead housing. It can be
selected by an operator for use as the primary tie-back method or can be
used when the left-hand threads inside the wellhead housing are damaged
and will not allow for make-up of the internal lockdown sub in the standard
tie-back system.

The external lock tie-back connector utilises a downward-looking funnel that aligns the 20" tie-back string with the wellhead housing. It has an
internal mechanical dog that mates with the external wellhead profile. It
can accommodate the same horizontal offset or angular deflection as the
standard tie-back adapter. The 13⅜" and 9⅝" tie-back subs used with the
external connector are the same as those used with the standard system.

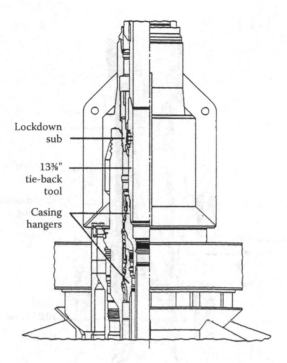

Lockdown
sub

13⅜"
tie-back
tool

Casing
hangers

FIGURE 5.32
Stabbing the tie-back adapter inside the lockdown sub. (Courtesy of GE Oil & Gas.)

Figure 5.33 shows an external lock connector with optional indicator rods. The rods give a positive indication, either to divers or remote TV, of the position of the cam ring and, hence the locking dogs. In this figure, running mode is shown on the right and locked mode on the left.

The full procedure for running the external lock tie-back adapter or external lock connector can be as follows:

1. The 20" external connector assembly on the bottom of the 20" tie-back casing is made up.

2. The 20" tie-back casing is then run through the jacket funnels with centralisers located at the funnel elevations.

3. The 20" external connector is stabbed onto the wellhead housing, using the weight of the casing to align the 20" tie-back casing with the wellhead.

4. The 20" torque tool is run on a drill pipe inside the 20" tie-back casing, landed on the lock nut and rotated slowly until the spring-actuated key finds the slot in the lock nut.

5. A low-pressure test is performed through the drill pipe to confirm the proper seating of the external connector over the wellhead housing.

Locked   Unlocked

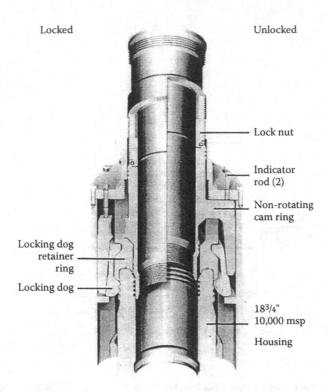

Lock nut

Indicator
rod (2)

Non-rotating
cam ring

Locking dog
retainer
ring

Locking dog

$18^{3}/_{4}$"
10,000 msp

Housing

FIGURE 5.33
External lock connector both in the running and locked mode. (Courtesy of GE Oil & Gas.)

6. The lock nut is rotated by turning the torque tool to the left. The lock nut, with left-hand threads, engages the threads in the connector body. Rotation of the lock nut drives the locking sleeve down, forcing the locking dogs into the external profile of the housing.

7. The 20" torque tool is then retrieved. The remaining steps are the same as those described for the standard tie-back adapter, beginning with Step 9.

5.2.2 Subsea Completion

Corresponding to the increase in the cost of platforms, there is a sharp increase in the number of wells being completed subsea in recent years. This is an indication that the reliability factor of subsea equipment available today in the market has gone up significantly due to which more and more operations are opting for the same.

Contrary to the belief that subsea completion is suitable only for high water depth areas where platforms are not feasible, it is found that most of the subsea completed wells are in water depth of less than 100 m.

Beside cost, some other technical advantages of adopting a subsea completion system are enumerated below:

(a) Allows economic exploitation of relatively small reservoirs or marginal fields, where a fixed platform may not be feasible.
(b) Allows production from infill locations in a field that has already been developed.
(c) The only proven method of completion in deep waters.
(d) Allows early production from a reservoir which is yet to be developed.
(e) Can be tied back to the platform in the future if the need arises.

A subsea completion system is basically an assembly of individual subsystems which can be utilised in many forms to suit the development requirements of a particular field. Depending on the requirement there are more than one type of completions.

5.2.2.1 Types of Subsea Completion

In general, there are two types of subsea completions:

1. Single-well completion or satellite completion
2. Multiwell completions

In cases where exploratory or developmental wells need to be completed for early production or where deviated wells drilled from a platform cannot reach the outer fringes of a shallow reservoir or in cases of injection wells, the satellite completions are made. In fact, a system proposed by British Petroleum, single well offshore production system (SWOPS), allows intermittent production to a dynamically positioned tanker through a flexible riser connected to the sea floor satellite tree.

But the second type, that is, the multiwall subsea completion, is more predominant where after drilling wells from floaters through subsea templates, subsea Christmas trees are installed on the template and then wells are produced through a production riser to a floating storage vessel or through flow lines to a remote tanker loading facility (e.g. SBM) or platform.

5.2.2.1.1 Satellite Completions

Satellite completions can be of two types: (1) *wet satellite completion* and (2) *dry satellite completion*.

There is a third type of satellite completion known as *submudline completion* which is nothing but another type of satellite wet completion. The terms 'wet' and 'dry' refer to the condition under which the satellite trees are installed.

'Wet' refers to the system where the Christmas tree is exposed to the sea around it whereas 'dry' refers to the system that encloses the wellhead and the production tree in a one-atmosphere inert gas-filled chamber. The sub-mudline completion system consists of placing most of the production tree in an extended conductor housing well below the ocean floor out of the way of icebergs, anchors and fishing nets. The upper part of the tree and the flow-line connection is housed in a low-profile protective cover.

A satellite completion system can be a simple type where a single-bore stacked valve Christmas tree, non-TFL type, with diver-assisted flow line connection and direct hydraulic control to a minimum number of valves is used. Conversely, it can be a very complex type also with features like multiple bores, TFL workover capability, diverless installation, multiflange flow line connection, electrohydraulic controls and monitoring systems and guidelineless reentry.

(a) Wet Satellite Completions

The major components of wet satellite completions are tubing hanger, subsea Christmas tree, control system and flowline connection system.

Tubing Hanger

Most tubing hanger systems are adaptable to a variety of tub-ing programs and production schemes. These tubing hangers lock and seal inside the uppermost casing hanger body in an existing wellhead installation. The tubing hanger and attached tubing strings normally are run as a complete assembly. Tubing hangers generally are available as either mechanical-set or hydraulic-set units. A significant advantage of the hydraulic-set tubing hanger is that, when run on the completion riser, it can be run, landed, sealed and all downhole operations performed in a single trip.

The hydraulic-set tubing hanger is run through the drilling riser and BOP stack on a running tool and completion riser. This completion riser is an important component not only for lower-ing tubing along with tubing hanger but also used for Christmas tree installation and for future workover.

Subsea Christmas Tree

The wet type Christmas tree (Figure 5.34) is made up of the following basic components:

1. Tree guide frame
2. Wellhead connector
3. Valve block(s)
4. Wye spools swab valve block

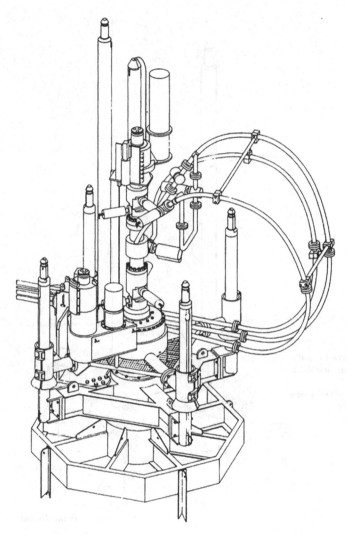

FIGURE 5.34
Wet type TFL Christmas tree. (Courtesy of GE Oil & Gas.)

5. Flow-line loops and crossovers

6. Tree manifold

7. Tree cap assembly

Tree guide frame – The tree guide frame (Figure 5.35) ensures positive guidance of the tree during installation. When the four guide funnels on the guide frame are stabbed over the guide-posts of the permanent guide base (PGB), proper rotational orientation is achieved.

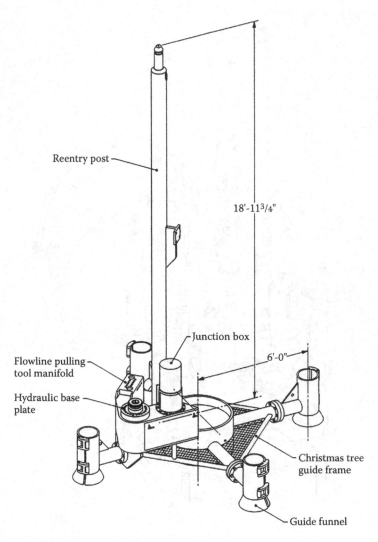

Reentry post

18'-11³/₄"

Junction box

Flowline pulling
tool manifold

Hydraulic base
plate

6'-0"

Christmas tree
guide frame

Guide funnel

FIGURE 5.35
Tree guide frame. (Courtesy of GE Oil & Gas.)

A *wellhead connector* – A hydraulically operated unit that locks and seals the tree directly to the wellhead housing. The connector is remotely operable, incorporating design features that allow it to combine its lock-and-seal integrity, essential during the life of the well, with a release-and-retrieve capability.

Valve block(s) – For non-TFL trees, it is common to include the master valves (for primary control of the well) and the swab valves in a composite valve block (Figure 5.36) allowing a more compact tree configuration. This valve block mounts directly

FIGURE 5.36
Composite valve block. (Courtesy of GE Oil & Gas.)

to the top of the wellhead connector and allows access to the matching tubing bores after the tree is installed. The valve block carries extension subs that remotely stab and seal in the tubing hanger. An orienting bushing bolted to the bottom of the valve block ensures accurate alignment with the tubing hanger.

Wye spool – For TFL* trees which are generally dual bore with an extra annulus access port, an additional component is required in the main body of the tree. This is a TFL diverter, more commonly referred to as a wye spool. The wye spool initiates the angle for the (minimum) 5-ft radius in both the service and production lines to accommodate TFL tools and service operations. It is mounted on the master valve block and allows both vertical and

---

* 'Through flow line' (TFL) technique or 'pump-down' technique requires two flow lines and at least two tubing strings as shown in Figure 5.37 to run a locomotive or piston unit along with flow line tools like scrapers, jars, stem, running and pulling tools and so on together with circulating fluids from a central location to the bottom of the well. It is a workover technique used mostly in highly deviated holes as well as in ocean floor completions.

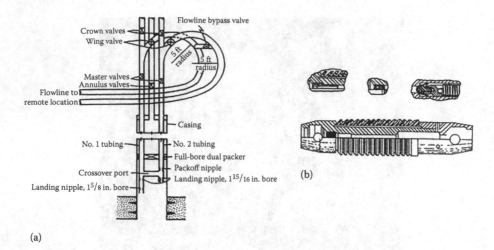

(a)

FIGURE 5.37
TFL or pump-down technique (a) Flow line hookup. (b) Locomotive or piston unit. (From Thomas O. Allen and Alan P. Roberts, *Production Operations*, Vol. 2, Oil & Gas Consultants International Inc., 1984.)

TFL access (for pump down tools) to the production and service tubing bores. Vertical access only is provided to the annulus bore.

Swab valve block (Figure 5.38) – Provides vertical access to the well. Operated only through the tree running tool, the failsafe closed (FSC) swab valves normally are closed when the well is producing. They are opened only during well servicing operations when vertical access to the tubing bores is required. The valves are identical to those used in the master valve block assembly.

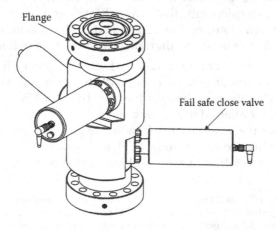

FIGURE 5.38
Swab valve block.

Flow-line loops and crossovers – The flow-line loops interface between the valve block and tree flange, providing flexibility for the necessary linear movement of the tree flange for engagement or disengagement of the flow-line connection.

Located in the flow-line loops are the wing valves. These provide individual fail-safe control for each flow line. When servicing the well vertically, the wing valves are closed and the swab and master valves are opened. A crossover line fitted with a fail-safe open (FSO) gate valve interconnects the flow line and service line. Additionally, with the wing valves closed, the crossover valve can be opened to allow pigging of the flow lines.

Tree manifold (Figure 5.39) – The uppermost fixed member of the subsea completion tree, mating directly to the top flange of the (composite) valve block. The manifold includes the following features:

1. Through bores installed using wire-line methods with internal preparations for receiving plugs
2. An external locking profile that accepts the tree cap or the tree running tool
3. A mounting plate for supporting the reentry post
4. Threaded ports for attaching the hydraulic lines controlling the tree functions
5. Receptacles for the hydraulic stab subs for either the tree cap or the tree running tool
6. Check valves, incorporated in the manifold, to isolate the hydraulic system from contamination when the stab subs are removed

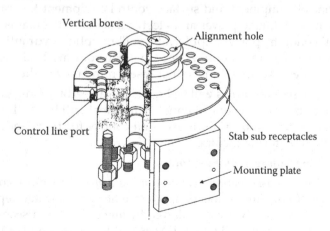

FIGURE 5.39
Tree manifold.

The manifold is the terminal point for all of the hydraulic control lines. When the subsea installation is being worked on from the surface, the through-bores provide communication with the well bore through the completion riser and tree running tool.

Tree cap assembly – When the well is completed and ready to be produced under direct hydraulic control from a remote location, the tree cap assembly is installed on the tree. The tree cap protects and seals the vertical bores of the tree. It serves as a hydraulic control-line connector in conjunction with the tree mandrel. It also serves as a conveniently retrievable package for the various hydraulic control components housed within.

Control System

Normally the tree functions are controlled by the control system which consists of the following sub systems:

1. Subsea controls

2. Control umbilical

3. Surface controls

The *subsea control system* basically consists of subsea control pods to achieve the following:

1. Direct hydraulic control for the well and the flowline

2. Electrical system for measurement of well head pressure and temperature

3. Electrohydraulic system with a sequenced hydraulic backup monitoring system

4. Variable hydraulic bean control for the well

The *control umbilical* serves as the link between the subsea control equipment and surface control equipment for transmission of hydraulic power and electrical signals. It contains electrical conduits, power supply cables, fibre optics, hydraulic tubes, other pipes and so on. A separate control umbilical bundle is provided as a spare line to increase redundancy of the system.

The *surface controls* will be housed on a nearby production platform as a master control panel. This panel will have a hydraulic power generation unit, individual control for trees and subsea data monitoring facilities.

Flow-Line Connection System

Diver-assist flow-line connection system – The most common type of flow-line connection to a satellite tree in water depths up to 600 ft is a diver-assist flanged connection. For a second-end connection where the flow line is laid to the tree with a flanged end, there are two options for making the connection. First, the diver can manoeuvre the end of the line into position making it up to the tree with a misalignment union. Second, if the line is large or difficult to move, a jig can be used to measure for a flanged jumper spool to be fabricated on the surface and then lowered into position for diver makeup.

Diverless flow-line connection systems – The diverless flow-line connection system is designed for tree and flow-line installation as independent functional operations. The flow lines can be laid and attached to the wellhead structure before or after running the tree.

(b) Dry Satellite Completion

In comparison to wet type completion, dry type completions are few but the special reason for interest in the dry tree stems from the expectations that production capability will move to water depths greater than those currently accessible to diver assistance. This is a technique to allow the use of surface-type Christmas trees and valves in a sea-floor application.

Dry completions are made possible by using a pressure-resistant vessel that surrounds the well equipment and encloses it within a bubble of air at atmospheric pressure. This means workers can hook up, repair and maintain the well equipment in a similar environment, as in the surface.

Workers are transported to and from the wellhead chamber by diving bell or submarine. The transport vehicles, which are a necessary component of the dry subsea wellhead system, must be able to lock on to a wellhead chamber, evacuating the water from the transfer trunk between the chamber and the transport vehicle, purging the atmosphere in the chamber, and pressurising the chamber and transfer trunk with breathable air before workers can move from the vehicle into the chamber. They must also purge air from the chamber and refill it with inert gas before returning to the surface. The complexity of the equipment required to service a dry wellhead chamber safely has been a major drawback to general acceptance of this production method.

A one-atmosphere chamber, commonly known as a wellhead cellar, is designed to attach directly onto the casing head of the well with a hydraulic connector of the same configuration as a BOP stack connector. Installation typically is made by running the complete wellhead cellar, loaded with the assembled tree components, on a special drill pipe running tool from a floating drilling vessel. After coupling the cellar to the wellhead, the drilling rig moves offsite and the service capsule (a specially equipped diving bell) or a special submersible with dry transfer capability descends and couples to the wellhead cellar. Assembly and hook-up of the tree and flow-line connections can then be completed by workers inside the protective chamber.

The basic components inside the subsea chambers are essentially identical to normal platform completion hardware since they, too, operate in the dry. Because a great deal of equipment must be installed in such a constricted space inside the pressure vessel, the packaging of the systems is critical. Naturally, the area taken up by items such as flanged connections and the accessibility required for their alignment and installation puts space at a premium. For these reasons, special hardware items are used exclusively in dry subsea completion installations.

5.2.2.1.2 Multiwell Completions

The major components of multiwell completions are subsea templates, subsea production manifold, Christmas trees production riser, flow lines and control system.

5.2.2.1.2.1 Subsea Templates   A subsea template is a large steel structure that is used as a base for various subsea structures such as wells, subsea trees and manifolds. It is the same as those described in Section 4.2.3.1, which are known as drilling templates. These templates can be either modular or unitised. The production template is made up of a drilling template with additional production equipment. This production equipment consists of the manifold, trees and control system. Depending on whether the manifold and the control system are integral with the template or independently fabricated and joined later, the subsea production templates can be classified as follows:

1. Unitised template with integral manifold
2. Unitised template with modular manifold
3. Modular template with modular manifold

A typical subsea template shown in Figure 5.40 is fabricated from a 30-in. diameter tubular. Mounted on the template are a three-point levelling system, six wellhead guide bases, a manifold base, two control pod bases (one electric and one hydraulic), four satellite flow-line connections and two sales line connection receptacles. Interconnecting piping and control lines are also mounted on the template.

The wellhead guide bases are modified from standard permanent guide structures to permit passage of a 36-in. diameter bit. Each guide base is equipped with remote post tops on each of the four posts and vertical stabs for connecting the flow line, annulus line, hydraulic lines and electrical cable from the template to the tree.

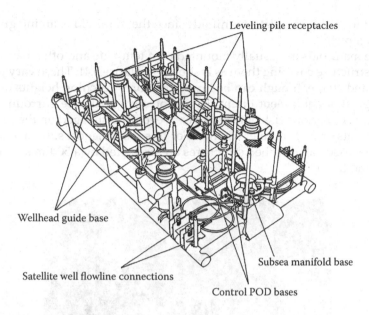

Leveling pile receptacles

Wellhead guide base

Satellite well flowline connections

Subsea manifold base

Control POD bases

FIGURE 5.40
Subsea production template.

The subsea manifold base provides a vertical stab for the flow lines and annulus access lines from each of the 10 (6 + 4) wells. It also has a receptacle for landing and cementing a 30-in. diameter pile to which the manifold is later attached with a hydraulic connector. Two receptacles are incorporated for receiving the vertical stab on the end of the 8-in. sales lines that connects to the tanker loading terminals. The control pod bases, located adjacent to the manifold base, are attachment points for hydraulic and electrical pods for controlling the 10 subsea trees.

The four satellite flow-line connections join the flow line, annulus line, hydraulic supply line and electric control cable from the satellite wells to the subsea template. These diverless flow-line connections have retrievable metal-to-metal seals.

The template-mounted piping connects the vertical stabs at the wellhead guide bases with the manifold base and connects the flow-line receptacles from the four satellite wells to the manifold base. The template-mounted control lines connect the guide bases and the flow-line connections with the control pod bases.

5.2.2.1.2.2 Subsea Manifold A subsea manifold is a system of headers, branched piping and valves used to gather fluids produced from individual subsea wells or to distribute injected fluids such as gas, water or chemicals. The manifold module provides the interface between the production pipeline, flowline and the wells. In Norwegian terms, the manifold is often referred to as the 'template/manifold', since the manifold is an integral part of the template.

These manifolds are usually mounted on a template and often have a protective structure covering them as is shown in Figure 5.41. These vary greatly in size and shape, though can be huge structures reaching heights of 30 m. Although this equipment often has a protective structure surrounding it, still there is a serious risk of fishing gear becoming snagged on the protruding elements or within inner cavities. For this reason, most subsea templates and other associated seabed structures are protected by a 500-m safety zone all around them.

FIGURE 5.41
Subsea manifold. (Courtesy of FMC Technologies/Wikimedia Commons/CC-BY-SA-3.0.)

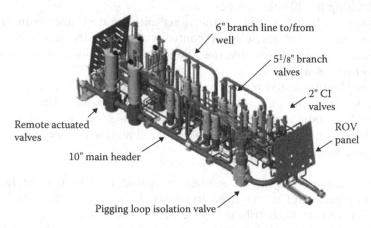

FIGURE 5.42
Manifold piping. (Courtesy of DNV GL Oil & Gas.)

Generally, more than one header – normally two headers – are advantageous in many respects. These increase the flexibility for future export options or tie-ins. Moreover, availability of production increases or chances of individual well testing, removal of hydrate plug and installing a pigging loop increases. The flowlines/piping connected to the header and other equipment in the manifold is shown in Figure 5.42. Each production flowline is equipped with a choke rotated by gate valves for service access.

5.2.2.1.2.3 Tubing Hanger and Christmas Tree The installation of a tubing hanger is carried out in the same manner as discussed for satellite completions.

The template trees are nearly identical to the satellite trees as discussed in satellite completions with the exception of flowline and control line connection. A vertical stab connection is made to the flowline and control line bundle as the tree is landed. The flow-line connection is latched with a hydraulically actuated connector.

5.2.2.1.2.4 Flowline and Production Riser Depending on whether the wells are produced to a nearby fixed production platform or any floating vessel anchored over the template, there can be either flowlines or production risers coming out of the manifold.

There can be one or two *flowlines* depending on well conditions and operational requirements of subsea wells. Their size depends on the flow rate, pressure drop and the tubing head pressure. To communicate with the tubing/casing annulus, another flowline is laid, which can help in monitoring the annulus pressure and for killing the well if required. This line can also be utilised for pigging, scraping or running TFL tools. These lines are not buried below the sea bed but rest on the sea floor itself. The pipes used nowadays are flexible ones made of laminated steel wires and other materials. These are mounted on a reel barge and are laid from that. The flowlines should be protected from any abrasion or physical damage and normally expansion loops are provided along the sea-bed to take care of the temperature changes.

Production risers are free hanging and flexible types which are attached to the floating vessel on its outer perimeter for easy access. These are lowered and connected to the manifold at the vertical stab provided on it. Here also more than one riser is used for the purposes mentioned above.

In some cases, the risers originate from a riser base at the bottom and the flowline connects the template and the riser base. There is a sliding 'L' spool on the riser base which accommodates any length discrepancy. It is important to note that the length of these lines and positional tolerance of the bases relative to the template are critical. Connection methods utilise come-alongs and Chinese fingers for manipulation of the pipe, together with swivel flanges and hydraulic bolt tensioners for make-up. The riser bases are installed and anchor piled to the seafloor.

5.2.2.1.2.5 Control Systems   The control systems comprise the same three subsystems as discussed in satellite completions; however, the connection of umbilical is made at the control pods located in the subsea manifold.

5.2.3 Subsea Completion Procedure

After the drillers have left the site after putting a corrosion cap on the top of the BOP which is placed over a PGB, the well needs to be connected to a floater again to hang the completion string from the tubing hanger to connect the Christmas tree and carry out the completion operations. For this purpose, a completion riser is used. The completion procedure is now enumerated for a wet satellite well.

First the landing of the tubing hanger on the wellhead is to be done. The type of hydraulic set tubing hanger shown in Figure 5.43 is advantageous because an unrestricted thorough bore connection can be maintained and control ports (within the hanger body) for downhole functions can be tied back to the surface through the completion riser. Also in this type of hanger, threaded inserts or internal profiles are incorporated for landing and setting tubing plugs.

The tubing hanger selected may provide for suspending a single string or multiple strings of tubing. The use of two or more strings enables production from separate zones and permits downhole maintenance using pump-down TFL tools. The tubing hanger illustrated is equipped with a 4-in. production bore and a 2 in. bore to provide access to the annulus between the tubing strings and the casing. This 2 in. bore can be used to monitor pressure in the annulus or for a gas lift.

Because of the relative complexity of the multiple-bore tubing hanger with downhole safety valve ports and the requirement for several individual hydraulic control functions to the hydraulic tubing hanger running tool, a simpler operation is achieved by running the tubing hanger on the *completion riser*. The completion riser, illustrated in Figure 5.44, encloses all tubing bore extensions, SSSV control-line extensions and hydraulic controls in an outer casing that provides a known profile for calculating riser stresses and top tension requirements.

The various tubing bore extensions, SSSV lines and hydraulic lines are spaced precisely to match the spacing on the tubing hanger running tool and tubing hanger. The pin-and-box end connections provide a means of fast makeup on the rig floor by activating the actuating screws carried on the pin section of each riser joint.

The completion riser is passed through a 13⅜ in. i.d. BOP stack and drilling riser system and the tubing hanger is installed.

Once the tubing hanger is landed and locked into place by applying hydraulic pressure downhole preparations and the system is successfully pressure-tested, blanking plugs are run on wire-line tools and set in the locking profiles of the tubing hanger. This is done so the well will be shut

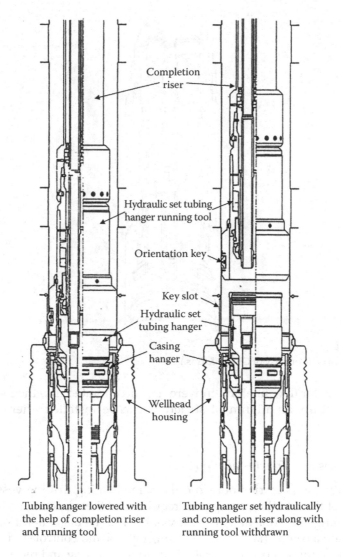

Completion riser

Hydraulic set tubing hanger running tool

Orientation key

Key slot

Hydraulic set tubing hanger

Casing hanger

Wellhead housing

Tubing hanger lowered with the help of completion riser and running tool

Tubing hanger set hydraulically and completion riser along with running tool withdrawn

FIGURE 5.43
Hydraulic set tubing hanger. (Courtesy of GE Oil & Gas.)

in effectively. Once the well is shut in, the tubing hanger running tool and completion riser can be retrieved and the BOP stack removed.

At this point in the completion sequence of the well, the operator is given the choice of running the tree or the flow lines first. Here, the tree will be run first.

The permanent guide structure for this system has the flow-line align-ment structure attached as a permanent member. In those instances when it is decided to produce the well after the wellhead has been installed and

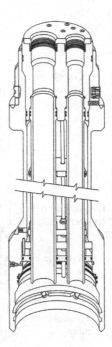

FIGURE 5.44
Completion riser. (Courtesy of GE Oil & Gas.)

preparations for flow line pull-in were not incorporated into the permanent guide structure, an alignment structure can be installed after wellhead installation.

5.2.3.1 Christmas Tree Installation

Tree installation is carried out from the same floating drilling vessel, since nearly all of the handling equipment required for these operations is utilised in the drilling operation and is already onboard. The major additional items required are a completion riser, tree running tool and control skid.

The tree running tool (Figure 5.45) is used for running and retrieving both the tree and the tree cap as individual units. The tool is run on a completion riser or may be run on a drill pipe or individual tubing strings in conjunction with a control hose bundle. The tool has an integral hydraulically actuated connector that latches onto the tree manifold or the top profile of the tree cap.

The tree normally is placed on a test stump on the installing vessel and is operated through functional cycles and pressure-tested prior to being deployed. The tree is run on the completion riser over guidelines and is landed and locked to the wellhead. It is again functioned and pressure tested. All downhole work is completed prior to retrieving the running tool and completion riser. When the running string is retrieved, the well bore is

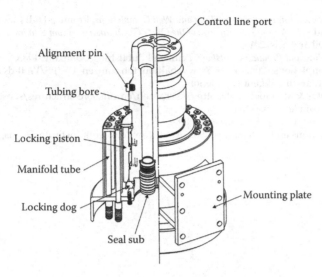

FIGURE 5.45
Tree running tool.

secured by the master valves and the tree is ready for installing the tree cap assembly, if required.

If no immediate completion operations are to be carried out, the tree cap assembly can be run in the same manner as the tree and is latched in place on top of the tree to seal off the vertical bores and complete the tree control package. For re-entry, the tree cap is retrieved and the tree running tool with completion riser is installed to gain downhole access to the well from the surface. All the completion operations are carried out through the completion riser.

In the case of multiwell completions, the procedure remains the same. Each wellhead system is supported in an individual receptacle on the template and those are completed separately with the help of completion risers run separately.

Bibliography

Allen, T. O. and A. P. Roberts, *Production Operations*, Vol. 2, Oil & Gas Consultants International Inc., 1984.

Completion (Oil & Gas Wells), 2015, http://en.wikipedia.org/wiki/completion_(oil_and_gas _wells).

ETA Offshore Seminars Inc, The Technology of Offshore Drilling, Completion and Production, Pennwell Publishing Company, 1976.

Frick, T. C. and R. William Taylor, *Petroleum Production Handbook*, Vol. I, SPE of AIME, 1962.

Gatlin, C. *Petroleum Engineering – Drilling and Well Completions*, Prentice-Hall Inc., 1960.

Gomes, J. S. and F. B. Alves, *The Universe of the Oil & Gas Industry – From Exploration to Refining*, Partex Oil and Gas, 2013.

Hall, R. S. *Drilling and Producing Offshore*, Pennwell Publishing Company, 1983.

Intelligent Completions, 2015, http://www.halliburton.com/en_US/PS/Well-dynamics/well-completions/intelligent-completions.

Subsea Templates & Manifolds, 2015, http://fishsafe.eu/en/offshore-structures/subsea-structures/subsea-templates-manifolds.

VETCO Templates and Tie-Back Systems, 1984.

Wireline Operations and Procedures, 1983, American Petroleum Institute, Dallas, TX.

6

Offshore Production

In the oil and gas industry, the term 'production' covers all the operations that are carried out *within the wellbore* to make the well fluid flow to the surface or the well head and the subsequent operations that are carried out *at the surface* to ultimately deliver oil and gas separately to the consumers as per their specifications. Down-the-hole operations like workover, well intervention, stimulations, artificial lift operations and so on are carried out exactly in the same way as is done in the case of onshore wells and hence have not been included here. The 'surface operations' though also are principally the same as onshore wells, the arrangement of the production system in the mid-ocean needs special attention. Hence, the subject of discussion will be the surface production system applied in offshore wells.

6.1 Outline of Surface Production System

The whole gamut of the surface production system includes separation of the well fluid into three phases, namely, oil, gas and water; processing of these phases into some marketable product(s) or disposing them in an environmentally acceptable manner; storing these products and finally transporting these to the respective consumers.

6.1.1 Oil and Gas Separation

Produced wellhead fluids are complex mixtures of different compounds of hydrogen and carbon all with different densities, vapour pressures and other physical characteristics. As the well fluid flows from the hot, high pressure petroleum reservoir, it experiences pressure and temperature reductions. Gases evolve from the liquids and the well stream changes in character. The gas carries the liquid droplets and the liquid carries gas bubbles. The physical separation of these phases is the main function of 'oil and gas separator.' In oil field terminology, a *separator* is a pressure vessel used for separating well fluids coming directly from oil or a gas well or a group of wells into gaseous and liquid components.

Separators are sometimes called *gas scrubbers* when the vessel is used for separating liquids from a gas stream or when GOR is very high, sometimes the term *trap* is also used in its place when the separators are used to handle flow directly from wells.

In any case, they all have the same configuration. These separators may be either 2-phase or 3-phase. The *2-phase* separators remove the total liquid from the gas and the *3-phase* separators also remove free water from hydrocarbon liquid.

The separators besides removing liquid from gas and gas from liquid also maintain optimum pressure on the separator and the liquid seal inside the separator.

For efficiency and stable operation over a wide range of conditions, a gas-liquid separator must have the following components (Figure 6.1):

1. *Primary separation section* for separating the bulk of the liquid from the well stream. It is desirable to remove quickly liquid slugs and large droplets of liquid from the gas stream to minimise turbulence and re-entrainment of liquid particles. This is accomplished by the tangential inlet which imparts a circular motion to the fluids.

2. *Liquid accumulation section* for receiving and disposing of the liquids collected. This section should be of sufficient volume to handle fluid surges that may occur in normal operation and should be arranged so that the separated liquid is not disturbed by the flowing gas stream. When large amounts of liquid are handled or severe fluid surges occur, it may be necessary to enlarge the liquid accumulation section.

3. *Secondary separation or gravity settling section* for removing the smaller liquid droplets. The major principle of separation in this section is gravity settling from the gas stream. Since the basic requirement for gravity settling is minimisation of turbulence, it is important to decrease the gas velocity right at the entrance to the separator. Properly positioned straightening vanes provide uniform gas flow throughout the separation section.

4. *Mist extraction section* for removing entrained droplets too small to settle by gravity. Entrained drops are those carried over from the secondary separation section, when the vapour velocity is greater than the rate of settling of the drops.

Mist extractors can be of various types like *vane type, wire-mesh type* and *arch plate type*. All these types utilise the physical principles like (1) impingement, (2) change of flow direction and (3) change of velocity which is self-explanatory from Figure 6.2. Some other types of mist extractors are also in use like *coalescing packs, cyclone type* and so on, which utilise centrifugal force as another physical factor.

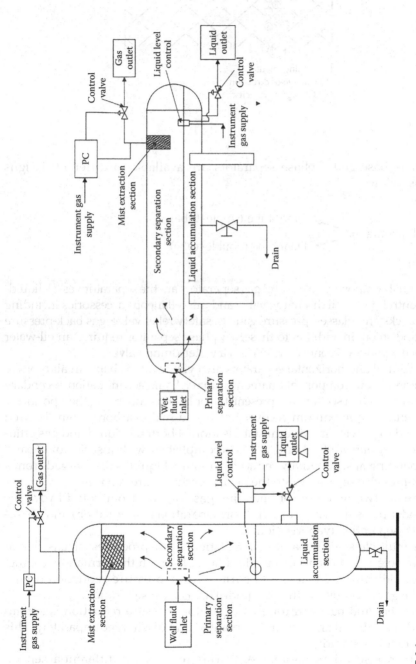

FIGURE 6.1
Components of vertical and horizontal separator.

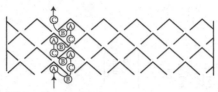

(A) Impingement
(B) Change of direction
(C) Change of velocity

FIGURE 6.2
Mist extractors.

Both 2-phase and 3-phase separators are available in different designs such as follows:

a. Vertical

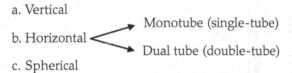

b. Horizontal

c. Spherical

Essential components of oil and gas separators are the separator vessel, liquid-level control (LLC), oil discharge valve and miscellaneous accessories including gauge cocks and glasses, pressure gauges, safety relief valve, gas back-pressure valve and so on. In addition to these, a 3-phase separator requires an oil-water interface liquid-level control means and a water-dump valve.

The vertical and horizontal separators are cylindrical in shape. In all types of separators all four components, namely primary, liquid accumulation, secondary and mist extraction sections are present, the only difference being their positions.

For obtaining maximum recovery of liquid hydrocarbons from the well fluid, and to provide maximum stabilisation of both the liquid and gas efflu-ent, *stage separation* of oil and gas is accomplished with a series of separa-tors operating at sequentially reduced pressure. Liquid is discharged from a higher-pressure separator into the next lower-pressure separator.

There are two processes of liberating gas (vapour) from liquid hydrocar-bon under pressure. These are: (1) flash separation (vaporisation) and (2) dif-ferential separation (vaporisation).

In a multiple-stage separator installation, both processes of gas libera-tion are obtained. When the well fluid flows through the formation, tubing, chokes, reducing regulators and surface lines, pressure reduction occurs with the gas in contact with the liquid. This is flash separation.

When the fluid passes through a separator, pressure reduction is accom-plished; also, the oil and gas are evaporated and discharged separately. This is differential separation.

The more nearly the separation system approaches true differential separa-tion from producing formation to storage, the higher the yield of liquid will be. Figure 6.3 shows the arrangement of separators in the stage separation

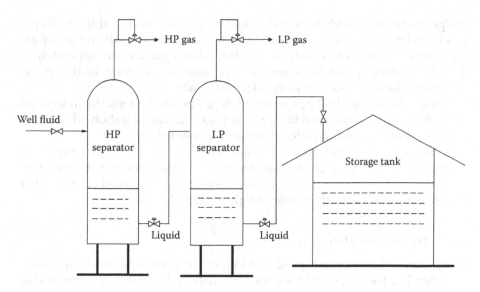

FIGURE 6.3
Three-stage separation.

process. Usually 3-stage or 4-stage separation is economical. A 3-stage separation means two separators and one storage tank and similarly a 4-stage separation means three separators and one storage tank.

The separators are always equipped with some safety devices. Some of these are (1) high-and-low-liquid level controls, (2) high-and-low pressure controls, (3) high-and-low temperature controls, (4) safety relief valves and (5) safety heads or rupture discs.

When wells are extremely of higher capacity, each well can have its own facilities for separation and metering, possibly also for treatment but as a common practice the well fluids are gathered at a particular station where stage separators are installed for separating the collected stream of well fluid. This gathering system can be of two types: (1) well-centre gathering systems and (2) common-line gathering system (Figure 6.4). In the first case,

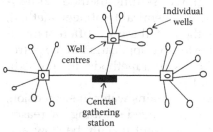

Well-centre gathering system

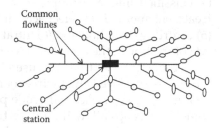

Common-line gathering system

FIGURE 6.4
Gathering systems – two types.

the produced well fluids from individual wells are collected at the well centres sometime called a group gathering station where gas, oil and water are
separated and then from each well centre, oil and gas are sent separately to
a central gathering station where further treatment is done. In this place,
sometimes facilities for storage of oil is also made.

In the second case, that is, common-line gathering system, the individual
wells flow through common lines and gather at a central station where after
separation of oil, gas and water, those are further treated.

In the case of gas wells, the gas stream is made to pass through a separator
or a scrubber to remove condensate and water in the same way as described
above in the case of an oil well. Condensate has the composition and other
characteristics very much similar to light crude oil.

6.1.2 Treatment of Oil

More than 80% of the crude produced in the world contains various amounts
of water. The treating of oil, therefore, consists of *dehydration* (i.e. removal of
water) and *desalting*, if required, by methods such as using a settling tank,
heater treaters, or electrical dehydrators and desalters such as water wash or
water dilution. The objective is to remove water and salt contaminants from
the oil to meet the processing requirements in the refinery. The crude oil supplied to the refinery should have water content less than 1% and salinity less
than 50 mg per litre (50 ppm).

Dehydration – This treatment is principally the treatment of emulsion. The
emulsion is defined as a mixture of two mutually immiscible liquids, one of
which is dispersed as droplets in the other and is stabilised by an emulsifying agent's film envelope around the droplet. The dispersed phase is known
as the internal phase and the liquid surrounding the dispersed droplets is
the external or continuous phase. Water-in-oil emulsion, that is, water as
dispersed phase and oil as continuous phase, is most common. Emulsifying
agents, such as formation fines, silts, asphaltenes, waxes and resins stabilise
the dispersion.

Methods employed in dehydrating crude petroleum emulsions may
be classified into six groups, such as: (1) gravity settling method, (2) heat-
treatment method, (3) electrical methods, (4) chemical-treatment method,
(5) centrifugal method and (6) filtration method. Of these six different methods of treatment, the first four have found extensive application in the petroleum industry; the fifth is less used than the first four methods; and the sixth
method is employed chiefly in experimental tests.

Gravity settling method – If crude petroleum contains water in suspension,
some, or perhaps all, will often settle out on prolonged standing, by reason
of the greater density of water. This separation will usually be slow and
incomplete and, in the case of thoroughly emulsified water, will perhaps
merely take the form of downward concentration of the suspended water
droplets, without actual coalescence, still permitting some water-free oil to

be skimmed off the top of the fluid. The lower portion of the fluid will, of course, contain a higher percentage of water than before gravitational settling occurred.

Gravity settling is often combined with other processes of treatment to accomplish final and complete separation of the two fluids, after the emulsion has been 'broken' by appropriate means. For example, certain chemical methods of treatment are effective in bringing about emulsification, but treatment by this method is followed by a period of gravity settling, during which the water is liberated and is permitted to settle out, leaving the supernatant oil practically free of water.

Heat-treatment methods – Application of heat to petroleum emulsions is often an effective means of accomplishing their dehydration and is, in many instances, used to assist other processes. Heat aids in bringing about separation of water from petroleum emulsions in three ways: (1) by reducing the viscosity of the oil, so that gravity may more readily operate in settling out the heavier water; (2) by effecting change in the interfacial tension relationships and colloidal properties of the emulsifying agent; and (3) if the temperature is carried high enough, by actually bringing about a change from the liquid to the gaseous state, the steam formed bursting the enclosing oil films about the water droplets. The equipment used for such treatment is known as heater treater. The most commonly used single well lease treater is the *vertical treater* as shown in Figure 6.5. Flow enters the top of the treater into a gas separation section. This section has an inlet diverter and a mist extractor.

The liquids flow through a downcomer to the base of the treater, which serves as a free-water knockout section. If the treater is located downstream

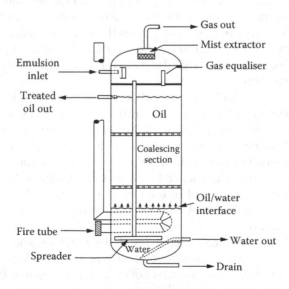

FIGURE 6.5
Vertical heater–treater.

of a free-water knockout, the bottom section can be very small. If the total wellstream is to be treated, this section should be sized for 3 to 5 min retention time for both the oil and the water to allow the free water to settle out. This will minimise the amount of fuel gas needed to heat the liquid stream rising through the heating section. The end of the downcomer should be slightly below the oil-water interface to 'water wash' the oil being treated. This will assist in the coalescence of water droplets in the oil.

The oil and emulsion rise over the heater fire-tubes to a coalescing section where sufficient retention time is provided to allow the small water particles in the oil continuous phase to coalesce and settle to the bottom.

Treated oil flows out of the oil outlet. Any gas, flashed from the oil due to heating, flows through the equalising line to the gas space above. Oil level is maintained by pneumatic or lever operated dump valves. Oil-water interface is controlled by an interface controller, or an adjustable external water leg.

For most multi-well situations *horizontal treaters* are normally required. The working principle of these treaters remains the same.

Electrical method – When brought within the field of influence of a high-potential alternating current, the dispersed water droplets of a petroleum emulsion coalesce to form larger water aggregates, which readily settle out under the influence of gravity. In this process, the emulsion is generally heated moderately to reduce viscosity and facilitate settling. Very complete separation of the water and oil is quickly effected at small cost by this method.

Theory indicates that the droplets of emulsified water, when passing between two electrodes upon which a high potential is imposed, become charged by induction. One side of each globule is negatively charged, the other positively and attraction between opposite charges causes the water droplets to align themselves, forming a chain between the electrodes. When this occurs, a discharge of electricity passes through the chain from one electrode to the other. As a result, the water globules comprising the electrical path coalesce and form one large drop of water, which readily settles from the oil under the influence of gravity.

Dehydration by the electrical method occurs in two steps. In the first step, the minute water globules coalesce, forming larger masses; and in the second, the larger masses settle from the oil under the influence of gravity.

Figure 6.6 illustrates a typical design of a horizontal electrostatic treater. Here an AC and/or DC electrostatic field is used to promote coalescence of the water droplets. It is experienced that such electrostatic treaters are efficient at reducing water content in the crude below the 0.5–1.0% basic sediment and water (BS & W) level. This makes them particularly attractive for desalting applications.

Chemical method – Chemicals used to break the emulsion known as demulsifiers are sold under various trade names, such as Tretolite, Visco and Braksit. Demulsifiers act to neutralise the effect of emulsifying agents. Typically, they are surface active agents and thus their excessive use can decrease the surface tension of water droplets and actually create more stable emulsions.

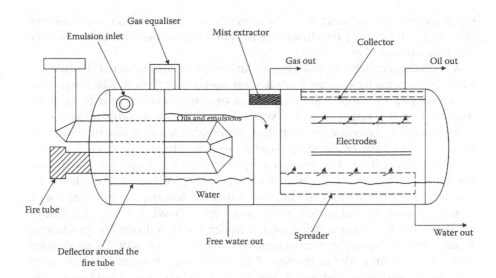

FIGURE 6.6
Horizontal electrostatic treater.

There are four important actions required of a demulsifier: (1) Strong attraction to the oil-water interface, (2) flocculation, (3) coalescence and (4) solid wetting.

When these actions are present, they promote the separation of oil and water. The demulsifier must have the ability to migrate rapidly through the oil phase to the droplet interface, where it must compete with the more concentrated emulsifying agent. The demulsifier must neutralise the emulsifier and promote a rupture of the droplet interface film which causes coalescence. With the emulsion in a flocculated condition, the film rupture results in rapid growth of water-drop size.

The way the demulsifier neutralises the emulsifier depends upon the type of emulsifiers. It would be unusual if one chemical structure could produce all four desirable actions. A blend of compounds is therefore used to achieve the right balance of activity.

The demulsifier selection should be made with the process system in mind. If the treating process is a settling tank, a relatively slow acting compound can be applied with good results. On the other hand, if the system is a *chemelectric* process where some of the flocculation and coalescing action is accomplished by an electric field, there is a need for a quick-acting compound, but not one that must complete the droplet-building action.

Centrifugal method – Centrifugal force, developed mechanically by rapid rotation of an oil-water mixture, is effective in bringing about separation of the two fluids. The effect is identical with that of gravity but is many times as powerful. Centrifuges capable of developing rotational speeds of 15,000 rpm develop a force equivalent to 13,000 times that of gravity. Efficiency in separation of the oil and water is dependent directly upon their relative densities – the greater the difference in specific gravities, the more effective the separation will be.

A centrifuge capable of bringing about complete separation of the two fluids from an emulsion must have a high speed of rotation. Machines thus far developed for large-scale continuous operation consist of a metal bowl mounted on the upper end of a vertical shaft that is rapidly rotated by a direct-connected electric motor. Belt-driven machines may also be secured if desired. The oil-water mixture, usually preheated to reduce viscosity, is fed into the bowl at its centre through a pipe that discharges into the bowl near the bottom. The selective action of centrifugal force then causes the water (the denser of the two fluids) to move outward toward the perimeter as it moves upward toward outlets in the cover of the bowl. A clear-cut cylindrical plane of separation is developed between the two fluids, the position of which depends upon the percentages of oil and water present. Clean oil overflows through an outlet in the top of the bowl near the centre, while water escapes through a second outlet near the outer perimeter of the bowl. Sand and other suspended impurities tend to follow the water, though the coarser solids generally remain in the bowl and must occasionally be removed by hand methods after stopping rotation and removing the cover.

Filtration method – In this method, processes are designed to separate emulsified oil and water by filtration under pressure through various porous mediums. Such filtering medium can be water-wet or oil-wet.

Water passes through such a water-wet filtering medium, and if the pore spaces of the filter are sufficiently small, the oil phase will be left on the upstream side of the filter. Conversely, an oil-wet filter cloth will pass oil but not water. Excelsior, sand, diatomaceous earth, glass wool and other filtering mediums have also been used with more or less success, often with the aid of the familiar filter press, so widely used in wax filtration in oil refining and for like purposes in other industries. In addition to their filtrative properties, the sharp angular edges and projections that such materials present are to some extent successful in altering the interfacial tension relationships between the water and the oil. 'Hay tank' is one such treater employing the filtrative principle.

Desalting – The water content of crude oil can be brought to the acceptable limit of less than 1% by electrostatic treatment; however, the salinity of the oil might be still higher than the accepted limit fixed by refinery.

Desalting operation is carried out mainly to reduce the salinity of crude oil. Desalting, which follows the initial dehydration, consists of

1. Adding dilution (or fresh) water to the crude.
2. Mixing this dilution water with the crude to dilute saline water.
3. Emulsion treating or second stage dehydration to separate the crude and brine phase.

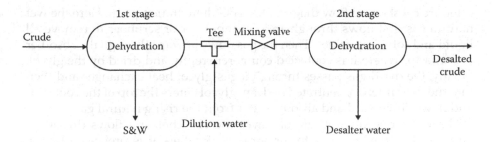

FIGURE 6.7
Desalting in two stages.

A schematic of 2-stage desalting is shown in Figure 6.7. In 2-stage desalting, the required water injection rate is usually 5–7% of the crude flow rate. The mixing tee mixes fresh dilution water and shears dilution water into droplets and disperses it throughout the crude.

6.1.3 Treatment of Gas

The major treatments made to the produced gases are *dehydration* and *sweetening*.

Dehydration – The major reasons for dehydrating natural gas are

1. Natural gas can combine with liquid or free water to form hydrate that can plug valves, fittings, or even pipelines.
2. When liquid water is present, natural gas becomes corrosive, especially if CO_2 and/or H_2S are also present.
3. Water vapour in natural gas can condense in the pipeline causing slug flow.
4. Water vapour increases the volume and decreases the heating value.

There are two methods of dehydration based on the principle of either absorption or 'adsorption,' and these are

1. Glycol dehydration
2. Solid bed dehydration

The 'glycol dehydration process' is an absorption process, where the water vapour in the gas stream becomes dissolved in a relatively pure glycol liquid solvent stream. Glycol dehydration is relatively inexpensive, as the water can be easily 'boiled' out of the glycol by the addition of heat. This step is called 'regeneration' or 'reconcentration' and enables the glycol to be recovered for use in absorbing additional water with minimal loss of glycol. The type of glycol commonly used is either di-ethylene glycol (DEG) or tri-ethylene glycol (TEG).

Figure 6.8 shows a flow diagram for TEG dehydration plant. Here the wet natural gas first flows through an inlet separator or scrubber to remove all liquid and solid impurities. Then the gas flows upward through the absorber or contactor where it is contacted countercurrently and dried by the glycol. Finally, the dried gas passes through a gas-glycol heat exchanger and then into the sales line. Concentrated or 'lean' glycol enters the top of the contactor and flows downward and absorbs water from the rising natural gas.

The water-rich or 'wet' glycol leaves the absorber and flows through a larger coil in the accumulator or surge tank where it is preheated by the hot lean glycol. After this glycol–glycol heat exchange, the rich glycol enters the stripping column and flows down the packed bed section and into the reboiler. Steam generated in the reboiler strips water from the liquid glycol as it rises up the packed bed and is finally vented from the top of the stripper. The hot reconcentrated glycol flows out of the reboiler into the surge tank where it is cooled. Finally, the 'lean' glycol is pumped through the glycol–gas exchanger and is fed back into the top of the absorber.

'Solid bed dehydration' systems work on the principle of adsorption. Adsorption involves a form of adhesion between the surface of the solid desiccant and the water vapor in the gas. The water forms an extremely thin film that is held to the desiccant surface by forces of attraction, but there is no chemical reaction. The desiccant is a solid, granulated drying or dehydrating medium with an extremely large effective surface area per unit weight because of a multitude of microscopic pores and capillary openings. A typical desiccant might have as much as 4 million square feet of surface area per pound.

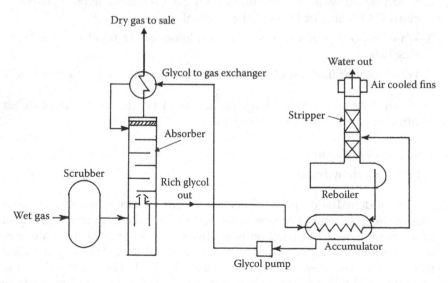

FIGURE 6.8
TEG dehydration unit – flow diagram.

Figure 6.9 shows a flow diagram for a typical two-tower solid desiccant dehydration unit. The essential components of any solid desiccant dehydration system are

1. Inlet gas separator.
2. Two or more adsorption towers (contactors) filled with a solid desiccant.
3. A high-temperature heater to provide hot regeneration gas to reactivate the desiccant in the towers.
4. A regeneration gas cooler to condense water from the hot regeneration gas.
5. A regeneration gas separator to remove the condensed water from the regeneration gas.
6. Piping, manifolds, switching valves and controls to direct and control the flow of gases according to the process requirements.

In the *drying cycle*, the wet inlet gas first passes through an inlet separator where free liquids, entrained mist and solid particles are removed.

In the *adsorption cycle*, the wet inlet gas flows downward through the tower. The adsorbable components are adsorbed at rates dependent on their chemical nature, the size of their molecules and the size of the pores. The water molecules are adsorbed first in the top layers of the desiccant bed.

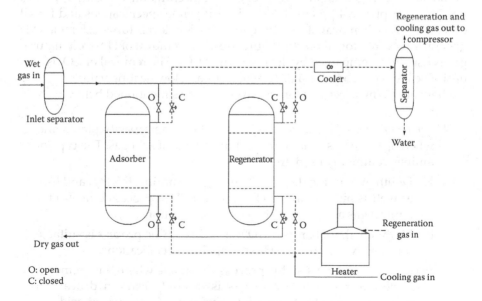

FIGURE 6.9
Solid bed dehydrator – flow diagram.

Dry hydrocarbon gases are adsorbed throughout the bed. As the upper layers of desiccant become saturated with water, the water in the wet gas stream begins displacing the previously adsorbed hydrocarbons in the lower desiccant layers. Liquid hydrocarbons will also be absorbed and will fill pore spaces that would otherwise be available for water molecules.

At any given time, at least one of the towers will be adsorbing while the other towers will be in the process of being heated or cooled to regenerate the desiccant. When a tower is switched to the *regeneration cycle*, some wet gas (i.e. the inlet gas downstream of the inlet gas separator) is heated to temperatures of 450–600°F in the high-temperature heater and routed to the tower to remove the previously adsorbed water. As the temperature within the tower is increased, the water captured within the pores of the desiccant turns to steam and is absorbed by the natural gas. This gas leaves the top of the tower and is cooled by the regeneration gas cooler. When the gas is cooled, the saturation level of water vapor is lowered significantly and water is condensed. The water is separated in the regeneration gas separator and the cool, saturated regeneration gas is recycled to be dehydrated. This can be done by operating the dehydration tower at a lower pressure than the tower being regenerated or by recompressing the regeneration gas.

Sweetening – Gas sweetening is processing of natural gas to remove acid gases such as hydrogen sulphide and/or carbon dioxide which form corrosive acids in the presence of moisture and ultimately lead to corrosion of metals. Hydrogen sulphide and carbon dioxide gases are found as a constituent of natural gas in many fields. The concentration of these gases may range up to 10%. Removal of hydrogen sulphide to a very low concentration, typically 1/4 grain per 100 scf (8 ppmw, 4 ppmv) is desired for pipeline specifications and for all other purposes. Removal of CO_2 to reasonably low levels (up to 2.5 mole%) is generally required for all practical purposes. The removal of H_2S from natural gas is usually accompanied by the removal of CO_2. The removal of CO_2 is also desirable if the gas contains moisture or if it has a low heating value.

There are many processes for gas sweetening as mentioned below:

1. *Amine process* – This is generally used for large capacity plants and where gas streams contain more than 0.075% of acid gas. The type of amines commonly used are

   a. Di-ethanol-amine (DEA) – By using this amine, the H_2S and CO_2 as well as mercaptans can be removed. This process is useful for natural gas as well as refinery gas applications.

   b. Methyldiethanolamine (MDEA) – This amine process is suitable for selective removal of H_2S to pipeline specifications.

2. *Physical solvent process* – This process is suitable where bulk removal and selective removal of acid gas is desired, heavy hydrocarbon quantity (% age) in feed gas is low, and partial pressure of acid gas in feed is greater than 50 psig.

3. *Hot carbonate process* – This process can remove CO_2 to very low specifications and H_2S to pipeline specifications.
4. *Batch process* – For example,

   Iron sponge process

   LO-CAT

   Chemsweet

   Molecular sieve

MDEA Process – Figure 6.10 shows a layout of an MDEA sweetening plant. Here the sour gas first passes through a knock-out drum where liquid carryover is separated. Liquid collected at the bottom of the drum is pumped out to the condensate treating section. Upon leaving the knock-out drum, the gas enters the amine absorber (tray type) where it comes into countercurrent contact with the aqueous solution of MDEA.

The absorber column normally contains 12 to 18 valve trays. The lean MDEA solution enters the column from the upper section at 45°C and gas enters at the bottom of the column. Different liquid feed nozzles on the column give the flexibility of allowing the correct feed plate to be selected to obtain the required results. The treated gas leaves the top of the absorber containing less than 4 ppm (vol) H_2S at a temperature of 40–44°C.

The treated gas enters a knock-out (KO) drum in which the liquid phase (condensed water, amine carry-out) is separated from the gas stream. The liquid from KO drum is sent to rich amine flash drum. The treated gas leaves from the top of the KO drum and is piped to triethyleneglycol

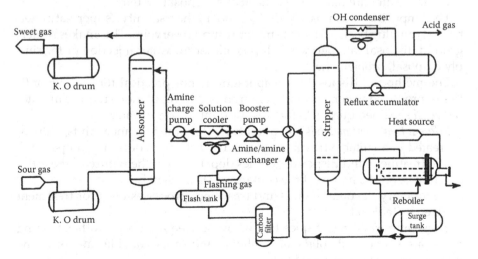

FIGURE 6.10
MDEA sweetening unit – flow diagram.

drying/dehydration unit. The rich amine solution flows from the bottom of the absorber to the rich amine flash drum and then to the rich/lean amine exchanger. MDEA solution stripping is accomplished in the regenerator.

Stripping vapor (primarily steam) is generated in the reboiler. The acid gases and the steam leave the top of the stripper and pass through a condenser, where a major portion of the steam is condensed and cooled.

The acid gases are separated in a separator and sent to flare or to processing. The condensed steam is returned to the top of the stripper as reflux.

6.1.4 Treatment of Produced Water

Petroleum reservoirs always contain some water along with hydrocarbons. This water is produced with oil and gas and contains some soluble inorganic compounds, such as salts of sodium, calcium, magnesium, potassium and some toxic metals. The composition and proportions of produced water in oil varies with both time and location. Produced water collected after separating oil is termed effluent or waste water. The water contains oil in the form of small droplets of varying sizes, along with salts and other chemical compounds. The impurities have an adverse effect on the environment if effluent is discharged into potable water sources or irrigation waters.

Different methods in vogue for the disposal of produced water include: (1) injection into permeable underground formations containing saline water, through disposal wells, (2) reuse for supplementary recovery operations like water injection, (3) disposal by evaporation and (4) surface disposal in rivers, lakes and so on. Before implementing any of these, environmental protection criteria have to be kept in mind because many of the impurities in produced water are harmful for the above disposal systems.

The impurities such as oil sludge and higher salinity (super saturated brines) can plug back permeable underground reservoirs. This makes underground disposal into permeable layers, either for water injection or for simply disposal, problematic.

The method of disposal by evaporation is not practical for large quantities of produced water. Similarly, environmental regulations require maintenance of specified quality before disposal into surface water sources.

Effluent treatment methods – In order to achieve the tolerance limits, effluent is treated in a suitable treatment plant. The selection, design and operation of the treatment and disposal facilities depend upon the nature of the effluent, that is, on the nature of the contaminants present in it.

A summary of contaminants and treatment processes used for treatment is given in Table 6.1.

These important techniques for removing free oil, grease, other floating materials, suspended solids, oily colloids and other colloidal matter in the effluent are briefly discussed below.

TABLE 6.1

Methods of Effluent Treatment

Contaminant	Unit Operation, Unit Process or Treatment System
Free oil, grease, other floating material	• Sedimentation, flotation • Filtration
Suspended solids, oily colloids and other colloidal material	• Chemical-polymer addition • Flocculation/coagulation followed by sedimentation
Biodegradable organics	• Activated-sludge variations Fixed film Trickling filters Rotating biological contactors • Lagoon variations/intermittent sand filtration/physical-chemical system
Heavy metals	• Chemical precipitation/ion exchange
Dissolved inorganics solids	• Ion exchange/reverse osmosis/electrodialysis

Gravity systems/sedimentation techniques – Here oil separates from water because of its lower density and then is skimmed at the water surface. Separation takes place in tanks, where retention time is high, sometimes up to several hours. Skimmed oil is recovered by gravity or with an oleophilic drum. Efficiency is limited by the size of oil particles (>150 microns). Gravity systems can accept any amount of oil in water, even a slug of pure oil. Solid particles also settle in these tanks from which they are periodically removed.

It is a typical primary treatment that can remove large quantities of oil but frequently does not reach the required standards. It is very efficient at regulating the flow rate and the oil content before a secondary treatment. Chemicals like flocculating agents are used for enhancing effectivity.

All these gravity systems fail to achieve the required standards but they are widely used in the oil industry as a primary treatment method.

Coalescing techniques – In this method, the main principle is to assist the collision of small oil particles to form large ones which rise more rapidly. Chemicals are used to give an oleophilic property to the coalescing medium. Different types of coalescing surfaces are plates, filter media (loose media), sponge-like foam and cylindrical cartridges (fixed media). Out of these, plate coalescers are more common.

Various configurations of plate coalescers have been devised, which are commonly called parallel plate interceptors (PPI), corrugated plate interceptors (CPI), or cross-flow separators. All of these depend on gravity separation to allow the oil droplets to rise to a plate surface where coalescence and capture occur. As shown in Figure 6.11, flow is split between a number of parallel plates spaced a short distance apart. To facilitate capture of the oil droplet, the plates are inclined to the horizontal.

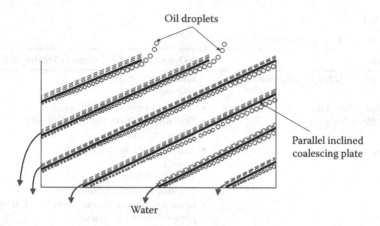

FIGURE 6.11
Gravity separation in parallel plates.

The first form of a plate coalescer was the *parallel plate interceptor* (PPI). This involved installing a series of plates parallel to the longitudinal axis of a horizontal rectangular skimmer (Figure 6.12). The plates form a 'V' when viewed along the axis of flow so that the oil sheet migrates up the underside of the coalescing plate and to the sides. Sediments migrate toward the middle and down to the bottom of the separator where they are removed.

Similarly in *corrugated plate interceptors* (CPI) the parallel plates are corrugated (like roofing material) with the axis of the corrugations parallel to the direction of flow. The plate pack is inclined at an angle of 45° and the bulk water flow is forced downward (Figure 6.13).

The oil sheet rises upward counter to the water flow and is concentrated in the top of each corrugation. When the oil reaches the end of the CPI pack it is collected in a channel and brought to the oil/water interface.

Cross-flow devices have modified CPI configuration for horizontal water flow perpendicular to the axis of corrugations in the plates. This allows the

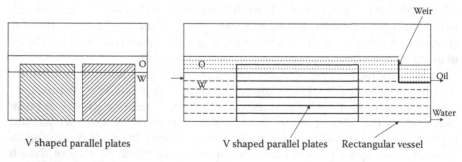

FIGURE 6.12
Parallel plate interceptor (PPI).

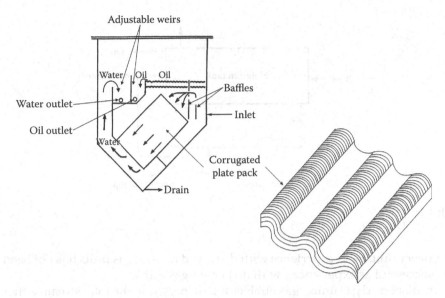

FIGURE 6.13
Corrugated plate interceptor (CPI) flow pattern.

plates to be put on a steeper angle to facilitate sediment removal, and to enable the plate pack to be more conveniently packaged in a pressure vessel.

Chemical treatment techniques – Chemicals are generally used in conjunction with the separation method, that is, gravity separation, coalescing, flotation, or filtration. The chemicals enhance the efficiency of these separation methods and different types of chemicals are used according to the technique used to remove oil from water.

Gas flotation techniques – This technique does not rely on gravity separation of the oil droplets. Two distinct types of flotation units have been used that are distinguished by the method employed in producing the small gas bubbles needed to contact the water. These are *dissolved gas units* and *dispersed gas unit*.

'Dissolved gas designs' take a portion of the treated water effluent and saturate the water with natural gas in a contactor. The higher the pressure the more gas can be dissolved in the water. Most units are designed for a 20–40 psig contact pressure. Normally, 20–50% of the treated water is recirculated for contact with the gas. The gas saturated water is then injected into the flotation tank as shown in Figure 6.14. The dissolved gas breaks out of solution in small diameter bubbles that contact the oil droplets in the water and bring them to the surface in a froth.

Dissolved gas units have been used successfully in refinery operations where air can be used as the gas and where large areas are available. In treating produced water for injection, it is desirable to use natural gas to exclude oxygen. This requires the venting of the gas or installation of a vapour

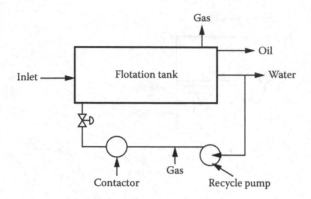

FIGURE 6.14
Dissolved gas unit.

recovery unit. Field experience with dissolved natural gas units has not been as successful as experienced with dispersed gas units.

In 'dispersed gas units,' gas bubbles are dispersed in the total stream either by the use of an eductor device or by a vortex set up by mechanical rotors. Figure 6.15 shows a schematic cross-section of a gas flotation unit with eduction. Clean water from the effluent is pumped to a recirculation header that feeds a series of venture eductors. Water flowing through the eductor sucks gas from the vapor space that is released at the nozzle as a jet of small bubbles. The bubbles rise causing flotation in the chamber forming a froth that is skimmed with a mechanical device.

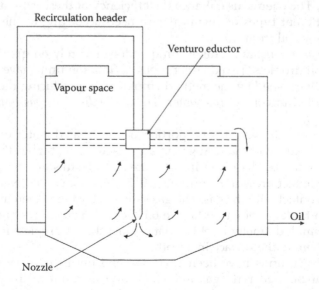

FIGURE 6.15
Dispersed gas flotation unit.

Filtration techniques – Here the main objective is to remove the residual suspended solids present in the effluent and the methods employed are layer, membrane and fibrous media filtration.

Filtration through sand, crushed anthracite coal or activated carbon is most common in the oil industry. A backwash system is always provided, requiring hot water or surfactants. This type of filtration is also known as 'granular media filtration.' The main advantages are that of compactness and high potential efficiency whereas disadvantages of this method are that backwash is necessary and plugging is a problem. Sophisticated automation can result in frequent downtime.

Once after the treatment the water meets the standard set by the EPA (Environmental Protection Agency), it can be disposed either in the surface like rivers or lakes or in the case of offshore into the sea water. In offshore, *disposal piles* are used which are large diameter (24–48 in.) open-ended pipes attached to the platform and extending below the surface of the water. These are designed to act as another treating experiment. Their main uses are to (1) concentrate all platform discharges into one location, (2) provide a conduit protected from wave action so that discharges can be placed deep enough to prevent sheens from occurring during upset conditions and (3) provide an alarm or shutdown point in the event of a failure causing oil to flow overboard.

Governing regulatory bodies require all produced water to be treated (skimmer tank, coalesce or flotation) prior to disposal in a disposal pile. Disposal piles are particularly useful for deck drainage disposal. This flow, which originates either from rain water or washdown water, typically contains oil droplets dispersed in an oxygen laden fresh or saltwater phase and gravitates to a low point for collection and either is pumped up to a higher level for treatment or treated at that low point. Disposal piles are excellent for this purpose. They can be protected from corrosion, they are by design located low enough on the platform to eliminate the need for pumping the water, they are not severely affected by large instantaneous flow rate changes (effluent quality may be affected to some extent but the operation of the pile can continue), they contain no small passages subject to plugging by scale build-up, and they minimise commingling with the process since they are the last piece of treating equipment before disposal.

The disposal pile should be as long as the water depth permits in shallow water to provide for maximum oil containment in the event of a malfunction and to avoid the appearance of any sheen. In the case of deep water, there is an alarm or shutdown signal to indicate the level of oil containment in a long length of pile. However, minimum pile length required is about 50 ft.

The *skim pile* is a type of disposal pile. As shown in Figure 6.16, flow through the multiple series of baffle plates creates zones of no flow that reduce the distance a given oil droplet will rise to be separated from the main flow. Once in this zone, there is plenty of time for coalescence and gravity separation. The larger droplets then migrate up the underside of the baffle to an oil collection system.

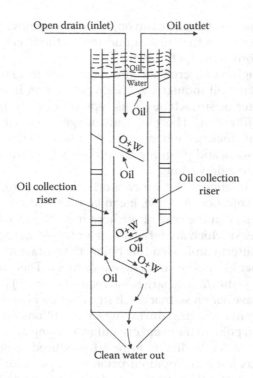

FIGURE 6.16
Skim pile.

Besides being more efficient than standard disposal piles, from an oil separation standpoint, skim piles have the added benefit of providing for some degree of sand cleaning.

6.1.5 Storage of Oil and Gas

The produced and treated oil and gas are required to be stored, mainly to meet the variation in demand and production and the demand during emergency situations like emergency shutdowns, during external aggression, or local disturbances and so on. A minimum of 10 to 20 days' production is stored to meet such eventualities.

Storage of oil – Several types of storage systems are in use all over the world. These include, tanks and underground storage. Sometimes long distance large diameter pipeline also works as a buffer storage system.

Presently most of the storage tanks are welded steel tanks although bolted steel tanks and riveted tanks are also used. These storage tanks can be classified as open top or roof type, closed fixed roof type and closed floating roof type (Figure 6.17).

The large tanks are designed and erected as per API standard 650. This standard covers materials, design, fabrication and erection.

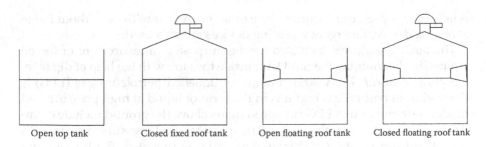

| Open top tank | Closed fixed roof tank | Open floating roof tank | Closed floating roof tank |

FIGURE 6.17
Different types of storage tanks.

Welded steel tanks are rugged and secured against leakage. Field welded tanks provide large capacities in a single-unit. These tanks are extensively used in oil fields.

In *fixed roof tanks*, the roofs are permanently attached to the tank shell. The tank is so designed that in case excess pressure builds up in the tank, the roof to shell connection will fail before any other joint fails and excess pressure will be relieved. The safety valve is also fit at the top for safety. These are very common tanks used extensively.

In *floating roof tanks*, the roof floats on the stored contents. This type of tank is used primarily for storage of volatile products near atmospheric pressure. Floating roof is designed to move vertically within the tank shell to provide constant minimum void between surface of the stored product and the roof. Floating roof tank may be either pan type or pontoon type.

The tank is protected against corrosion by providing internal coating of any one or a combination of paints, coal tars, epoxy resins, rubber linings or galvanisation. The external coatings are provided against weather exposure, and in some cases cathodic protection is also considered to control corrosion.

Some of the merits of the floating roof tank are minimum evaporation loss and minimum fire hazard but the demerits are that the initial investment is more and maintenance is costlier than a fixed roof tank. These are not suitable for storage of petroleum liquids of RVP (Reid vapour pressure) greater than 0.34 bar.

The tanks are generally fit with pressure/vacuum relief devices, platforms and ladders, gauging devices, manways and other connectors. These are also equipped with sumps at the bottom, inlet and outlet nozzles, temperature gauges, pressure gauges, vents and blow downs.

In the case of a tank battery, firewalls or dikes are provided to isolate the tank from surrounding adjacent areas, and from other tanks. The dike walls may be of earth, steel, concrete or solid masonry.

Because of the diurnal variation of temperature, the vapours get accumulated at the top of the liquid inside the tank which requires venting and this is accomplished by a breather valve, a pilot operated relief valve, a pressure relief valve, a pressure vacuum valve or an open vent with or without flame arresting device. Emergency venting devices are also used.

The tanks are always calibrated for the purpose of measurement of the oil volume by determining the fluid level inside the tank with the help of dip tape.

Storage of gas/LPG – A large amount of liquefied petroleum gas (LPG) is derived from natural gas and it is in the form of liquid at high pressure and moderate temperature. LPG storage systems above the ground include cylindrical pressure vessels, that is, bullets, spherical steel pressure vessels, with or without refrigeration and insulation, fully refrigerated, flat bottom and low pressure steel tanks.

Bullet storage – These tanks are long, cylindrical, usually horizontal, pressure vessels. LPG is stored at ambient temperature and design pressure of 250 psig. Most bullet tanks have capacities under 300 m³ (2000 barrels). Typical dimensions are 14 ft diameter and 175 ft length. The main drawbacks are low storage capacity and large surface area requirements.

Spherical steel pressure vessels – The spherical tanks provide maximum storage capacity for the same weight of steel used for construction, compared to other types. This type of vessel is extensively used for storage of volatile products under pressure. At ambient temperature, the LPG is stored in this vessel which withstands the pressure appreciably higher than vapour pressure of the liquid at the highest liquid temperature.

Fully refrigerated flat bottom tank – This type is widely used for storage of a large volume of LPG in excess of 50,000 barrels. It is designed to operate at near atmospheric pressure and fully refrigerated conditions. The roofs are either spherical or ellipsoidal, and the vessel may be either single or double walled.

6.1.6 Transportation of Oil and Gas

The movement of oil and gas starts at the well itself. The well fluid comprising oil, gas and water needs to be moved from the wellhead to a common point where separation of individual phases has to take place. Then the separated oil and gas need to be sent to the storage facility from where they need to be sent to the consumers. There are different modes of transportation of crude oil and gas like rail/road tanker, inland river and lake barge, ocean-going tank ship or tanker and pipeline out of which the most important and common mode of transportation is pipeline because of its many advantages over other modes.

In oil fields, transport of crude oil and natural gas by pipelines from wells up to refinery or primary consumer point consists of three different types of pipelines: (1) *Flow lines* which vary from 2–4″ in diameter and are laid from individual wells to a gathering station, (2) *collector lines* which carry oil and gas separately for further treatment to a central collecting station and these vary from 8–10″ in diameter and (3) *trunk lines* which carry the treated oil and gas as a finished product from the collecting station to the refinery or consumer points. Even the line carrying further processed products from refineries like gasoline, diesel, LPG and so on to the consumer stations are also known as trunk lines. While flow lines and collector lines are laid over lengths of 10–30 km within the oil field, the trunk lines are *cross-country pipelines* which run over several hundred kilometres and they can be as large as 56 in. in diameter.

Steel pipe or the tubular used in pipeline construction is commonly called *line pipe* as distinguished from drill pipe, casing pipe, or tubing. These line pipes conform to the API Specification 5L. These are bevelled on both ends so that they can be joined by welding. No threading is found in line pipes. In its design, three parameters are normally determined; those are the material, that is, the grade of the pipe, diameter and wall thickness. The pressure drop calculations are made by using an energy balance equation. The flow equation for gases is modified by incorporating the compressibility of the gas and the Reynolds' number and friction factor relationship.

In designing a pipeline system, the location of the pump or compressor in between the origin and destination of the line must be determined as well as the capacity and arrangement of different pumps or compressors used in each station. Depending on the requirement, the pump or compressor stations can be categorised as *originating station, booster station* and *injection station*. The number of booster stations varies widely depending on the length of the line, ground profile and the capacity of the originating station which is located at the beginning. The injection stations are used at a location where some chemicals need to be injected into the line.

The types of pumps or compressors commonly used are either reciprocating type or centrifugal type. For pumping liquid (i.e. oil or products), multistage centrifugal pumps are preferred because of its higher throughout and maintaining a required pressure. The reciprocating pumps are used for hydro-testing purposes where raising of pressure with a little throughput is possible. Similar to pumps, compressors for transporting gas can also be either reciprocating or centrifugal type. Reciprocating compressors generally operate at slower speeds and are used where higher pressures are required. It produces a pulsating flow that can damage the equipment and piping. In contrast, the output from a centrifugal compressor is smooth; as a result, vibrations are minimised which makes them preferable for using in offshore platforms. By connecting a number of centrifugal compressors in series, the compression ratio can be increased to that of a reciprocating one.

To run the pumps or compressors, proper prime mover selection is a must and the important criteria for selection are horsepower output and efficiency. Further availability of power to run the prime mover and the cost involved are also other criteria. It can be either electrically powered, if electricity is available, or else it can be a diesel operated or even a gas turbine generator.

In onshore pipeline, it is buried under the ground at a depth of 1 m from the ground level to the top of the pipe. This height is known as 'cover' and 1 m is the common value although depending on the location, it can be more than that. This pipeline traverses, following the topography of the land, through barren land, populated area, jungles, swamps, marshy areas and crossing roads, railway lines, rivers, water-logged areas, rocky areas and so on. This pipeline, before being lowered to the bottom of the trench, is radio-graphed (x-rayed) at every joint and then coated and wrapped at the exterior to protect it from leaking and from corrosion. After having lowered to the bottom of the trench, the top of the pipe is covered with soil, up to the ground level by backfilling. Once the whole line is laid, the complete line is hydraulically tested before commissioning. As a further protection from corrosion, cathodic protection (CP) of the whole line is done by installing CP stations in between just like booster pump stations, from where the required amount of current is discharged by the rectifier to the anode bed for nullifying the electrochemical cells being formed.

Once the pipeline is commissioned and put into operation, its maintenance becomes very important. It is a very common practice in pipeline to clean it time-to-time from inside of the debris, waxes, scales, water and so on, that get deposited. The operation to carry out this is known as 'pigging.' A *pig* is described as a free-moving piston that is inserted into a pipeline to perform certain functions. Among these functions, other than pipeline maintenance are (1) interface separation of fluids, (2) removal of obstruction and (3) pipeline monitoring. It should also be noted that pigging becomes necessary even during construction and commissioning and during inspection. The pigs (Figure 6.18) can be of various types as shown in Chart 6.1.

The normal propellants used in pushing the pig at a certain flow rate are water, air, or nitrogen. For launching the pigs at the inlet end of the pipeline and simultaneously receiving it at the outlet end, launching station or launchers and receiving station or receivers are used. Figures 6.19 and 6.20 show a pig launching and receiving station. By carefully looking at these two figures, it can be easily found that one is the mirror image of the other. The

FIGURE 6.18
Different types of pigs. (Left: Courtesy of Harvey Barrison/Wikimedia Commons/CC-BY-SA-2.0. Middle and right: Courtesy of Wilfra/Wikimedia Commons/Public domain.)

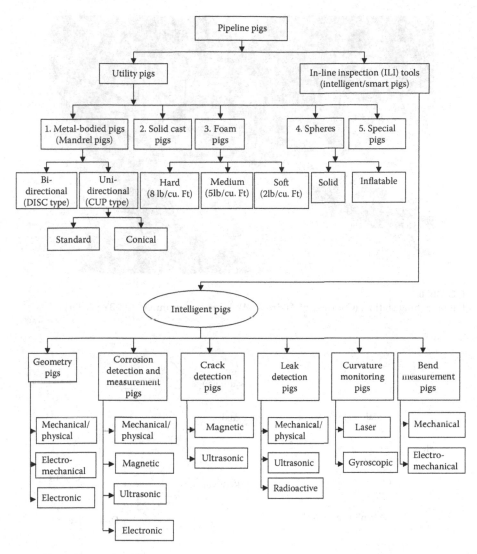

CHART 6.1
Classification of pipeline pigs.

pig launcher or receiver is simply a length of pipe at least two sizes larger than the main line equipped with a flanged closure at one end and an eccentric reducer at the other, flanged to match up with the existing pipeline. The pig launcher will be equipped with the necessary connections for propelling the pig and monitoring pressure and so on. Similarly, the pig receiver will be equipped with the required connections for vents, drains, pressure recorder and so on.

FIGURE 6.19
Pig launching station. (Courtesy of Audriusa/Wikimedia Commons/CC-BY-SA-3.0.)

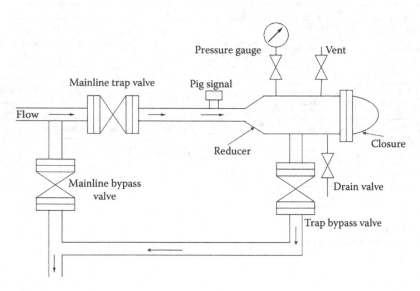

FIGURE 6.20
Pig receiving station.

6.2 Offshore Production System

Knowing the major surface oil and gas operations, the next thing is to look into the aspect of how the arrangement of the production systems involved in these operations is made in offshore water. To be more precise, it is the extension of the knowledge gained during drilling operations. Following a similar approach, the offshore production system can be either fixed or floating type based on which these are classified as fixed production system and floating production system. In Chapter 3, the classification of these two types based on the structural variation of different units has been made. Here these two types will be dealt with from a surface production system point of view.

6.2.1 Fixed Production System

This system includes fixed production platforms and also the offshore terminals like SBM and so on.

The fixed production platforms can be divided into three types:

1. Well/satellite platform
2. Process/well-cum-process platform
3. Water injection platform

This classification is based on the gathering system described earlier in Section 6.1.1. Here the well platforms are small unmanned fixed jacket type platforms which normally accommodate 4–6 wells. The well fluids collected from each of these 4–6 wells are sent through a single flowline to a process platform where fluids from other well platforms are also collected. For better understanding, Figure 6.21 gives a true representation of such platforms in an Indian offshore field. In its north NA, NB, NC… are the well platforms and in the south SA, SB, SC… are the well platforms. BHN and NQ in the north and BHS and SH in the south are the process platforms. It also shows the flowlines network. In the periphery of this offshore field, injection well platforms WI-2, 3 are also shown and for treatment of injection water, injection platforms WIN and WIS are also shown. Now the topside of each of these platforms are described below.

6.2.1.1 Well Platform

The function of a well platform is to gather produced fluid from wells, to test liquid and gas flow rate of individual wells after separation, and to transfer well fluids to the processing platforms through subsea flow lines.

Structurally well platforms are four-legged jackets with corrosion protection, having a superstructure consisting of a main deck, cellar deck and

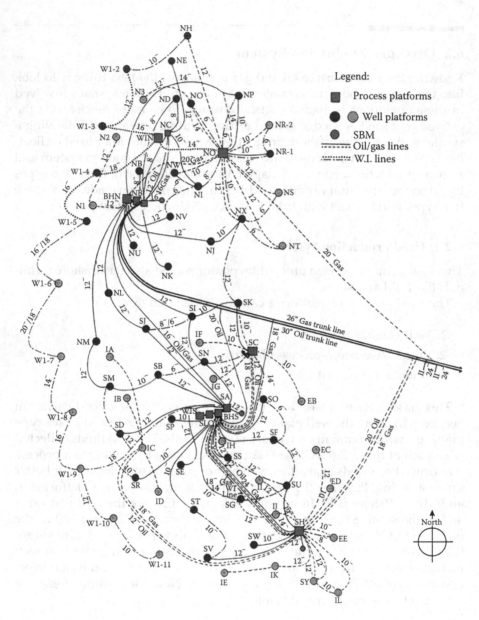

FIGURE 6.21
Layout of production platforms in Indian offshore field. (Courtesy of ONGC.)

helideck. It has 4–12 well slots with a conductor within the jacket framing. The 2-level boat landing facility is also provided. For mechanical lifting, pedestal cranes and monorail hoists are provided. Though these platforms are unmanned, in case of an emergency for the personnel to stay back, a small bunk house facility is also available.

The *topside* facilities are as follows:

The Christmas trees are located in the cellar deck which are provided with the shutdown valve (SDV) located after the wing valve and the sub-surface safety valve (SSSV) located about 150–300 m from the tubing hanger in the tubing. The flow arms are fit with pressure switches high and low (PSH and PSL), and the wells are controlled through a wellhead control panel. The panel operates pneumatically. In case of fire and emergency shutdown, this panel gets activated and the wells close. With telemetry and telecontrol facilities, the well can be opened or closed by giving commands from the process platform, where a central control room remains always operational. The SDV is pneumatically operated and the SSSV is operated hydraulically.

The wells are connected to a manifold, which consists of two headers, namely, a test header and a group header. The test header carries fluid to the test separator and the group header is connected to the riser of the flow line connecting the well platform to the process platform. The fluid from the test separator (oil, gas and water) after testing are recombined and joined at a point before the riser.

The test separator is a high-pressure 3-phase horizontal separator as described earlier. It has a positive displacement (PD) meter for measuring oil and water rate and an orifice meter to measure the quantity of associated gas.

The platforms are equipped with gas or dual fuel generator and switchgear room. In addition, they have solar panels. The instrument gas system utilises gas stream after conditioning.

The fire water pump is a diesel engine driven vertical centrifugal pump. Utility gas is stored in volume bottles for starting-up.

A small chemical injection pump with tanks is provided for injecting chemicals like corrosion inhibitors, demulsifiers and so on.

Process safety devices are installed at strategic locations like on the well flow line, manifold, riser, instrument gas skid which gets actuated in case of some upset conditions and shuts down the system or the platform depending on the condition.

These are also equipped with fire and gas detection systems, consisting of gas detectors and fusible loop. A fire and gas panel gets actuated in case of eventuality and shuts down the platform.

The firefighting system includes dry chemical powder (DCP), foam, portable extinguishers and water.

The life saving devices like life jacket, life buoy, life raft and pilot ladders are available on this platform. Navigational aids are also provided.

The *linkage* between each well platform and the process platform is through the riser and the flowline which are separate for each well platform. The riser here (commonly called production riser) connects the group header at the topside and the flowline at the seabed. In such a platform the riser is free standing, vertical tubular supported within the jacket frames. The riser is equipped with a pig barrel for pigging the flow line connecting the well to the process platform. Refer to Section 6.2.2.2 for further understanding about the linkage between subsea and topside.

6.2.1.2 Process Platform

This particular platform is equipped with processing, treatment and pumping facilities which are the primary operations. The other supporting facilities include living quarters, utilities, control and communication systems. Most of the process platforms are attached to at least one well platform (Figure 6.22). Process platforms are normally eight-legged platforms and do not have any wells. The produced fluid from different well platforms is sent for processing and transportation.

The process facilities include the necessary equipment to:

1. Separate oil, gas and water.
2. Treat and pump oil to a subsea off-take pipeline.
3. Clean up and dispose of the produced water.
4. Treat and compress gas to a subsea offtake pipeline.
5. Condition associated gas for fuel.
6. Dispose of excess gas by flaring.
7. Generate power.
8. Make potable water.

FIGURE 6.22
Well cum process platform. (Courtesy of ONGC.)

At the process platform, the well fluids join at the inlet manifold and from there it is sent to the inlet separators. The inlet separators are high-pressure separators. The fluid comes out of these vessels and is sent to the low-pressure separator. All these separators are 3-phase and horizontal. From the low-pressure separator, the crude flows to the surge tank. The water separated out from the separators is sent to the produced water conditioners, which are usually corrugated plate interceptors (CPI). The treated water, after meeting the specified limit of oil content, is directly disposed in the sea or sump caisson.

The gas produced from the high-pressure separator is sent to a scrubber, which is normally a vertical vessel. The gas after scrubbing is sent for compression and dehydration and dispatched to the onshore facilities. The low-pressure and excess high-pressure separator gas are sent to respective flare knockout drums, wherein the gas is scrubbed of liquids, before sending to the flare. The flare is installed on a tripod and is bridge connected. The knocked-out liquid is sent to the sump caisson.

Apart from PPD, corrosion inhibitors are also dosed depending on the requirement, therefore, a chemical injection facility with tank, transfer pump and dosing pumps is provided.

The main power generation is by the diesel/gas engine driven generators, but for emergency power, the emergency generator is diesel engine driven. Fuel gas is tapped from the high-pressure separator and passes through the fuel gas conditioning skid, where the liquids are knocked off, before it is sent as a fuel to the generators.

The platform has the central control room, which controls the entire process operations.

The platform is also equipped with safety devices like gas, smoke, heat, UV (ultra-violet) detectors and the firefighting system includes DCP, halon, foam, water and portable extinguishers (Figures 6.23 and 6.24). Normally, there are two fire water pumps, diesel engine driven, vertical turbine type.

The life-saving devices include life jacket, life buoy, life boat, life raft, scrambling net and pilot ladder (Figures 6.25 and 6.26).

Utility, control and communication system available on offshore installations include the following:

Diesel fuel system – The diesel fuel system is provided to supply diesel to turbo-generators, utility generator, pedestal cranes and fire water pumps. Under normal conditions the turbo-generators use fuel gas. But diesel fuel is used during start-up or if the fuel gas supply fails. The diesel fuel is stored in the pedestals of the pedestal cranes.

Jet fuel system – The system is designed to provide clean and metered fuel to the helicopter fuel tanks. The unit has an air eliminator and strainer.

Dry chemical fire extinguishers | Halon system

Fire hose reel | Fire hydrant | Sprinkler system

FIGURE 6.23
Firefighting system. (Top left and middle: Courtesy of Dante Alighieri and Firetech 117/ Wikimedia Commons/CC-BY-SA-3.0. Top right: pochi-2002/Wikimedia Commons/Public domain. Bottom left: Denis Apel/Wikimedia Commons/CC-BY-SA-2.0. Bottom middle: Brandon Leon/Wikimedia Commons/CC-BY-SA-2.0.)

UV detector | Thermal detector | Gas detector | Fusible plug

FIGURE 6.24
Fire detection system. (Courtesy of N.K. Mitra.)

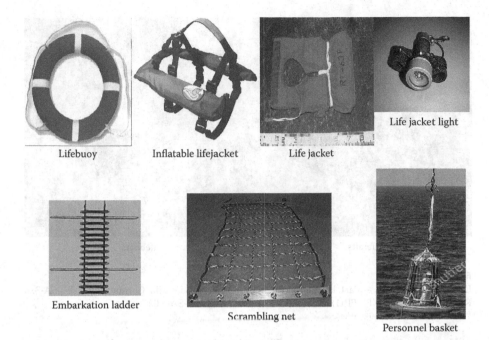

Lifebuoy   Inflatable lifejacket   Life jacket   Life jacket light

Embarkation ladder   Scrambling net   Personnel basket

FIGURE 6.25
Life saving devices – Part I. 1. Lifebuoy (Courtesy of B M Machinery Co. Ltd). 2. Inflatable life jacket (Courtesy of Ericlmccormick/Wikimedia Commons/CC-BY-SA-3.0). 3. Life jacket (Courtesy of Accident Investigation Board Finland/Wikimedia Commons/CC-BY-SA-3.0). 4. Life jacket light (Courtesy of Peter Southwood/Wikimedia Commons/CC-BY-SA-3.0). 5. Embarkation ladder (Courtesy of www.offshore-technology.com). 6. Scrambling net (Courtesy of www.mid-continents.com). 7. Personnel basket (Courtesy of Shutterstock).

Instrument, utility and starting air system – The instrument and utility air system supplies dry air for instrumentation, fusible fire loops and the operation of pneumatic valves. The system also supplies non-dried air for general utility use. There must be two air compressors, one as a standby.

The starting air system provides high-pressure air for starting the diesel powered fire water pumps.

Potable water system – The potable water system consists of water makers which produce fresh water from sea water, potable water storage tanks, water distribution pumps, potable water vessel, potable water header and distribution piping. The potable water for human consumption is generated by the reverse osmosis (RO) plants.

Utility water system – Utility water system is provided to supply sea water for water maker, chlorinator, wash down hose reel, pressurisation of fire water headers, toilet flush and flushing of sewage treatment unit. A wash down pump acts as a standby for the utility water pump.

Liferafts   Lifeboats

FIGURE 6.26
Life saving devices – Part II. (Left: Courtesy of Poe/Wikimedia Commons/CC-BY-SA-3.0.
Right: Courtesy of NPL FP11/Wikimedia Commons/CC-BY-SA-3.0.)

Hypochlorite generation system – The chlorinator produces sodium hypochlorite by electrolysis of sea water. Hydrogen gas is generated as a by-product of the electrolysis of sea water. The system includes a hydrogen removal tank. The gas is vented to the atmosphere, as it is below the 4% explosion limit of hydrogen in air. Sodium hypochlorite is distributed to the casing of subsea pumps to prevent the algae growth on the pump, casing and the strainer.

Material handling – The equipment normally available on platform for the purpose of moving material and equipment and assisting in maintenance are pedestal crane, electric monorail hoists, manual hoists and manual trolley hoists.

The cranes transfer personnel, equipment and supplies between barges or supply boats and the platform.

Cooling water system – The cooling water system is provided to supply cooling water to the various pumps on the platform. The system forms a closed circuit which consists of a cooling water tank, cooling water pumps and water cooler. Potable water is used as the cooling medium and the make-up water for the cooling water system is manually added to the cooling water tank.

Fuel gas system – The fuel gas system is provided to supply adequate, clean and dry gas to the combustion gas tubing of generators and process gas compressors. The system also supplies the purge gas for high-pressure flare and low-pressure flare systems. The fuel gas is treated in the conditioning skid before supplying to the consumers.

Vent system – A vent header is provided on the platform for the collection of low pressure gas from various equipment. The collected gas from each piece of equipment is scrubbed in the vent scrubber and then released through the vent boom to the atmosphere. Two flame arrestors are provided on the inlet line of the vent boom to avoid any flashback into the vent system. Carbon dioxide cylinders are provided to snuff out any possible fire in the vent boom.

Flare system – High-pressure and low-pressure gases, released from various production vessels and equipment are collected and burned at the flare tip, for reasons of safety and pollution control. The liquid is knocked out from the gas by passing through knock-out drums. HP and LP flare headers are constantly purged with a metered quantity of gas from the fuel gas system.

Waste heat recovery system and hot oil circulation system – The hot exhaust gases from the turbo-generators are routed through the waste heat exchangers and then discharged to the atmosphere from exhaust ducts. The waste heat recovery from the gases warms a circulating hot oil which in turn exchanges heat with different facilities on the platform.

Communication systems – The different types of communication systems used in a process platform are as follows:

Radio system – It consists of

1. Life boat radio equipment to provide for emergency distress communication from life boat to a shore radio station or any other lifesaving agency.
2. VHF–FM marine transceivers to provide platform to platform and to ocean vessel communication within the production field.
3. VHF–FM walkie talkie sets to provide operational communication with nearby stations.
4. VHF–AM aero transceivers to provide communication between the platform and helicopters working in the areas.
5. Non-directional beacon (NDB) to provide a continuous code radio signal to assist helicopters working in the area in locating the offshore installations.
6. HF–SSB transceivers and radio teletype to provide both voice and teletype communication between platform and shore.

Page party communication system – The system is a complete intraplatform as well as an interplatform page party communication system with an emergency tone generator. The system is common talking, so that any handset user may take part in any conversation. Speakers are used only in the page mode and emergency alarm mode.

Closed circuit television system – The CCTV system provides a high-resolution camera system with remote controlled pan, tilt and zoom capability and high-resolution monochrome monitors. It is suitable for low light operation. It is useful in monitoring helicopter landing, boat operation at any external unknown approach.

Remote telemetry unit – Remote telemetry unit in combination with remote platform monitoring console (RPMC) assists in operation of well platforms from the central control room and acquisition of data about equipment.

Private automatic branch exchange and telephone system – The private automatic branch exchange (PABX) and telephone system provides an internal switching, operating and interfacing system for the telephone network. The system is totally automatic for unattended operation.

Satellite communication system – Communication through satellite has become very handy and contacting anyone, anywhere has become a matter of seconds only. The satellite phone or mobile phones are meant for communication within and outside any organisation or operator company.

Support system – Any process platform where a network of offshore operations is carried out is well supported by various support services day in and day out. The important agencies rendering valuable services can be categorised as:

Offshore supply vessel (OSV): The OSVs transfer workers and materials to platforms by waterways. They also provide services in attending to well closures on unmanned platforms and temporary deck platforms at odd times. They render useful service during firefighting and platform evacuation.

Multi-support vessel (MSV): The MSVs, also called *dive support vessel* (DSV), provide a wide range of services to offshore networks. Besides handling firefighting operations and emergency situations, they carry out surface as well as underwater inspection, maintenance and repair of the equipment and pipelines.

Helicopter: A helicopter is the fastest mode of transportation in offshore. Besides transferring workers and materials, they help in evacuating personnel from the platform in emergencies and search and surveillance operations.

6.2.1.3 Water Injection Platform

This platform can be divided into (1) injection water process platform and (2) water injection well platform.

Injection Water Process Platform

In offshore, the processing of water for injection is done by a semi-closed system which comprises the following:

1. Sea water lifting
2. Coarse filtration
3. Fine filtration
4. Deoxygenation
5. Dosing of chemicals
6. Pumping of treated water to injection well

A process flow schematic is shown in Figure 6.27. The general process description of such a platform is given below.

Sea water lifting – The major components are sea water lift pumps and chlorinators. The raw sea water is lifted by two sea water lift pumps (SWLP) from −30 MSL and there are two chlorinators that produce NaOCl (sodium

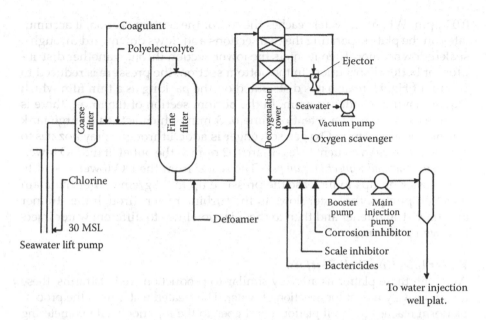

FIGURE 6.27
Water injection process platform – flow diagram.

hypochlorite) by electrolysis of sea water. The produced NaOCl is injected at the intake of the SWLP.

Coarse filtration – The coarse filters remove 98% of all particles of 80 microns or larger. About three such filters are generally used. The filter has a single cylindrical metal filter element mounted internally. The water pressure forces the water out through the element in the outer annulus of the coarse filter and thereafter goes to the fine filter. Each filter has a backwash filter channel which is held firmly against the filter element by a rotating element. An automatic sequencer closes or opens the valve and puts the motor on or off. The measurement of differential pressure across the filter gives an indication of the requirement of backwashing.

Fine filtration – The fine filters are designed to remove 98% of all particles of more than 2 microns. To aid in the filtration process coagulant and polyelectrolytes are added before the water enters the fine filter. About 8 fine filters in parallel are generally used. These are vertical pressure type, down flow filters with dual filtering media of anthracite and garnet. Normally, a filter is backwashed after 48 h and takes about 30 min. A timer sequencer controls the backwash cycle. In case the differential pressure of any filter exceeds 2.0 kg/cm², the filter will be taken up for backwash on priority.

Deoxygenation tower – The filtered water is sent to the DO tower after dosing defoamer. The water enters the tower through a distributor above the top section. In the top section, most of the dissolved gases (O_2, N_2, CO_2, Cl) and water vapour are taken out and their concentration is reduced from 7 ppm to

0.02 ppm. When the water reaches the top of the bottom section, it accumulates on the plate separating the two sections and flows downward through a
sealed downcomer. Then it enters the lower section through another distributor for better distribution. In the bottom section, the pressure is reduced to
20 mm of Hg. Here also it is distributed on the packing as a thin film which
aids in having a better vacuum. In the bottom section of the tower, there is
storage capacity with a retention time of 3 min, which acts as a surge tank
for the booster pumps. Oxygen scavenger is added through spray nozzles to
reduce the oxygen content to less than 0.02 ppm at the outlet of the DO tower.

Booster and main injection pumps – The water from the DO tower is sent to
the booster pumps to increase the pressure up to 9 kg/cm². After the main
injection pumps, the water flows to the turbine meter through the strainer
and flow straightener and then to the injection lines to different water injection well platforms.

Water Injection Well Platform

Although these platforms are very similar to production well platforms, these
are exclusively meant for injection of water. The treated water from the process
platform reaches the well platform and goes to the injection well via metering
devices, which are meant for finding out the injection rate of each well. The facilities in this platform are water injection manifold skid, well shutdown panel, diesel storage tank, diesel generator, launcher and receiver, pedestal crane, monorail
hoists, solar panels, battery room, telemetry room, switch gear room, power
package for operating shutdown valves, firefighting (dry chemical skid, hose
reels, portable extinguisher), lifesaving equipment (life raft, buoys, jackets) and
navigational lights. A hydraulic power package is used for the operation of the
shutdown panel. The hydraulic power package is operated by the injection water
pressure. The power can be generated by the diesel engine driven generator.

6.2.1.4 Offshore Terminals and Offloading Facility

Whenever transportation of the processed oil is to be made in the absence of
any pipeline, the only other mode possible is by tanker. These tankers, normally called shuttle tankers or export tankers, have to be berthed near some
process platforms and that too in the middle of the water where no port, har-
bour, or any protected shelter is possible. Hence, a new kind of port loading/
unloading facility called 'offshore terminal' has come into prominence.

Offshore terminals now play an important role in the handling and transportation of crude oil, petroleum products and other liquid cargo. The
immense growth in the volume of petroleum traffic alone over the years has
demanded tremendous increases in the cargo tonnage capacity of the shipping industry. There has been a spectacular trend toward using larger tankers for transporting crude oil.

As a result, the developments that followed were all directed toward
shifting port facilities to offshore areas where the natural water depth was

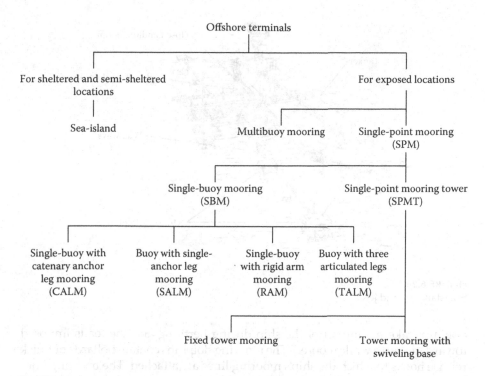

CHART 6.2
Classification of offshore terminals.

adequate to allow plying and manoeuvring of the super tankers; no dredging was involved, installation time was short and the cost economical. Offshore terminals may be classified as sea-islands, multibuoy moorings (MBM) and single-point moorings (SPM). Chart 6.2 gives a comprehensive picture of the various available offshore mooring systems.

Each type has three major components, namely, (1) a means to hold the tanker in position, (2) a means to transfer cargo to/from the tanker's manifold from/to a manifold on a loading/unloading platform or on the sea-bed and (3) a pipeline between the manifold on the sea bed and the process platform or the shore. The various types of offshore terminals differ mainly in their means to hold the tanker in position, which eventually determines the usefulness and effectiveness of the terminal in a specific environment.

6.2.1.4.1 Sea Islands

The sea island or fixed jetty berth is basically a conventional jetty berth constructed offshore to serve as a deepwater terminal. The application of this type of berth is limited to sheltered or semisheltered offshore areas where environmental conditions are not severe.

The main components of such a berth are the breasting dolphins, the mooring dolphins and a loading/unloading platform (Figure 6.28). The breasting

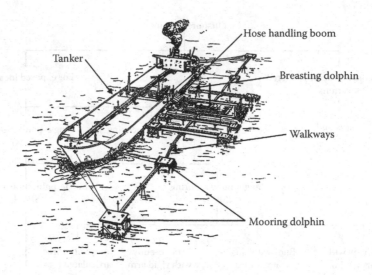

FIGURE 6.28
Sea island or fixed jetty.

dolphins take the impact of the ship during berthing and the loads imposed toward the berth while moored. The mooring dolphins contain bollards or quick release hooks to which the ship's mooring lines are attached. The mooring lines hold the ship in a fairly fixed envelope in space and permit the use of loading arms mounted on a central platform to connect the ship's manifold with operations. The prevalence of any cross-wind or cross-current would render both the ship and berth liable to severe damage due to excessive berthing impact.

6.2.1.4.2 Multibuoy Moorings

The multibuoy mooring or conventional buoy mooring (MBM or CBM) evolved for application in exposed locations. The movement of the mooring buoys and flexibility of the mooring lines provide adequate elasticity to the system for tankers to be held at berth even in rough sea conditions.

In this system 5–7 moored buoys are installed, generally in a semicircular pattern around the desired position of the stern of the tanker (Figure 6.29). Because the mooring will permit berthing in one general direction only, the buoys are laid out in such a way that the vessel will be oriented with its longitudinal axis parallel to the prevailing wind, wave and current direction. The connection between the submarine pipeline end manifold and the tanker's manifold is made by means of underwater hoses picked up by the tanker after berthing is complete. The advantages of this type of berthing system are

1. Oscillatory motions and the heaving of ships do not induce appreciable loads in the mooring lines as the displacement involved is easily absorbed by the elasticity of the mooring lines and mobility of the anchored buoys.

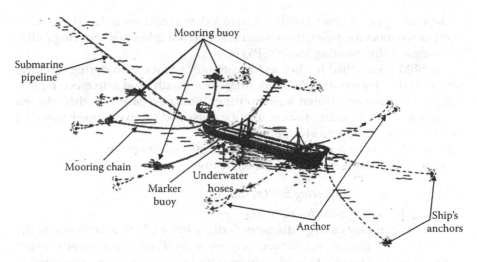

FIGURE 6.29
Multibuoy mooring system.

2. This system needs little capital investment as the buoys required are small in size and the connection between the tanker's manifold and the submarine pipeline is a simple arrangement.

3. Maintenance requirements and costs are moderate.

4. The time required for construction and installation is quite short.

The disadvantages of the system are

1. Berthing operations are cumbersome and time consuming, and at least 2 launches and 1 or 2 tugs are required for berthing and mooring the tanker.

2. Due to limitations on the number and sizes of hoses that can be installed, the loading/unloading rate is slow.

3. Since tankers can be moored in a fixed heading, any deviation in the directions of wind, waves and current from the designed heading will make the terminal inoperative.

4. This type of terminal is not suitable for tankers above 100,000 dwt.

5. The life of the system is short.

Single-Point Moorings

The single-point mooring system (SPM) has a single floating buoy or a fixed tower to which the tanker is moored using a bowline. The tanker can freely move around the terminal taking up the most convenient position with respect to wave, wind and current. This enables the terminal to operate in severe environmental conditions with the least force being imposed on the terminal.

Depending on whether a buoy is employed or a fixed tower has been envisaged as the mooring point, the system is called a single-buoy mooring (SBM) or a single-point mooring tower (SPMT).

The SBM is classified further, as the catenary anchor leg mooring (CALM), single-anchor leg mooring (SALM), single-buoy with rigid arm mooring and buoy with three articulated legs mooring. Similarly, based on their design the single-point mooring towers may be classified as fixed tower mooring and tower mooring with swivelling base.

Many SPMs have been installed and operated successfully.

Some of these systems are described below.

6.2.1.4.3 Single-Buoy Mooring System (SBM)

Catenary Anchor Leg Mooring (CALM)

This system employs a large-diameter floating buoy which is anchored to the sea-bed by 6–8 chains and anchors (Figure 6.30). The buoy is an all-welded water-tight steel vessel of circular shape. The anchor chains are connected to a circular skirt welded to the outside of the buoy body. The mooring point and oil handling system, which are located on a turntable mounted on the top of the buoy, can swing freely as the tanker swings around the buoy. The tanker's manifold is connected by floating hoses to pipe arms leading to a centre swivel on the buoy, which is connected to the submarine pipeline by underwater hoses.

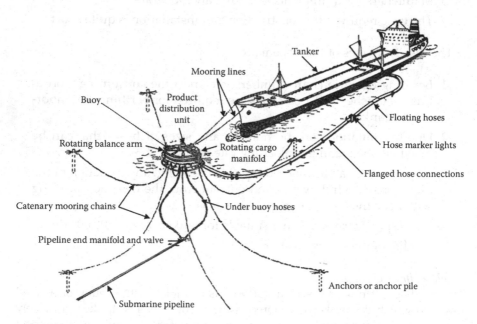

FIGURE 6.30
Single buoy with catenary anchor leg mooring (CALM).

The main advantages of this system are

1. It involves moderate capital investment.
2. The terminal can be installed in a short time. The actual time required for installation may be as little as 3 months provided the hardware is available at hand.

The main disadvantages of this system are

1. The high rate of wear of underwater and floating hoses and anchor chains due to wave action and so on, involves high maintenance costs. It is necessary to regularly inspect and tension anchor chains and frequently replace hoses and other vulnerable components.
2. Accidental over-riding of the buoy by the tanker may result in serious damage to hoses and chains.
3. The system has a short service life and the buoy may require major overhauling necessitating dry docking once every 3 or 4 years.

Almost 90% of the SBMs designed and installed thus far belong to this category.

Single-Anchor Leg Mooring (SALM)
This system employs a smaller size buoy than the previous type and uses a single-anchor leg of rigid tubular section articulated at the base and connected to the buoy by a pretensioned chain (Figure 6.31). The base, which is a steel framework, is secured to the sea-bed by means of piles. The tubular section also serves as the riser pipe and is provided with a swivel joint at the top to which underwater hoses are connected. The bottom end of the tubular section is connected to the submarine pipeline through a universal joint. The tanker is moored to the mooring point provided on the buoy, and the crude oil or other liquid cargo is loaded/unloaded using floating hoses connected to the underwater hoses.

The main advantages of the SALM system are

1. Tankers can manoeuvre close to the SALM with less concern for damage to the system because the buoy can be ruggedly built, and the fluid, anchor swivel and hose connections may be placed below the keel of the tanker.
2. If necessary, tankers may drop anchor without the danger of fouling the anchor of the buoy, since the anchor chain is located directly beneath the buoy.
3. Hose wear is reduced since the swivel unit is located underwater. The underwater hose length is reduced.
4. The SALM is less expensive than the CALM in deep water.

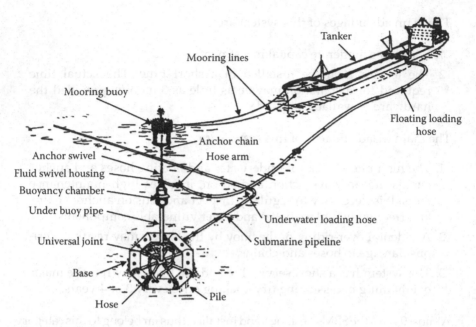

FIGURE 6.31
Buoy with single anchor leg mooring (SALM).

The main disadvantages of this system are

1. The underwater swivel, hoses and articulated joint at the base are potential sources of trouble and may require frequent inspection and maintenance.
2. The repair or overhauling of any mechanical component would require its recovery and reinstallation which would put the terminal out of commission for some time.
3. Chain wear is high due to large movement of the buoy.
4. The safety features claimed to be inherent in this system are not strictly valid in shallow waters.

Buoy with Rigid Arm Mooring (RAM)
In this system a large-diameter buoy is used, but instead of chain anchors the buoy is held by a rigid arm (Figure 6.32). The arm is connected to a base at the sea-bed through ball and swivel joints which enable the whole system to rotate around a vertical axis as well as to ride up and down with the waves. The main advantage of this is the elimination of anchor chains and long submarine hoses. However, the short length of the underwater hose and swivel at the base are likely to require frequent inspection and maintenance.

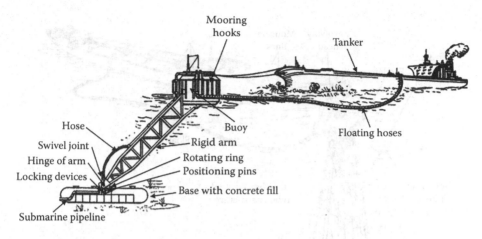

FIGURE 6.32
Buoy with rigid arm mooring (RAM).

Single-Buoy with Three Articulated Legs Mooring (TALM)
This system employs a buoy as in the conventional SBM, but the buoy is connected to the anchor base by three articulated legs. The legs are attached to the buoy and anchor base via cardon joints which permit the bottom portion of the legs to swivel and the buoy to move in a horizontal direction. As in many other systems, floating and submarine hoses are used for transferring cargo through a submarine pipeline.

The advantages claimed are that the buoy remains in a horizontal position for all conditions of sea and external loads, mooring loads are reduced, and with anchor chains being eliminated, maintenance is reduced.

Out of these four types of SBM, the CALM and SALM types are more common.

6.2.1.4.4 Single-Point Mooring Tower (SPMT)

Fixed Tower Mooring
In this system a fixed tower is built on the sea-bed and a turntable is mounted on its top. The turntable carries the mooring point and swivel similar to the catenary leg SBM system. The manifold on the turntable is connected to the submarine pipeline by means of a steel riser pipe, thus eliminating underwater hoses. One design (Figure 6.33) employs an open braced tower with floating hoses to connect the tanker's manifold with the turntable. Another design uses a mono-tubular tower instead of the open-braced construction.

A third design uses an open-braced tower of slightly different construction with an additional submerged arm carrying pipelines from the tanker's manifold to the tower replacing the floating hoses.

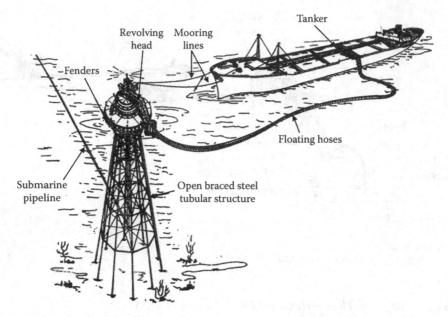

FIGURE 6.33
Fixed tower mooring (SPMT).

The principal advantages of this system are

1. Steel riser pipes replacing the underwater hoses enable faster cargo transfer rates to be achieved.

2. Maintenance requirements are reduced due to the absence of submerged hoses, underwater swivels and chains.

3. The fixed tower, cathodically protected, may be designed for a life of even up to 30 years.

4. The provision of a submerged rigid loading arm instead of floating hoses further reduces maintenance problems and it has an improved berth occupancy coefficient due to higher loading/unloading rates and higher weather operating limits.

The major drawbacks of the system are

1. In case of overriding by the tanker, serious damage may result, both to the tanker and to the tower. Once such damage occurs, it normally requires a long time to rectify it and put the system back into commission.

2. Normally the towers are fixed on the sea-bed by means of piling. When the sub-bottom soil conditions are not favourable, piling becomes difficult, time-consuming and costly.

Tower Mooring with Swivelling Base
This system consists of a large-diameter tubular section hinged to the base. The swivel and mooring arrangements are installed on the top. The tower remains in a vertical position due to the forces of buoyancy. The SARUS tower installed in the Gulf of Kutch is an example of this type.

Offshore terminals mainly SBM and SPMT can be operated in moderate to severe sea conditions because of their unique way of mooring a tanker, allowing it to weather-vane and take up the most convenient position with respect to wind, wave and current. However, the berthing of a tanker at a terminal requires the assistance of a mooring launch, and therefore, berthing operations are limited by the sea conditions beyond which the mooring launch cannot operate. These limits generally are: (1) wave heights of 2–2.5 m, the lower value being applicable for local wind generated waves and (2) wind velocity of not more than about 50 km per hour.

Once berthed, the tanker can remain moored and discharge cargo at conditions worse than above. Cargo transfer may continue so long as the movement of the floating hoses is not violent and the relative motions of the tanker and hoses do not threaten the snapping of the hose strings. Normally, the following are thought to be the limiting conditions for the transfer of cargo: (1) wave heights up to 3.5 m; (2) wind velocities up to 75 km per hour; and (3) current velocities up to 2–2.5 m per second.

6.2.1.4.5 Tanker Offloading System
For offloading crude from the tanker, the system comprises the following:

1. Pumping system/crude transfer pumps
2. Stripping system

Pumping System
For crude offloading conventional centrifugal cargo pumps are commonly used in tankers. There are basically three types of centrifugal cargo pumps:

1. Horizontal split-case cargo pumps, preferably with an external bearing arrangement
2. Vertical split-case cargo pumps
3. Barrel-type cargo pump

As space is at a premium, especially on a tanker, by far the majority of cargo pump arrangements are vertical, thereby reducing the size of the pump room required. Nowadays the preferred arrangement of vertical-type cargo pumps is the vertical overhung, barrel-type pump unit. From the point of view of the case of onboard maintenance, the barrel-type cargo pump has been developed. The overhung impeller-type cargo pump, referred to as the barrel type, normally has the same basic hydraulic design as an equivalent-sized, conventional vertical/horizontal split-case cargo pump. Indeed, it is often possible to use the same design of impeller. The suction/discharge of the barrel type pump can be either of the straight-through arrangement or at a 90-degree suction/discharge angle. Both connections are in the main bottom half casing.

Barrel-type cargo pumps are normally fit with mechanical seals, and it is also possible to fit a split seal. The make and type of mechanical seal fit are normally as per the owner's preference. With this design of cargo pump, the impeller can be situated lower in the cargo pump room than the impeller of a vertical, conventional split-case type and thus improving the suction conditions to the impeller. The barrel-type pump is designed with special attention to easy onboard maintenance.

The materials of construction for both types of cargo pump are selected in accordance with the type of products that the pump will handle.

Stripping Systems
It means collecting liquid residue in ship tanks at the last stage of discharging in a safe and efficient way. The primary requirement of the offloading system is that the cargo pumps have to be capable of stripping the tanks free of cargo to the specific level. Vertical centrifugal pumps will be able to pump the cargo down to near the centreline of the impeller. To go lower than this, other low-suction screw pump eductors or stripping pumps will have to be used. Stripping systems are required for the barge-based and tanker-based FPS systems.

6.2.2 Floating Production System

In the beginning of offshore oil and gas operations, all the production systems in offshore were of fixed type but as the search for oil and gas moved to deeper waters and away from the shore, the use of floating production systems (FPS) gained prominence. With the advancements in subsea technologies and because of other advantages as listed below, the use of FPS has increased manifold. Some of the advantages of FPS over fixed types are

1. Lower capital cost compared with fixed platform
2. Reduced time from discovery to production (i.e. early production)
3. Can be relocated and reused in another field
4. Ability to operate in deep and ultra-deepwater
5. Can be used for reservoir testing in different locations
6. Possible use in earthquake prone or ice-infested areas

Different types of floating production units like TLP, FPSO, SPAR and FPS have been discussed in Chapter 3. All these have four common elements, out of which topsides and risers are relevant in the context of production operations. On the topside the decks along with different production modules and other modules are placed whereas the risers form a linkage between the sub-sea and the topside.

6.2.2.1 Topsides

The different types of topsides, depending on the way these are built, have been discussed in Chapter 4. Out of these the 'modular deck' type has been considered here and the following relevant modules are discussed.

1. Process system module
2. Utilities and support system module
3. Safety system module

The other modules as discussed in the context of drilling in Chapter 4 remain common. A general layout of topsides and the important parameters have also been discussed in the same chapter. However, certain common guidelines followed for preparing layout of production equipment on an FPS are given below in order of priority.

1. Isolating quarters and helideck on windward side
2. Placing vent or flare on leeward side and locating cranes
3. Separating ignition sources from fuel sources wherever possible
4. Locating rotating machinery for access to cranes
5. Putting utilities and water handling equipment near quarters
6. Optimising placement of equipment to minimise piping

Different types of layout of oil production facility on a 2-level deck system are available in API Recommended Practice (API RP 2G 1974). Both these layouts have certain merits as well as demerits. For any system, the layout is to be tailor-made keeping in mind certain basic principles as mentioned above.

6.2.2.1.1 Process System Module

As discussed in the outline of the surface production system (refer to Section 6.1.1) the processing system comprises separation, treatment, storage and transportation. The basics of each of these operations are discussed there. Thus, a process system module consists of the following main modules: (1) oil module, (2) gas module and (3) produced water module. There are some auxiliary modules like (4) main oil line pump (MOL) module, (5) process gas compressor (PGC) module and (6) venting and gas flaring module.

In the *oil module* separated oil from a high-pressure (HP) separator goes to low-pressure (LP) separators via heat exchangers for the next phase of processing and thereafter oil is fed to the surge tanks via heaters for further lowering of water content. A surge tank is again a separator with similar construction. Surge tanks can operate in two modes, that is, 'pressurised mode' and 'atmospheric pressure' mode. In 'pressurised mode' surge tanks are operated at 4–5 kg/cm^2. It is used when oil is to be sent via pipelines. Operating at this pressure helps in reducing loss of lighter fractions into flare. When the oil is to be sent to tankers, it has to be stabilised; in other words, all the gases are to be knocked off. In this case, the surge tanks are operated at atmospheric pressure. It is evident that there are bound to be stabilisation losses in this process. Stabilised crude gets offloaded from the floating production systems to shuttle barges or tankers. The rate of crude transfer is dependent on the size of the shuttle tanker and the loading time. Custody transfer of crude can be done in two ways, that is, by tank gauging and 'positive displacement (PD) meters'. Metering has an accuracy of ± 0.025% of total capacity, while automatic tank gauging, with a servo-operated system, has a reported accuracy of ±1 mm of the actual level. A centralised control panel (CCP) has a flow rate integrator to know the quantity of oil dispatched.

In the case of emulsified crude, consideration should be given to installing an electrostatic treater or chemical injection facility. This possibility, however, is to be considered in the engineering phase. An exact treatment procedure has to be tried out with crude samples, under laboratory conditions.

In the *gas module*, the high-pressure gas is fed to the compression system by a common manifold called a compressor inlet manifold from where gas either goes for delivery (normally with FPS, this gas delivery option for use by external customers is not available) or for getting used as fuel gas (for running turbine, compressors) or being used as reinjection gas for gas lift purposes (gas is injected back into the reservoir to enhance oil production). But prior to the delivery or reinjection, this HP gas passes through a dehydration process in a dehydration module. Low-pressure gases from LP separators or from surge tanks are flared. Gas is measured at the outlet of the separator using an orifice meter. Orifice in the gas line gives the gas flow-rate. The CCP also has a gas flow integrator. Further, if we have to handle sour gas (H_2S) or acid gas (CO_2), it has to pass through a sweetening module.

In the *produced water module*, the water separated in the separators and surge tanks is sent to the produced water conditioner wherein the oil is skimmed off and the treated water is either dumped in the sea or in the sump caisson. Sometimes, this produced water is also used for water injection. And accordingly this produced water passes through a series of treating, de-oxygenation and conditioning processes prior to injection.

Environmental and state/national regulations require that produced water from the separators be treated to reduce the oil content in the produced water to a specified level, say 35 ppm, or less, before it can be safely disposed to the sea. The oil quantity in water to be disposed at sea depends upon the region in which FPS is operating as different regions have different guidelines with respect to the maximum contamination of effluents like the North Sea (25 ppm), Canada (25 ppm), China (15 ppm) and S.E. Asia (50 ppm). For Indian offshore water, it has to be less than 50 ppm.

The *main oil line pump (MOL) module* consists of the main oil line pumps which are horizontal, split case, multistage skid mounted units and is used to transfer the crude oil either to the hull of the tanker or for offloading crude oil to the shuttle tankers. These pumps are equipped with the most advanced control systems and operating mechanisms. The pumps have inlet and outlet shutdown valves (SDV), which are motor operated and known as MOV. The electric motor is directly coupled with the pump shaft whereas the gas generator is coupled to the gear-box and thereafter to the pump shaft.

Normally these MOL pumps have a standby provision; a provision by which the standby pump can takeover automatically in case of failure of one pump.

In *process gas compressor (PGC) module* the process gas compressors are used to compress the separated gas to a specific discharge pressure for either its transmission or for its use somewhere, internal or external to the FPS. The compressors are powered by gas turbine generators. Due to its light weight and because of a wide range of operation and usage of natural gas as fuel, gas turbines become a popular choice.

In *venting and gas flaring module* the excess HP and LP gases get disposed through vents or flares. Vent and gas flaring are of three types:

1. Atmospheric vents
2. High-pressure gas flaring
3. Low-pressure gas flaring

In atmospheric vents, the vent header is atmospheric. The gas from the pig receivers, launchers, separators (depressurising), surge tank (depressurising), fuel gas conditioning (depressurising), process water separator, skimmer vessel, sump caisson and so on, can be released through the vent boom. The gas passes through a knock-out drum (KOD) before it is vented. The vent booms are equipped with CO_2 snuffing unit to extinguish. Some of the vessels like chemical storage tanks and diesel storage tanks have independent vents. The process gas compressor packages also have independent vents to depressurise the system.

Apart from the atmospheric or cold vent there are two headers, high pressure and low pressure, through which the gas is sent to flare. The high-pressure vessels like separator, surge tank, compressors, dehydration skid, fuel gas skid and so on, are connected to the high-pressure flare header. The liquid carried over by the gas gets knocked off in the HP flare KOD before it is sent to the flare.

The low-pressure system also has flare KOD. Gas from the surge tanks during stabilised mode and glycol flash drum is sent to the flare through this system. The liquid drained gets collected in the skimmer vessel. Since the flow through this header is normally less, this header gets purged continuously with fuel gas.

The objective of the flare is to burn the combustible gases. Normally there are three types of flares: (1) pipe flare, (2) tulip flare and (3) fin flare.

The *pipe flares* are simple in construction and are just open-ended pipes. Normally these are equipped with windshield, pilot burners and thermocouples. The high-pressure gas comes out from the centre pipe and low-pressure gas through the annulus.

The *tulip flare* utilises a skin adhesion effect known as 'coanda effect,' the principle of which is simple. When high-pressure gas is ejected from the annular slot, it changes direction and follows the profile of the flare tulip entraining air. The gas entrains sufficient air approximately at the point of maximum diameter of the tulip for combustion to begin at the outside of the gas film. Combustion takes place from outside to inward, thus there is always a protective layer of unburned gas between the flame and the tulip preventing flame impingement on the tulip metal. At a lower flow rate due to subsonic flow, the combustion characteristics approach those of a conventional pipe flare wherein the desired cooling effect reduces on the flare metal surface.

The *fin type* is like a pipe flare, but instead of the gas coming out of the pipe it is routed through holes in the fins welded around the circumference of the tip. This is meant for better combustion. But in case liquid hydrocarbon is carried over, then it will cause high flare temperature and can damage the fins.

To avoid liquid going to the flare, a high liquid level switch is provided in the KOD, which when actuated causes shutdown of the processing system. The high-pressure and low-pressure flare lines are interconnected with valves and bursting disc. In case the pressure becomes excessive, then the disc ruptures and the pressure gets released through the other header. The flares are also equipped with pilot burners. A flame front generator or pallet launching unit (PLU) is provided on the FPS to ignite the flare if it gets extinguished.

On a given floating production system, there exist various alternatives for gas disposal like remote flaring, onboard incinerator, conventional flare boom and ground flare.

Remote flaring is preferred in barge type FPS used in shallow water. Because of its remoteness from the storage and production facilities, the potential fire hazards and health hazards are reduced. It is desirable to have the wind blow the hazardous gases away from the production facilities and personnel. At a minimum distance of 500 m, the gas will be sufficiently diluted to be nonhazardous even if the wind direction is reversed. In shallow waters, the remote flare may be mounted on a fixed structure. Two sub-sea lines are required, one for HP gas and the other for LP gas, which terminate at a buoy or a tripod.

The *onboard incinerator* is firebrick lined. It uses about 25% excess air to burn the gas and additional air to quench the exhaust gas. However, this is not preferred because of the dimension it takes and process it utilises.

Conventional flare booms are normally installed on semisubmersible FPS. The boom length would have to be about 230 ft (70 m) to ensure an acceptable level of heat radiation on deck. Two cantilever flare booms will be required to account for the changing direction of the wind. A shuttle barge will be moored alongside to offload the crude. The cantilever flare booms will interfere with the offloading operation. Hence, it is proposed to offload crude from the platform via a pipeline and buoy located away from the floating platform; the restriction of having an open flame near offloading operations does not apply.

A *ground flare* has found wide acceptance on weather vaning FPSO or tanker type FPS. It is normally mounted vertically on the deck, in a location where it is downwind of the living quarters. It is normally supplied with forced draft air at the base, to provide a short, nonluminous flame. The burner and the flame are completely shrouded by a refractory-lined vertical stack to reduce radiation levels on the deck. The stack has openings around the base to provide natural cooling of the refractory lining. Since the tanker is to be stern-moored to an SPM, the living quarters are located in the deck house at aft; the ground flare is therefore to be located near the bow.

6.2.2.1.2 Utilities and Support System Module

The following utility and support systems are generally available on any floating production system either as individual modules or a few combined together as a single module:

1. Diesel fuel system
2. ATF refuelling system
3. Instrument, utility and starting air system
4. Potable water system
5. Utility water system
6. Hypochlorite generation system
7. Material handling system
8. Cooling water system
9. Fuel gas system
10. Hot oil circulation system
11. Communication systems
12. Power generation system
13. Chemical injection system
14. Crude oil washing system
15. Purge gas (inert gas) system
16. Ballast water treatment system
17. Sewage system
18. Helideck
19. Heating, ventilation and air conditioning equipment
20. Drain system

The systems listed from No. 1 to 11 have already been discussed before in the context of Fixed Production System (Section 6.2.1.2). The remaining systems are briefly discussed below.

Power generation system – The FPS is always equipped with sufficient power generations, which supply power to the living quarters, process pumps, air compressors, lighting, battery chargers and so on. Power generation system normally comprises gas turbine generators, utility and emergency diesel generators, switchgear rooms and UPS and is ably supported by a fuel gas system. Normally the power generation is centralised and all the units except the gas compressor are fed power from a turbine generator.

Chemical injection system – Pour point depressant (PPD), demulsifier and oil and gas corrosion inhibitors are injected into the crude oil and gas streams. Oxygen scavenger and bactericide injection systems are also used. The system consists of injection pumps, storage tanks for chemicals and mixing facility.

Crude oil washing (COW) system – For the tanker-based FPS, complies with the requirements of IMO Tanker Safety and Pollution Prevention (TSPP 1978) and is used for the sludge control and cleaning of cargo tanks. The system serves as a fixed deck-mounted and submerged tank-cleaning machine in all cargo tanks.

Purge gas system – Used to prevent air from being drawn into the storage tanks so that safe explosive limit is ensured inside the storage vessel. This can be achieved by connecting the vacuum side of the breather valve to the inert gas system.

Ballast water treatment system – Treats the water from the shuttle tankers at the FPS's loading facility. The International Maritime Consultative Organisation proposes a maximum oil content from tanker deballast water to be 15 ppm or less. The ballast water is generally offloaded from the shuttle tanker and loaded onto an FPS vessel. Produced water tank compartments at FPS are to be used to accommodate this ballast water from a shuttle tanker. Oil removed from these compartments will be recycled back to the main separation system.

Sewage system – Includes sewage treatment/effluent treatment plants which are installed to chemically and physically treat the sewage from the living quarters and the waste water from the galley/kitchen before being discharged into the sea to meet the statutory environmental regulations for offshore installations.

Helideck – Installed at the FPS to allow the landing of state-of-art helicopters and accordingly the dimensions of the helideck are determined. The landing area is to be readily accessible from the accommodation and the take-off and approach path is to extend over an area of 210°. This area must be clear of obstructions. The landing area should be delineated by alternate yellow and a safety net is to be provided at the edges. A rope net with suitable anchorage points is to be provided on the upper surface.

Heating, ventilation and air conditioning equipment (HVAC) includes air handling units, chiller handling units and pressurising units. For efficient functioning of different control systems in process control rooms, generator control rooms and for safety purposes, all closed and living quarters are kept pressurised and a temperature of around 25°C is maintained through the HVAC system. Besides, all other residential rooms, recreation rooms, galleys, offices and so on are temperature, humidity and pressure controlled.

Drain system – Consists of three types of drains. Those are deck drain, closed drain and condensate drain. The deck drains are open drains. Water used for cleaning decks and other spillages on the decks is dumped into the sump caisson through the deck drain header.

The closed drain header is a low-pressure header and the liquids thus drained also go to the sump caisson/skimmer vessel.

The condensate drain header is a high pressure drain header through which liquids from a dehydration system, gas compressor, KODs and so on are drained to the surge tank.

6.2.2.1.3 *Safety System Module*

The main objectives of any safety system in an FPS are

1. To prevent any undesirable event that could lead to release of hydrocarbon.
2. Shut-in hydrocarbons to a leak or overflow, if it occurs.
3. Accumulate and recover hydrocarbon liquids and disperse gases that escape from the process.
4. Prevent ignition of released hydrocarbons.
5. Shut-in the process in the event of fire.
6. Prevent undesirable event that could cause release of hydrocarbons from equipment other than that in which the event occurs.

Safety in any hydrocarbon processing facility is very much interrelated with fire and firefighting systems. Hence, these two systems are compulsory on any floating production system, either in all comprehensiveness or in some exclusivity depending upon the process, systems and field requirement. Following are the details.

1. Safety systems
   i. System and equipment safety system
   ii. Emergency shutdown (ESD) and fire shutdown (FSD) system
   iii. Personnel safety system and personal protective equipment (PPE)
2. Fire and firefighting system:
   i. Firefighting systems
   ii. Fire detection system
   iii. Fire suppression system

System and Equipment Safety
The various systems and equipment are fitted with different safety devices like shut down valve (SDV), pressure switch high (PSH), pressure switch low (PSL) and so on. A safety analysis function evaluation (SAFE) chart is made for the given FPS, which relates all the equipment and the sensing devices, shut down devices and emergency support systems to their function. The system in general provides two levels of protection, that is, primary and secondary to prevent or to minimise the effects of an equipment failure within the process.

Emergency Shut Down System
ESD is a system of manual controls located on an FPS which when actuated will initiate shut down of all wells and other process systems. ESD system provides a means for personnel to initiate process shut down of an FPS when an abnormal condition is detected. In case of actuation of ESD, all process operations will stop, SDVs, MOVs and control valves will go to the fail-safe position, blow down valves (BDVs) will open and vessels will get depressurised. The emergency shut down system consists of a pneumatic loop, kept pressurised at a given critical pressure specific to the chosen FPS, say for example, 40–50 psig and goes all around the FPS. When actuated, it initiates shut down of the complete facility. This ESD either needs to be actuated manually or should get actuated automatically if emergency so arises or if any vulnerable process upsets might get witnessed.

Fire Shut Down System (FSD)
Similar to an ESD system, a fire shut down (FSD) system is also provided at the FPS. It is a system of manual control, in addition to control from various fire sensing devices. This system consists of a pneumatic loop running throughout the FPS. The loop is normally kept pressurised and comprises fusible plugs. In case of fire, the fusible plug melts resulting in loss of air pressure in the loop and actuates the FSD system. In addition to this and ESD pull buttons, FSD pull buttons also exist on the platform at strategic locations, which when actuated leads to FSD.
   The following situations lead to FSD on the FPS:

   1. Actuation of UV detector

   2. Actuation of thermal detector

   3. Actuation of smoke detector

   4. Loss of air pressure in FSD loop by melting of fusible plug

   In case of FSD, total facilities (process and utilities) on the FPS come to a halt. The sprinkler system actuates and thus the pressure of firewater header reduces which in turn starts the firewater pump. Power for the pump in this situation is being supplied by an emergency generator. The ESD/FSD stations should be conveniently located at different places on the decks but should be protected against accidental activation.

Personnel Safety Systems
Safety of personnel is the most important factor in the operation of any industry. General safety rules are practiced and enforced for all personnel on board the FPS as summarised below:

1. All personnel in the open deck area shall wear safety helmet and shoes.
2. Each person on board shall know where the safety and fire suppression equipment is located and how to operate it.
3. Smoking shall be permitted only in specified areas.
4. Safety belts shall be properly and firmly tied up while working at higher elevation.
5. Helideck rules shall be pasted at each entrance to the helideck and should be rigidly followed.
6. Escape routes are to be prominently displayed at each strategic location.

All floating production systems are provided with adequate lifesaving equipment and are to be maintained, tested and kept ready for instantaneous use. The purpose of life-saving equipment is to provide safe means of survival in an emergency situation offshore. Such equipment is life boats, life raft, life buoys, life jackets, personnel baskets, fire blankets, breathing air apparatus and fire suits. Figures 6.25 and 6.26 show all these safety items.

Personnel Protective Equipment (PPE)
All persons on board the FPS must use proper personnel protective equipment while working. Specific PPE is also required to be worn for hazardous operations. The commonly used PPE available at FPSs (Figure 6.34) are cotton dungarees, safety helmets, safety shoes, hand gloves, ear muffs/plugs and goggles.

Fire and Firefighting System
These systems are designed to ensure the early detection of fires, by way of automatic or manual methods, and of giving early and effective alarm. The fire detection and protection systems are engineered as per the approved codes of practice and should be in accordance with the ABS Guide for Building and Classing Industrial Systems, and SOLAS 74 (Safety of Life at Sea 1974). The firefighting system is shown in Figures 6.23 and 6.24.

It is important to know some basics related to fire like fire triangle, hazardous area classifications and so on. Fire is an uncontrolled exothermic chemical reaction in which air/oxygen, inflammable material and heat energy are

FIGURE 6.34
Personnel protective equipment. 1. Safety shoe (courtesy of Francis Flinch/Wikimedia Commons /CC-BY-SA-3.0). 2. Goggles (courtesy of Lilly_M/Wikimedia Commons/CC-BY-SA-3.0). 3. Gloves (courtesy of Pixabay/public domain). 4. Ear muffs (courtesy of Santeri Viinamaki/Wikimedia Commons/CC-BY-SA-4.0). 5. Safety helmet (courtesy of John Tallent, MSA/Wikimedia Commons/CC-BY-SA-4.0).

subjected beyond ignition temperature. This means three elements, namely, fuel, air, or oxygen, and source of ignitions are involved in occurrence of fire, thereby constituting a 'fire triangle.'

Fire is classified into four types:

'*Class A' fire* – These fires are due to ordinary combustible materials such as wood, cloth, paper and plastics. Examples of such materials commonly used in the oil field are: constructive materials and wood decking scaffolding, fibre ropes, clearing rags, tarpaulin and so on.

'*Class B' fire* – These are the fires due to flammable liquids, gases, greases, oil and condensate, residue from stored hydrocarbons, welding and cutting gases, paints, lubricating and hydraulic fluids and so on.

'*Class C' fire* – These are the fires that involve energised electrical equipment where electrical nonconductivity of an extinguisher is of importance. When electrical equipment is de-energised, fire becomes a Class A or B fire.

'*Class D' fire* – These are the fires of combustible metals such as magnesium, zirconium, sodium and potassium.

To determine the type of electrical installation suitable for use in different hazardous atmosphere, the hazardous areas have been classified into three zones, namely, zone 0, zone 1 and zone 2, according to the degree of probability of the presence of hazardous atmosphere.

Zone 0 hazardous area means an area in which a hazardous atmosphere is continuously present and any arc or spark resulting from failure of electrical apparatus in such an area would almost certainly lead to fire or explosion.

Zone 1 hazardous area means an area in which a hazardous atmosphere is likely to occur under normal operating conditions. Such conditions are likely to occur at any time at oil and gas wells and production facilities, which therefore require the most practicable electrical protection.

Zone 2 hazardous area means an area in which a hazardous atmosphere is likely to occur only under abnormal operating conditions, which may be caused only by the simultaneous occurrence of a spark resulting from an electrical failure and a hazardous atmosphere arising through failure of a control system.

An electrical spark near a flammable substance can cause serious fire hazard. A fire or explosion can occur if a flammable atmosphere and a source of ignition (spark) exist together. To avoid this, special electrical equipment is to be used depending on the area. In a zone 0 area, it is preferable to avoid all electrical equipment; otherwise, only intrinsically safe type equipment can be used. In the case of zone 1, only 'certified flameproof' electrical equipment can be used. If the area comes under the category of zone 2, 'non-sparking' types of electrical equipment can be used.

Fire can be controlled in any of the following ways:

Cooling method – where fire is controlled by 'removal of heat from the surface of fire.'

Starvation method – where fire is controlled by 'isolation of fuel from fire.'

Smothering method – where fire is controlled by 'cutting off air/oxygen from the surface of the fire.'

Fire Detection Systems

Fire detection is accomplished through:

1. Ultraviolet (UV) detection system
2. Thermal and smoke detection system
3. Gas detection system
4. Fusible plug loops

UV Detection System
UV detectors are utilised to give the earliest possible detection of rapidly developing gas and oil fires. Detectors are sensitive to radiation over the range of 1850–2450 angstrom and insensitive to light. The UV detectors are not affected by wind, rain, humidity, or extremes of temperature or pressure and are suitable for both indoors and outdoors. They are housed in explosion-proof stainless steel enclosures. On the UV controller panel, the probable source or type of fault can be displayed and provide a location code number for the information of maintenance personnel. When coincident fire signals are received from two detectors, the FPS audible alarm and central control panel annunciator sound is activated along with causing an FSD.

Thermal and Smoke Detection System
The thermal detection system detects an increase in temperature caused by a fire, initiates an alarm signal, and actuates the automatic extinguishing system for that particular area. When *thermal detectors* are located in an enclosed room protected with a Halon 1301 system, the signal is transmitted to the fire and gas panel, which in turn produces audible and visible alarms, sends a shutdown signal to the HVAC unit, and actuates the Halon system. The *smoke detectors* detect both visible and invisible particles of combustion by means of an inner reference and outer sample chamber. The coincident action of two smoke detectors in one protected area actuates the Halon system. Action by only one detector will initiate an alarm only. When detectors operate in a room with a sprinkler system, audible and visible alarms are activated, but the spray systems operate only when the fire melts the fusible sprinkler head, which then permits water spray on the fire. All detectors are intrinsically safe, explosion-proof type.

Gas Detection System
A gas detection system is provided to detect flammable and combustible gases before they reach a concentration level that would cause a fire or explosion. They are the first line of defence in prevention of human injury and equipment damage.

Fusible Plug Loops
Firewater header is always maintained at a predetermined pressure, say, for example, 10 kg/cm^2 and all open deck areas in the complex are covered with a spray system. A fusible plug loop is a ring of pressurised air tubing, which contains fusible plugs at fixed intervals. This loop covers all the open deck process areas. When these plugs come in contact with heat, they melt down and pressure of this loop is vented to the atmosphere. The fall in the pressure of the fusible plug loop activates fire shutdown. Fire shutdown closes down all process systems, generates audio-visual alarms, opens the deluge and starts the fire water pump.

Fire Suppression and Fire Fighting Systems
Some of the systems used for fire suppression and for firefighting are (1) fire water system, (2) spray system, (3) sprinkler system, (4) foam and water hose reel, (5) dry chemical firefighting system, (6) Halon system, (7) firefighting vessels and (8) fire extinguishers.

The *fire water system* primarily consists of equipment to pump and distribute water for firefighting purposes. The water for this system is salt water taken from the sea. The pump takes suction from the sea and starts automatically upon the command of the FSD from the fire and gas panel. Firewater sprinkler/deluge loops located all over the open areas spray water over the equipment and vessels. Foam water hose reels are located at various locations throughout the platform so that all areas can be reached from at least two hoses. These hose reels deliver an aqueous film forming foam (AFFF or light water) with seawater to assist in extinguishing spill fires.

A *spray system* consists of a deluge valve, open type spray nozzles, fusible plugs and fire shutdown valves. The spray system is actuated automatically or manually. Automatic operation takes place by coincident action of any two UV detectors and by action of fusible plugs. Manual operation takes place by pulling FSD, by pressing the remote manual switch for the deluge valve in the fire and gas panel. Manual operation also takes place in case the automatic deluge valve cannot be operated by the above procedure and in that case the bypass valve is to be opened.

The *sprinkler systems* are installed to extinguish class A fires, which are likely to occur in the living quarters, workshops and storerooms. The sprinkler system includes manually operated isolation valves, which normally are locked open. If a fire occurs in a room of the living quarters, the smoke detector will be actuated by combustion gas and particles and fusible metal of the sprinkler head will be melted by the heat generated by fire. Smoke detectors will transmit the signal to the main fire and gas panel. Salt water will be permitted to exit only through the sprinkler nozzles, which have been melted away by the heat, to extinguish a fire. Reduced water pressure will result in the operation of the firewater pump.

AFFF is a two-dimensional medium in a *foam and water hose reel system*, which acts by sealing the static surface of fuel and preventing evolution of flammable vapor. On the fuel surface, AFFF breaks down rapidly, releasing the film forming solution, which then spreads on the surface of the fuel. AFFF is very effective for use on oil spill fires, but is not effective on gas pressure fires. At least two foam/water hose reels, which consist of an AFFF concentrate storage tank, eductor, hose reel and hose with nozzle, are installed to reach any location on the deck. A foam/water hose reel can discharge either a 6% foam solution or water, only by changing the valves. It can also discharge a straight jet stream by turning the grip handles of the hose nozzle.

Dry chemical powder (DCP) is important because it is the most powerful three-dimensional medium available. Powder is effective as an airborne cloud of small particles, which inhibits the chemical reaction of fire. Dry chemical extinguishing systems are considered satisfactory protection for flammable or combustible liquids, combustible gases and electrical hazards.

For operating, the desired amount of hose length is reeled off, the cylinder valve is opened, the chemical is directed at the base of the fire using a sweeping motion to cover the fire area and thus the fire gets extinguished.

Halon 1301 systems are generally used on floating production systems to protect areas where there is an electric fire potential, such as electrical rooms, control rooms, turbine generator enclosures and switchgear rooms, since water is an electrically conductive fluid. Halon 1301 agent is a colourless, odourless, nonflammable and electrically nonconductive gas. It is considered nonhazardous to personnel when exposed for brief periods at low concentrations (less than 7% by volume). Halon 1301 extinguishers extinguish fire by isolating oxygen from fuel and electrical circuit isolation. As Halon 1301 is not very environmentally and equipment friendly, worldwide operators are switching to FM-200.

Big fires are difficult to fight. MSV and *firefighting vessels* have firefighting capability for fighting large fires. MSV are dynamically positioned and normally have 4–5 remote-controlled firefighting pumps.

Portable *fire extinguishers* are designed for small fires and are used in close proximity of burning materials. Various types of extinguishers are available which are suitable for different classes of fire like Class A, B, C and so on.

6.2.2.2 Linkage between Subsea and Topside

The surface production facility on the topside of an FPS requires a feed of well fluid from its underneath, that is, the well head or the wet Christmas tree at the seabed. So, Figure 6.35 shows the arrangement between the seabed and the topside. The well fluid flows from the wellhead through a *jumper*

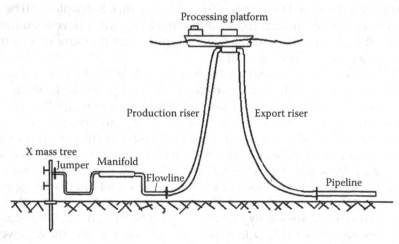

FIGURE 6.35
Arrangement between seabed and topside.

to a manifold from where it moves through a *flowline* to a *production riser* and gets delivered to the FPS where processing takes place. Subsequently the processed and treated oil and gas moves through an *export riser* to a *subsea pipeline* and then to the consumers at shore. To provide a connecting medium for monitoring and controlling the wells and these subsea items from the topside, *umbilicals* and the *control system* play a very important role. The foregoing discussions are on each of these items.

6.2.2.2.1 Jumper and Flowline

A jumper is a prefabricated section of steel pipe or a length of flexible composite line especially configured to make a structural, mechanical and pressure-tight connection at each end. The shape of the jumper is like the letter 'U' with its horizontal base much longer than the two vertical sides but its exact length, orientation, relative offset and angles between the termination points can be determined only after the wells and manifolds are in place.

Flowline and pipeline are the same except they differ in their sizes and also for the purpose they are used. This is discussed in detail in Chapter 8.

6.2.2.2.2 Production Risers

The production risers and export risers are again one and the same so far as their type, shape and configuration are concerned. The production risers used with fixed or compliant platforms are of two types, namely *attached risers* which are clamped to the outside structure of the platform and *pull tube risers* which are actually extensions of the flowline or pipeline dragged through a pull tube permanently fixed in the platform structure (Figure 6.36).

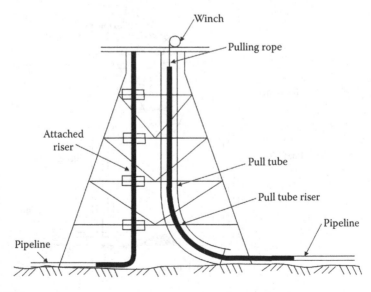

FIGURE 6.36
Attached riser and pull tube riser.

The production risers used with the floating production units are of four types, namely,

1. Top-tensioned risers (TTR)
2. Steel catenary risers (SCR)
3. Flexible risers
4. Free standing or tower risers

Top tensioned risers (TTRs) are long flexible circular cylinders used to link the seabed to a floating production unit specially TLP and SPAR. These are straight risers which run down from the platform and terminate at a bottom junction structure (Figure 6.37) and are subjected to steady current with varying intensity and oscillatory wave flows.

TTRs are almost exclusively associated with dry trees and hence do not compete with flexible risers or SCRs. Under the varying influence of waves and currents, there is vertical displacement between the point of connection with the bottom junction structure and the point of connection with the TLP or SPAR. To compensate this motion within allowable stresses in the riser, a motion compensator similar to the one used with the drilling riser is integrated into the top tensioning system.

TLP and SPAR riser systems are virtually identical below the floater. The differences are in the manner of supporting and tensioning the riser at the top. TLP risers are supported by the hull buoyancy. The tension is provided at the load ring, which is supported by tensioners. Rollers at the

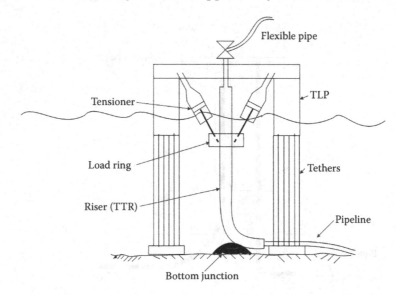

FIGURE 6.37
Top tensioned riser.

TLP deck centralise the riser and accommodate the angle between the riser and the hull.

SPAR risers are supported by 'air cans' or 'buoyancy cans,' not by the SPAR hull itself. These buoyancy cans can be 'integral' or 'non-integral' type. The integral air cans are attached to riser joints whereas non-integral ones are deployed separately.

Flexible risers are the most common type of production riser which may be deployed in a variety of configurations as shown in Figure 6.38 depending on the water depth and environment. The flexible risers are associated with wet trees. Because of its beneficial bending characteristics, flexible pipe is used more as the primary risers. These have traditionally been limited by diameter and water depth, yet they are a tough competitor with SCR and the choice between a flexible riser and SCR is not clear cut. All the configurations shown in Figure 6.38 involve providing buoyancy modules or cans at selected locations on the flexible pipe or using anchored buoyancy modules to give a desired shape such as *steep S* or *wave* and *lazy S* or *wave*. There are various other riser configurations under development today for use in deepwater and ultra-deepwater.

Steel catenary riser (SCR) is a cost-effective alternative for oil and gas export and for water injection lines on deepwater fields, where the large diameter flexible risers present technical and economic limitations. Centenary riser (Figure 6.35) is a free hanging riser with no intermediate buoys or floating devices. A typical profile of an SCR is shown in Figure 6.39. The SCRs are designed by analysis in accordance with the API codes (API RP 1111 and API RP 2RD) or the DNV codes. For detail designing and so on, Reference 7 may be cited.

Free standing or *tower riser* is a typical example of a riser used first in offshore Angola. There three steel column towers each 4200 ft long were anchored to the sea floor. Total water depth was 4430 ft and each tower contained four production risers along with four gas lift risers, two injection risers and two service line risers. At the top of each columnar tower, buoyancy tanks (Figure 6.40) of 15–26 ft in diameter and 130 ft long are provided. The upward force generated by the buoyancy tank holds the

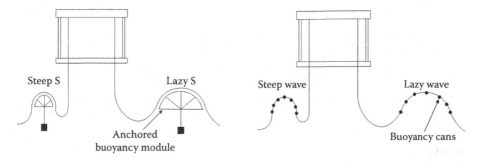

FIGURE 6.38
Flexible risers.

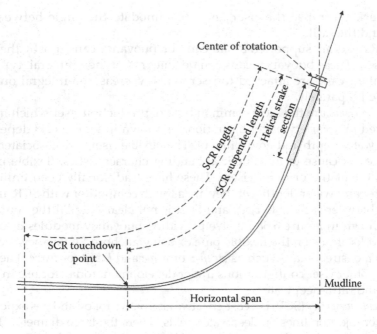

FIGURE 6.39
Profile of steel catenary riser (SCR).

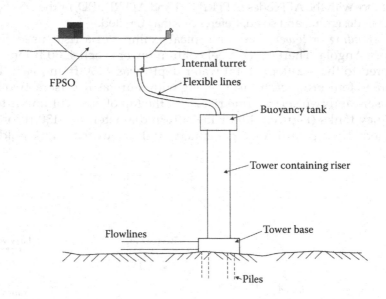

FIGURE 6.40
Free standing or tower riser.

tower vertically stable. Flowlines from subsea wells connect to the tower base and at the tower top flexible lines from FPSO connect with different risers within the column.

6.2.2.2.3 Umbilicals and controls

Umbilical is defined as a bundle of pipes comprising a number of reinforced hoses connecting sub-sea wells to surface structures and conduct the flows necessary to keep the system alive.

Umbilicals provide the connecting medium for electrical, hydraulic, chemical injection and fibre optic lines between the topside facilities on the host platform and various subsea items like the manifold, termination assemblies (Figure 6.41), subsea trees and controls.

The number and the character of umbilicals vary according to specific system needs and development plans. Umbilicals may be single function, for example, a hydraulic line only. They are more commonly multifunction integrated umbilicals that provide hydraulic tubes, electrical lines and tubes that carry chemicals to the manifolds and trees. Some umbilicals have thermoplastic tubing for low-pressure chemical injection service, for injecting chemicals like corrosion inhibitor, hydrate inhibitor and so on.

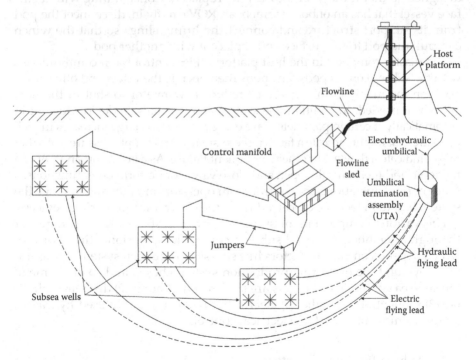

FIGURE 6.41
Cluster well system.

The electrical conductors in the umbilical transmit signals from instrumentation on the subsea components (temperature, pressure, integrity checks) back to the control centre. Electrical power can move through an umbilical to operate solenoid valves on the subsea control pods, which in turn control hydraulic pressure supplied to the subsea valves on the manifolds and trees. In some cases, power is also supplied to operate well bore or subsea pumps.

Combining multiple services into a single umbilical can save manufacturing and installation costs. However, distance, depth and weight may require multiple umbilicals.

The subsea end of an umbilical has a termination head that connects to a distribution structure, the umbilical termination assembly or UTA. From there the umbilical services are distributed to trees or any other miscellaneous equipment in the vicinity by flying leads, sort of subsea extension cords.

A *sub-sea control system* has the ability to monitor and control wells and manifold functions from the topside of the floating unit with confidence and safety. At the seabed trees and manifolds have control pods, modules that contain the electro-hydraulic controls, logic software and communication signal devices. Redundancy is built into the systems to provide alternate ways to retrieve and transmit data and commands. Most control pods are designed so that if they fail they can be replaced. Collaborating with a surface vessel that has an onboard winch, an ROV can fly in, disconnect the pod from its support structure and connect the lifting slings so that the winch can pull the pod to the surface and replace it with another pod.

The master computer in the host platform/FPS control room communicates with the subsea control pods. The pods then operate the valves and other functions on the manifold to increase or reduce flow rate or to shut in the flow entirely, if needed. In addition, the system contains failsafe devices to shut in automatically if certain parameters are exceeded, such as large increases in well pressure that might indicate a fault in the system or an abrupt pressure drop that might indicate a flowline or other component failure. As a last resort an ROV can fly down and mechanically open and close valves or perform other functions.

For normal operations, the control room operators monitor and manage the system based on feedback received from the electronic monitoring devices.

The various components of sub-sea control systems are held together through a member of varied sub-sea control logic options, the choice of which depends on various factors like sub-sea production system configuration, lay-out, type of floating production system chosen and so on. Some of the sub-sea control logic options are (1) direct hydraulic, (2) discrete hydraulic, (3) sequential hydraulic, (4) electro-hydraulic, (5) multiplexed hydraulic, (6) ultrasonic control and (7) electrical control.

6.2.3 Subsea Production System

In the subsea production system it is mainly the way of collecting well fluid from different wells to a common point where the commingling takes place from where a single line carries these fluids to a processing platform either fixed or floating. In Section 5.2.2.1, different types of subsea completions were discussed and based on that different subsea system configurations are possible which are discussed below.

6.2.3.1 Subsea System Configuration

The key elements of a subsea system are the wells and their locations, the flowlines, manifolds and umbilicals whose layout will depend on the (1) nature of the seafloor, that is, undulating or smooth, (2) water depth contour or the topography, (3) flow assurance problems, that is, variation in temperature at the sea bed, (4) number of flowlines and pigging requirement and (5) type of host platform, that is, fixed platform or a floater.

Naturally a number of probabilities exist. For example, a single well may be connected through a single flowline and single umbilical to a host or a number of wells production is gathered at a single manifold via jumpers or flowlines and the commingled fluid is then transported through a single flowline from this manifold to the host. Thus, the first probability is known as *satellite well system configuration*, and the second case is known as *cluster well system configuration*. There is a third system configuration that is *template well system configuration*.

In the case of the *satellite configuration*, the single well is directly connected via a long flowline to a host and it provides for independent operation. This configuration does not put any field development restriction and because of less development time can put the well on production quickly. But the major problem is the length and number of flowlines which are expensive and increases the risk of damage by anchor and trawler/fishing gear.

When the satellite wells become closely spaced, they become *cluster wells* and these are clustered around a central manifold (refer to Chapter 5) where the fluid from individual wells gets commingled. This manifold is connected by a flowline and an umbilical to the host platform. Figure 6.41 is an example of a cluster system where the umbilicals from individual subsea wells known as *flying leads* are shown getting terminated at a single structure known as umbilical termination structure or umbilical termination assembly (UTA). Sometimes the flying leads may terminate in a junction box on the central manifold itself from where a single umbilical will connect with the host. The wells in this system are spaced approximately 10 m from the central unit.

The third type of configuration, that is, the *template well system* configuration, becomes visible in the case of multiwell completion (Section 5.2.2.1). The concept of an underwater template system is seen as a replacement for the platform, capable of supporting a substantial number of wells in very deep water by making full use of the advantages such a system can pose. A subsea production template is shown in Figure 5.40 (Chapter 5) where it can be seen that there is a facility for even connecting satellite well flowline which are later connected to the common manifold placed over the template. In this configuration, no well flowlines are required and it has comparatively reduced the number of surface facilities. It covers a limited ocean floor area and the protective devices for fishing gear and so on are less expensive here. The only disadvantage of this system is that the production can start only after all wells are drilled and completed.

Figure 3.46 (Chapter 3) shows a much bigger and more complex system that includes all these types of configurations.

6.2.3.2 Selection of Subsea System for Different FPS

Depending on the type of floating production system, for example, barge, tanker or semisubmersible, the subsea production system needs to be selected. The choice in general will be based upon better compatibility, improved operational flexibility, least complexity and reduced cost.

Normally in the case of barge-based FPS, satellite well system configurations are preferred. The reason is that the barge-based FPSs are normally used for small field developments of one to four wells or for single well extended test systems. Naturally this small number of wells does not warrant the use of manifolds or templates.

For tanker-based FPS although any of the three configurations may be possible, satellite well options are preferred owing to less complexity of the sub-sea systems like template, manifold and so on and mainly from the cost point of view. It maximises production availability as sub-sea equipment failure or well work over result in only one well being shut in, therefore having a minimum of lost production due to maintenance or equipment failures.

In the case of semisubmersible-based FPS either a cluster system or a template system may be used. It is always better to place the sub-sea wells immediately beneath the semisubmersible vessel, thereby getting benefit from the features that are inherent to a semisubmersible FPS. This type of well placement enables the rig to perform well maintenance work (e.g., wire line operation, work-over operation) also. In this case, out of the two options, it will be always be preferable to use the template system configuration because the capital cost is less, there are fewer flexible risers, there is greater deck load capability and there is a quick disconnect system.

Bibliography

API RP 2RD, Recommended practice for design of risers for Floating Production Systems (FPS) and Tension Leg Platforms (TLPs), 1998, American Petroleum Institute.

Arnold, K. and M. Stewart, *Surface Production Operations*, Vol. 1 & 2, Gulf Publishing Company, 1991.

Chakrabarti, S. K. *Handbook of Offshore Engineering,* Vol. II, Elsevier Publisher, 2008.

Katz, D. L. et al., *Handbook of Natural Gas Engineering,* McGraw-Hill Book Company, 1959.

Lake, L. W. *Petroleum Engineering Handbook,* Vol. III, Society of Petroleum Engineers, 2007.

Leffler, W. L., R. Pattarozzi, and G. Sterling, *Deepwater Petroleum Exploration & Production – A Nontechnical Guide,* Pennwell Corporation, 2011.

Malhotra, A. K. *An Introduction to Ocean Science and Technology,* National Book Trust, India, 1980.

Mitra, N. K. *Fundamentals of Floating Production Systems,* Allied Publishers Pvt. Ltd. India, 2009.

Mukerjie, R. K. et al., *Technical Manual for Production Operations,* IOGPT, ONGC, India, 1994.

Uren, L. C. *Petroleum Production Engineering,* McGraw-Hill Book Company Inc., 1939.

7

Offshore Storage

The oil and gas separated at the processing platform, fixed or floating, are normally sent to the shore through separate pipelines, but where no pipeline exists, the oil is stored in tanks offshore and the gas is either flared or sent back down a riser for reinjection in the producing reservoir or some other nearby subsurface structure. Part of the gas separated is used to drive pumps that transfer crude from the platform to the storage tanker.

Oil produced from offshore wells can be stored in the following manner.

7.1 Above Water Storage

7.1.1 On Platform Storage Tank

Some oil processing platforms (e.g. Gulf of Mexico) are provided with storage facilities in addition to separation and treating equipment. This becomes at times necessary because of the location of fields far away from the shore where transportation by pipeline becomes very costly.

The amount of storage varies with the number of well platforms served by the processing platform. About five days' storage is normally required for efficient operation. With this much storage, a producer can usually maintain his monthly allowable producing rate despite inclement weather conditions which may often prevent the dispatch of crude by barges to the shore.

7.1.2 Floating Storage

In the foregoing discussions (Chapters 3 and 6) on 'Floating Production Systems,' FPSO, FPS and SPAR have been discussed in detail. Hence, in the present context vessels from small barges to large tankers are considered as the main storage units for storing crude oil in offshore water. The large tanker installations are normally considered permanent storage facilities while barges usually provide temporary or emergency storage only.

7.1.2.1 Barge Storage

Although anchored barges have been used for oil storage before, they were an operational headache due to the breaking of anchor chains during rough weather. To solve this, a floating buoy arrangement for mooring the barges was developed. Details on single-buoy mooring (SBM) are available in Chapter 6 (Section 6.2.1.4.3).

The storage barge connection is such that it can weathervane about the buoy according to the sea conditions. Crude is produced from a platform to a submarine pipeline which is connected to the buoy by a hose. Oil reaches the barge from the buoy via a floating hose connection. Figure 7.1 shows oil being transferred from the storage unit to a small oil transport barge or transfer barge.

It was found in a typical location of the Gulf of Mexico, where a 17,000 bbl capacity, 200′ × 54′ storage barge was tied to a floating buoy (SBM) with a bridle composed of two 120′ long nylon ropes. The total cost of installation of this system including the cost of the tug and transport barge was much less as compared to a pipeline, which in this case was laid to do the same job.

FIGURE 7.1
Barge storage.

Besides, a pipeline would have required months to put it in operation. The SBM permits production of wells as soon as they are completed.

7.1.2.2 Tanker Storage

A tanker is defined as a ship designed to carry liquid cargo in bulk. The tanker can be used for storing liquid cargo, mainly crude oil, either by mooring to a floating buoy (SBM) or anchoring conventionally in relatively sheltered waters near the shore. Generally, SBM is capable of handling large vessels in high seas.

The design of a tanker considers the particular trade for which it is intended. A high rate of loading and discharging is desirable; pumping capacity and size of pipelines are important in this respect. The safety factor must be borne in mind with the provision of a fire smothering installation and cofferdams at the ends of cargo spaces, ventilating pipes to tanks and so on. Ships intended for the carriage of heavy oils would have steam heating coils fitted in the tanks. Figures 7.2, 7.3 and 7.4 show the features of an oil tanker, a product tanker and an LNG tanker.

The cargo space is generally divided into three sections athwartships by means of two longitudinal bulkheads and into individual tanks by transverse bulkheads.

The maximum length of an oil tank is 20% L (L is length of vessel) and there is at least one wash bulkhead if the length of the tank exceeds 10% L or 15 m. Tanks are generally numbered from forward, each number having port, centre and starboard compartments. Pump rooms are often located aft so that power may easily be supplied to the pumps from the engine room, but ships designed to carry many grades of oil at once may be fitted with two pump rooms placed to divide the cargo space into three sections. The system of pipelines used in a tanker is such that great flexibility is possible in the method of loading or discharging, and different parcels of cargo may be completely isolated from one another during loading and subsequently during discharge. In some cases, a small, separate line is used for stripping the last few inches of oil from each tank.

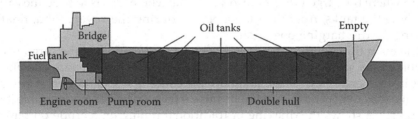

FIGURE 7.2
Features of an oil tanker.

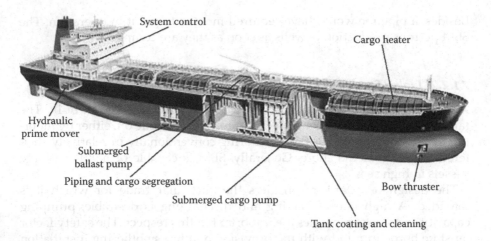

FIGURE 7.3
Features of a product tanker.

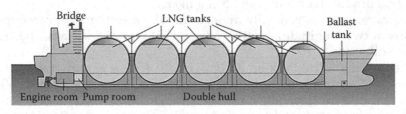

FIGURE 7.4
Features of an LNG tanker. (Courtesy of Tosaka/Wikimedia Commons/CC-BY-SA-3.0.)

The loading line system is the basic element of the cargo handling equipment on an oil tanker. Treatment or handling of cargo includes all transport of the cargo, ballast handling, loading, discharging, internal cargo transferring, tank cleaning – either with cargo (COW) or water, cargo heating and so on.

On a traditional crude oil tanker, the vessel is equipped with an efficient line system for loading the cargo on board and discharging the cargo ashore. When discharging the cargo ashore, the cargo goes via the vessel's pump room where the cargo pumps are located. The whole idea is to keep the cargo safely in the tanks, from the time it enters, during the voyage, and, finally, during the discharging operation.

The main thing with cargo in such a closed system is that the cargo is not visible at any stage of the operation. Fixed checklists provide safe operations and instruments show where and how the cargo flows. On different vessels, the line system in principle is similar, but each vessel has its own peculiarities.

Figure 7.5 shows the drawing of traditional piping on a crude oil tanker. The vessel is fitted with four centre tanks (CT) and five pairs of wing tanks (WT) in the port (p) and starboard (s) side for cargo.

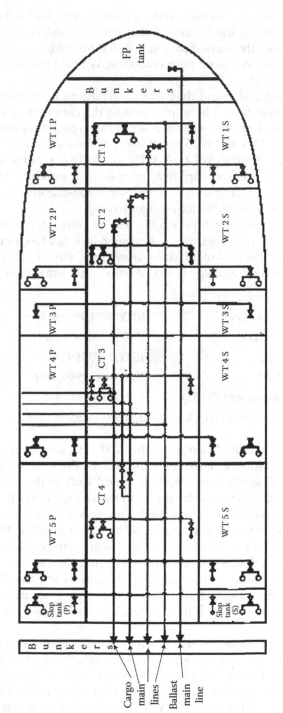

FIGURE 7.5
Piping in a crude oil tanker.

The cargo main lines are located in the vessel's centre tanks. The location of these lines will be on the bottom of the vessel generally called 'bottom lines,' which are usually supported about 4–6 ft above the vessel's bottom. Crossover valves, two valves on each crossover, connect the bottom lines to each other.

From the drawing it is found that, from the bottom lines, there are lines that lead to each cargo tank. These lines end on the cargo tank's suction bell mouth. Each bottom line serves its own set of cargo tanks; for example, bottom line no. 1 serves CT1 and WT5 p/s. Bottom line no. 2 serves WT1 p/s and CT4. Bottom line no. 3 serves WT2 p/s, CT3, and WT6 p/s and so on.

Tankers range in size and capacity from several hundred tons, which includes vessels for servicing small harbours and coastal settlements, to several hundred thousand tons, for long-range haulage.

Tankers used for carrying liquid fuels are classified according to their capacity. Though in earlier days, the 'afra' system (average freight rate assessment) was used to classify tanker sizes, nowadays, tanker sizes are classified as per 'dead weight tonnage (dwt).' Accordingly, tankers are classified as below:

General purpose tanker	10,000–24,999 dwt
Medium range tanker	25,000–44,999 dwt
Large range 1 (LR1)	45,000–79,999 dwt
Large range 2 (LR2)	80,000–159,999 dwt
Very large crude carrier (VLCC)	160,000–319,999 dwt
Ultra large crude carrier (ULCC)	320,000–549,999 dwt

Tankers are also classified as per the 'hull,' that is, single-hulled tankers and double-hulled tankers. In *single-hulled tankers*, the hull is also the wall of the oil tanks, and any breach results in an oil spill. In *double-hulled tankers*, a space is available between the hull and the storage tanks to reduce the risk of a spill if the outer hull is breached. This space is used to carry water ballast when the ship is not carrying an oil cargo. In practice, the addition of an extra hull should prevent such a ship from suffering a catastrophic breach of the hull. A double-hulled tanker is generally safer than a single-hulled tanker in a grounding incident, especially when the shore is not very rocky. However, double hulls are not a complete solution, and they are at greater risk of explosion if petroleum vapour collects in the space between the hulls.

Over and above these classifications, there is a very common term used in the shipping industry known as 'super tankers' which stand for the world's largest tanker ships having deadweight tonnage (dwt) of above 250,000 and capable of transporting around 2 or 3 million barrels of oil. In the shipping industry, very-large crude carriers (VLCC) and ultra-large crude carriers (ULCC) are now commonly designated as super tankers.

Due to their size and mass, super tankers have very poor manoeuvrability. The stopping distance of a super tanker is typically measured in miles. When operating close to the shoreline, they are vulnerable to running aground, largely because at slow speed it is impossible to control their movements. When this happens, there are chances of oil spills and accidents.

The term dead weight tonnage (dwt) used for classifying the tankers is defined below as per SOLAS (Safety Of Life At Sea 1974).

Deadweight is the difference in metric tons between the displacement of a ship in water of a specific gravity of 1.025 at the load waterline corresponding to the assigned summer freeboard and the lightweight of the ship. Deadweight (often abbreviated as dwt for deadweight ton) is the displacement at any loaded condition minus the lightship weight which includes the crew, passengers, cargo, fuel, water and stores.

Lightweight, similarly as per SOLAS 74, is the displacement of a ship in metric tons without cargo, fuel, lubricating oil, ballast water, fresh water and feed water in tanks, consumable stores together with passengers and crew and their effects. The term 'ton' and 'tonnage' needs to be understood clearly and hence a little background is given below.

Early cargoes carried in ships used to be wine, which was transported in wooden casks called tuns. A tun was a cask of wine that could be carried in a cart pulled by two horses, as shown in Figure 7.6. In 1350, in England, an import levy of 2 shillings per tun of wine was granted to the crown. This import payment became known as tunnage. Ships paid tunnage ostensibly to pay for the protection of trade at sea, and all English kings from Henry VI to James I received this tax. In chartering ships, it became common practice to describe a ship in terms of its tunnage, that is, the number of tuns it could carry. Eventually the tun was fixed by weight in Great Britain as being equal to 20 hundredweight or 2240 pounds. Today, many calculations in marine work still use a weight measure of a ton equal to 2240 pounds.

FIGURE 7.6
Cask carried in a horse cart.

In the United States, this is called the long ton to differentiate it from the short ton, which is equal to 2000 pounds.

In marine work people have their own way of talking about tonnage. Tax is levied on a ship according to the size of the ship. The size of a ship was once determined by multiplying the length times the width times the depth of flotation. This calculated volume in cubic feet was divided by a number that varied with the amount of taxes that the taxing authorities wished to collect. Eventually the distraught ship owners succeeded in making the authorities agree that the number should be 100. Thus, a ton, when used to describe the tonnage of a vessel (including modern drilling units) equals 100 cubic feet and is actually a volume measure.

The total volume of a vessel up to a height specified by law is called the *gross tonnage*. The term *net tonnage* is used to describe the volume of a vessel that can carry cargo. Net tonnage, however, is not determined by actually measuring volume. Instead, it is determined by subtracting from the gross tonnage allowances for space occupied by the crew's quarters, engine room, fuel tanks, and other areas not used for carrying cargo. The word *allowances* is important: depending upon the agency making the determination, volumes greater than the actual volume of these areas may be used. Some of these allowances are intended to encourage the building of safer ships. The remainder after allowances have been subtracted from the gross tonnage is the net tonnage. Once a ship is built and the tonnage is measured for registration, the resulting *registered tonnage* is fixed. It never changes unless some physical change is made to the ship or drilling unit.

A 38,000-ton tanker used in the Persian Gulf (Figure 7.7) permanently moored to a buoy receives crude through an 8″ floating hose. As required, export tankers visit the field and tie up to the buoy alongside the storage vessel. Crude is then transferred and sent to market. The unique single buoy

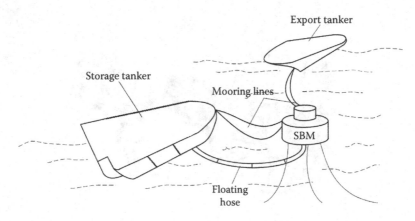

FIGURE 7.7
Tanker storage.

mooring system and special fendering which permit the ships to tie up together even in high seas make the operation feasible. The mooring system allows the vessels to swivel about the buoy in a complete 360° circle so that the wind and sea direction have no appreciable effect on the operation. This installation has withstood 60 knot winds and 25-foot seas without damage.

7.2 Over and Under Water Storage Tank

This type of storage unit is unique in certain offshore fields and is not commonly found in every field. In an offshore field where large productions are to be handled and the existing floating storage vessels are unable to handle the complete production per day, this extraordinary special type of tank is put into operation. One such example is the Arabian Gulf's Fateh oil field where 500,000 bbl capacity underwater oil storage tank was installed. This field's daily production was 100,000 barrels of oil for which over and above two floating storage vessels, namely, Majmaa 1 and 2, Dubai–Khazzan type storage tank were commissioned. Figure 7.8 shows this storage unit in Fateh field. The other two examples are Tenneco Oil Co.'s water tower type tank placed in the U.S. Gulf Coast of Louisiana, and Elf-ocean type tank used off the coast of Western France.

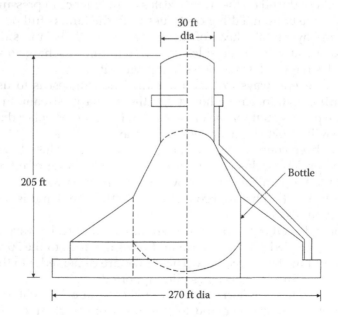

FIGURE 7.8
Dubai–Khazzan type tank.

7.2.1 Dubai–Khazzan Type Tank

A half-million-barrel capacity underwater oil storage tank of unusual design has been successfully set in 158 ft of water off the Shaikhdom of Dubai. The unit is shaped like an inverted funnel. It is 205 ft high; 270 ft in diameter; has no bottom and weighs 15,000 tons (Figure 7.8). Constructed onshore in the Trucial State of Dubai, the unit was floated in an excavated pit, then towed through a dredged channel to the open sea. After a 58-mile tow, it was sunk adjacent to Fateh field producing facilities.

The underwater storage unit has an open bottom, an internal diameter of 250 ft and straight sides of 31 ft. A conical transition connects the roof to a 30-ft diameter shaft that extends 45 ft above the water surface.

The straight shell portion is 4 ft wide with concrete placed between the two steel shells. The concrete adds ballast and lowers the centre of gravity, important during towing and submergence.

An 80-foot diameter pressure vessel or bottle is in the centre of the structure. Its main function is to assist in the submerging operation. Once on bottom, openings are made to allow communication of water or oil from the outer volume to the inside of the bottle. Hence, the bottle is also used for storage. The roof is heavily stiffened to transmit wave loads to the shell and then to the supporting system.

The structure operates on the water displacement principle and will always be full of oil or water or a combination of the two. After separation and treating (if necessary), oil is routed to storage by flowing into the central riser.

The weight of the oil on the water creates an imbalance in pressure forcing water out of the bottom. Filling continues until the tank is full of oil, and at that time the hydrostatic head of the oil on the inside balances the hydrostatic head of water on the outside. Deep well pumps discharge oil from the unit. As oil is removed, water flows in replacing it.

There are several ways to gage the tank. The simplest is to use meters for measuring both incoming liquid and the discharge stream. In addition, capacitance probes monitor the oil/water interface. Probes placed at various elevations will actuate high- and low-level alarm systems.

There has been concern about creating an emulsion at the oil/water interface. However, lab work shows that vigorous intermixing of oil and water is required to form an emulsion. Movement of the oil/water interface in the tank is very slow (1 fph with production of 100,000 bpd) and is not enough to create an emulsion.

The roof and shell of the structures are designed to resist wave loads created by the so-called 100-year storm and transmit them to the 30 skirt piles placed around the tank periphery. These piles are cemented into the sea bed and to the structure, so it becomes pile supported.

Piles are used to resist horizontal wave forces, support weight of the structure when it is full of water, and to help resist uplift when it is full of oil.

Pile design considers vertical wave forces, both up and down, in combination with the tank being full of water or oil.

Corrosion protection is provided by self-sacrificing anodes on the inside and outside of the unit to protect both surfaces.

7.2.2 Water Tower Type Tank

This unit in which oil is stored above and below water is being used by Tenneco in Vermilion Block 245 field of Louisiana. The 30,000-barrel capacity unit, the only one of its type, has the largest capacity of any underwater storage installation used in U.S. Gulf waters. Essentially, the structure consists of a 'doughnut' shaped hull, which rests 5 ft below the sea bed, surmounted by three legs which support an above-water tank (Figure 7.9). Shear cans attached below the lower hull prevent horizontal movement of the unit during rough weather. Weight prevents vertical movement. The ring-like hull shown separately in Figure 7.9 has a capacity of 24,000 barrels. The figure also shows the 6000 barrels above the water tank that is installed on top of the structural members (legs) which protrude above the water's surface from the ring.

Configuration of the facility is similar to a municipal water tower design, and is considered feasible for use in water depths up to 200 ft. Units with capacities to 500,000 barrels also are possible.

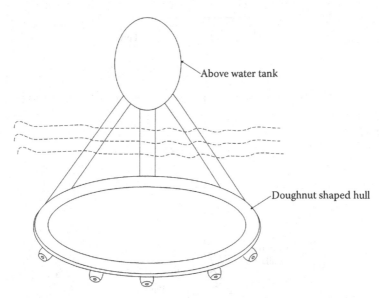

FIGURE 7.9
Water tower type tank.

Oil is delivered to the above-water tank from a nearby production platform and eventually overflows into the bottom hull through one of the supporting legs, displacing seawater. Salt water pumped into the lower hull will displace crude back to the surface for off-loading. Production rates to 10,000 bpd have been handled by this unit and it takes about 6 hours for off-loading 15,000 bbls of crude.

7.2.3 Elf-Ocean Type Tank

The Elf-ocean oil storage and production platform makes possible the development of an oil well too far from shore or of too small an output to make a marine pipeline economic. Oil is stored in the column and unloaded onto a tanker at intervals. This platform (Figure 7.10) consists of a cylindrical shell articulated on the base and used as the storage tank. Part of the shell is divided into ballast compartments which are used as a reserve for buoyancy to provide stability. The engine compartments and crew accommodations are all located in the upper part. In the case of one well only, the platform is set down exactly on top of the well. The base is then set over a template so that it is precisely over the completion. Oil from other wells or from under-water storage units can be brought to the platform by marine pipelines. To compensate for variations of vertical loads, one universal joint is provided. A regulating pump can partially flush the buoyance ballast.

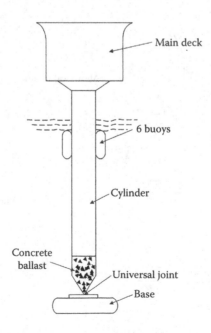

FIGURE 7.10
Elf-ocean type tank.

To unload the oil into a tanker, a rotating joint is located on the upper part of the platform. The tanker is itself moored to the rotating head of the platform and is thus always on the leeward side (away from the direction of the wind) of the platform.

The 23-ft diameter cylindrical offshore production platform, Elf-ocean, extends 408 ft from the sea floor to its heliport. The unit, which oscillates with wave action, was successfully installed by the French oil company ELF and C.F.E.M. in the Bay of Biscay.

Chief components of the unit are

1. A rectangular 69-ft × 79-ft × 13.9-ft deep base plate.
2. A universal joint that connects base and cylinder.
3. The 23-ft diameter × 355-ft long cylinder.
4. Six buoyant stabilising tanks, 56.7 ft long × 14.7 ft in diameter, attached near the upper end of the cylinder.
5. A 24.6-ft high × 34.4 ft diameter living quarters/work deck.

When the sea is calm, the platform does not move and no force is exerted on the rotating joint. When the sea is rough, the platform moves to and fro in phase with the waves. This results in a very light horizontal drag force much less than would be exerted on a fixed platform of equivalent capability.

The *cylindrical hull/tank/shell* is supported by steel wind rings of different grades and variable thicknesses according to stress concentrations. Wind rings are stiffened with horizontal bracing every 4.9 ft and vertical bracing every 10°. A 3.3-ft diameter central conduit and two 4.9-ft diameter pipes extend from the top to the bottom of the cylinder. These tubes are used for testing simulated well heads and other oil field equipment.

The cylinder is partitioned into sections by nine bulkheads from bottom to top.

The floating tanks or *buoys* fastened on the cylinder provide the stability and uplift torque to overcome stresses imposed by waves, wind, current and top loads.

The *universal joint* that connects cylinder to base plate consists of (1) a lower plate that includes two laying guides and hydraulic locks for connection, (2) an upper plate welded to the cylinder on the end shell ring and (3) a ring joining the two plates with two pairs of bearing necks.

The *superstructure* or top part of the unit contains living quarters, work areas, helicopter deck and so on. A central passage allows a 5-ton traveller and hydraulic handling crane to work through the three vertical conduits that traverse the unit from top to bottom.

The *rectangular base* has to provide a secure mooring point for the column. It is designed to be floated to location, sunk on the site and to moor the universal joint. It consists of four 13.9-foot diameter cylindrical buoys cross-braced with lattice girders that contain the universal joint connection stage.

7.3 Submerged Storage Tank

Underwater stationary tanks for storing large amounts of crude oil consist mainly of a hemispherical chamber sitting on the sea bed. Because of its shape, it is known as a 'cupola' type tank.

7.3.1 Cupola Type Tank

These are dome-shaped tanks described as a 'cupola' fastened to an annular base (Figure 7.11). The annular base acts as a float during towing. It is ballasted during immersion and transmits the wave forces to the pile foundation. It also helps in storing the oil within the cupola. These tanks may be designed with capacities up to 600,000 bbls. They operate on the water displacement principle. A 600,000 bbl tank has an outside dimension of 300 ft and a height of 70 ft. Depending upon the size, the tank may be built in a dry dock and towed to its location. The immersion of such a tank is done with two small barges equipped with floats at the end of articulated arms. During immersion, the tank has no movement and there is no risk of damage near the bottom. The procedure followed is a safe one. Such tanks may be easily recovered and relocated elsewhere.

These tanks can be located in water depths greater than 130 ft and where the wave heights range between 30 and 70 feet. For very deep waters, the Elf-ocean type of articulated device can be fitted to the top of the submerged tank and tankers can be loaded from it.

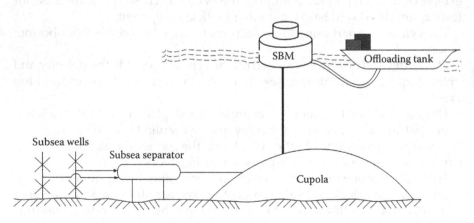

FIGURE 7.11
Cupola type tank.

Bibliography

Buoyancy, stability, and trim, Petroleum Extension Service, 1985.

Different types of tankers, http://www.slideshore.net/Ruranha/different-types-of-tankers, 2015.

Mitra, N. K. *Fundamentals of Floating Production Systems*, Allied Publishers Private Ltd. India, 2009.

Offshore Handbook, Vol. 2, World Oil, Gulf Publishing Company, 1971.

Offshore Handbook, Vol. 3, World Oil, Gulf Publishing Company, 1973.

Offshore Handbook, World Oil, Gulf Publishing Company, 1969.

Oil tanker design and equipment, http://tankerfleet.ru/into/oil-tanker-design-equipment.pdf 2015.

8

Offshore Pipeline

Regarding transportation of oil and gas through a pipeline, a general description of it covering various aspects like types, design, laying, maintenance and so on with reference to the onshore areas has been made in Chapter 6. In the last three decades, oil and gas production in offshore areas has shown a marked increase, resulting in a growing network of underwater pipelines. As these pipelines have moved into deeper water, new subsea technology has been gained and improved techniques of laying pipelines in offshore water have developed. Previously in Chapter 6, it was found that in the case of both the fixed or floating production system, oil, gas or both are transported to shore or an offshore terminal via submarine pipelines. These submarine pipelines are classified into four types depending on the line function, pipe sizes and operating pressure.

These classifications are flowlines or intrafield lines, gathering lines or interfield lines, trunk lines and loading (unloading) lines.

A *flowline* connects a well to a platform or subsea manifold. Usually the line has a small diameter and may be bundled. Flow inside of it may be at high pressure such that the fluid flows through the line without pump or compressor.

A *gathering line* connects from one (multiwell) platform to another platform and is usually a small- to medium-sized diameter line but can be large diameter, too. The line may be a bundled oil, gas, condensate or two-phase flow. The range of operating pressure is usually between 1000 and 1400 psi. Flow in the lines is done by booster pumps or compressors which are often installed on the platform.

A *trunk line* handles the combined flow from one or many platforms to shore. The line is usually of large diameter and can either be oil or gas. Booster pumps or compressors must be provided at intermediate platforms for very long trunk lines. A trunk line is usually a common carrier, carrying product owned by many producers.

The *loading (unloading) lines* usually connect a production platform and a loading facility or a subsea manifold and a loading facility. The lines can be small or large diameter and carry liquid only. Connection may be from a shore facility to an offshore loading or unloading terminal. Loading lines are usually short, ranging from 1–3 miles long. The loading facility may be temporary, such as an early production facility, to provide limited product shipment until a gathering or a trunk line can be completed. The loading line can be used with a permanent loading facility for small reservoirs and in remote areas.

Similar to the onshore pipeline design, here also the pipe parameters are carefully evaluated and the most common parameters evaluated are internal pressure, flow properties, hydrodynamic forces, vortex-induced pipe oscillation, pipeline-soil stability, pipe buckling, effects of large soil movements, effects of sea bed irregularities and scour and erosion of the bottom soil. Since the pipeline remains filled and often buried under the mud line, the hydrodynamic forces become more prominent during construction than during operation. The details of designing subsea pipeline are available in the references cited and some of the parameters evaluated are common to onshore pipeline which follow strictly the API (API5LX) and ASME (B31.4 & 31.8) codes.

The fact that makes it different from onshore pipeline lies in the way such pipes are laid below sea water causing development of installation stresses and the technology involved to overcome those in such methods. The second aspect that is important for consideration here is the flow assurance of the fluids (oil and gas) flowing through these lines and the third aspect is the maintenance issues, mainly corrosion and its monitoring. Subsequent discussions are on these three aspects.

8.1 Laying

The methods of laying offshore pipeline are many and the type of pipe considered here is the steel pipe conforming to the API specification and not the flexible one. Following are the different pipe laying methods:

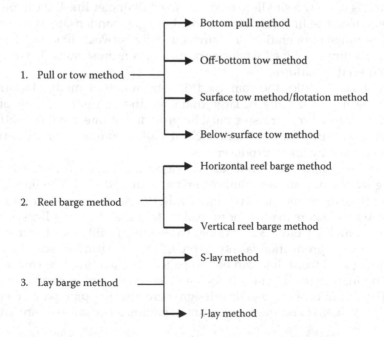

8.1.1 Pull or Tow Method

In this method, long sections of pipe are welded together onshore and pulled into the water a section at a time. The method is particularly applicable to crossing narrow, deepwater channels, but it can also be used for moderate-length offshore pipelines. The sections are pulled by winches in the case of a narrow crossing or by a tow vessel in the case of an offshore line.

This type of offshore pipeline construction has several advantages; for example, much less expensive offshore construction equipment is required, much of the work is done onshore using conventional techniques, and the time the operation is exposed to severe offshore weather is reduced. Large diameter lines (30 in.) up to 20 miles long, have been installed using this method. Typically, though, projects using this technique involve 10–15 mile of 16-in. diameter to 24-in. diameter pipeline.

The tow method has been recommended in a number of applications.

1. Near shore in shallow water where a lay barge operation is not possible.
2. For bundles of several pipelines or very large-diameter lines that are difficult to handle by a lay barge.
3. Where difficult or dangerous manoeuvring by lay barge is required.
4. In deep water, where the capacity of lay barge tensioners, stinger, or positioning system is exceeded.
5. In the Arctic, where heavy ice cover exists.
6. Where only a short installation season is available because of high sea states or other environmental conditions.

There are four tow methods as mentioned before. A brief description of each of these follows:

8.1.1.1 Bottom Pull Method

When the bottom pull method is used for narrow crossings, the pipe is fabricated onshore into one or more sections, launched into the water and pulled along the bottom and into its final position by means of a winch.

The launchway consists of a graded area with dollies, upon which the pipe is placed, mounted on a track. The winch may be located on the far shore, in the case of a river crossing, or upon a barge firmly anchored in the case of a long underwater line.

Often the pipeline must be divided into several sections because of limited space on the launchway. When this is the case, the pull must be interrupted for tie-ins. Figure 8.1 illustrates this method.

When the bottom tow method is used for moderate length offshore line, the pipeline – or bundle of pipelines – is towed along the ocean floor by a tug

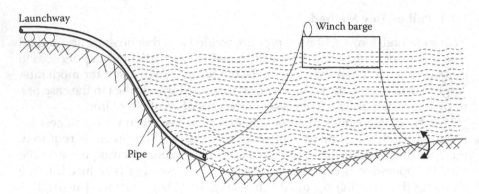

FIGURE 8.1
Bottom pull method.

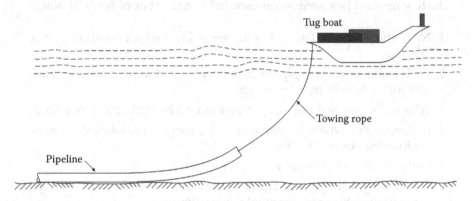

FIGURE 8.2
Bottom tow for offshore line.

(Figure 8.2). The ocean floor where the pipeline is towed should be relatively flat and free of obstacles.

In the bottom-tow design, the tow route is one of the basic design factors. Route considerations affect coating design for abrasion criteria, stability during tow, tow vessel size and optimum length of towed segments. Multiple route surveys may be required to identify an acceptable towing corridor. Route surveys and installation-site surveys should include a detailed investigation of variations in soil conditions to be encountered, bottom currents, bottom contours and identification of obstructions within the corridor. Route surveys should include the near-shore and surf zones at proposed make-up sites in the same detail as the deepwater corridor.

8.1.1.2 Off-Bottom Tow Method

In this method, the pipeline is floated at a height a little above the ocean floor by adjusting buoyancy with weights and floats. It requires both a primary

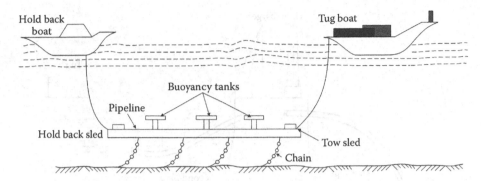

FIGURE 8.3
Off-bottom tow.

tow vessel and a small hold-back vessel. Buoyancy tanks are attached to the pipe string at specified intervals, and the tow and hold-back sleds are positively buoyant. Lengths of chain are suspended from the buoyancy tanks. During the tow, the weight of the chain raised off the seafloor balances the buoyant thrust, and the pipe string is supported at a predetermined design height off the seafloor. This method is illustrated in Figure 8.3. When the pipeline is in position, the floats are either released to surface or flooded, allowing the pipeline to sink to the seabed. There is less chance of damage to the pipeline with this method than when towing along the seabed.

8.1.1.3 Surface Tow Method/Flotation Method

In the surface tow method (Figure 8.4), the pipeline is towed to its site while buoyed near the water's surface with floats or pontoons. In addition to the primary tow vessel, a second vessel is usually needed for control of the floating string in the surface-tow technique. This hold-back vessel can generally

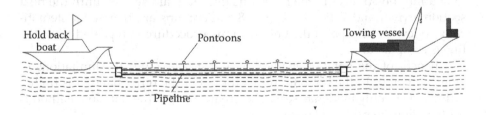

FIGURE 8.4
Surface tow for offshore line.

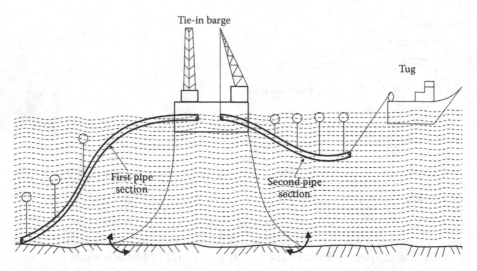

FIGURE 8.5
Flotation method.

be smaller than the primary tow vessel, as it is only required to exert a relatively small tension force on the string. However, it may be desirable to use two or even three vessels of the same size to provide redundancy during a long tow. Lowering of the pipeline in shallow water can be done by releasing the floats or pontoons in one step. In deep water, floats can be released successively to allow the pipeline to settle to the bottom in an S-curve configuration. Environmental conditions – wind velocity and wave height – have a significant effect on this technique while the pipeline is in the floating position.

An extension of this surface tow method is known as *flotation method* where pipe joints are first welded into a number of long strings onshore. Pontoons are attached to provide buoyancy and section by section the strings are towed into position. A barge holds one end of a laid section until the next section arrives and is tied-in (Figure 8.5). Pontoons are released systematically to lower the pipe to the bottom. This procedure is repeated until the line is completed.

A benefit of the flotation method is that it overcomes the limitations of length inherent in the bottom pull method. On the other hand, it is highly vulnerable in only moderate seas. The greatest use of the flotation method is for long lines in protected waters.

8.1.1.4 Below Surface Tow or Mid-Depth Tow Method

As illustrated in Figure 8.6, this method uses flotation devices to support the pipe string below significant wave action. Spar buoys are generally used to limit the amount of surface motion transferred to the pipeline. It is a controlled underwater flotation method that overcomes some of the deficiencies

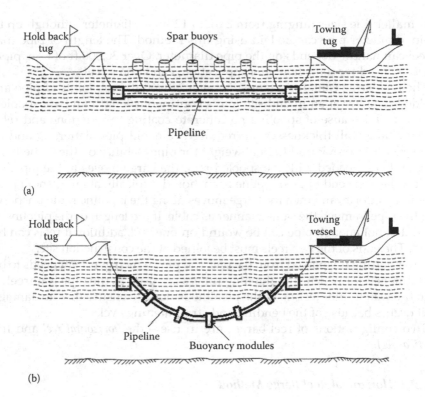

FIGURE 8.6
(a) Below-surface (b) mid-depth tow in offshore pipeline.

of the surface flotation method. This method offers the advantages of the off-bottom tow method – the operation is not significantly affected by sea floor conditions and smaller tow forces are needed than those required for the bottom tow method. Here the line is towed at a depth of 5 m or so below the water surface and is kept under tension with one boat towing another acting as a stern drag. Upon arrival at the site the line is lowered to the seafloor by cutting off every second spar, by a diver or ROV, while keeping the line under tension. When the line is on the seafloor, the remaining spars are removed. This method is used for laying flowlines.

All of these four techniques of pull or tow method must be carefully evaluated to determine which is the most appropriate for a specific pipeline construction project.

8.1.2 Reel Barge Method

The reel barge method of laying submarine pipelines uses a continuous length of pipe coiled onto a reel or drum. If pipe is not of large diameter and pipe line is relatively short, it can be laid by the reel method. It is used primarily for smaller line sizes ranging from 2 in. to 12 in. in diameter although up to 18-in. diameter pipe can be laid using this method. The length of pipe that a reel can handle depends on the pipe diameter. Over 30,000 ft of 6-in. pipe, the type used for some flow line application, can fit on a reel.

The reel barge contains a large diameter (40–60 ft) reel. Joints of pipe are welded together at an inland site into a continuous length and are spooled onto a reel. Because of spooling no concrete coating can be done and relatively heavy wall thicknesses are required to avoid pipe flattening and in some cases, to provide additional weight for pipe stability on the seabed.

After being loaded with the spooled reel, the barge moves to the pipeline laying site. One end of the pipeline is anchored, typically at an offshore production platform, and then the barge moves along the pipeline route unspooling the pipe in much the same manner as cable. If the length of the pipeline is such that not enough pipe can be wound on one reel, additional reels can be used. The ends of the two reels must be joined at the construction site.

After getting unspooled from the reel, the pipe which gets bent needs to be straightened for which straightening rollers are used as the pipe is unreeled into the water. Prior testing of the pipeline is done to ensure that no damage to it occurs because of the bending and straightening cycle.

Two configurations of reel barges are in use – the *horizontal reel* and the *vertical reel*.

8.1.2.1 Horizontal Reel Barge Method

Here a horizontally mounted reel is used and the pipeline installation is accomplished by unspooling and straightening the pipe as the barge moves forward. The horizontal reels lay pipe with an S-lay configuration (Figure 8.7), which is similar to the conventional lay barge operation. In such cases, pipe laying can take place at relatively high speeds; dynamic positioning can be

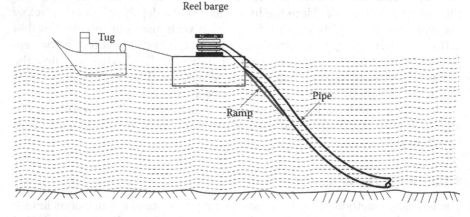

FIGURE 8.7
Horizontal reel barge method.

used as opposed to a spread mooring system. To limit the sagging of the pipe while being laid, tension is applied at the barge and made to come out of the barge by resting on a stinger or ramp.

8.1.2.2 Vertical Reel Barge Method

This method is advantageous in deep water applications, since it can be loaded for discharge from the top and not require a stinger (Figure 8.8). Santa Fe's reel ship *Apache*, for example, can lay pipe with diameters up to 16 in. from a reel mounted vertically on a ship and because of the large size of pipe, it is used on an adjustable deck-mounted ramp to control the pipe entry angle.

8.1.3 Lay Barge Method

Lay barges use essentially the same method of laying pipelines as in onshore fields, except that for a specific operation, a specific area on the barge is earmarked and more or less the same spread system is followed as in onshore construction. Those areas are known as the work stations. The pipe is delivered to the lay barge in single or double length joints (20- or 40-ft length) by a cargo barge and some lay barges are capable of handling even 80-ft joints, which are stacked at one side on its deck. In offshore pipeline, normally the yard coated pipes with a bare portion left at each end for permitting welding are used and after joining by welding, these bare portions are field coated on the lay barge. This coating is meant for protecting the pipe externally from corrosion. Additionally, offshore pipelines are also coated with a layer of concrete primarily to provide 'negative buoyancy' (i.e. weight needed to keep the pipe on the seafloor) and to resist damage during the laying and trenching.

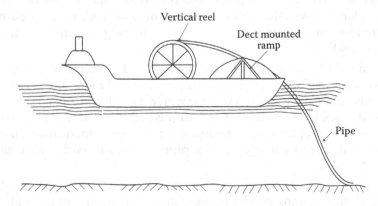

FIGURE 8.8
Vertical reel barge method.

A typical lay barge is a large floating vessel often up to 400-ft long and 100-ft wide. The deck area provides pipe joint storage prior to their being welded together on a long working ramp, producing a finished pipeline ready for lowering to the sea bottom.

The vessel provides living quarters for the workmen and all necessary welding, inspection and field joint coating equipment. An elaborate mooring system serves to position and move it systematically down the pipeline right-of-way. Auxiliary tugs assist in handling anchors while supply barges transport pipe and other materials. Depending on the configuration of the pipe during laying either S-shaped or J-shaped, the arrangement on the barges also changes. Thus, two different methods evolved and these are discussed separately below.

8.1.3.1 S-Lay Method

Most S-lay barges are equipped with 5–10 work stations. Figure 8.9 shows a typical 10-station lay barge where the sequence of operation is as follows:

Station 1 – *Pipe supply* – The appropriate number of 40-ft pipe joint is brought to the worksite. In the case of double joint lay barges, two joints are prewelded on board to create 80-ft joints.

Station 2 – *Transfer to bevel station* – Conveyor systems move pipe joints to the bevelling station.

Station 3 – *Pipe bevelling* – Both ends of the pipe are chamfered to give a V-shape with the help of a pipe facing machine. This helps to prepare a butt joint for welding two pipes.

Station 4 – *Transport to line-up station* – Transverse carriages are utilised to move the bevelled pipes into position in the line-up station along the firing line.

Station 5 – *Ready rack* – Here the pipe is lined-up and the internal line-up clamp moves through the pipe to the open end. Line-up clamps are used for aligning two pipes and preparing a fit-up for starting the welding.

Station 6 – *Line-up station and weld station 1* – This is the beginning of what is called the firing line, which includes all remaining work stations before the pipe goes into the water. The pipe is moved from the ready rack to weld station 1 and aligned ready for welding. The root bead and hot pass welds are deposited using automatic internal and external welding machines. The pipe is then moved to subsequent weld stations.

Station 7 – *Weld stations 2–6* – The external weld passes, that is, filler bead and capping bead, are completed in the remaining weld stations (2–6) using external welding machines.

FIGURE 8.9
S-lay barge. (Courtesy of CRC Evans.)

Station 8 – *NDT station* – The nondestructive tests like radiographic and ultrasonic tests are carried out on each completed joint at this station to ensure that the finished weld meets the defect acceptance criteria.

Station 9 – *Pulling the pipe* – During operation, each workstation on the line is indicated by a red light in the control room. As a station completes its task, the light turns to green. When all lights on the line are green, the vessel advances 40 or 80 ft by pulling on its anchor lines or utilising dynamic positioning thrusters. Once the pipe is pulled, all lights return to red and work continues. The performance of each station is recorded for improvement actions.

Station 10 – *Coating stations* – Field joint coatings are applied to guard against corrosion. The number of coating stations varies according to parameters such as parent coating type, pipeline operating temperature and insulation requirements. Typically, 2 or 3 stations are used for anticorrosion and/or insulation coatings. On concrete-coated pipelines, one more station may be needed to infill the annulus between concrete ends to allow smooth passage of the joint over the stinger rollers.

The next step is the *plunge,* that is, the newly welded, inspected and coated pipe is moved into the water via the stinger (described later), which is buoyancy controlled to maintain a smooth curve profile to the target water depth to minimise the stresses on the pipeline during laying.

As the work at each station is completed, the barge moves forward one joint length and a new joint of pipe is added to the assembly line.

The basic elements of the pipelaying system are shown in Figure 8.10. These include:

1. The lay barge
2. The mooring system
3. The pipe tension devices
4. The stinger
5. The suspended pipe span

The suspended pipe span, characteristically an 'S' shaped curve, is divided into a lower portion, the sag-bend region (concave up), and an upper portion, the over-bend region (convex up). The transition point divides the sag-bend from the over-bend.

The stinger provides support for the over-bend while tension devices serve to control the pipe in the sag-bend. The tension devices, located between various workstations on the barge, develop and maintain a restraining force between the pipe and the lay vessel as the barge moves forward on its anchors.

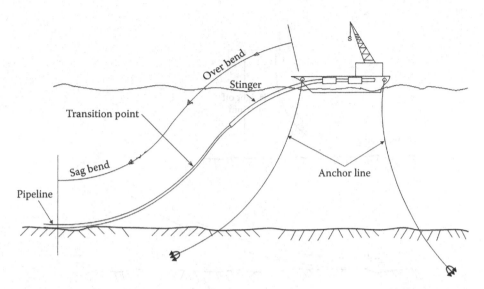

FIGURE 8.10
Elements of pipeline in an S-lay system.

During laying continuous control of the suspended pipe span is the most crucial aspect and malfunction of any portion of the pipe handling system usually leads to catastrophic failure of the pipeline.

The mechanics of the suspended pipe span can best be discussed by starting at the bottom and considering the *sag bend*. First, let us consider lifting the end of the pipe off the sea bottom by exerting only lifting force on the end of the pipe as shown in Figure 8.11a. As the end is lifted in this manner, a sag bend will develop which for small displacements can be analysed by ordinary beam theory. The curvature of the pipe is dependent on its submerged unit weight, W, and the flexural rigidity, EI, where E is the modulus of elasticity and I is the moment of inertia of the cross-section.

As the end of the pipe is lifted off the bottom it may be noted that, at first, curvature increases rapidly and then approaches the limiting value when the height of the pipe is a distance off bottom approximately equal to C, the characteristic length of the pipe.

$$C = \frac{EI^{1/3}}{W}$$

Under these conditions, bending strain is usually beyond the acceptable limit for most practical pipelaying conditions and horizontal pipe tension must be used to maintain bending within permissible limits. Therefore, in addition to the vertical lifting force, by applying a horizontal force (tension) to the pipe as shown in Figure 8.11b bending stresses in the sag bend can be controlled.

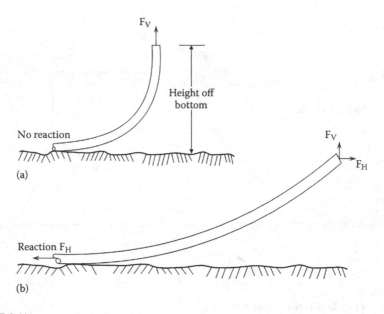

FIGURE 8.11
Sag band configuration under various lifting conditions. (a) Vertical lifting force and (b) vertical and horizontal forces.

Let us now consider the support of the upper portion of the 'S' curve, that is, the *over bend* of the suspended pipe span. Thus far, the sag bend has been discussed as if the pipe terminated at the transition point where there are no bending moments. Let us consider maintaining the transition point under constant conditions of position, slope and force components, although extending the pipe some distance upward. Its result will be that the pipe extension will assume a reverse or convex curve approaching a nearly horizontal attitude tangent to the lay barge ramp. Analysis of the over bend shows that it must be supported by upward reactions spaced at frequent intervals along the length of the pipe. This support could be provided by multiple pontoons of uniform positive buoyancy spaced at frequent intervals along the pipe.

In practice, the most efficient and manageable way to provide the required buoyancy support of the over bend region is with a buoyant curved stinger – a stinger of a curvature equal to the permissible bend of the pipe and of an arc length necessary to accommodate the required change in slope of the pipe (Figure 8.10). Since the slope of the transition point increases with water depth, the stinger length (arc length) is appropriate for only one depth. The choice is usually made to design the stinger for the deepest water expected and to tolerate unnecessary stinger length at more shallow depths.

The net positive buoyancy required of the stinger also varies with water depth, since the horizontal distance to pipe touch down changes with water depth. Therefore, the buoyancy of the stinger must be adjustable over a range determined by the unit weight of the pipe to be laid and the horizontal distance to the touchdown point.

A novel design of stinger known as *articulated stinger* has developed which is especially suited for the horizontal tension method of sag bend control. The stinger consists of several segments connected in series by special hinge joints providing a limited degree of free vertical and restrained lateral and torsional flexibility. The segments have adjustable buoyancy compartments which can be ballasted to provide the required pipe support. Vertical flexibility, in combination with applied pipe tension, increases the water depth capability and minimises the length of stinger required. The articulated stinger is shown in Figure 8.12 in a variety of configurations to indicate its shape in shallow, intermediate and deep water. The number of segments is dictated by the depth of water and the permissible pipe bend radius of curvature.

Control of the *suspended pipe span* during laying operations is maintained by proper buoyancy of the stinger and control of the horizontal tension in the pipe. Pipe tension is induced by tension devices positioned along the lay barge between workstations. As shown in Figure 8.13, one design employs several large pneumatic tyres which contact the outer surface of the pipe to develop a forward thrust. Some designs have caterpillar treads which grip the pipe. Most tension systems are power driven and controlled to take-in or give-up pipe as required, maintaining pipe thrust within limits.

In summary, the pipe handling system consists of the pipe tensioning machinery, the stinger and the barge positioning or anchoring system. These

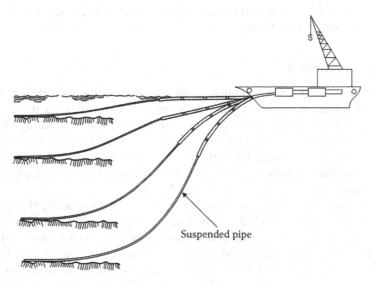

Suspended pipe

FIGURE 8.12
Shape of articulated stinger at different water depths.

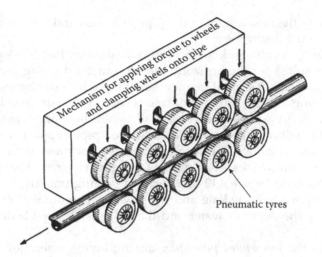

FIGURE 8.13
Pipe tensioner.

components must work in constant unison for the safe handling of the suspended pipe span. Too little pipe tension not only increases the bending strain in the sag bend region, but also reduces the amount of pipe weight imposed on the stinger. With constant buoyancy there is a compensating rise of a rigid stinger or an increased curvature of the articulated stinger. For an extreme reduction in horizontal tension, the upper segments of the articulated stinger will rise to the water surface, lose buoyancy and therefore tend to reduce the excessive lift on the pipe span. This is a very significant advantage of the articulated stinger.

8.1.3.2 J-Lay Method

To overcome the problems related to excessive weights and stresses on the overbend when using an S-lay barge equipped with a stinger in very deep water, a method is being developed in which the pipe is welded together in a vertical position on the lay barge and is allowed to enter the water vertically below the vessel. A tower-like structure much like a derrick on the stern side or over the side where pipes are stacked would support the pipe on the barge. This technique is called the J-lay method.

On the vertical-lay barge, pipe would be suspended vertically (Figure 8.14). A new joint would be added, aligned with a line-up clamp and welded in the vertical position. The pipe would then be lowered so the next joint could be added. Weld inspection and coating stations could be included at different levels on the vessel. Realistically in the case of a tall tower, only one welding and coating station can be used, so the pipe arrives on transport barges in long, prewelded, coated lengths up to 240 ft. The pipeline would

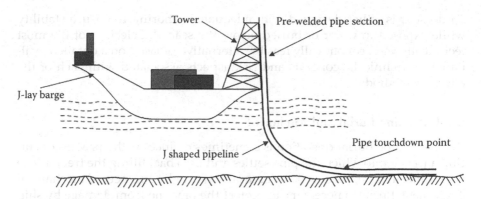

FIGURE 8.14
J-lay barge and J-lay method.

exit vertically through a moonpool or from the stern side and has no over-bend. Sometimes this method utilises a hinged ramp (Figure 8.15), inclined only slightly from vertical on which a triple- or even quadruple-jointed pipe segment is joined by advanced method of welding.

Not only would total tension required on the pipe during laying be less than when laying from a horizontal position, but the bulk of the tension force would be in the vertical direction and would not have to be resisted by the vessel anchors. In fact, the small magnitude of this force might make the use of dynamic vessel positioning practical for vertical lay vessels. Forces on equipment suspending the pipeline during laying would be well within the capacities of derrick towers.

Whether this approach to deepwater pipelaying is used depends on the economics of a particular project. Nevertheless, it does offer a potential alternative in extreme water depth.

Once the pipeline is laid by any of the methods mentioned above, it is essential that the pipeline remains stable on the sea bed during its lifetime.

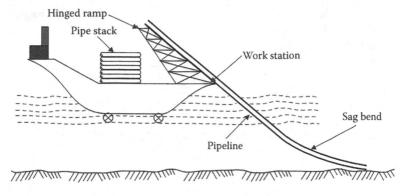

FIGURE 8.15
J-lay method with hinged ramp.

To do so it is either designed with adequate anchoring to ensure stability while exposed in water or buried below the seabed. Selection of the most technically and economically feasible alternative is based on a detailed evaluation of technical, economic and risk aspects associated with each of the alternate methods.

8.1.4 Pipeline Burial and Anchoring

Burial – A pipe burial operation is sometimes defined as the process of cutting a trench into which the pipe settles and then backfilling the trench. The backfilling in this case may be by natural action or by engineered mechanical equipment. Burial is necessary to protect the pipeline from damage by ship anchors, fishing gear and natural hazards. But the only drawback to burial is that in the presence of backfill, finding a leak and repairing it becomes difficult.

Pipe trenching may be done using three different modes: (1) *pretrenching*, where a trench is made prior to pipe installation mostly in shallow-water; (2) *simultaneous trenching*, where trenching occurs during the installation process and (3) *post-trenching*, where trenching is done subsequent to the pipeline installation.

The two most common approaches to burial are jetting and plowing or mechanical cutting. There is also another method known as fluidisation. A brief outline of these methods is given below.

In *jetting*, a jet sled (Figure 8.16), consisting of a frame on which pumps, jets and associated equipment are mounted, is lowered to the ocean floor from a jet barge (Figure 8.17) over the pipeline laid on the sea floor. As the jet barge pulls

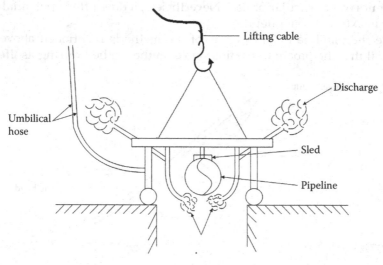

FIGURE 8.16
Jet sled.

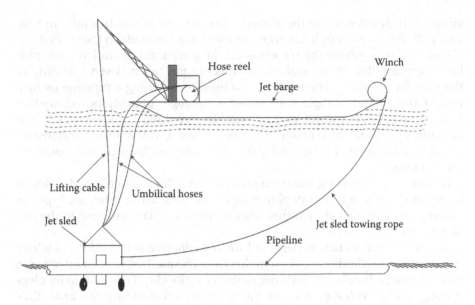

FIGURE 8.17
Jet barge illustration.

the sled over the pipeline, its powerful pumps directed under the pipeline force soil from beneath the pipe and allow the pipeline to settle into the resulting ditch. The displaced soil then covers the pipe as the jet sled moves along the pipeline. The type of soil on the ocean floor has a great effect on how deep the pipe can be buried in the seabed by jetting. Where the soil type changes frequently, the pipeline may be left suspended between two relatively hard soil areas. These unsupported spans can cause undesirable stresses on the pipe.

To provide more effective burial under conditions in which jetting is difficult, underwater *plows* were developed. Three different methods of plowing exist:

Preplowing, where a trench is cut in which the pipeline is later installed. Here the plow is towed by surface vessel. This method has been used to cut a trench in 130 m water depth for a 36-in. pipeline which was then installed by the bottom tow method. Preplowing is often feasible for pipelines to be installed by bottom pull or by bottom tow methods.

This method is not favoured for laying and plowing pipe in deepwater, since it is difficult to control the position of the pipe touchdown point in deepwater. Also, the trench will have to remain open until the pipe is placed in it. In some cases, bottom currents may cause material to fall back into the trench before the pipe is lowered into the trench.

Simultaneous plowing is where a trench is made in a combined operation with pipe installation. This technique has been used with lay-barge installation in shallow-water depths where a plow was attached to the end of a stinger which extended to the seabed. This method is effective only in relatively shallow water applications since the stinger extends to the seabed.

Postplowing is where the trench is cut in a separate operation after pipe has been installed at the seabed. This technique can be used with any of the pipe lay methods. The method was used for plowing a pipeline using a draw barge to draw the plow. The main advantages of postplowing are that it can be used in deepwater applications, has a very high production rate and can quickly protect the pipeline after installation. Careful design and instrumentation are needed to minimise the possibility of damage to the pipeline during plowing.

In mechanical cutting some underwater trenching machines developed by several companies are deployed at the seabed using different types of cutters and propulsion facilities. This method is relatively new to the offshore industry.

The *fluidisation method* is designed for noncohesive soil conditions where conventional methods have been ineffective. The method is most effective in sand and slightly cohesive sediments, such as silty clay. Fluidisation involves forcing a large volume of water into the soil surrounding the pipe, thus reducing the soil density and allowing pipe to settle in the soil. The main advantage of this method is that, during fluidisation, the pipe is immediately covered with sand and full pipe protection is achieved. The main disadvantage of the method is that it is effective only in sandy soils, and considerable variations in soil type are normally encountered along a pipe route. Also, fluidising equipment is large and bulky and requires large volumes of water. Hose-handling complications are inevitable during operation.

Among the various burial methods discussed, the selection of any one will depend upon many factors like soil type, pipe size and weight, water depth, production rate, sea state, trench stability, soil disposal and power consumption.

Anchoring – Pipeline stabilisation, or anchoring, can be defined as a system designed to maintain the pipeline in a desired position relative to the surrounding environment and subject to the various forces acting on the pipeline. Because of the many demands on the anchoring system, the anchors can take many forms and combinations.

Anchors currently available for pipelines consist of two basic types: *density* and *mechanical* (Figure 8.18). The *density anchor* simply consists of weight added to the pipeline to increase the average density or negative buoyancy to some acceptable level that will be stable under prevailing conditions. These anchors are usually concrete and take the form of either bolt-on weights, set-on weights, or a continuous concrete coating. In contrast, mechanical anchors are usually fabricated from steel and not designed to add weight. They maintain a minimum hold-down force on the pipe when properly installed in the soil. Because the holding power of mechanical anchors is much greater than their own weight, they are significantly more efficient, by weight, than density anchors.

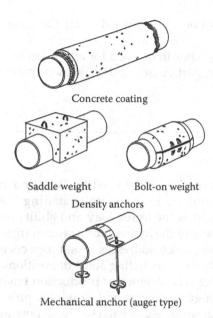

Concrete coating

Saddle weight   Bolt-on weight
Density anchors

Mechanical anchor (auger type)

FIGURE 8.18
Density and mechanical anchor.

Continuous concrete coating is described as a coating of concrete completely encompassing the pipe. It can be applied either in a coating yard with special equipment or on the job site.

Another type of density anchor is the *bolt-on concrete weight*. Bolt-on weights are built in two halves and designed to be clamped on the pipeline. The two halves are held together with long bolts.

The most economical form of density anchor is the *set-on weight*. These weights, sometimes known as *saddle weights* also, are shaped like inverted 'U' and they are set on the pipeline after the pipeline is in the ditch. The weights are designed with the centre of gravity as low as possible. The legs are designed 2–3 in. longer than the diameter of the pipe. This is to prevent the weight from rolling off the pipe and to enable the ditch bottom to take the load of the weight.

Mechanical anchors differentiate from density anchors in that they derive their holding power from the shear strength of the soil. They are inserted into the soil and attached to the pipeline. They are usually made of steel and are either pile auger or expanding type.

The most commonly used type of mechanical anchor is the *auger type*. This anchor consists of a round steel plate shaped like an auger and attached to the end of a long steel rod. The other end of the rod is threaded for attachment to the pipeline.

This system consists of two anchors and a strap shaped to fit the pipeline. It consists of installing an anchor on each side of the pipeline and attaching the strap to both anchors. The formed strap fits snugly over the pipeline securing it in place.

The exact type of anchor to be used for any specific location will depend on the soil conditions, particularly soil resistance and the sea-bed profile.

8.2 Flow Assurance

The term 'flow assurance' was first used by Petrobras in the early 1990s in Portuguese as 'Garantia do Escoamento', meaning literally 'guarantee of flow'. Thus, it is defined as the technology and ability to ensure the successful and economical flow of the hydrocarbon stream from the reservoir to the point of sale. Flow assurance addresses broad aspects of the problems and solutions of flow distortion, including solid depositions of waxes, hydrates, asphaltenes and scales. The deepwater production environment and transportation through longer pipelines and flowlines provides conditions that brought about flow assurance challenges. These flow assurance challenges are as illustrated in Figure 8.19. For effective subsea production, it is important to identify the potential for and quantify the extent of any solid deposition in the system. Some flow assurance challenges such as wax and hydrate

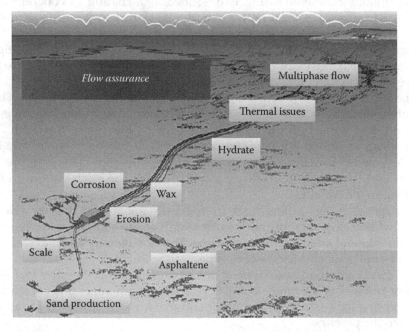

FIGURE 8.19
Flow assurance challenges. (Courtesy of www.diva-portal.org)

formation and deposition are very common in subsea pipeline because of the favourable subsea pressure and temperature conditions. The asphaltene and scale deposition also cause problems. A brief discussion on each of these problems follows.

Wax problem – As a solid organic phase, mainly long-chain n-alkanes (C16-C80+), wax comprises a mixture of components. In oils, wax crystallises within the fluid as the fluid temperature drops to below the cloud point. However, it only deposits on the pipe wall when the wall temperature is below the cloud point and colder than the bulk fluid. The wax crystallisation process develops in three stages:

1. Nucleation, at which the first nuclei appear
2. Growth, at which mass is transported from the solution toward the nuclei
3. Agglomeration, at which the developed crystals join together and bigger crystals are formed

Wax precipitation can occur in the reservoirs, the production column, the flowlines and in the surface production equipment.

Cloud point is the temperature at which paraffin wax begins to crystallise, also known as wax appearance temperature (WAT) and is identified by the onset of turbidity as the temperature is lowered. In other words, cloud point is the highest temperature at which the wax crystal forms. The cloud point of the particular crude is dependent on the oil composition and is affected by small amounts of high molecular weight paraffin in the crude. As the temperature falls below the cloud point, wax crystals begin to precipitate from the oil phase.

Pour point is the lowest temperature at which the liquid ceases to flow. Therefore, the crystals of waxes or paraffins start to solidify as their temperature reaches the pour point.

Wax deposition is controlled by many factors besides the fluid properties. The key controlling factors are the temperature difference between fluid and pipe wall (thus, the heat flux), the concentration gradient and the mass transfer resistance determined by fluid properties and flow rate, and so on.

Pressure has different effects on the wax formation in a single phase system and a multiphase system. In a single-phase oil system, since the wax phase is denser than the oil, an increase in pressure slightly increases the wax deposition tendency. In a multiphase system, an increase in pressure drives the light ends of the mixture into the liquid phase and tends to decrease the cloud point, therefore, tending to reduce the amount of wax formed at a particular temperature.

Wax gelation is less common in steady state than is wax deposition. This can have greater impact if, during production system shutdowns, fluid temperatures cool below the fluid pour point, thus allowing the formation of a 'candle' or solid wax column. The melting point of wax deposits is normally about 20°C higher than the cloud point. Figure 8.20 is a schematic of a wax plug removed from a subsea pipeline.

However, wax precipitation and deposition in the offshore environment poses a great challenge such that remediation costs become very significant or massive. Thus, this is a key issue for the oil industry in analysing the economics of waxy crude production in cold environments.

Hydrate problem – Hydrates are crystalline compounds formed by chemical combination of natural gas and water under pressure and temperature considerably above the freezing point of water. These natural gas hydrates mostly in gas pipelines may block the pipelines, facilities and instruments. These can cause flow and pressure monitoring errors, reducing gas transportation volume, increasing pipeline pressure differences and damaging pipe fittings. The nature of a hydrate plug removed from a subsea pipeline is as shown in Figure 8.21.

Hydrates are of clathrate structure wherein guest molecules are entrapped in a cage-like framework of the host molecule without forming a chemical bond. They look and behave like ice although they are formed at a temperature higher than the freezing point.

There are four major conditions necessary for hydrate formation. These, as shown in Figure 8.22, include:

1. Water as the liquid phase condensing out of the hydrocarbon.
2. Hydrate formers. These are small gas molecules such as methane, ethane and propane (gas composition).
3. The right combination of low temperature and high pressure.

FIGURE 8.20
Wax plug. (Courtesy of www.diva-portal.org)

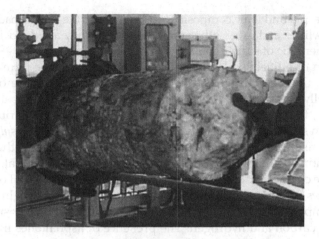

FIGURE 8.21
Hydrate plug. (Courtesy of www.diva-portal.org)

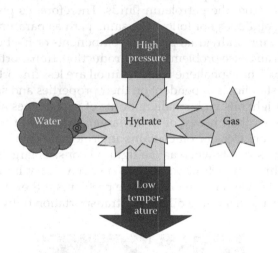

FIGURE 8.22
Hydrate formation conditions. (Courtesy of www.diva-portal.org)

Hydrate formation is favoured by low temperatures and high pressures typically 20°C and 100 bar.

Scale problem – Scale is a precipitation of inorganic salts (minerals) in production equipment such as carbonates and sulphates of barium, strontium, or calcium. Scale may also be salts of iron like sulphides, carbonates and hydrous oxides. Oilfield scales are deposited from oilfield brines when there is a disturbance in the thermodynamic and chemical equilibrium that may result in a certain degree of supersaturation. The disturbance in thermodynamic and chemical equilibrium can be a change in pressure,

temperature, pH and ionic composition. They reduce the flow capacity of pipelines or flowlines and cause other problems when they are deposited. Figure
8.23 is a schematic of scale deposits on a pipeline cross-section.

Formation of oil field scale can be in one of two ways: the formation water
(brine) may undergo change in conditions such as temperature or pressure.
This generally gives rise to *carbonate scales*. Also, when two incompatible waters,
for example, formation water rich in calcium, strontium and barium and sea
water rich in sulphate, mix together. This generally gives rise to *sulphate scales*.
The mechanism of *mineral scales* formation is dependent upon the degree of
supersaturation of the water with respect to a particular mineral, the rate of
temperature change and changes in the pressure as well as the pH of the water.

Asphaltenes problem – Asphaltenes are high molecular weight polycyclic
organic compounds with nitrogen, oxygen and sulphur in their structure, in
addition to carbon and hydrogen. The presence of asphaltenes in the petroleum fluids is also defined as a fraction of petroleum fluid or other carbonaceous sources such as coal. They are soluble in benzene and deposit by
addition of a low-boiling paraffin solvent. Asphaltenes do not crystallise
upon deposition from the petroleum fluids. Therefore, its phase transition
from liquid to solid does not follow the same path as paraffin. It is also not
easily separated into individual purified components or fractions.

Asphaltenes can cause problems in oil production, transportation and processing facilities. The asphaltene contents in oil are less important than their
stability. Their stability is dependent on their properties and solvent properties of the oil. Light oils with small amounts of asphaltenes are more likely
to cause problems than heavy oil with larger amounts of material in the
asphaltene fraction. Heavier oil contains intermediate components that are
good asphaltene solvents whereas the light oil consists largely of paraffinic
materials in which, by definition, asphaltenes have very limited solubility.
Asphaltenes in heavier oils can also cause problems if they are destabilised
by mixing with another crude oil during transportation or by other steps in

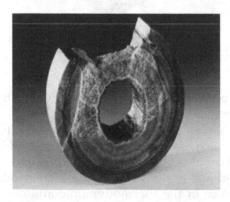

FIGURE 8.23
Scale deposit inside a pipe. (Courtesy of www.diva-portal.org)

oil processing. In general, asphaltenes cause little or no operational problems because most asphaltic crude oils contain stable asphaltenes.

Out of the four issues mentioned above, wax and hydrates consume a considerable amount of time for the flow assurance team. So, major focus is given on these two so that their formation can be controlled.

Generally, there are three basic methods of controlling hydrates, waxes and asphaltenes: thermal, chemical and mechanical. They may be used alone or in combination. Thermal methods encompass heat conservation using a passive insulation system, an active addition of heat or an active system of insulation and heating. The active heating can be fluid heating, electric heating, or thermal-chemical exothermal heating.

Chemical methods include both inhibition and dehydration for hydrate prevention and the use of long chain polymers to maintain waxes and asphaltenes in the fluid.

Mechanical means, such as pigs, may be used for 'prevention' if operated on a planned frequency.

Wax remedial treatments often involve the use of solvents, hot water, the combination of hot water and surfactants, or hot oil treatments to revitalise production. Eight methods are available for removal of wax, paraffin and asphaltenes: hot fluid, solvents, dispersants, crystal modifiers, a combination of the above, mechanical means (scraping), SNG (or NGS, nitrogen generating system), thermo-chemical cleaning and microorganisms.

There are chemicals available that can be tailored to work with a particular crude oil composition, but tests should be carried out on samples of the crude to be sure that the chemical additives would prevent the wax deposition.

Removal by means of a hot fluid works best for downhole and for short flowlines. The hydrocarbon deposits are heated above the pour point by hot oil, hot water, or steam circulated in the system. This practice, however, has a drawback. The use of hot oil treatments in wax-restricted wells can aggravate the problem in the long run, even though the immediate results appear fine.

The combined hot water and surfactant method allows the suspension of solids by the surfactant's bipolar interaction at the interface between the water and wax. An advantage of this method is that water has a higher specific heat than oil, and therefore usually arrives at the site of deposition with a higher temperature.

Solvent treatments of wax and asphaltene depositions are often the most successful remediation methods, but are also costlier. Therefore, solvent remediation methods are usually reserved for applications where hot oil or hot water methods have shown little success.

In the case of crystal modifiers (pour point depressant [PPD])/dispersants, paraffin wax crystal modifiers are chemically functionalised substances that range from polyacrylate esters of fatty alcohol to copolymers of ethylene and vinyl acetate. These crystal modifiers attack the nucleating agents of the hydrocarbon deposit and break down and prevent the agglomeration of paraffin crystals by keeping the nucleating agents in solution.

Dispersants do not dissolve wax but disperse it in the oil or water through surfactant action. They divide the modifier polymer into smaller fractions that can mix more readily with the crude oil under low shear conditions.

NGS, introduced by Petrobras in 1992, is a thermo-chemical cleaning method. The basic concept of NGS is related to the irreversible fluidisation process of the organic deposits. Such a process is caused by the simultaneous actions of the temperature increase of the fluid and paraffin, the internal turbulence during the flow and the incorporation of the organic solvent into the deposits. The heat is generated with nitrogen simultaneously by the chemical reaction between two inorganic salts in aqueous saturated solution. The NGS process combines thermal, chemical and mechanical effects by controlling nitrogen gas generation to comprise the reversible fluidity of wax/paraffin deposits.

Pigging is the oldest way of cleaning out a flowline which has been discussed next.

To *combat the formation of hydrates* there are three ways currently in use:

1. Preservation and application of heat by using insulation and supplemental heating
2. Use of inhibitors
3. Dehydration or removal of enough water from the stream so that a hydrate will not form (glycols)

Inhibition of the hydrate formation process is a commonly used practice. There are two kinds of inhibition – thermodynamic and kinetic. Thermodynamic inhibition prevents hydrate formation by adding a third active component into a two-component/intermolecular interaction and thermodynamic equilibrium between molecules of water and gas, and to modify the hydrate formation temperature.

Kinetic inhibitors are adsorbed on the surface of hydrate micro-crystals, and microdispersed droplets of water are absorbed in the flow of a fluid. Unlike thermodynamic inhibitors, kinetic inhibitors do not lower the hydrate formation temperature, but preclude the process of hydrate formation. Kinetic inhibition is a temporary inhibition and is effective in producing and transporting hydrocarbons.

Glycol dehydration may be more efficient, but requires expensive downstream process (recovery) facilities. Triethylene glycol is commonly used to dehydrate produced gas.

In a situation where there is a high potential for constriction or plugging, an effective insulation system is used for flowlines where different heat retention alternatives are tried. One such alternative is *pipe-in-pipe flowline*. The space between the pipes is partially evacuated and then filled with insulating material such as a polymer. The result behaves like an elongated thermos bottle, eliminating most of the heat loss from the well fluids as they pass through. These efficient insulation systems allow the well fluid to be transported great distances while maintaining almost all their original heat, preventing the formation of both hydrates and wax.

Other common approaches to heat retention are as follows:

1. Insulating with a material such as a polymer that can withstand the flowline installation process.
2. Burying the already insulated or even bare pipe flowline so that the surrounding soil acts as an insulator.
3. Pipe-in-pipe bundles, use of flexible pipes that have inherent insulating characteristics due to their composition.

Where heat retention techniques are not enough, the pipeline can also be heated from an external source. Direct electrical heating applies an electrical current along the flowline, where electrical resistance to the current causes direct heating of the flowline. In another technique, hot fluids are circulated inside a pipe bundle. The heat permeates and transfers to the produced fluids.

8.3 Pigging

Earlier it was mentioned that one of the means of controlling flow assurance problems was the mechanical one, that is, by using pigs about which already a brief introduction was given in Chapter 6. Here major emphasis will be given on pigging offshore with a thrust on differences between onshore and offshore pipeline pigging.

As is known, pigging requirements will vary with the type of pipeline system in question and the stage of development, that is, construction, commissioning, maintenance, inspection and so on. In the beginning, the classification of subsea pipelines has been mentioned, which are flowlines, gathering lines, trunk lines and loading lines. Sometimes a part of the loading line connected between the subsea manifold to the main export system is called a *spur line*. As far as the pigging requirements are concerned, these will vary for the above-mentioned lines.

The *pipeline system*, according to Pipelines Inspectorate's definition, is the entire system between the pig launcher and up to the first isolation point of take-off. So, pigging is always referred to as a particular pipeline system.

8.3.1 Pigging during Construction and Precommissioning

Pigging during construction and precommissioning offshore is carried out to enable flooding, cleaning, gauging, dewatering and drying to be carried out. For a completed pipeline system, pigging requirements are essentially

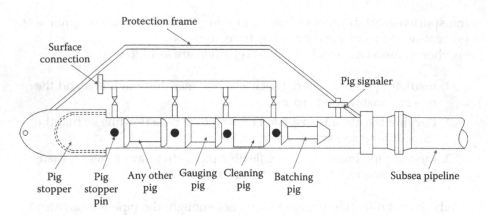

FIGURE 8.24
Laydown launcher.

identical to those for an onshore line. Here also a pig launcher and receiver are used. An example of a laydown launcher configuration is shown in Figure 8.24.

Typically, the procedure will comprise sequentially launching the bi-directional batching pig followed by a cleaning pig and a gauging plate. The pigs are propelled using water, usually filtered and treated with biocides and corrosion inhibitors. The successful recovery of an undamaged or acceptable-worn gauging plate, together with a successful hydrotest will certify it for commissioning.

Other than cleaning, gauging, dewatering and drying, one more aspect of removing welding spheres or bladders becomes necessary in case of tying-in two subsea lines. Subsea welded pipeline connections are currently performed hyperbarically, that is, within a controlled atmosphere at ambient pressure. This necessitates the use of a habitat which can be filled with gas. To prevent water entering the habitat via the pipeline ends, inflatable welding bladders or spheres are inserted into the line as shown in Figure 8.25. The sphere must be sufficiently inflated to withstand tidal pressure variations. On completion of the tie-in, the spheres are swept through during cleaning by a cleaning pig prior to pressure testing. During cleaning long pipelines, pigs undergo significant wear, contributing to the potential for jamming.

To overcome cleaning difficulties, a recent development is the use of pigging gels. Used in conjunction with conventional cleaning pigs, the gel collects and suspends the debris ahead of the cleaning pigs, so preventing jamming and reducing wear. The gel pig train configuration in Figure 8.26 shows the principle of the mechanism.

Before commissioning, the line should be dried though drying of the oil line serves no purpose and the water permits direct pressuring of the system. For gas pipelines, drying may be necessary to prevent corrosion, to meet

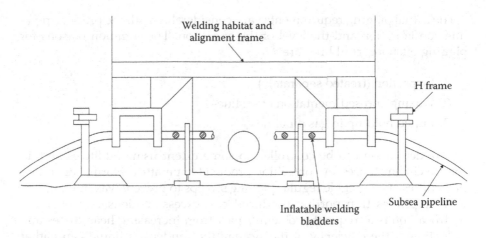

FIGURE 8.25
Use of welding bladders in hyperbaric welding.

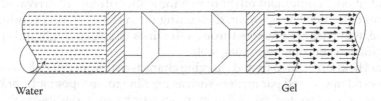

FIGURE 8.26
Gel pigging.

sales gas specification and to prevent hydrate formation unless suppressants such as glycol or methanol are injected into the streams.

Traditionally, pipeline drying has been with methanol or glycol swabs between pigs, usually performed in accordance with standard procedure. After drying, the lines have been batch-filled with nitrogen or sales-quality gas if available. Provision for handling and disposal of such drying chemicals offshore can be expensive, and in the case of methanol is potentially hazardous. This has led to the development of vacuum drying which has already proven successful for offshore lines, reducing overall pigging activities and employing comparatively small-sized equipment. Indeed, the method is now understood to be used onshore.

8.3.2 Pigging during Operation

8.3.2.1 Trunkline Systems

Offshore pipelines are used for produced hydrocarbons that have undergone varying degrees of treatment, and give rise to the need for routine pigging programmes.

The actual pigging requirements vary considerably with the product, pipeline configuration and the level of the operation. The common reasons for pigging offshore trunklines are

1. Inspection (treated separately)
2. Waxing and sedimentation of oil lines
3. Liquid hold-up in gas pipelines

Wax deposition can be controlled to some extent using additives and, in certain circumstances, by controlling product temperature. Normal operating procedures often include regular pigging sweeps to prevent wax build up.

For pipelines transporting produced gas, excess condensation of liquid hydrocarbon is a frequent problem. Apart from increasing head losses and reduction of the efficiency of the system, the condensed liquid can gather to form slugs which can arrive at the receiver facilities at high velocity and damage equipment. The arrival of large slugs also creates the need to provide additional equipment offshore to handle the infrequent arrival of large volumes of liquid. The operator, depending upon the operating condition of the line, will pass sphere pigs through the lines to sweep out liquids before accumulations become unmanageable.

Liquid hold-up behaviour and slugging characteristics are difficult to predict, and depend upon fixed parameters such as pipeline route topography and riser configuration, as well as on variation in operating conditions. Sweeping routines must therefore be developed by trial and experience taking due consideration of the cost of sphering, the efficiency of the system and the size of slugs received. Sphering costs can become significant, particularly for unmanned satellite platforms where remotely controlled launchers are employed.

A *flowline* system needs special mention here because the procedures discussed thus far relate to larger-diameter systems. Although sometimes installed from lay barges, flowlines are increasingly being installed from reel vessels. Furthermore, J-tube entry into platforms is used more and more, so precluding the need for riser/pipeline tie-ins. These procedures have many significant effects on pigging practices during construction.

In a reel barge, the reeling and straightening processes involve plastic deformation of the pipeline. This necessitates the use of line pipe possessing a small diameter/thickness (D/t) ratio (typically less than 20), and the reeling and straightening processes are more rigorous than the subsequent laying of line on the seabed. Moreover, with the line installed within the platform from commencement of laying, dewatering can be readily achieved in the unlikely event that damage occurs. In fact, for smaller flowlines the reel vessel could probably recover the line without dewatering. Thus, no provision for pigging may be necessary until the point of cleaning and filling the line. For very small lines, except through-flowlines (TFLs), pigging may indeed be waived altogether, cleaning being by flushing.

The *spur lines* can also require pigging. In the past, this has normally been restricted to the passage of spheres through the spur which have been burst against bars at the tee connection. Recently, however, the industry has witnessed the development of a piggable subsea wye or assembly, designed to enable future pipeline tie-in and to permit the passage of all types of pigs including inspection vehicles. The wye assembly is purpose-designed for the product in question and is installed complete with support and protection covers, and all required valves.

8.3.3 Pigging for Inspection

Although internal inspection tools have been developed for onshore systems, for offshore inspection the same tools cannot be used.

Thus far, as practicalities of internal inspection of offshore pipelines are concerned, apart from pig availability in certain sizes, the major difficulties arise over space, handling and wall thickness. The implications of bend geometry are equally significant.

Offshore platforms are congested; thus, to avoid potentially expensive modifications in service, the facility to deliver, manoeuvre and launch large inspection tools should be considered at the earliest stages of design. The difficulties of providing launcher/receiver facilities several metres long together with handling facilities for the heavy inspection tools cannot be overestimated.

Although having demonstrated reasons for detailed internal inspection of offshore pipelines, it must be pointed out that in many instances appropriate inspection cannot be performed, since currently available inspection devices do not have the capacity.

It is believed that developments of inspection tools will continue and that eventually most offshore inspection needs will be met.

8.4 Corrosion and Maintenance

The life of any pipeline depends on how efficiently it is maintained for which the first and foremost consideration is the corrosion, its monitoring and protection from it. So first, a brief discussion on corrosion of pipeline is made and other maintenance issues are dealt with later.

8.4.1 Corrosion and Its Monitoring

Corrosion can be simply defined as nature's way of returning refined metal to its natural state. Metals usually exist in their natural state as compounds and are thermodynamically stable. When we extract a metal and do not substantially change it (i.e. by alloying), the thermodynamics become unstable. If exposed to the environment this metal will corrode, reverting to its natural state as a compound.

The important thing from the standpoint of corrosion is not the formation of the compound, nor even the nature of the compound formed, but going into solution of the metal in the first place. This is the initial step, the one that must be taken before any of the others can follow, and the one that will completely arrest the corrosion process if it can be brought to a halt.

The corrosion may be of various types:

1. Sweet corrosion (due to CO_2)
2. Sour corrosion (due to H_2S, S)
3. Microbiological corrosion (aerobic and anaerobic bacteria)
4. Oxygen corrosion
5. Electrochemical corrosion

Type 3 and 4 can be included in electrochemical corrosion because they are one of the reasons for the formation of a concentration cell. Types 1 and 2 are generally found in the inside of the pipe because it depends upon the carrying fluid which may contain CO_2 and H_2S dissolved in it. So, these are the main reasons for internal corrosion of pipeline.

Thus far as *external corrosion* is concerned, it is broadly the *electrochemical corrosion* although electrolytic corrosion also falls into this category.

Every metal has some tendency to go into solution in the presence of water – and it must be realised that there is some water always present in even the driest soil. What to talk of an offshore environment, where presence of water vapour or moisture in the air is very common. The process by which the smallest possible part of a metal goes into solution may be simply described as follows.

As an integral part of the metal structure, the atom of metal exists as an atom without an electric charge; the only way it can enter solution is by becoming an ion, and it can only do this by obtaining a positive electric charge from some source; wherever there is water, it is to some extent ionised; that is, separated into positively charged hydrogen ions (H^+) and negatively charged hydroxyl ions (OH^-)

$$H_2O \rightarrow H^+ + OH^-$$

The metal atom can seize a positive charge from a hydrogen ion, and thus become a metal ion, and go into solution.

$$Fe + 2H^+ \rightarrow Fe^{++} + H_2$$
$$Fe^{++} + 2(OH)^- \rightarrow Fe(OH)_2$$
$$\text{(crust)}$$

Thus, the initial step is of primary importance, for if it can be prevented, the corrosion can be stopped.

Different metals differ in their readiness with which they can hold charges; thus, some metals can take charges away from others. Iron, for example, easily and rapidly takes positive charges away from copper ions; so, a piece of iron placed in a solution of copper sulphate goes rapidly into solution, and metallic copper (the ions having lost their charges cannot remain in solution) plates out. The relative tendency of different metals to hold charges is tabulated in the electromotive series of metals given in Table 8.1.

The difference between the potentials given in the series for two metals is an approximate indication of the voltage which will be developed between pieces of the two when placed in a solution that is electrically conductive; that is, which contains ions. The exact voltage actually developed depends upon the kind of solution, the concentration, the temperature and several other

TABLE 8.1

Electromotive Series of Metals

Metal	Ion Formed	Standard Electrode Potential
Potassium	K+	+2.92
Calcium	Ca++	+2.87
Sodium	Na+	+2.71
Magnesium	Mg++	+2.40
Aluminium	Al++	+1.70
Beryllium	Be++	+1.60
Manganese	Mn++	+1.10
Zinc	Zn++	+0.76
Chromium	Cr++	+0.56
Iron (Ferrous)	Cd++	+0.44
Cadmium	Cd++	+0.40
Cobalt	Co++	+0.28
Nickel	Ni++	+0.23
Tin	Sn++	+0.14
Lead	Pb++	+0.12
Iron (Ferric)	Fe+++	+0.04
Hydrogen	H+	0.00
Copper (Cupric)	Cu++	−0.34
Silver	Ag+	−0.80
Mercury	Hg++	−0.80
Gold (Auric)	Au+++	−1.36
Gold (Aurous)	Au+	−1.50

Note: Standard electrode potential is determined by placing an electrode of the pure metal in 'standard' solution of its own ions and measuring the potential difference between it and a standard hydrogen electrode to which is assigned the arbitrary value of zero. The standard solution contains an ion concentration of one mole per 1000 g of water and the standard temperature for making the determination is 25°C.

factors. If a connection is then established between the two pieces of metal, a current will flow and one of the metals will be attacked (Figure 8.27a). This is the one from which the current flows into the solution; it is known as the anode. The other metal, called the cathode (or sometimes noble), will not be attacked by the process. *The anode corrodes, the cathode does not.* The amount of corrosion depends upon the total amount of current flow; that is, upon the current value and the time (I.t). The current flow, in turn, is determined by the voltage developed and by the total circuit resistance (I = V/R). Further developments, as corrosion proceeds, usually change the potential (by changing the nature of the solution and the nature of the metal surface) and the circuit resistance (by changes in the solution and by films formed on the surfaces). The potential difference, then, can only give an indication of the initial tendency to corrode.

This mechanism just briefly described is an explanation of *galvanic* or *bi-metallic corrosion*, where two different metals are connected together and placed in conductive soil or water. It has already been pointed out, however, that electrode potentials are influenced by factors other than the nature of the metal. This means that it is possible to have a corroding cell, both electrodes of which are made of the same metal, and it is this type of cell which is by far the most common cause of underground corrosion.

One type of cell, in addition to the bi-metallic already described, is the *concentration cell* (Figure 8.27b). In this the difference lies in the solution or electrolyte, it contains a different concentration of dissolved substances at two points, in contact with a single piece of metal. In general, the metal in contact with the more concentrated solution will be the anode (corroding) while the other will be the cathode. The single piece of metal forms both electrodes and the connecting circuit. Underground or subsea, this type of cell occurs when a structure, for example, a pipeline, lies in contact with two different soils. It can also occur when contact is made with a mixed soil, in which case the anodes and cathodes are likely to be very close together.

Another kind of cell responsible for much underground corrosion is the *differential oxygen* or *differential aeration cell* (Figure 8.28a). This occurs when part of the structure has better access to air than another; the aerated portion

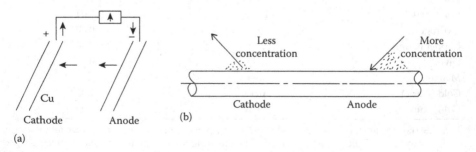

FIGURE 8.27
Illustration of (a) galvanic corrosion and (b) concentration cell.

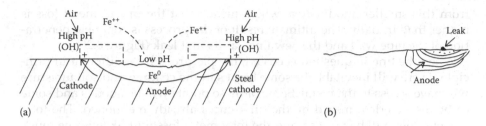

FIGURE 8.28
Illustration of oxygen corrosion and development of leakage. (a) Differential aeration cell and (b) leakage due to corrosion.

will be the cathode, while that from which the oxygen is excluded will be the anode. This type ordinarily results in attack on the bottom of the pipe; the top surface has some access to air through the soil above it, while it is almost totally excluded from the bottoms.

Step by step reactions are as follows:

1. Oxidation at the anode where the metal is eaten away or destroyed.

$$Fe^\circ \rightarrow Fe^{++} + 2e$$

2. Reduction at the cathode where oxygen combines with water and gains electrons to form the hydroxyl ion.

$$O_2 + 2H_2O + 4e \rightarrow 4OH^-$$

3. The reaction of the ferrous ion with the hydroxyl ion to form ferrous hydroxide (rust) which costs the cathode.

$$Fe^{++} + 2(OH)^- \rightarrow Fe(OH)_2$$

When a new pipeline is constructed, there arise hundreds and thousands of cells from these various differences – cells of varying size and strength. They overlap, they add and subtract voltage in a bewildering complexity. However, at virtually every point on the surface of the pipe, there is either a net flow of current to the surface of the metal, in which case that area is a cathode and nothing much is happening, or there is a net flow from the surface of the metal, in which case it is an anode and corrosion is taking place.

As time passes, changes occur, invariably some of the weaker anodes first become neutral and then cathodes; until, as time passes, there is smaller and smaller anodic area. This is not a gain; the total current flow must now pass from this smaller anodic area, which means that the rate of metal loss is higher than initially. The ultimate result of the process is complete penetration of the pipe wall and the new line has its first leak (Figure 8.28b).

When the line in question is coated, the picture is little different in principle. There will inevitably be some breaks in the coating, and soil moisture will have access to the metal. Some of these points will be anodes and some cathodes – as determined by the differences already mentioned. The total current flow will be far less, and the total metal loss will likewise be much smaller; it may happen, however, that at the worst anode the penetration rate will be greater than on the bare line. Coated lines are by no means immune from corrosion.

Besides the currents generated by the naturally occurring cells described, underground structures are subject to damage by currents from other sources such as high tension line, old existing pipeline, DC machinery etc., which happen to be passing through the earth. This is known as *electrolytic corrosion* (Figure 8.29). Some of these can be incepted by the structure and where they leave the surface for the soil, corrosion occurs.

Condition/Corrosion Monitoring

Corrosion measurement is the quantitative method by which the effectiveness of corrosion control and prevention techniques can be evaluated and provide the feedback to enable corrosion control and prevention methods to be optimised.

Corrosion monitoring techniques can help in several ways:

1. By providing an early warning on corrosion inducted failure.

2. By studying the correlation of changes in process parameters and their effect on system corrosivity.

3. By diagnosing a particular corrosion problem, its causes and the rate controlling parameters.

4. By evaluating the effectiveness of a corrosion control/prevention technique such as optimal chemical application.

5. By providing management information on maintenance requirement and ongoing condition of plant/pipes.

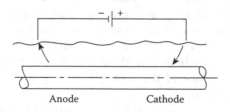

FIGURE 8.29
Electrolytic corrosion.

Internal Corrosion Monitoring
Internal corrosion monitoring techniques provide measurement of metal loss/corrosion rate in the internal wall of a pipeline. The techniques are as follows.

Metal Corrosion Coupons (Weight Loss Measurement)
Weight loss tests are the most common of all corrosion rate measurement tests. Corrosion coupons are strips of mild steel of various sizes with a hole in one end so that coupons can be mounted on an insulating (plastic) rod and inserted in a pipeline through a threaded fitting. Arrangements can also be made to insert coupons into a pipe or vessel under pressure through a valve.

After preparation, usually by sandblasting and degreasing, coupons must be accurately weighed before exposure. Corrosion of the coupon must be prevented while it is stored or being transported to and from the test location. Grease, oil and fingerprints on the uncorroded coupon will prevent the coupon from corroding properly when exposed. After being exposed to the corrosive fluid for two to four weeks, the coupon must be removed and cleaned of all corrosion products without any attack of the metal by the cleaning technique.

Weight-loss area of coupon and exposure time are used to calculate corrosion rate, which is reported in mils per year (MPY) of metal loss (one mil equals 0.001 in.). Pitting penetration rate and physical appearance of the coupon should be included in reporting corrosion of coupons.

Corrosion Probes: Electrical Resistance (E/R) and Linear Polarisation Resistance (LPR)
In the electrical resistance method, a corrosometer is used principally in a gas stream because it does not have to be submerged in water to function.

The corrosometer measures electrical resistance by using Wheatstone bridge. This consists of fair resistance, with two of the resistances in the probe and two in the instrument box. In this probe resistance a small strip, wire, or tube of steel or other metal is exposed to the corrosion environment. Due to corrosion, the cross-section reduces which changes resistance of the probe. By taking a reading of the instrument and characteristic corrections for the particular probe, the corrosion rate may be determined to the nearest micro inch.

In the linear polarisation resistance method, the corrosion rate meter uses a three-electrode probe that is inserted into the system. The corrosion rate meter measures corrosion current and corrosion rate because metal loss is directly proportional to current flowing from the test electrode. For the test electrode to change its potential, external current is applied from the auxiliary electrode with respect to the reference electrode. The applied current is related to metal loss and corrosion rate. By this a very low corrosion rate is also detected. The instrument gives a direct reading in MPY, which represents instantaneous corrosion rate.

Microbial Activity (SRB Count)

This is used to identify the presence of sulphate reducing bacteria (SRB). These anaerobic bacteria which consume sulphate from the process stream generate sulphuric acid which is very corrosive to pipeline steels.

Iron Count

This is a chemical analytical method to measure the iron content increase during the flow of a liquid in a pipeline.

The weight loss coupons and probes are installed in the pipeline through corrosion monitoring access fittings shown in Figure 8.30. This is an intrusive technique where the probes are inserted into the pipeline. An access fitting is a heavy-schedule machined metal tube that is attached to the pipeline by welding. A 45-mm hole is trepanned through the base of the fitting to allow access into the pipe. The fitting has internal and external threads and is sealed by an internal threaded plug that holds the weight loss coupons or corrosion probes. A steel outer cap is fitted to provide a secondary seal and to prevent damage to the external threads.

For installation, the fittings are usually fabricated into spool pieces that are installed in suitable locations in the topside section of the pipeline or associated pipe work.

After installation, the fitting is sealed using a solid plug. The seal between the solid plug and the corrosion-monitoring fitting is achieved using a soft, large face gasket and a secondary O-ring seal. The fine threads of the plug are not the sealing element, and Teflon tape is not required to establish a seal. External plastic caps should be fitted to protect the external threads on the access fittings. A plastic cap will fail during the hydrotest if the installation procedure of the solid plug or access fitting is faulty. The plastic caps will be

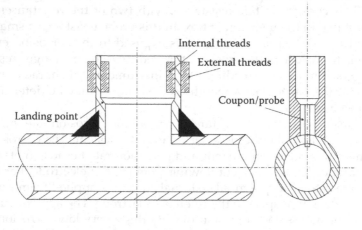

FIGURE 8.30
Corrosion monitoring access fitting.

replaced with steel or alloy caps when the fittings are commissioned on the operational pipeline.

Beside these intrusive techniques, there are some nonintrusive techniques like hydrogen patch probes, thin layer activation (TLA) – irradiated spools and so on. The other two methods, that is, SRB count and iron count fall, under fluid sampling techniques.

External Corrosion Monitoring for Offshore Pipelines
ROTV-ROV-CP Surveys

Equipment ROTV (remote operated tow vehicle) with side scan sonar, altimeter, underwater camera and forward-looking sonar and ROV (remote operated vehicle) shown in Figures 8.31 and 8.32 assess the external condition of submarine pipelines in the following aspects:

a. Bathymetry free-span and exposures. Burial depth of pipelines.

b. Physical condition; pipeline crossover, present position, condition of trenches.

c. Cathodic protection survey to know about coating condition and CP-current leakage.

d. ROV survey in exposed/damaged/critical section.

There are several ways to protect corrosion which are based on causes for that corrosion. The common ways of protection are by selecting metals close together in the galvanic series, by using inhibitors, by using protective coating, by electrically insulating, by cathodic protection and so on. The basic *principle of cathodic protection* is as follows:

FIGURE 8.31
Remote operated tow vehicle (ROTV).

FIGURE 8.32
Remote operated vehicle (ROV).

In any kind of electrochemical cell, whether the current flow is caused by self-generated potentials due to various kinds of differences, or is impressed from an external source of power, the anodes corrode and the cathodes do not (there are some exceptions, but only when certain metals other than iron or steel are involved). This basic principle is the one involved in cathodic protection; the entire surface of the threatened structure is made to act as a cathode and thus does not corrode. However, if there is to be a cathode, somewhere there must be an anode, a terminal from which the current can flow to the electrolyte, which in the case of a pipeline is the soil or surrounding water in which it lies. The anodes may be made of a metal which is naturally anodic to iron; generally, *magnesium and zinc* are used. These types of anodes are known as *'sacrificial anodes'*, which are commonly used during construction of the pipeline as a temporary protection.

In other cases, the anodes may be some other substance – carbon or graphite rods, junk iron, steel or high silicon cast iron; in which case, an external source of driving voltage is needed, usually a rectifier. These types of anodes are known as *energising anodes*. The details of design are many and varied; the principle is simple.

Another point of view for the description of cathodic protection is based upon the initial corrosion step – the conversion of an atom of metal into an ion, by its seizing a positive charge. If the surface of the metal is connected to a source of negative electricity, and negative charges are made available there in sufficient quantity, then any stray positive ion which comes that way will be neutralised; there will be no positive charges available for the formation of metal ions, and the metal cannot go into solution at all. By thus stopping the very first necessary step in corrosion, the entire process is prevented. This too, is cathodic protection, merely looked at from another point of view.

During laying of offshore pipeline, sacrificial anodes are fitted externally. A bracelet of anodes, made from special alloy metals, are fitted externally on the pipeline in a certain interval of length. Design life of these anodes is generally more than the design life of the pipeline.

8.4.2 Sub-Sea Pipeline Maintenance

Subsea pipelines, one of the most important links in the chain of operations that carries millions of tons of oil and natural gas from offshore producing fields to consumers, need special attention for an uninterrupted supply. This can only happen when a good maintenance practice is adopted. Beside uninterrupted flow of fluid, there are other objectives of pipeline maintenance and those are minimisation of operating cost, risk mitigation and pipeline life extension, safety and environment protection.

The operating subsea pipelines are very much vulnerable to both external as well as internal threats.

The *external threats* are

a. Seabed scouring due to current leads to washing off the bottom of pipeline and as a result a long span of line becomes free and unsupported. This may lead to damage of the pipeline.

b. Lateral shifting of the pipeline due to strong underwater currents specially in the case of nonanchored lines.

c. Physical damage on the outer surface of the pipeline due to dragging of anchors of other vessels and so on.

d. Damage to the pipeline crossing where two or more lines cross each other.

e. Deterioration of cathodic protection system.

f. Encroachment in bay areas.

The *internal threats* can be one or some of the following:

a. Transporting corrosive fluids

b. Generation of undesirable velocities, which may cause deposition of solids in the form of waxes, scales and so on

c. Growth of bacteria, mainly sulphate reducing bacteria (SRB)

d. Hydrogen embrittlement

e. Stress corrosion cracking

f. Transporting waxy crude in an oil pipeline

g. Formation of hydrate in case of a gas pipeline

h. Settlement of water in either an oil or a gas pipeline in a typical sea-bed profile

So, to plan a maintenance program, the first thing is the inspection and already we have discussed both internal as well as external inspection tools. After inspections, the observations are to be reported. Hence, the first activity is IRR, that is, inspection and reporting requirements. Then comes the type of activity to be carried through the pipeline after approval from the standing committee. Chart 8.1 shows the different maintenance activities and their interdependence.

The recommended frequencies of all the pipeline activities will vary depending on the type of operation. For example, maintenance pigging including intelligent pigging is carried out annually. Corrosion monitoring like SRB count in oil trunk line is carried out monthly whereas in water injection lines it is done once in two months. The LPR probe inspection is carried out on a continuous basis. The H_2S monitoring for oil trunk lines is done weekly. Similarly for bay pipelines, patrolling in urban areas varies from weekly to monthly depending on the fluid they carry. For cathodic protection monitoring the transformer units are maintained weekly whereas anode beds are attended to annually.

Thus, a maintenance schedule is prepared for every operation and is strictly adhered to. Today, maintenance of pipeline is not considered an isolated activity; rather, it is integrated in a concept of 'pipeline integrity management.' Pipeline integrity should ensure that a pipeline is safe and secure. It starts with a good design and construction, but it involves pipeline inspection, management and maintenance. Thus, the key to pipeline integrity is to prevent, detect or mitigate defects in the pipeline.

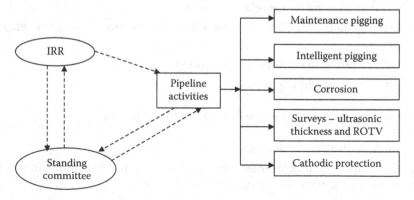

CHART 8.1
Maintenance activities.

Bibliography

Palmer, A. C. and R. A. King, *Subsea Pipeline Engineering*, Pennwell Corporation, 2004.

Design and Application Manual, Pipeline Engineering & Supply Co. Ltd., 1999.

Flow assurance challenges, 2015, http://www.diva-portal.org

Flow assurance: Maintaining plug-free flow and remediating plugged pipelines, 2001, http://www.offshore-mag.com/articles/print/volume-61/issue-2/news/flow-assurance

Gerwick, Jr., B. C. *Construction of Marine and Offshore Structures*, CRC Press LLC, 2000.

Dean, H. *Oil/Gas Pipelining Handbook*, Energy Communications Inc., 1975.

Kennedy, J. L. *Oil and Gas Pipeline Fundamentals*, Pennwell Publishing Company, 1993.

Leffler, W. L., R. Pattarozzi, and G. Sterling, *Deepwater Petroleum Exploration & Production*, Pennwell Corporation, 2011.

Mitra N. K. *Fundamentals of Floating Production Systems*, Allied Publishers Pvt. Ltd., 2009.

Mousselli, A. H. *Offshore Pipeline Design, Analysis and Methods*, Pennwell Publishing Company, 1981.

Tiratsoo, J. N. H. *Pipeline Pigging Technology*, Gulf Publishing Company, 1987.

9

Diving and ROV

Divers and remotely operated vehicles are both integral to any underwater or subsea operation that is carried out for all the petroleum operations discussed thus far. Both carry out some common operations although some are an exclusive specialty of either of these, for example, in deep or ultradeep water ROVs can operate safely whereas in shallow water divers become more useful and cost-efficient. Sometimes diving is also supported by a fleet of ROVs in the offshore oil industry. In any case, the support from either a diver or an ROV or both is a must for working in offshore and hence the nature of work of each of these in an underwater environment needs to be understood without which the subject of 'offshore petroleum operation' will remain incomplete.

9.1 Diving

Diving in general is defined as plunging into water in a lake, pool, or sea but professional diving is defined as a method of entering into water by a person enabling him to perform work in the ocean or on the ocean floor. There are two basic ways people can work underwater. The first is *pressurised* (or *hyperbaric*) diving in which the diver goes directly into the ocean and is therefore continuously subjected to increasing pressures with depth. The second method is *unpressurised* (or *isobaric*) diving in which the diver goes underwater in an unpressurised sealed diving bell, submarine, or other device that is capable of withstanding external pressure at different depths.

In the second case, the diver is protected from the inherent high pressure underwater environment whereas in the first case he or she is completely exposed to the underwater environment which can lead to many physiological as well as psychological problems. To understand it better, the basic principle of diving is dealt with below. Knowing the principle, the different methods of diving, the diving equipment and different diving services are also dealt with subsequently.

9.1.1 Principle of Diving

The basic principle of diving is based upon underwater physics and underwater physiology, a break-up of which is shown in Chart 9.1.

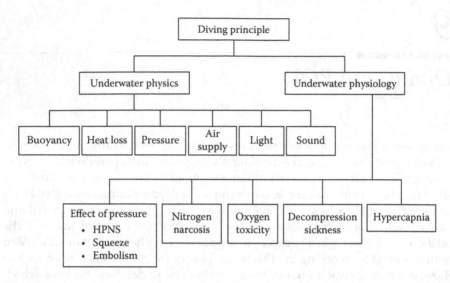

CHART 9.1
Principle of diving.

9.1.1.1 Buoyancy

Upon entering the underwater world, one of the most immediately notice-able differences is the tendency to rise or float. This elemental water force is known as buoyancy.

This weightless condition underwater can be a problem. Weighted div-ing shoes and belts are worn to compensate for this loss of weight. The equipment will give a diver negative buoyancy, enabling him or her to remain stationary at any desired level between the surface and the ocean bottom.

Positive buoyancy, which allows a diver to rise, is achieved by venting air into a standard diving suit. A diver must be extremely cautious when working with positive buoyancy because of the danger of 'blowing up,' or rising to the surface uncontrollably. During the ascent the decreasing pres-sure expands the diver's suit and gives him or her more buoyancy and more speed. These conditions can rupture the diving suit and cause the diver to crash back down toward the bottom. In this instance the risk of a body squeeze is greatly increased.

9.1.1.2 Heat Loss

Most working dives take place in water temperatures between 45°F and 60°F. The human body suffers rapid heat loss in these temperatures because (1) the specific heat of water is 1000 times greater than that of air and (2) the thermal conductivity of water is 20 times that of air. Heat loss can create hazardous situations where diving safety is concerned. A diver working in shallow depths for 16 hours may be able to turn his skin red after surfacing with the aid of a hot shower, but his body, or core, temperature may remain well below the safe level of 98.6°F.

Hypothermia, the scientific term for heat loss to key organs like heart, brain, kidneys and liver, is a direct result of overexposure to chilling temperatures. The body combats this heat loss by constricting the blood vessels, which slows blood circulation to the extremities. The body's natural insulation of fatty tissues also curbs heat loss to some extent. The problem the diver has is restricted blood flow. Without blood circulating through the hands, the diver begins to suffer from numbness. Since the diver is working in near-zero visibility much of the time, he or she relies heavily on their sense of touch. When numbness in the hands becomes apparent, the diver's productivity and safety are greatly reduced. Divers working in cold waters usually wear standard wet suits equipped with hot water systems to protect them from the chilling underwater environment. Vigorous exercise, cold-water training and proper diet also aid a diver who must work in cold water. A diver not prepared for such assignments is vulnerable to oxygen toxicity, nitrogen narcosis and decompression sickness.

9.1.1.3 Pressure

The pressure a diver is most interested in is the liquid pressure surrounding him or her. The pressure of water is directly proportional to its depth. Water pressure is the same at a given depth both inside and outside an open vessel. For a diver this is not true because of free air spaces like lungs, sinuses and middle ear. If these cavities trap any gas while a diver is descending or ascending, the change in pressure will cause a pressure differential that could cause severe problems.

9.1.1.4 Air Supply

When a person dives below the surface, he or she must take a portable gas supply with him or her, either in tanks on his or her back or through an umbilical line to the surface. His demand for this air supply is subject to change with changes in depth. This change is governed by the basic gas laws that are Boyle's law, Charles's law and a combination of these two laws, known as the Henry's law.

Some problems encountered in diving will show the usefulness of gas laws specially to calculate the effect of change in volume, pressure and temperature of the air inside the breathing apparatus law.

9.1.1.5 Partial Pressure and Dalton's Law

A diver uses various mixtures of gases, and to make them effective he or she must understand how these different mixtures behave as pressure changes occur. For this discussion, air shall be considered to be 20% oxygen and 80%

nitrogen. Other common mixtures of gas used in diving are helium and oxygen; nitrogen and oxygen (in different percentages than in air); and oxygen, helium and nitrogen.

In any mixture of gases, each gas exerts its share of the total pressure, known as partial pressure. Dalton's law states that the total pressure exerted by a mixture of gases is the sum of the pressures that would be exerted by each of the gases if it alone were present and occupied the total volume.

As the diver goes deeper under the surface, he or she is exposing himself to greater atmospheres of pressure, and oxygen will be 20% of the pressure if air is the breathing medium. At 15 atmospheres (about 462 ft underwater), the partial pressure of oxygen will be 3 atmospheres, or 44.1 psi. Thus, the partial pressure of a gas at sea level atmospheric pressure may be insignificant but can become extremely dangerous as pressure increases. Breathing 44.1 psi of oxygen can cause a toxic, possibly fatal reaction, and for this reason gases are mixed to prevent such events from occurring.

9.1.1.6 Light

The physical properties of light change dramatically when exposed to the underwater environment. As light rays strike and penetrate the water's surface, the water begins to absorb different light rays. The red wavelength is the first to be absorbed, followed by the yellow spectrum at a slightly greater depth.

Besides being absorbed, light is also diffused, or scattered, by the water and objects floating or living in it. Although the diffusion effect may be on a small scale, some objects become completely obliterated. Red objects appear to be black and disappear as part of the water background. A diver's field of vision is reduced considerably.

Wearing a face mask also creates changes in the properties of light. By looking through a glass face mask, a diver is observing light rays that are refracted, or bent, as they travel through the water into the eyes of the diver. This refracting effect magnifies the diver's view. Until visual perception is adjusted, objects will appear to be 1.3 times larger than their actual size. This magnification actually improves close-up underwater tasks. However, this small advantage is the only one a diver is afforded in the underwater environment.

9.1.1.7 Sound

The underwater world is not the tranquil, silent world many believe it to be. Sound travels four times as fast in water as it does through air. Number, volume and speed of the multidirectional sound sources are enough to confuse even the most veteran of divers. The speed of the sound is too rapid for the ears to distinguish its directional source. In the viscous medium of water, sounds cannot be distinguished as they can on the surface.

9.1.1.8 Underwater Physiology

The human body, a highly complex and sensitive system of cells, tissues, fluids and bone, functions in a normal fashion when at sea level pressure. In different environments, like the ocean, the body must adjust to different pressures in order to survive.

9.1.1.9 Effect of Pressure

Within the present diving depth capabilities of humans, human tissues are insensitive to the increased pressures. However, for a diver to be relatively insensitive to pressure changes, his or her breathing gas must have access to all body cavities, such as the lungs, the middle ear and the sinuses. Trapped gases in these free air spaces are compressed by increasing pressure of water depth and by compliance of the cavity walls. No significant pressure differential can exist between these spaces and the outer environment; otherwise, immediate tissue damage occurs. This fact is of the utmost importance because of our rapid descent capability. Experienced divers are usually able to descend at rates greater than 100 ft/min.

Although rapid descents have been simulated, an experienced diver will not risk his or her safety and health, or even his or her life, by racing to the bottom. Rapid descents below depths of 600 ft are a main factor in initiating the high-pressure nervous syndrome (HPNS) discovered only recently.

HPNS is a newly discovered phenomenon of deep-sea diving using helium, and the medical community is continuing to gather data on the effects and causes of this reaction. Rapid descent seems to be one cause that may bring about the symptoms. Experiments on test animals have shown symptoms of fatigue, muscle tremors and convulsions. Divers have suffered from fatigue and tremors, but convulsions have been limited to lab experiments with animals.

9.1.1.10 Squeeze

If the openings to the middle ear or the sinus spaces are closed during descent, as may happen to divers with head colds, slight pressure increases may cause severe pain. This pressure phenomenon is called middle ear or sinus squeeze, depending on the location of the pain. Squeeze is a common occurrence in diving operations.

A reverse squeeze occurs if these same cavity spaces are closed off during ascent. For example, maximum outward pressure on the eardrum may rupture the membrane. Immediate relief is felt by the diver. Unfortunately, if the diver is not wearing a diving helmet, water may enter the ear, causing a feeling of vertigo.

Thoracic squeeze, or compression of the lung cavity, will occur if the breath is held during descent. The increasing pressure compresses the lungs. If this pressure is not equalised, the lungs will collapse and fill with blood and tissue fluid.

9.1.1.11 Embolism

If the breath is held during ascent, the formation of an air bubble in the bloodstream may occur. This blockage in the bloodstream is known as an embolism and can create serious hazards to safe diving. Breath-holding over-expands the lungs as a diver rises to the surface. This over-expansion can cause ruptured air sacs and blood capillaries in the lungs. Once these sensitive tissues are ruptured, air is forced into the pulmonary capillary bed and air bubbles are carried to the left chambers of the heart. These bubbles are then pumped into the arteries. Any bubble too large to pass through an artery will lodge in the bloodstream and form an obstruction, thus depriving other tissues of blood. Consequences of an embolism depend on the area or organ that is affected. Often the brain is affected, and unless the diver is rapidly recompressed to release the bubble and allow the blood flow to continue, death can result.

9.1.1.12 Nitrogen Narcosis

The purpose of any inert, or carrier gas is to dilute the flow of oxygen to the lungs, thereby preventing oxygen toxicity. In air and in shallow diving, nitrogen serves this purpose. However, breathing high partial pressures of nitrogen causes a narcotic, euphoric feeling. This detrimental effect, called nitrogen narcosis, eliminates the use of nitrogen as a breathable gas below depths of about 190 ft.

If nitrogen narcosis occurs, a feeling of drunkenness or euphoria takes place. Thought processes slow down, and a diver's judgment and reaction time may be impaired. His or her working ability may also be affected, although some divers may be capable of withstanding higher doses of nitrogen than others. The narcotic effect usually wears off as a diver ascends, and by the time he or she reaches a shallower depth and a clearer state of mind, the diver may not remember if the assigned tasks were accomplished. Because of these limitations, another inert gas is usually substituted for nitrogen in deep dives. The most common substitute is helium.

However, recent experiments at depths greater than 600 ft have shown evidence of HPNS.

9.1.1.13 Oxygen Toxicity

Breathing extremely high partial pressures of oxygen can lead to a reaction known as oxygen toxicity. Experimental evidence indicates that divers may suffer from two different forms of oxygen toxicity. The first and most serious kind affects the central nervous system (CNS), causing convulsions similar to those brought about by an epileptic seizure. Total loss of muscle control occurs, and the diver suffering from the effects of CNS oxygen toxicity may become unconscious. These dangers eliminate the use of pure oxygen as a breathing medium below 50 ft in commercial operations.

The second type of oxygen toxicity involves the pulmonary muscle and can occur at depths even more shallow than those that cause the CNS oxygen toxicity. Symptoms include pains in the chest area, coughing and painful inhalation. Fortunately, these symptoms are rapidly reversed when air is used as the breathing medium instead of pure oxygen.

9.1.1.14 Decompression Sickness

The term 'decompression' in diving derives from the reduction in ambient pressure experienced by the diver during the ascent at the end of a dive or hyperbaric exposure and refers to both the reduction in pressure and the process of allowing dissolved inert gases to be eliminated from the tissues during compression.

Any dive beyond the safe no decompression zone (about 30 ft) carries with it the risk of decompression sickness, or the bends, so named because the bending joints of the body are most often affected. Nitrogen or helium bubbles in the bloodstream or tissues result from breathing air or mixed gases at increased pressures and not allowing sufficient time for the inert gas to dissipate. If this decompression procedure is not carefully executed, a diver may suffer numbness, weakness, pains in the chest or limbs, paralysis, residual bone disease, blindness or loss of hearing. If the CNS is affected, death may occur if the proper treatment is not administered promptly.

The bends affect different areas of the body in different ways. Knees and shoulders are common sites for bends pain. Problems in some tissue areas can be more acute than in others, but as in many other diving instances, an individual's own physiological makeup is one of the determining factors in the seriousness of the case. The nervous system is extremely sensitive. Bubbles in the brain or spinal cord can cause permanent damage.

Decompression sickness is usually noticeable within 30 minutes after a diver has surfaced but can occur during ascent. Bends can also occur after long delays, from 24 to 32 hours.

If the bends are suspected after a diver's ascent, he or she will be assisted to a recompression chamber by diving tenders. The surface crew will help remove the diver's suit (Figure 9.1), and the timekeeper will prepare the decompression charts and data sheets for monitoring recompression. During recompression, the chamber pressure will be monitored, the diver's vital signs checked and communication maintained using electronic equipment. Once recompression is completed, the diver will leave the chamber and be directed to the medical station on board. All vital signs will be thoroughly examined to assure that the diver is no longer suffering from the illness. The bends are a serious threat to any diver and have been responsible for many diving deaths. Treatment by recompression is required if the bends are suspected.

FIGURE 9.1
Diver being helped by surface crew to remove his suit.

9.1.1.15 Hypercapnia

Hypercapnia, or excess carbon dioxide in the body, is caused by inter-
rupting the elimination of CO_2 from the body. It often results from an
excessive CO_2 partial pressure in the breathing mix. Hypercapnia causes
blood vessel dilation and increases the flow of blood to the brain. This
increased blood flow can cause oxygen toxicity and expose the diver to
loss of muscle control.

All bodily tissues are susceptible to hypercapnia, but brain tissues are the
most sensitive. Symptoms of CO_2 excess include loss of motor coordination,
confusion and ultimately, unconsciousness. Hypercapnia rarely leads to
death and when given fresh air to breathe, the diver is usually able to recover
rapidly. However, he or she may still suffer from headaches and nausea after
initial recovery.

9.1.2 Diving Methods

There are three types of diving offshore, typically defined by the depth at
which the dive takes place:

1. Air diving
2. Saturation diving
3. Hard suit diving

9.1.2.1 Air Diving

It is a type of diving in which the diver's breathing medium is a normal mixture of oxygen and nitrogen. This is sometimes known as surface supplied
dives that generally run to a maximum depth of 50 m at which the diver

is exposed to the pressure. After this depth, the risk of oxygen toxicity is greatly increased and a mix of breathing gases is required.

In this type of diving, the diver's air, power and communications are supplied by an umbilical which runs between the surface vessel and the divers helmet. This umbilical is handled by the tenders onboard. The diver will also wear SCUBA (self-contained underwater breathing apparatus) type breathing bottles, which act as a backup. This way there will be a 'bail out' in place if the umbilical gets trapped or damaged and needs to be cut loose.

Because surface supplied diving equipment includes communication capability with the surface, this adds to the efficiency of the working diver. Breathing gas is supplied from the surface, either from a specialised diving compressor, high-pressure cylinders, or both. In commercial and military surface-supplied diving, a backup source of breathing gas is always present in case the primary supply fails.

The types of jobs performed on a surface dive are the same as would be done on a saturation dive. Some examples are: (1) underwater burning and welding, (2) survey work, (3) equipment installation and (4) operation and monitoring of subsea machinery.

When returning to the surface, much of the time the diver will go straight into a decompression chamber. In deep recreational dives, the diver will make calculated stops during the ascent to properly decompress. In a working environment, the divers can use the chamber to decompress thoroughly and in a controlled environment such divers may have to get back in the water again the next shift. Figure 9.2 shows how such a diver runs off the side of a vessel to get into the water.

FIGURE 9.2
Diver running off the side of a vessel.

9.1.2.2 Saturation Diving

Saturation diving is a diving technique that allows divers to remain at great depth for long periods of time.

'Saturation' refers to the fact that the diver's tissues have absorbed the maximum partial pressure of gas possible for that depth due to the diver being exposed to breathing gas at that pressure for prolonged periods. This is significant because once the tissues become saturated, the time to ascend from depth, to decompress safely, will not increase with further exposure.

Saturation divers work at depths in excess of 50 m for weeks at a time. At times the depth can become 500 m or so. An interesting item about the diver's equipment is that while working in the water, warm water is pumped into the suit to keep the diver warm.

On dry land, the air we breathe consists of approximately 21% oxygen and 78% nitrogen, with 1% of other trace gases. However, it can only be used in diving to a depth of around 40 m, after which nitrogen narcosis can occur, which can obviously become very dangerous in a working environment. So, breathing gas mixtures such as Heliox (hydrogen and oxygen) and Trimix (oxygen, nitrogen and helium) are mixed onboard the ship by the life support technicians (LSTs) and used by the divers for breathing. These mixtures are prepared to avoid the risks of nitrogen narcosis and oxygen toxicity. The gases need to be blended at specific ratios, depending on the working depth of the divers.

While not on shift, the divers live in a controlled habitat onboard the vessel known as the *saturation system* (Figure 9.3), which typically comprises a *living*

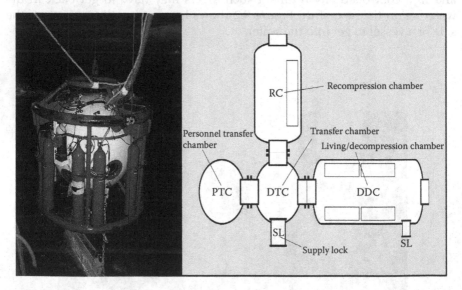

FIGURE 9.3
Diving 'saturation system.' (Left: U.S. Navy photo courtesy of Chief Photographer's Mate Andrew McKaskle; https://zh.wikipedia.org. Right: Courtesy of Pbsouthwood/Wikimedia Commons/CC-BY-SA-3.0.)

chamber, transfer chamber and submersible decompression chamber, which is commonly referred to in commercial and military diving as the *diving bell*. The system can be permanently placed on a ship or ocean platform, but is more commonly capable of being moved from one vessel to another by crane.

The entire system is managed from a control room, where depth, chamber atmosphere and other system parameters are monitored and controlled. The diving bell is the elevator or lift that transfers divers from the system to the work site. The bell will generally have one diver inside that tends the working divers tether and stands by to assist if required.

The divers will work a 12-hour shift at depth, dividing the time in the water. Once back on board, the time off will vary with the number of dive teams in saturation, but it will be at least 12 hours.

At the completion of work or a mission, the saturation diving team is decompressed gradually back to atmospheric pressure by the slow venting of system pressure, at an average of 15 m per day. In an emergency, this rate could be increased.

9.1.2.3 Hard Suit Diving

A third type of diving takes place from within a hard or *atmospheric* suit (Figure 9.4). Consisting of a hard and bulky suit fitted with small thrusters and control arms, the diver inside is kept at a steady pressure of 1 atmosphere;

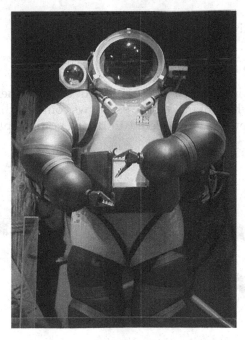

FIGURE 9.4
Hard suit. (Local travel photo. Courtesy of Wikimedia Commons/CC-BY-SA-2.0.)

the same as we all experience on the surface. This allows the team to lower the diver down to the working depth where the job is performed, then bring up to the surface with no special breathing gas or decompression time required.

Divers can work at depths of more than 700 m with no ill effects such as decompression sickness or nitrogen narcosis; more than twice that of a saturation diver. Oceanworks International makes the atmospheric diving system (ADS) line of hard suits.

One of the advanced designs now being used in offshore diving operations is the 1 atmospheric diving suit and is supplied to offshore oil industry under exclusive rights of oceaneering. The pressure inside the suit is always at 1 atmosphere and the operator inside the suit can descend to 1000 ft in 5 min. The diver can surface in the same amount of time and needs no decompression. The suit weighs 1100 lb out of water but only 65 lb when submerged. The disadvantage is the platform restricts an operator's vertical mobility and the diver has to perform in near zero visibility.

9.1.3 Diving Equipment

The equipment used in today's diving operations covers a wide range and is broadly classified as

1. Air diving equipment
2. Diving bell
3. Saturation diving system

9.1.3.1 Air Diving Equipment

These are the equipment that a diver will put on his or her body and carry with him or her while diving. Figure 9.5 shows common diving gear.

Following equipment is considered air diving equipment.

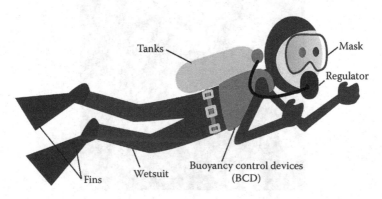

FIGURE 9.5
Common diving gears. (Courtesy of https://pixabay.com)

9.1.3.1.1 *Self-Contained Underwater Breathing Apparatus (SCUBA)*

Scuba equipment utilises high pressure cylinders to store compressed breathing air. The air is supplied to the diver by means of a demand regulator which delivers the air at a pressure equal to the surrounding water pressure. The diver's exhalation is exhausted into the water.

The advantages of scuba are horizontal mobility (because the system does not require any connections to the surface), portability as a result of its small size and weight, and a high degree of depth flexibility and buoyancy control. The disadvantages of the system are a depth limitation of about 60 ft (130 ft under ideal conditions), a limited gas supply, and an exertion limitation (caused by the mechanics of the demand breathing regulator). A serious operational disadvantage is caused by the fact that there is usually no communication between the scuba diver and topside personnel.

Scuba is seldom used in commercial operations because most underwater work does not require horizontal mobility, but almost always requires an unlimited breathing gas supply and good two-way communications for safety reasons.

9.1.3.1.2 *Lightweight Equipment*

Air diving or surface-supplied diving equipment that employs swim fins or boots, face masks or helmets and protective clothing is known as *lightweight gear*.

Swim fins and boots – Fins and boots are two of the diver's most fundamental needs. The diver will need a good pair of swim fins for propulsion. Working divers will have two pairs of fins, one pair for wearing with rubber booties and the other pair for short duration skin dives. Boots are often used in commercial dives especially when walking along the ocean bottom.

Face masks or helmets – Full face mask or a separate multipiece or combination of two are used in supplying air to the diver. Full face masks have a demand regulator mounted on the mask itself.

Several different companies manufacture commercial face masks that are similar in nature and function. Some masks contain an oral-nasal breathing piece inside that reduces carbon dioxide buildup. Some are constructed of fiberglass with stainless steel and chrome-plated fittings and feature a dual exhaust valve that minimises the exhalation effort. An umbilical cord from the surface to the mask supplies a breathing medium and communications to a diver.

Lightweight diving helmets can be worn with either a wet or a dry suit and feature a neck ring that will mate with the neck ring on either suit. The communication system is contained inside the helmet, permitting the diver to carry on verbal communication with personnel at the surface or in the bell. Noise-canceling devices increase the communication efficiency of the helmet (Figures 9.6 and 9.7).

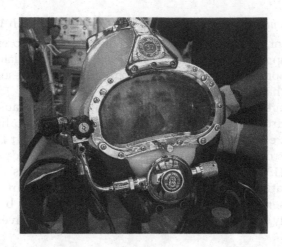

FIGURE 9.6
Diving helmet with communication system. (Courtesy of the U.S. Navy/Wikimedia Commons/ Public domain.)

FIGURE 9.7
Helmet used in saturation diving. (Courtesy of Shutterstock.)

Lightweight diving helmets enable a commercial diver to go to the maximum depths possible with surface-supplied air. Helmet design gives a diver more freedom of movement, and swimming is much easier than it would be with deep-sea diving gear. Another advantage that lightweight gear has over deep-sea dress is that a diver usually needs little or no assistance in donning the helmet.

Protective clothing – Divers wear either a wet or a dry diving suit.

1. *Wet suit* – Usually made of neoprene material, it is widely used in diving activities as a form of protective clothing. Although not as effective as a dry suit, cost and application have made the wet suit highly popular among sport divers. A small amount of water is

allowed to enter between a diver's body and his wet suit. Heat from his body warms this thin layer of water and provides protection for short duration divers.

2. A *dry suit* is designed to accommodate a layer of insulation between the diver and the suit. The usual form of insulation is thermal underwear. This gives the diver maximum warmth and protection from water. Many dry suits are inflatable through a low-pressure supply hose that has a quick disconnect fitting. The exhaust valve on the suit allows rapid suit depressurisation for situations that require such action.

3. *Buoyancy control equipment* – A flotation vest is a common type of buoyancy control equipment used in sport diving. Carbon dioxide cartridges are fired when inflation of the vest is necessary. The vest should automatically keep a diver afloat in the proper position and should be capable of keeping him or her alive at any depth. Commercial divers use weight bells and weighted shoes to achieve negative buoyancy. Although these items can be worn separately, they are often used with a deep-sea diving suit. The weight belt is worn around the waist and a diver should be able to quickly remove it if necessary.

9.1.3.1.3 Deep Sea (Hard Hat) Equipment

Deep-sea or hard hat gear consists of a metal hard hat, a watertight suit (dry dress), a weight belt and weighted boots. As with lightweight gear, an umbilical from the surface provides the air supply and communications. The advantages of deep-sea gear are maximum physical protection, good working leverage (because the diver can make himself heavy underwater) and good temperature protection. The disadvantage of the equipment is its bulk and weight. The depth restriction and surface support equipment are the same as for lightweight gear.

9.1.3.1.4 Mixed Gas Diving Equipment

Below 190 ft of depth, divers switch from air as a breathing medium to mixed gas. This type of diving combines the mobility of scuba swim gear with the safety and communications advantages offered by surface supplied equipment. A means of partially recirculating the diver's exhaled air is incorporated into a mixed gas system to conserve on the expensive premixed gases that must be used.

Recirculation of the exhaled air requires carbon dioxide scrubber systems. The backpack carbon dioxide scrubber and the helmet scrubber are two frequently used systems.

In a carbon dioxide scrubber system, the premixed gas from the surface enters the backpack and is introduced to the helmet through the right-hand interconnecting hose (Figure 9.8). A diver breathes the breathing medium in a normal fashion and exhales.

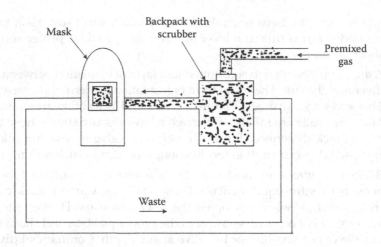

FIGURE 9.8
Carbon dioxide scrubber system.

As the diver exhales, pressure inside the helmet builds up. By the time half of the exhalation is completed, the exhaust valve on the left-hand side of the helmet opens and gas containing a large amount of carbon dioxide is released into the water. As exhalation is completed, the exhaust valve closes and the remaining gas, about 80% of the diver's exhalation, is recirculated to the backpack.

Gas brought back to the backpack enters the left side and travels laterally through carbon dioxide absorbent materials, which may be soda lime or baralyme. As the gas passes through the absorbent materials, an exothermic chemical reaction takes place and the material absorbs all the exhausted carbon dioxide from the diver's helmet. Backpacks under normal operation will effectively scrub the recirculated gas up to 7 hours.

The temperature of the canister containing the carbon dioxide remover is important in determining the effectiveness of the entire system. The scrubber system works on an exothermic, or heat-producing, reaction. If the gas or the canister is too cold, the system will be inhibited in its performance. An external heat source is sometimes used to retain heat so that the removal of CO_2 can be accomplished. Close monitoring of the gas mixtures by the surface team is essential.

9.1.3.2 Diving Bell

The effectiveness of air diving reduces in deeper dives. Hence, in deep waters the diving bell evolved. The bell serves not only as transportation to a deepwater work site, but also as a decompression chamber on the return to the surface.

In a bell divers can be transported to a work site with the atmosphere at normal surface pressure. Their decompression time is thereby reduced, since they will be exposed to high pressures for a shorter time than if they were compressed before or during descent. Once at the site, the chamber can be pressurised to the surrounding water pressure. After work on the bottom is completed and the divers are inside the bell, their decompression process may begin. If the decompression time is short, they may continue decompression inside the bell until it has been completed. If a lengthy decompression time is required, they may transfer under pressure to a deck decompression chamber (DDC). Many designs enable the bell to be mated to a deck decompression facility either from the top or from the side. The bell is usually designed to transport three persons – generally two divers and a tender, or a diver apprentice, although some bells may hold up to four persons. Gas can be supplied from the surface through umbilicals, or the bell can be run independently with a 50-bottle gas supply contained inside. The bell support system consists of a self-contained carbon dioxide removal unit, metabolic oxygen supply, emergency gas supply oxygen analyser, communications system and internal and external lights. An umbilical line from the bell to the surface provides power, communication, a mixed-gas breathing supply, heat and surface monitoring equipment services. Divers are connected to the bell in much the same fashion, usually with an umbilical that provides similar services.

The normal method of operation is for the two bell occupants to enter and seal the bell hatch while it is on deck. All systems are checked, including life support, manipulators and thrusters. The bell is then lowered to a stable shallow depth, and all onboard equipment is checked. After this inspection, the bell is lowered to the work site. At the site it is aligned visually and then attached, in most cases, to the subsea equipment for working leverage and stability. Once stable, the bell's load line to the surface is slacked off so that the bell will not absorb any motion due to the movement of the sea at the surface.

Next, the bell is pressurised to the working depth environment, and one diver enters the water through the egress hatch. The alternate diver inside the bell constantly monitors the movements and work of the diver outside. When the working diver has been exposed to a certain amount of inert gas, he or she reenters the bell and the alternate takes over the task. This technique is used until the divers must surface or until the job is completed.

After all tasks have been accomplished, the bell is raised to the surface and landed on deck. Control and monitoring of all the services of a bell diving system take place in a *control van*, a building that houses electronic equipment necessary to carry out the monitoring function. The control aboard provides weather protection and noise shelter for the control crew.

Launching, mating and recovering a diving bell require a means of handling the system on board the rig. A handling system must be designed and configured for the particular diving system, the conditions under which it will operate, and the rig on which it will be installed.

9.1.3.3 Saturation Diving Systems

A typical saturation system is made up of the following basic components.

9.1.3.3.1 Diving Bell

The bell is used to transport the divers from the topside habitat to the work site and return. It must be capable of mating with the living chambers.

9.1.3.3.2 Deck Decompression Chambers (DDCs)

Divers often complete their decompression times in a DDC, which usually consists of one or two pressure vessels that provide a comfortable living area during the decompression time.

The living chambers must provide space for sleeping, relaxing, work preparation, eating and sanitary facilities. The systems are usually outfitted with transfer locks, TV, radio, air conditioning and heating and all life support functions.

9.1.3.3.3 Operational Motor/Control Module

Generally, this is a van containing communications and the equipment required to monitor and control the in-water dives.

9.1.3.3.4 Decompression Control/Monitor Module

This is a van that contains all of the controls and monitor equipment needed to operate the deck chamber complex and life support functions.

9.1.3.3.5 Support Equipment

Support equipment for saturation systems can vary, but most systems designed for full-scale deepwater diving will include the components like water supply system, air compressor, diver heating unit, hydraulic power unit, bell handling system (winch, tuggers) and gas storage cylinders.

Saturation diving begins to make economic sense in water depths greater than 300 ft for projects requiring quite a few hours of working bottom time and in water depths greater than 200 ft when a relatively large number of working hours is required. For instance, several different jobs in 200 ft of water may prove to be more economical if done with the saturation technique.

9.1.4 Diving Services

The actual tasks that working divers undertake are similar to jobs done on land. Common assignments include welding, inspection, surveying, demolition and maintenance. Today's commercial diver is an underwater marine technician, who is well trained, experienced and competent in basic mechanical engineering. Additionally, he or she is an able rigger, welder and underwater observer.

9.1.4.1 Surveys and Inspections

This is one of the most common assignments and may range from a very simple bottom survey to ensure that there are no obstructions prior to setting a platform to very complex inspections to determine structural damage.

Observation bells and submarines are useful tools for survey and inspection work. Also, television, video tape and photography provide permanent records. Ultrasonic equipment with strip recorders can be utilised to check wall thicknesses and critical weld areas with a record being maintained for future reference.

Surveys and inspections can be made quite accurately if the need exists. Limiting factors are poor visibility, depth, temperature, strong currents and rough weather.

9.1.4.2 Maintenance and Repair

Routine maintenance of subsea installations is a common diving requirement. The diver can inspect for scour, leaks, coating damage and other discrepancies. Miscellaneous repairs to docks, piers and other shore facilities will be performed by the diving team. These workers are skilled craftsmen who can usually perform a multitude of duties other than diving.

Underwater repairs of every conceivable nature can be achieved. Damaged pipelines and platforms can be repaired by utilising wet or dry welding techniques, or mechanical devices. The depth and scope of work will dictate the equipment, method and cost of a particular repair. It would be safe to say that almost any repair can be made to existing installations in depths up to 600 ft. Dry welding is usually practical, though costly. Mechanical connections are proving to be reliable and will be used more and more.

The placement of concrete and repairs to concrete structures can be done with great success. Underwater cleaning and painting has reached the point where good results can usually be obtained.

9.1.4.3 Pipeline Construction

Diver support of pipeline construction operations is one of the more important roles played by today's diver. This role will expand in importance as operations move to deeper water depths. The divers are used to inspect the bottom for obstructions prior to laying the line. During the lay operation, the divers swim the stinger to ensure proper alignment of the pipe and to profile the stinger position. Quite often divers manipulate ballast valves to change the profile of the stinger.

The divers also check to ensure that coating damage is minimal and that the pipe is laying properly. During the burying phase, divers attach the jetting equipment and/or hand jet the line to the proper depth. Inspection divers are employed to check the line for proper installation.

Underwater connections and riser installations rely on the diver to a great extent. Familiarity with pipeline terminology and procedures is a prerequisite for this type of work. Preplanning and pre-engineering of subsea connections and riser installations will go a long way toward reducing the costs and complexities.

9.1.4.4 Platform Installation

The basic task for a diver associated with platform installations is to check the bottom prior to setting the jacket. The role of the diver becomes more important when a platform is being installed over an existing wellhead. In this case, the diver assists in guiding the jacket into the proper position.

9.1.4.5 Harbors – Port Facilities

The major role of the diver in new port and harbor development is in the area of inspection. However, if rock is encountered, divers are often assigned the task of blasting away rock overburden or outcroppings.

9.1.4.6 Offshore Drilling Support

Probably, the one area where the contribution of the diver is most significant is in support of offshore drilling operations utilising subsea completion equipment. This is especially true for foreign exploration operations, which are often done in remote or isolated locations. To a large extent, the diver can be compared with a local firefighter.

Basically, divers are utilised for the following tasks:

1. Checking the bottom and setting of base plate.
2. Stabbing guide wires for stack and for TV system.
3. Assisting in stabbing conductors and marine risers.
4. Inspecting BOP stack (when required) to check rams, kill and choke lines and various hydraulic fitting and components. Locating and repairing leaks.
5. Restabbing broken guide wires and clearing debris from the stack.
6. Manually operating components if hydraulic failures occur.
7. Assisting in stack recovery.

A tremendous amount of rig down time can be saved by having a diver inspect or repair in lieu of having to pull and rerun the stack. Rig divers must be thoroughly trained and familiar with whatever subsea drilling equipment is being used. They must understand the inner workings of a

BOP stack and be aware of what can be done to correct malfunctioning parts.

9.1.4.7 Underwater Welding and Habitats

Hot taps, pipeline repairs and structural repairs can be performed with a high degree of effectiveness. A habitat that is large enough to accommodate two diver-welders and the structural member to be repaired is lowered to the work site, sealed around the pipeline and then dewatered with compressed air or some other inert atmosphere. Welding carried out inside the habitat is called 'dry welding'.

The two most common welding techniques known as TIG and MIG carried out inside water are known as 'wet welding' and require both training and skill to achieve code grade results. In addition, special equipment and gas shielding are required. A method that requires much less training and special skills and compares closely with standard stick welding in the dry utilises manual metallic arc 'stick' electrodes with special covering to stabilise the arc under hyperbaric atmospheres.

9.1.4.8 Demolitions

The use of demolitions underwater has great applicability. Bulk and shaped charges can be designed for a multitude of jobs. Demolitions are used in salvage work, pipe trenching, harbor and port blasting and in removing conductors, wellheads and caissons below the mudline. Although setting underwater explosives is not an everyday task, it is one that only divers can do. Drill pipe casing, for example, is sometimes cut with explosives known as shaped charges and these must be placed on the casing by the diver. Divers doing this job must have field experience under live conditions.

9.1.4.9 Submersible Work Vehicles

There are a number of small submarines on the market today. Depending upon the job, they can be very useful and effective. Most are expensive, but under certain circumstances they can provide the most economical means of accomplishing a task.

Beyond diver depths, subs can be extremely valuable. Subs are used for salvage work, inspections, surveys and scientific assignments.

All the above services described give a clear indication that to do these jobs highly skilled divers are required and the divers of the present generation to do operations in the offshore oil and gas industry are given special training to learn the skills that include a broad technical background in the area of offshore oil drilling, production and pipeline transportation. Today's divers are highly competitive and full of confidence. Most divers are healthy individuals who realise the inherent dangers of their occupation and they are highly paid professionals.

9.2 Remotely Operated Vehicle (ROV)

A remotely operated vehicle (ROV) is the common accepted name for tethered underwater robots in the offshore industry. ROVs are unoccupied, highly manoeuverable and operated by a person aboard a vessel. They are linked to the ship by a tether, a group of cables that carry electrical power, video and data signals back and forth between the operator and the vehicle. High power applications will often use hydraulics in addition to electrical cabling. Most ROVs are equipped with at least a video camera and lights. Additional equipment is commonly added to expand the vehicle's capabilities. These may include sonars, magnetometers, a still camera, a manipulator or cutting arm, water samplers and instruments that measure water clarity, light penetration and temperature. Exactly who to credit with developing the first ROV will probably remain clouded; however, there are two who deserve credit. The PUV (programmed underwater vehicle) was a torpedo developed by Luppis-Whitehead Automobile in Austria in 1864; however, the first tethered ROV, named POODLE, was developed by Dimitri Rebikoff in 1953. ROVs gained in fame when U.S. Navy CURV (cable controlled underwater recovery vehicle) systems recovered an atomic bomb lost off Palomares, Spain in an aircraft accident in 1966, and then saved the pilots of a sunken submersible off Cork, Ireland, the Pisces in 1973, with only minutes of air remaining. This created the capability to perform deep-sea rescue operations and recover objects from the ocean floor. Building on this technology base, the offshore oil and gas industry created the work class ROVs to assist in the development of offshore oil fields. More than a decade after they were first introduced, ROVs became essential in the 1980s when much of the new offshore development exceeded the reach of human divers. During the mid-1980s, the marine ROV industry suffered from serious stagnation in technological development caused in part by a drop in the price of oil and a global economic recession. Since then, technological development in the ROV industry has accelerated and today ROVs perform numerous tasks in many fields. Their tasks range from simple inspection of sub-sea structures, pipeline and platforms to connecting pipelines and placing underwater manifolds. They are used extensively both in the initial construction of a sub-sea development and the subsequent repair and maintenance.

Two of the first ROVS developed for offshore work were the RCV-225 and the RCV-150 developed by Hydro Products in the United States. Many other firms developed a similar line of small inspection vehicles. Today, as oil exploration migrates into deeper and deeper waters, ROVs have become not only capable, but also highly reliable. With ROVs working as deep as 10,000 ft in support of offshore oil and other tasks, the technology has reached a level of cost-effectiveness that allows organisations from police departments to academic institutions to operate vehicles that range from small inspection vehicles to deep ocean research systems.

It was once thought that something thrown into the ocean was lost and gone forever; however, organisations such as Mitsui and JAMSTEC in Japan have ended that thought. With the development of their ultra-deep ROV Kaiko, they have reached the deepest part of the ocean – the Challenger Deep in the Mariana Trench, at 10,909 m.

The ROV is an unmanned, free-swimming underwater vehicle that has six degrees of freedom and can be operated and controlled from a surface station. It is propelled and controlled by several thrusters, which have speed controllability and reversibility. The controls are often guided by the information from many sensors, which collect data on the ambient environment and supply it to a host computer. The data is filtered, processed and used as an input to computer software which generates corresponding control signals to enable the ROV to perform the manoeuvers.

ROV carries as payload various instruments, which collect data and/or perform tasks for which the vehicle is deployed. The telemetry and power supply from the surface station is often sent through an umbilical cable. In some cases, onboard batteries provide power and the telemetry through an acoustic link.

9.2.1 Types of ROVs

ROVs can be classified as follows:

1. 'Work' class ROVs
2. 'Observation' or 'general' class ROVs
3. 'Mini' and 'micro' class ROVs
4. 'Special purpose' ROVs

Work class ROVs are very large in size and operated by a crew. The crew consists of a supervisor, pilot and in some instances a co-pilot. Generally, the members are very experienced with great knowledge in electronics, mechanics and hydraulics. These work class ROVs are used for deepwater trenching, cable burial, repair jobs and the recovery of larger objects. These big heavy ROVs are lifted in and out of the water by cranes. Work class ROVs are an essential world tool, making today's underwater jobs less of a challenge.

These systems are designed to do complicated work underwater at various depths. Most of these can work in water up to 3000 ft, and some up to 10,000 ft. These ROVs range in weight from 220–4410 lb (1000–2200 kg) with typical payload capacities in the 220–600 lb (100–272 kg) range. The vehicles themselves tend to be large and powerful and may be deployed from a cage called a tether management system. The vehicles have two or more multi-function manipulator arms. This class of ROV (Figure 9.9) has the best track record in subsea drilling support.

A typical work class (also called open frame) ROV for the offshore industry consists of a frame that supports hydraulic pumps, thrusters, all ancillary equipment (cameras, sonars, etc.) and the electronic control equipment. The mass is distributed to achieve balance in air and water. The submerged weight is compensated by buoyancy packs made of Syntactic foam fitted to the upper part of the frame. It is fitted with a five-function grabber arm, used to position the ROV steadily in one position, and a seven-function manipulator is then used to perform assigned robotic tasks. The manipulator is similar to those found in various industrial applications on-shore, having a number of joints, a rotating wrist and a hand-like claw. For some applications, the ROV is equipped with a special tooling kit designed to locate and lock on to a docking panel whence various valves and controls can be activated.

Light work class ROVs, typically less than 50 hp (propulsion) are able to carry some manipulators. Their chassis may be made from polymers such as polyethylene rather than the conventional stainless steel or aluminum alloys. They typically have a maximum working depth less than 2000 m.

Heavy work class ROVs typically less than 250 hp (propulsion) have the ability to carry at least two manipulators. With new requirements to perform

FIGURE 9.9
Work class ROV. (Courtesy of Brennanphillips/Wikimedia Commons/Public domain.)

subsea tie-in operations in deepwater installations and to carry very large diverless intervention systems, this class of ROV is becoming increasingly large, powerful and capable of carrying and lifting large loads – thus the term 'heavy work class vehicle' has been adopted by the industry. These vehicles can weigh more than 10,000 pounds and resemble a minivan in size. Three-thousand-meter depth capable systems are now commonplace, with at least one system capable of 6000 m. A cable and flow line burial system powered by four electro hydraulic units totalling 1000 hp is in use today, and at least one ROV that can lift and manoeuvre 16,000 pounds has been built. Cameras, lights, sonars and other sensors necessary to operate at great depths are readily available. Manipulators capable of lifting hundreds of pounds are commonly installed on these vehicles. In open frame of work class ROVs, the emphasis is more on the design and fit of the robotic tools than on the hydrodynamic design of the frame. The framework along with thrusters acts only as a vehicle.

Trenching/burial vehicles typically ranging from 200–500 hp (propulsion) have the ability to carry a cable laying sled and work at depths up to 6000 m in some cases.

Observation or general class ROVs are much smaller in size but perform many underwater tasks, specifically in areas where work class ROVs cannot go, or where they just might be too much. Tasks include pipeline inspections, search and rescue operations, ship inspections, treasure recovery, port inspections and so on. In many cases, general class ROVs can be deployed and controlled by just a couple of people (Figure 9.10). This can make jobs easier and less expensive.

This general class ROV with typically less than 5 hp propulsion occasionally has small three-finger manipulator grippers, such as on the very early RCV 225. These ROVs may be able to carry a sonar unit and are usually used on light survey applications. Typically, the maximum working depth is less than 1000 m though one has been developed to go as deep as 7000 m.

Mini and Micro class ROVs are very small in size and weight. Today's mini class ROVs (Figure 9.11) usually weigh around l5 kg and the micro class ROV

FIGURE 9.10
General class ROV. (Courtesy of Shutterstock.)

FIGURE 9.11
Mini class ROV. (Courtesy of Ph. Saget/Wikimedia Commons/CC-BY-SA-3.0.)

can weigh as little as 3 kg. Essentially one person could take the complete ROV system out with them on a small boat, deploy it and operate it with no problems. This can be very handy in many applications; and with these systems being significantly lower in price, they make for a good alternative to divers specifically in places where a diver might not be able to physically enter such as a sewer, pipeline or small cavity. Occasionally both micro and mini classes are referred to as *eyeball class* to differentiate them from ROVs that may be able to perform intervention tasks.

Special purpose ROVs are those ROV systems that have been designed to perform specific underwater jobs ranging from searching for the wreckage of the Titanic in 13,000 ft (3900 m) of water to trenching telecommunication cables. Some of them can work in water depths as great as 20,000 ft (6060 m). These systems may or may not be one of a kind. These ROVs are often used as intervention and inspection vehicles for offshore applications, particularly at depths that are beyond the limit of saturation diving.

9.2.2 ROV System

A standard ROV system consists of three major hardware sub-components, namely the ROV vehicle, tether management system (TMS) and the launching and recovery system (LARS). Besides, there are other ancillary systems like sensors and controls, power supply and power distribution unit and umbilical.

9.2.2.1 ROV Vehicle

It is a free-swimming vehicle with six degrees of freedom, capable of moving forward, backward, laterally, and vertically up and down by means of suitably located thrusters. It also carries payload in the form of scientific instruments, cutters, manipulator tools and so on for use at required depths.

9.2.2.1.1 Sensors and Controls

The ROV is fully controlled from a surface station based on the sensory feedback on the system. The sensors provide the information on the depth, position, angular and axial velocities and accelerations, altitude from the seabed, images of the environment and so on. These sensory data are used as the input for the control software, which generates the necessary signals to perform the intended tasks. In addition, there are often many gadgets and systems like manipulators, homing and docking system, which makes the vehicle more versatile. To accomplish the user-specified tasks, dedicated sensors like camera, sonar and so on are often installed on the ROV.

9.2.2.1.2 Power Supply and Power Distribution Unit (PDU)

There is a generator to provide independent power for the ROV system. Sometimes rig power can be unreliable or generates power peaks which can damage the ROV system's electronics. Hence, ROV systems require a dedicated generator or a motor generator which gets power from the rig and delivers clean power to the ROV system.

9.2.2.1.3 Umbilical

It is a central armoured cable that runs between a tether management system and the surface.

9.2.2.2 Tether Management System (TMS)

ROVs that operate at greater depths (in excess of 1000 m) are often equipped with a TMS or top-hat for ease of deployment and retrieval. TMS is known as the vehicle garage. It reduces the umbilical drag by separating the cable expanse between the vehicle and the surface into two segments. These are (1) the main umbilical from the surface to the tether management cage and (2) the neutrally buoyant tether between the cage and vehicle. This TMS also acts as a depressor so that the major portion of the umbilical remains erect and is not pulled along when the ROV moves. This greatly reduces the power consumption of the ROV. When the required depth is reached, the ROV slips out of the TMS and during retrieval, it goes back to the TMS. The TMS also prevents the ROV from being damaged due to collision during launch/recovery. The TMS contains the following sub-systems, namely:

1. Underwater winch to pay out cable when the ROV is launched.
2. Underwater slip ring to connect the cable in the winch to the main umbilical.
3. Homing device to allow the ROV to find the location of the TMS and enter.
4. Docking arrangement to allow the ROV to be physically held within the TMS.

9.2.2.3 Launch and Recovery System (LARS)

The LARS consists of a skid-mounted crane (Figure 9.12), which can be in the form of an A-frame, Jib-boom, knuckle-boom and so on, with suitable docking head and guidance system so as to launch and recover the sophisticated TMS-ROV system safely. An umbilical winch is a part of the launching/recovery unit, which is located on the ship. The ROV with the TMS system is often very heavy and it needs to be lowered down or hauled up from the working depths. This requires a lot of power depending upon the size and weight of the system and it is delivered by the winch. Most heavy-duty winches are hydraulically operated with speed controllability and reversibility. The umbilical is the component through which the ROV maintains physical contact with the mother ship. In most ROVs, the cable supplies the power to the ROV. The ROV transmits the signal from the sensors through the cable to the host computer on the ship. Based on these signals, the control software determines the necessary control signals that should be downloaded through the umbilical to the ROV so that the required manoeuvres are carried out. The umbilical also acts as a mechanical means for handling the ROV in or out of the water. In many cases, the LARS also includes a heave

FIGURE 9.12
Launch and recovery system (LARS). (Courtesy of http://wikioceannetworks.ca/)

compensation system which reduces the dynamic loading factor on the umbilical and therefore gives the flexibility of working in a higher sea state.

9.2.2.4 System and Frame Construction

Conventional ROVs are constructed with a large flotation pack on top of a steel or alloy chassis to provide the necessary buoyancy. Syntactic foam is often used for the flotation. A tool sled may be fitted at the bottom of the system and can accommodate a variety of sensors. By placing the light components on the top and the heavy components on the bottom, the overall system has a large separation between the centre of buoyancy and the centre of gravity; thus providing stability and stiffness to do work underwater.

Electrical cables may be run inside oil-filled tubing to protect them from corrosion in seawater. Thrusters are usually located in all three axes to provide full control. Cameras, lights and manipulators are on the front of the ROV or occasionally in the rear for assistance in manoeuvring.

Most of the work class ROVS are constructed as described above; however, this is not the only style in ROV building. Specifically, the smaller ROVs can have very different designs each geared toward their own task. One company's ROV even has wings that allow the vehicle to move more efficiently while being towed and/or operating on thruster power in high currents.

The frame of the ROV provides a firm platform for the fixing of the necessary mechanical, electrical, propulsion components including special tooling such as sonar, cameras, lighting, manipulators and scientific sensor and sampling equipment. ROV frames have been made by many different materials from plastic composites to aluminium tubing. In general, the materials used are chosen to give the maximum strength with the minimum weight, since weight has to be offset with buoyancy which is very critical.

The ROV frame also has to comply with international regulations concerning load and lift path strength. In general, the local regulation has to be investigated before designing an ROV frame for use with an ROV.

9.2.3 Functions of ROV

Following are the tasks performed by a work class ROV (WROV) in the offshore oil and gas industry.

1. Diver observation – to ensure diver safety and aid while being used in conjunction with diving.
2. Platform inspection – to monitor the effects of corrosion, fouling, locating cracks, estimating biologic fouling and so on.
3. Pipeline inspection – to follow underwater pipelines to check for leaks, determine overall health of the pipeline and ensure the installation is acceptable.

4. Surveys – both visual and acoustic surveys are necessary prior to installing pipelines, cables and most offshore installations.

5. Drilling support – everything from visual inspection, monitoring installation, operational support and repair when necessary using multiple manipulators.

6. Construction support – a natural follow-up to drilling support. Tasks can become more complex with the use of manipulators, powered tools and cutters.

7. Debris removal – offshore platforms can become a 'trash dump' underwater. ROVs provide a cost-effective method of keeping the area clean and safe.

8. Platform cleaning – it is one of the most sophisticated tasks using manipulators and suction cups for positioning 100 hp system driving brushes, water jets and other abrasive devices.

9. Subsea installations – to support the construction, operation, inspection, maintenance and repair of subsea installations, especially in deep water.

10. Telecommunications support (inspection, burial, or repair) – from towed plows that bury cables for protection from trawlers and anchors to sophisticated vehicles that can locate, follow, retrieve and rebury subsea telecommunication cables.

11. Object location and recovery – search, location and recovery of lost objects.

A typical case of ROV's assistance in offshore drilling operation is presented below.

The first ROV operation is to perform a bottom survey of the drilling location. The equipment used for this is the scanning sonar and the TV camera. During this survey, debris, boulders and so on may be found. In such cases, the ROV is used to remove this debris from the location. The next ROV intervention is to assist during the lowering of the guide base and the installation of guide lines. All this work is done by use of the ROV manipulators. Once the guide base is installed, the next ROV intervention is to control the cement return. Cement may cover parts of the guide base or Christmas tree components and has to be removed. This removal is done by water jetting, utilising the ROV mounted water pump. The type of pump used will depend on the amount to be removed. During normal drilling operation, no ROV intervention is required. The ROV and the crew are kept in standby on the drilling platform. A typically unplanned intervention is to change a broken guide wire or to remove a seal ring. Guide wire change out is performed by the ROV, that is, removing the remaining piece of broken guide wire by use of a wire cutter and then installing a new guide wire. For the seal ring, a special

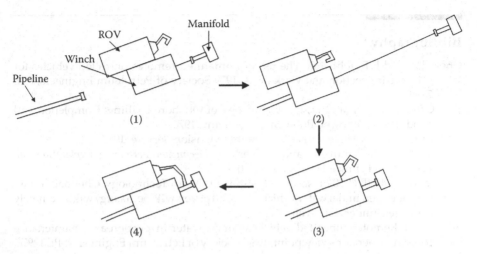

FIGURE 9.13
ROV attaching a subsea pipeline to a manifold.

purpose-built tool is used. At the completion of the drilling operation, the ROV performs a bottom survey.

Also, a *task of attaching a subsea pipeline to a subsea manifold* performed by an ROV is schematically presented in Figure 9.13. The sequence of operation is listed as below:

1. The ROV hooks its winch to the manifold.
2. Then it backs up by unwinding the winch and holds the pipeline end.
3. After that the ROV winches the pipeline into place, aligns it with the manifold hub and moves it into final position.
4. Lastly, the ROV's articulated arm or the manipulator closes the clamp, inspects the seal and tests the integrity.

Though all the above listed tasks are performed by ROV, it cannot work by itself without being operated from the surface, that is, the vessel deck. It is the operator who maneuvers the vehicle with a joystick watching screens on a console that relay information from sonar and the camera on the ROV as well as the transponders on the seabed or equipment being serviced. These operators are highly skilled who manoeuvre the vehicle under the water where many obstacles like equipment, flowlines, jumpers, flying leads, umbilicals and so on are present. So, these operators get training on simulators to establish their proficiency. Their job is similar to that of an air traffic controller, although complexity in operation is more in the case of ROV operators.

Bibliography

Chew, J. J. and J. R. Johnston, The use of unmanned remote-controlled vehicles for offshore inspection and work tasks, 1979 Society of Petroleum Engineers, SPE 7769.

ETA Offshore Seminars Inc., the Technology of Offshore Drilling, Completion and Production, Pennwell Publishing Company, 1976.

Haggard, R. *Diving & Equipment*, Petroleum Extension Service, 1982.

Leffler, W. L., R. Pattarozzi, and G. Sterling, *Deepwater Petroleum Exploration and Production*, Pennwell Corporation, 2011.

Manual on ROSUB 6000 of National Institute of Ocean Technology, Chennai, India.

Remotely operated underwater vehicle, 2016, http://en.wikipedia.org/wiki/Remotely -operated-underwater-vehicle.

Simpson, J. Remotely operated vehicle as underwater inspection and maintenance tools: An operator's viewpoint, 1984. Society of Petroleum Engineers. SPE 12976.

10

Health, Safety and Environment

Health, safety and environment (HSE) are of major concern in any industry. However, the history of the oil and gas industry, in addition to the inherent difficulties and dangers in exploration and production (E & P) activities, make it much more susceptible to questions raised by different affected parties like local authorities and environmentalists. The HSE component has now become an integral part in the management of oil and gas companies. From the beginning of the E & P activities, there had been many incidents and accidents which had severe impact to the environment, ecology and humans and which drew the attention of the whole world community.

The most recent example is the famous Macondo disaster which took place on 20th April, 2010 in a deepwater drilling platform operated by BP on the deep offshore of the Gulf of Mexico. The blowout occurred due to operational, technical and regulatory problems, causing an explosion that killed several people. Over three months, the uncontrolled well resulted in the largest marine oil spill in the history of the oil and gas industry. It affected the marine flora and fauna along with the tourism and fishing industries. The harm caused to the environment is still not quantified.

Whenever there is talk on *health* in HSE management, it has altogether a different connotation. According to a statement made by Pericles, an eminent Athenian politician in the year 430 BC, which is still very much pertinent even today, 'Health is more than the absence of disease! It is a moral, mental and physical well-being that allows the individual to face any challenge in life!' It is not enough to cure or avoid illness, rather, it is fundamental that the people are provided with an atmosphere that ensures their well-being. In terms of issues related to health, this refers to the response to accidents and illnesses, non-exposure to harmful environments such as smoke, noise and so on, ergonomic conditions and the well-known stress. For any operating company, it is the first and foremost duty to look after the well-being of its people.

In case of *safety* in HSE management, it is the policies and procedures adopted to ensure safety in a very large and complex operation such as E & P. Earlier safety was perceived as a set of indicators for lost time and accidents and was not treated so seriously as a common objective for the company but recently companies have started to assess how best to define their safety measures which is given the highest priority and are followed very seriously step by step. The aim is now to establish general objectives and leave to the staff the task of defining and implementing the best ways of achieving these objectives within certain parameters.

The *environment* is undoubtedly one of the aspects that confers greater visibility to the oil and gas industry. The type of accidents that are common in E & P activities can be broadly grouped into three categories:

1. Oil spills, both on land and at sea
2. Gaseous emissions to the atmosphere
3. Changes in the Earth's surface, from the rainforests to the deserts, as well as residential areas

The impact of a serious accident or any such action may extend over years, decades, or even centuries like the Hiroshima or Nagasaki nuclear tragedy in Japan and Bhopal gas tragedy in India. With the growing importance of the oil and gas industry in the world energy scenario, it is crucial to address the environmental problems in a systematic manner and transform this industry into an effective contributor to the sustainable development of the world we live in.

These three components are now being addressed separately limiting the discussion to within offshore oil and gas operations.

10.1 Occupational Health Hazards and Risks and Their Control and Mitigation

Occupational health is about *protecting the physical and mental health of workers* and ensuring their welfare in the workplace. It deals with the relationship between someone's state of health and his or her job. This is very wide ranging, but a priority is to prevent ill health arising out of conditions at work. Other important aspects include ensuring fitness to perform a job safely; providing first-aid and emergency medical services; health education and promotion; and rehabilitation after illness or injury.

All these aspects are relevant to the offshore oil and gas industry, and across this industry, each year many more people become ill as a result of their work than are killed or injured in industrial accidents. Most diseases caused by work do not kill, but they can involve years of pain, suffering and discomfort for those affected. They include respiratory disease, dermatitis and hearing loss caused by excessive noise.

Though health and safety overlap to a great extent, it is important to recognise that health issues are often much less clear-cut than safety issues. For example, the effects of being hit on the head by a falling brick are immediate and visible. There is no doubt what caused them. However, if the same brick is ground to a very fine powder which is inhaled over a long period, the possible result (lung disorder) may not be seen for years. It may be difficult to tie the result back to the brick dust. So, health problems may not be as obvious as a safety failure such as a structural collapse, a machinery accident or fire and explosion. Most people may never see cases of occupational ill health while at work. They may miss the connection between the effect and its cause. So, it is even more important to adopt a proactive approach to managing health issues.

In the offshore industry, the management of health issues is integrated into the installation's safety management system (SMS). This ensures that health objectives are built into existing control systems, such as permits to work and audits. Even more effective is to assess potential occupational health problems at the design stage. It is much easier to design out a health problem, or design inadequate controls, than to put right a problem once everything is in place. The discussion on occupational health is made under four categories:

1. Identifying *hazards* to health (a hazard is anything with the potential to cause harm, such as chemicals, noise, or vibration).
2. Assessing the *risks* posed by the hazards identified (a risk is the likelihood, great or small, that someone will actually be harmed by the hazard in the circumstances in which it is met).
3. Introducing measures to ensure that the health risks are adequately *controlled* (i.e. that they are reduced to the lowest level that is reasonably practicable) and checking that the measures remain effective.
4. Making arrangements to *mitigate* the effects if the control measures fail or if problems arise which could not have been foreseen.

The main health hazards found in the offshore oil and gas industry can be divided into the following five broad groups:

1. Chemical hazards
2. Physical hazards such as noise, vibration, or radiation
3. Biological hazards such as microorganisms
4. Ergonomic hazards such as manual handling
5. Psychosocial hazards

Tables 10.1–10.5 give examples of each of these groups of hazards that can be found in offshore. The tables describe briefly the possible short-term (acute) and long-term (chronic) adverse health effects of each example.

10.1.1 Health Risk

There is an important difference between *hazard* and *risk*. A hazard is something that could do harm (such as a toxic chemical), but the risk is the actual chance that someone will be harmed by the chemical. Risk depends on the

TABLE 10.1

Chemical Hazards

Agent	Example	Source or Use	Potential Health Effect
Toxic substances	Hydrogen sulphide	Sour gas production, bacterial activity in stagnant water	Acute – respiratory irritation, chemical asphyxiation
	Benzene	Component of crude oil, concentrated in gas dryer vent emissions	Acute – central nervous system effects Chronic – anaemia, leukaemia
	Methanol	Gas drying and hydrate control	Acute – narcotic effects, damage to the optic nerve
Corrosive substances	Hydrochloric and	Well stimulation	Acute – skin, eye and respiratory irritation, burns, ulceration, tissue destruction
	Sodium hydroxide (caustic soda)	Drilling fluid additive	Acute – skin, eye and respiratory irritation, burns, ulceration, tissue destruction
	Sulphuric acid	Wet batteries, regenerant for reverse osmosis water makers	Acute – skin, eye and respiratory irritation, burns, ulceration, tissue destruction
Irritant substances	Manufactured mineral fibre	Thermal insulation and construction materials	Acute – skin irritation, dermatitis Chronic – possible lung cancer risk
	Cement dust	Oilwell cementing	Acute – skin, eye and respiratory irritation, dermatitis
	Sodium hypochlorite	Injection water treatment	Acute – respiratory irritation
Sensitising substances	Isocyanates	Two pack paints	Acute – respiratory and skin allergic reactions after sensitisation
	Glutaraldehyde	Biocide for water treatment systems	Acute – respiratory and skin allergic reactions after sensitisation
	Terpenes	Degraded d-limonene based degreasers	Acute – degradation fumes may cause sensitisation
Possible carcinogens	Asbestos	Thermal insulation and construction materials (encountered during removal)	Chronic – asbestosis, bronchial carcinoma, mesothelioma
	Polycyclic aromatic hydrocarbons	Used engine oils	Chronic – possible skin carcinomas

Source: Oil Industry Advisory Committee.

TABLE 10.2

Physical Hazards

Agent	Example	Source or Use	Potential Health Effect
Noise	Machinery noise	Engine rooms, compressor rooms, drilling brake, air tools, machine shop (metal grinding)	Acute-temporary threshold shift Chronic – noise-induced hearing loss
	Relief valve noise	Actuation of relief and emergency blowdown valves	Acute – physical damage to inner ear, including burst eardrum
	HVAC noise	Accommodation ventilation systems	Chronic – sleep disturbance, irritation
Vibration	Hand tool vibration	Needle gunning	Chronic – hand-arm vibration syndrome
Ionising radiation	Sealed sources	Radiography, well logging, densitometers, interface instruments	Acute – radiation burns, radiation sickness Chronic – blood disorders, leukaemia, carcinomas
	Naturally occurring radioactive material	LSA scale in tubulars and process plant	Chronic – possible lung disorders
Non-ionising radiation	Ultraviolet	Arc welding, sunshine	Acute – sun burns, keratoconjunctivitis Chronic – melanomas
	Infrared	Flares	Acute – sun burns, blistering Chronic – cataracts
	Electromagnetic fields	Transformers, power cables	Acute – induced body currents, static shocks
Thermal extremes	Heat	Wellhead area, near the flare, on the monkey board under certain conditions	Acute – sweating, fatigue, skin burns, heat stress
	Cold	Open modules in winter	Acute – shivering, hypothermia

Source: Oil Industry Advisory Committee.

TABLE 10.3

Biological Hazards

Agent	Example	Source or Use	Potential Health Effect
Water-borne bacteria	*Legionella pneumophila*	Cooling systems,+ domestic water system	Acute – Legionnaires disease
Food-borne bacteria	*E. coli*	Contaminated food	Acute – gastro-intestinal disorders, food poisoning

Source: Oil Industry Advisory Committee.

TABLE 10.4

Ergonomic Hazards

Agent	Example	Source or Use	Potential Health Effect
Manual handling	Moving equipment and containers	Pipe handling on drill floor, sack handling in sack store, manoeuvring equipment in awkward locations	Acute – back spasm, torn ligaments Chronic – slipped disc, back pain
Workstations	Offices with display screen equipment	Poorly designed office furniture and poorly laid out workstations	Acute – headaches, muscular discomfort Chronic – upper limb disorder

Source: Oil Industry Advisory Committee.

nature of the hazard and the extent to which workers are exposed to the hazard and for how long.

The relationship between hazard and risk, is given as below:

Health risk = severity of the hazard × potential for exposure

Severity of the hazard means the capability of the agent to do harm.

For example, (1) the toxicity of a chemical; (2) the sound pressure level at the operator's ear and the frequency of a noise source; (3) the intensity and penetrability of ionising radiation and (4) the mass of a load.

In some cases, the severity of the hazard may be greater for some workers who are more susceptible to its effects. For example, they might be sensitised to certain chemicals. This means it is necessary to introduce extra measures to protect them. First, though, one should deal with the risks to most people.

Potential for exposure means the opportunity given to the agent to do harm. For example, the form in which a chemical is used (e.g. powder, granules); the length of time of each period of exposure; the frequency of the exposures and the effectiveness of the controls in use.

So, when a health hazard with enough severity to cause harm is on a work-site and there is a potential for exposure of a worker to the hazard, there is a risk to the health of the worker.

It may be possible to identify a risk, but it may not be big enough to cause concern. It is not possible to remove all risk from everything being done. Instead, the law requires that the level of risk is kept as low as is reasonably practicable. There is guidance on what this means for different circumstances, but it recognises that a minimum level of risk can be acceptable.

The level of risk is indeed by risk assessment. This does not have to involve complex calculations. The law requires quantitative risk assessment (QRA)

TABLE 10.5

Psychosocial Hazards Related to Work

Agent	Example	Source or Use	Potential Health Effect
Stressors connected with work organisation and job content	Organisation and job design	Ambiguity of job requirements, unclear reporting relationships, over-supervision, under-supervision, constant paperwork	Acute – worry, interrupted sleep, excessive smoking and eating, fatigue, argumentativeness and acrimony
	Work planning issues	Work overload or underload, unrealistic targets, lack of clear planning, poor time management, atmosphere of uncertainty due to constant change, poor communications, inadequate training	Chronic – possible serious physical and mental ill health, particularly if pressures are intense. Stress has been associated with high blood pressure, heart disease, thyroid disorders and mental problems including anxiety and depression
	Relationships at work	Personality clashes, blame culture, age and background differences, constant personnel changes, bullying, sexual or racial harassment	
	Work environment	Noise, air quality, overcrowding, ergonomics	
	Working hours	Inflexible or overdemanding work schedules	
	Working and living in a hazardous environment	Mistakes can be catastrophic, vulnerable to mistakes of others, responsible for the safety of others	
Offshore location stressors	Artificial society	Male dominated society, rigid work and social hierarchy, limited leisure and exercise options, no alcohol	
	Lack of privacy	Shared cabins, same faces, same routine, feeling of claustrophobia and imprisonment	
	Living on the job	Difficult to turn off in leisure time Difficulty of escape in an emergency, Reliance on helicopter travel, adverse weather	

Source: Oil Industry Advisory Committee.

techniques to be applied only to certain major offshore risks, but not to occupational health risks. Guidance on risk assessment is available in detail from the HSE guidebook.

Health risk assessment is a continuing and evolving exercise. Assessments should be kept up to date. When a task changes, for example, by introducing new control measures, the assessment is repeated for the new circumstances.

10.1.2 Controlling Risks

Following are the generally accepted hierarchy of control measures to control any situation:

1. Elimination or substitution
2. Engineered controls
3. Procedural controls
4. Personal protective equipment

Following is an explanation of this hierarchy in more detail.

10.1.2.1 Elimination or Substitution

Stop using the offending process, substance or equipment or use it in a different form.

For example, use less toxic degreasers and rig washes; stop using carbon tetrachloride in laboratories; use mercury-free downhole samplers; replace noisy pumps and compressors with less noisy ones (when reasonably practicable to do so, such as during a major refurbishment) and use smaller sacks for mud additives.

10.1.2.2 Engineered Controls

Redesign the process or equipment, if reasonably practicable, to eliminate the release of the hazard; enclose the process or equipment to capture and absorb the hazard or release it in a safe place or dilute to minimise the concentration of hazard on release.

For example, install exhaust ventilation over shale shakers and mixing hoppers; install acoustic hoods and enclosures; use fume cupboards in laboratories and use automatic pipe handling apparatus on drilling rigs.

10.1.2.3 Procedural Controls

Institute work systems and procedures so that work is performed in a way that limits exposure to hazards.

For example limit work periods in hot environments; limit access to noisy modules; classify hydrogen sulphide areas and develop good housekeeping and chemical rotation procedures.

10.1.2.4 Personal Protective Equipment

Select, provide and ensure the use of appropriate protective equipment.

For example, respiratory protective equipment, such as facemasks and breathing apparatus; personal protective equipment, such as gloves, goggles and aprons; and hearing protective equipment, such as earmuffs and earplugs.

Once control measures are installed, it is to be ensured that they are used properly by:

1. Working procedures, work packs, platform instructions, permit to work systems and other means of managing work on the installation or barge.
2. Educating the workforce on the hazards and risks involved in their work and how the control measures will protect their health.

Both methods demand that the workforce receives adequate information, instruction and training. These are required by law. Good ways of doing this include providing workers with safety data sheets and workplace posters; making presentations at health and safety meetings; and one-to-one training given by supervisors or specialists.

Putting in place a regime of maintenance, examination and testing for each control measure will help to ensure that their performance does not drop off with time.

Having put an occupational health programme into practice, its monitoring and review is necessary which includes workplace monitoring, personal monitoring, performance monitoring and audit.

A health surveillance system is put into practice to look for early signs of ill health caused by work, if any. It includes keeping a health record for each worker at risk. If possible, this record should continue from job to job.

Health surveillance is mandatory in some cases, such as work with certain chemicals or with ionising radiation. It may also be appropriate, under the Management of Health and Safety at Work Regulations 1999, where there is a high level of exposure to other hazards.

10.1.3 Mitigation

The main aim of an occupational health programme is to prevent workers' health being affected by their job. But it is also necessary to act quickly if anything goes wrong, to minimise any ill effects. For example:

1. If health surveillance shows that a worker's exposure to lead is approaching the agreed limit, remove the worker from exposure before any harm is done.

2. If symptoms of minor ailments are detected, take action to prevent them from becoming major health problems.

While in offshore, workers cannot visit their ground facilities if a health problem arises.

Instead, medics and first-aiders should be provided, trained and equipped to deal with the full range of health problems that may arise. From time to time a worker may develop a serious health problem or suffer a serious injury. There should be arrangements to transport the sick or injured worker to shore promptly and receive medical attention. Acting quickly and efficiently will help to prevent the condition getting worse and improve the worker's chance of recovery.

Finally, workers who have recovered from illness or injury may still have difficulty in adjusting again to work, especially after a long period. They may need assistance or advice to hasten the day when they return to being a valued and productive part of the industry's workforce.

10.2 Safety Offshore

Safety is defined as 'control of accidental loss or hazards'. It involves constant awareness of critical work hazards through a constant improving system. Compared to onshore operations, the offshore operations call for greater awareness regarding safety because of the extra parameter, that is, the marine environment.

Regarding safety systems, this has been discussed in Chapter 6 (Section 6.2.2.2.1.3), which are common to both drilling and production. However, while working either in fixed or floating units, there are certain other aspects that need to be considered from a safety point of view.

Safety offshore has two main aspects. Firstly, the safety of the installation and secondly workers' attitude toward safety. The method of dealing with safety problems and accidents offshore are different because of the situations that occur offshore and nowhere else; for example, lack of space, helicopter and sea transport, rescue equipment and so on. The work is performed in a marine environment by workers who are often not used to working offshore. Then there are installations that are not designed to be continuously manned, where workers are required to do their job independently with a high degree of self-control and self-discipline. The attitude to safety must be reflected daily in the example the supervisors set for their staff. This is the only way by which the confidence of workers will develop into conviction.

In case of any emergency, workers may be required to abandon an offshore installation and take to the life boats or enter the water. Abandonment of an offshore installation even in controlled circumstances and relatively calm sea conditions is a hazardous operation. An even greater threat to life exists if abandonment takes place in adverse sea and weather conditions or during a blowout or fire. To help overcome these problems, personnel must receive training in the use of life-saving appliances and in the application of personal survival techniques.

Brief and common guidelines are given in the following sections to help the reader understand the factors mentioned above.

The guidelines are based on safe practices adopted by the offshore petroleum industry the world over. The safety provisions contained in the following documents have been referred to for this purpose.

1. IMO Code for Mobile Offshore Drilling Units (Resolution A-414. Xl dated 15th November, 1979)
2. International Convention for the Safety of Life At Sea (SOLAS) 1974
3. The Merchant Shipping (fire appliances) Rules 1969 under the Merchant Shipping Act 1958
4. API RP-14C for offshore production platforms
5. API RP-14G for fire prevention and control on offshore production

10.2.1 Safety in Offshore Drilling

While drilling from an offshore platform, the places or activities where safety is to be ensured are as follows:

Mud pumps and pits – These are located in between decks in an enclosed space. Gas cut mud from the well can cause hazard unless the ventilation is adequately maintained and there is continuous monitoring for CO_2 and H_2S. Proper ventilation systems should be provided along with suitable sensors for detecting the presence of CO_2 and H_2S.

Cranes and winches – All the safe practices followed for crane operations during lifting and handling heavy equipment with stringent inspection and maintenance schedule need to be followed here. Air winches should be operated only by persons who are adequately experienced.

Adequate means of communication between crane operator and his signalman should be provided; walkie-talkie sets may be used for this purpose.

No crane should be operated during the time a helicopter is landing or taking off. Suitable announcement should be made over the public-address system to this effect.

Walking surfaces, ladders, guard rails and so on – The passages most frequently used during normal operations should be marked by safety signs.

Emergency routes should also be marked but in a different manner. Except the helicopter landing deck, all decks, platforms, bridges, ladders and hatches (flush deck openings) should be surrounded by rigid and securely fixed guard rails.

Blow-out prevention and control – In offshore operations, the consequences of blow-out are much more severe than it is onshore. Therefore, the means of detection of kicks as well as their control are much more elaborate. The following safety practices regarding blowout need to be followed.

1. At the flow line a device called a flow sensor should be provided to detect any change in the rate of flow of mud. In case of any sudden increase in the rate of flow, it gives an automatic alarm at the driller's control panel and at the geologist's instrument cabin.

2. Continuous monitoring of mud volume going into and coming out of the hole, gas content in mud and rate of flow of mud should be made and indicated in an instrument panel by means of sensors.

3. It is absolutely necessary to use a 'trip tank' for measuring the amount of mud being pumped into the hole as the drill pipes are pulled out. This is a small calibrated tank with a measuring device consisting of a float connected to a weighted marker which can be clearly seen by the driller. The trip tank should be used together with a trip sheet in which the amount of mud required to fill the hole for any given number of stands pulled out is indicated. Thus, any variation in the calculated measured volume of mud is instantly detected.

4. In exploratory wells, a diverter system should be used when drilling below 30″ conductor casing. In development wells, the diverter system should be used unless geological information indicates an absence of shallow gases up to the depth of next casing. The diverter should be provided with two outlets for connection to the choke and kill line manifold.

5. In a well, after the surface casing is installed the blow-out preventer stack should be installed and maintained before resuming drilling.

6. There are three control panels for the BOP stack. One on the drill floor near the driller's stand, another at the accumulator (koomy) and the third in the tool pusher's cabin in the living quarters. The accumulator unit should be located outside the safety perimeter. The control panel should be equipped with pressure and flow indicator and suitable markings for closed and open positions.

7. The testing for surface control equipment, choke and kill manifolds, intermediate and production casing should be done as per the procedures laid out in the safety manual.

8. The different methods of well control should be carefully studied by the drillers so that they are familiar with the techniques of well control. Of course, the actual method to be adopted will depend on the condition of the well when it kicks.

9. Training in well control is a must for drilling personnel. Blow-outs occur because of one or more of the following reasons.

    a. The fact that the kick was detected too late

    b. Incorrect initial reaction

    c. Unsuitable surface equipment

In the process of controlling a kick, there are three main phases, firstly detecting the kick, secondly closing the well and thirdly regaining control of the well. In this process, the risk is highest during the first stage, that is, before closing the well. It is therefore important to train all categories of drilling personnel to react correctly in the event of kick and to remain constantly alert to detect the signs of a kick as early as possible.

Hydrogen sulphide – The risk constituted by hydrogen sulphide (H_2S) gas begins with drilling operations and is present in all the succeeding operations. It is highly toxic, heavier than air (1.2 times), smells like rotten eggs (only at low levels), burns with a blue flame to form sulphur dioxide, which is a corrosive gas, forms explosive mixtures with air, and is readily soluble in hydrocarbons and water. Hydrogen sulphide is hazardous to humans and at 10 ppm (0.001%) concentration in air, a person should not be exposed for more than 8 hours. If the concentration is more than 10 ppm, persons must use breathing apparatus because at higher concentrations, exposure even for a few minutes could prove fatal.

Another hazard is due to corrosion caused by H_2S on the surface of steel which leads to embrittlement cracks and failure of equipment.

H_2S detection and alarm system – Comprises H_2S sensors located at predetermined points, for example, in air conditioned or ventilated areas or on gas-carrying equipment (well nipple, shale shaker, mud pits, drillers stand, etc.).

It is a pure alarm system with two warning stages and cannot trigger an emergency shutdown alone. In case of high level H_2S alarm, the *visual warning signs* should be displayed to alert helicopters and vessels near the drilling rig.

Since level H_2S is heavier than air, it is likely to settle down at lower levels particularly in still air or in light winds and cut off the natural *escape route* to the boat landing. This situation gives rise to the following requirements.

1. Sufficient staircases on the upwind side of prevailing winds for escape route up-the-stairs or down to the lifeboat.

2. *Muster stations* for operating personnel in the event of a gas alarm, areas in the open on the upper deck which can be kept free of H_2S by the wind.

Forced air *ventilation* to disperse any accumulation of H_2S should be provided by fans (bug blowers) at the shale shaker, working platforms and control rooms.

Control of H_2S kick may be achieved either by pumping it away or circulating it out. The actual method to be adopted will depend upon the condition of the well, but in general all persons on the drilling floor shale shaker area, mud pump and tank should put on self-contained breathing apparatus when the kick is to be circulated out.

10.2.2 Safety in Offshore Production

The main objectives of the safety system in production platforms, fixed or floating, are

1. To prevent undesirable events that could lead to a release of hydrocarbon
2. Shut-in the process or affected part in the process, to stop the flow of hydrocarbon to a leak or overflow if it occurs
3. Prevent ignition of released hydrocarbons
4. Shut-in the process in the event of a fire

The system provides two levels of protection, that is, primary and secondary, to prevent or to minimise the effects of an equipment failure within the process.

Once a production platform with a suitably designed safety system is installed, the focus should be on the maintenance of the system so that the designed level of protection is available at all times.

Following are the guidelines for different safety systems in production operation.

Process safety system – This consists of primary and secondary protection provided by sensors and safety devices against undesirable events, as shown in Table 10.6. These sensors and safety devices should be periodically checked.

Well control safety system – In order to protect wellhead installations of producing wells, sub-surface safety valve (SSSV) is installed in a well, below the wellhead, designed to prevent uncontrolled well flow when actuated, in case the wellhead is sheared off or otherwise damaged. The SSSV valve is controlled from the surface by hydraulic means and hence is known as surface control sub-surface safety valve (SCSSV).

A surface safety valve (SSV) is also installed on the wellhead to shut down the well when actuated. These valves should be regularly checked and maintained in good working order.

TABLE 10.6

Protection Provided by Sensors and Safety Devices

Sl. No.	Undesirable Event	Primary Protection	Secondary Protection
1.	Over-pressure – pressure in excess of maximum allowable working pressure.	Pressure safety high (PSH) sensor to shut off any flow.	Pressure safety valve (PSV) to release pressure.
2.	Leak-accidental escape of fluid from pressure component to atmosphere.	Pressure safety low (PSL) sensor to shut off inflow and a flow safety valve (FSV) to minimise back flow.	Emergency shut down system
3.	Liquid over flow-discharge of liquid from gas outlet.	Level safety high (LSH) sensor to shut off inflow.	Emergency shut down system
4.	Gas blow by-discharge of gas from process component to liquid outlet.	Level safety low (LSL) sensor to shut off inflow or the liquid outlet.	Safety device on downstream component.
5.	Under pressure – pressure in a process component less than the designated collapse pressure.	Vent or gas make-up system.	Another vent or pressure safety valve (PSV)
6.	Excess temperature – temperature above that for which a process component is designed to operate	Temperature safety high (TSH) sensor to shut off fuel supply	Temperature safety high (TSH) stack/media sensor

Emergency shutdown system – The ESD system (discussed in Chapter 6) provides a means for personnel to manually initiate platform shutdown when an abnormal condition is detected.

The *ESD stations* should be conveniently located but should be protected against accidental activation. The stations are generally located at helidecks, exit stairways, boat landings, muster stations, near driller's console during drilling and work-over and near the main exits of living quarters. ESD buttons should be located on decks.

In the following *emergency conditions*, the ESD system should be actuated by any person on the spot:

1. Fire on a platform
2. Leakage in main oil line
3. Blow-out
4. Leakage of gas in excess of 60% of lower explosive limit (LEL)

In the following cases, the ESD system should be actuated only if the Field Production Superintendent so decides:

1. Cyclone and severe weather conditions
2. Fire in a nearby installation
3. Oil spills around the installation
4. Collision involving the installation

Once every six months, the ESD system should be function-tested under the direct personal supervision of the Field Production Superintendent.

Gas detection system – Accumulation of combustible gases in the atmosphere in offshore platforms could create a threat to safety. The precautions are adequate ventilation, particularly of enclosed areas, installation of gas detectors at locations like air-conditioned or ventilated areas or on gas carrying equipment (e.g., wellhead installation, gas compressors).

Pollution prevention – In offshore installations one of the primary considerations is prevention of pollution of marine environment. Disposal of any contaminant which may pollute the marine environment should receive careful consideration. The contaminant could be liquids or solids containing liquid hydrocarbons, relatively high concentration of caustic or acidic chemicals, raw sewage and inedible garbage.

For this *containment of spill oil* and disposal of solid waste are the two important steps to be taken which is discussed in the next sections.

10.2.3 Common Safety in Drilling and Production

The safety systems that are common in both offshore drilling and production platforms are

1. Fire safety system
2. Lifesaving appliances
3. Helicopter transportation
4. Transfer of personnel and material
5. Safety in communication and electrical system
6. Safety in use of explosives and radioactive substances
7. Safety in case of emergency

Regarding (1) and (2), that is, the fire safety system and lifesaving appliances, these have been discussed in Chapter 6. The rest of the systems are dealt with below.

Helicopter Transportation

Helicopters are an operating necessity in offshore areas. In emergency as on other occasions, the helicopter is the main means of transport. Personnel are regularly transported to offshore installations by helicopters.

Though helicopters are quite safe for travel, the chance of an accident or fire involving a helicopter cannot be ruled out. It is therefore essential that safety rules are strictly observed in the operation of helicopters.

The helideck should have a non-skid surface like a nylon rope net. Adequate drainage facilities should be provided to prevent accumulation of liquids on deck. There should be no obstruction on the helideck itself and within 3 m of its perimeter, the closest super-structure above the height of the helideck should be provided with a red obstruction light. The perimeter of the helideck should be provided with a safety net of sufficient strength designed to prevent any person from falling overboard the deck. The outer edge of the net should not rise above the edge of the helideck.

The deck should have both a main and an emergency personnel access route. For a safe landing, the helideck should be painted with day markings and also be fitted with night lights. Also, the name of the offshore installation in fluorescent paint should be displayed on every offshore installation.

Proper firefighting arrangement should be made on the helideck.

All the requisite rescue tools for use in the event of an accident involving the helicopter should be provided at a site easily accessible from the helideck.

The *refuelling facilities* should be in a place that is isolated from any source of ignition and protected from possible impact from landing helicopters. All the precautions during fuelling should be taken to prevent a fire.

The *helicopter pilot* is responsible for the safe conduct of the flight and he alone decides whether the flight can take place. The pilot's authority includes command over his crew and passengers. Every passenger in the helicopter must abide by the instructions of the pilot.

Every helicopter should be provided with adequate number of safety equipment like safety belts, ear muffs, sea survival packs and life rafts.

All the workers going offshore as visitors should be made aware of safety rules for flying by helicopter over the sea.

Communication is very important; radio communication should be established between the installation and the helicopter before it lands on the installation or takes off from it.

At every offshore installation, a person should be designated as helideck attendant to carry out his assigned duties for safe landing or take-off of the helicopter.

Transport of restricted articles like explosives, compressed gases, flammable liquids and radioactive material which are required for use in oilfield operations are regulated under the IATA Restricted Articles Regulations.

The emergency procedure to be adopted in case a helicopter is to ditch in the sea should be carefully written down and the passengers should be instructed by the pilot about the emergency alarm as also the procedure to be followed.

Transfer of Personnel and Material

Off-loading of materials and workers from supply vessels to offshore installations and vice-versa in a heavy swell is always hazardous. It requires quite an exceptional level of coordination between the captain of the supply vessel and the platform crane operator. The latter only has a small 'time window' in which to start hoisting the load to avoid it being hit by the heaving deck of the vessel. This is the reason that the use of a personnel basket to transfer workers requires extreme care.

So far as the transfer of personnel is concerned, it can be done in two ways: (1) direct transfer and (2) transfer by personnel basket.

In fine weather, personnel should transfer directly on the boat landing platform by wearing life jackets. Boat landing platforms should be provided with ladders made of synthetic fibre rope and there should also be a 'swing rope' which passengers/crews should hold on to.

If sea is rough or wind does not permit vessels drawing alongside or if there is no landing platform, personnel should be transferred by unsinkable basket (see Figure 6.25).

As far as transfer of material is concerned, it should be undertaken after necessary permission has been obtained from the master of the vessel. The loads are placed on the landing net and are correctly slung before picking it up and stowing. The transfer should be done under the supervision of two supervisors, one each on the deck of the installation and on the vessel. These two supervisors should be able to communicate through walkie-talkie sets or megaphone.

Safety in Communication and Electrical System

An efficient communication system is a vital requisite for the safety of offshore installations and the personnel on board, not only during normal operations but also during unfavourable weather conditions and emergency situations. It is also of utmost importance for emergency medical treatment, as in the case of a critically ill patient, it is only by reliable radio-communication that the doctor or a specialist onshore can provide advice to the medical officer on the installation.

Different types of communication equipment have been discussed in Chapter 6. It is necessary that all such equipment be provided with all safety devices and those should respond and work perfectly at the time of necessity without any failure. All the precautions required to operate such equipment in a vicinity of interference by any other electronic device must be taken.

All sort of emergency communication arrangements should be made so that messages can be sent promptly from the installation to any other place like shore base, helicopters, offshore supply vessels, rescue vessels and so on.

It is important to maintain this equipment properly so that it remains always in working condition. The batteries should be fully charged daily. Any high voltage system should be provided with protective devices like circuit breakers and so forth.

In case of an electrical system, there are additional electrical equipment in offshore installation like central air conditioning and refrigeration plant, galley (kitchen) equipment, cooking appliances and laundry equipment. Moreover, the generation of electrical power is made in the same platform itself. Hence, extreme caution is needed with regard to equipment inside closed spaces, where any explosion or fire can have serious consequences. A strict adherence to the safety rules laid down by certifying agencies like American Bureau of Shipping or others should be made while designing and constructing all the electrical equipment and appliances.

There are different hazardous areas classified as Zone 0 to Zone 2 (refer to Chapter 6) and accordingly the safety devices like flameproof or intrinsically safe equipment should be used. Details on this will be available in any electricity rule.

Every offshore installation should be provided with a self-contained emergency source of electrical power, that is, either a generator or a storage battery sufficient to carry the required load. These emergency sources should automatically take over in case of failure of the normal source. Even if required, suitable arrangement must be provided for selective shut down of electrical systems with respect to the ventilation system, non-essential electrical equipment and so on.

Over and above the safety devices, precautionary arrangements must be made so that no chance is taken for any hazard from electrical equipment.

Safety in Use of Explosive and Radioactive Substances

Generally, the use of *explosives* is made during a 'perforation' operation. So, naturally starting from its transportation to the platform up to storage and handling, safety guidelines must be followed.

The explosives should be transported with proper packaging and with utmost care in a cargo which should display a red flag. Detonators should be carried in a locked compartment removed from other explosives.

For storage, only such quantities of explosives and detonators should be stored which are necessary to carry out immediate operations. Their containers should be locked and placed in different areas.

The perforation job should be carried out in daylight hours only and all other safety rules regarding handling and usage of explosives should be applicable here as well.

As far as radioactive substances are concerned, these are used during logging operations, NDT operation, pipeline leak detection or some other testing purposes.

The radioactive source container should be kept inside a suitably constructed chamber secured to the body of the installation which is located far away from the living quarters, the working areas and areas through which personnel frequently pass, so that no one is affected by its radiation. From storage to handling has to be made under the complete supervision of an authorised agency like Bhabha Atomic Research Centre in India.

Safety in Case of Emergency

Despite all prevention measures, there remains an element of risk in offshore installations. Emergency procedures are therefore designed to indicate the duties, equipment and instructions so that necessary action can be taken in an uncontrolled situation. Personnel unfamiliar with emergency procedures can make the situation worse, whereas those who are well trained take the right steps to prevent it from getting worse. The procedures must be checked through training and exercises. If these exercises show that the platform personnel do not have the necessary capabilities and practices in the use of the equipment and procedures more instructions and training will be necessary. The exercises also reveal whether the procedure is suited to the situation at hand. If this is not the case, the procedure should be revised. If the total approach to any given situation is correct and the guidelines are checked during exercises, the very knowledge that one has reviewed the various situations and the guidelines will generate much greater confidence in platform personnel to tackle any emergency situation.

It is important that all the personnel on offshore platforms are familiar with alarms and code of signals. They are aware of their duties to be performed at the time of emergency. For this, drills need to be regularly held. For example, fire drills should be carried out once every week. In relation to landing, swinging out and lowering of life boats (propelled)/survival craft, the drills should be undertaken once every two months.

Some of the duties considered as emergency duties which should be assigned to specific persons are closing of wells, valves, vents, fire doors and ventilators, water-tight doors, stopping of machinery and launching of life boats and life rafts. These emergency exercises should be carried out once every month. Beside all these exercises and drills, all the workers including visitors should be handed over essential instructions regarding safety printed on a small booklet or a card.

Detailed emergency procedures in the case of fire, abandonment of platform, H_2S emission and so forth are available in detail on every offshore platform and all workers must be familiar with these.

The recovery of survivors from the water is achieved by two principal means.

1. Rescue or pick-up boats
2. Rescue ships

The pick-up boats are robust, powered boats whose speed is over 26 knots which enables them to reach the survivors quickly, pull them on board and go where large vessels cannot approach.

The role of rescue ships is to recover persons from the sea and to provide first-aid treatment. Life-belts and life-rafts which are also secured to the rescue ships help the survivors to hang onto those from where the persons are brought on board the ship.

10.3 Environmental Pollution and Control

Environmental pollutions are linked to health issues in humans, animals and plant life. There are 7 kinds of pollution, which are as follows:

1. *Air pollution*, which means contamination of air by smoke and harmful gases. The probable causes are exhaust fumes, burning of fossil fuels, harmful off-gassing and radiation spills.

2. *Water pollution*, meaning contamination of any body of water like lakes, ground water, oceans and so on. The possible causes are raw sewage running into lakes or streams, industrial waste spills, radiation spills, illegal dumping of substances and biological contamination such as bacteria growth.

3. *Land pollution* means degradation of Earth's surface by improper disposal of waste. The probable cases of improper disposal are litter found on roads, illegal dumping in natural habitats, oil spills on land, use of particles in farming, damage and debris caused by mining activities and radiation spills.

4. *Noise pollution* is the loud sounds harmful or annoying to humans and animals. Examples are sounds caused by airplanes, helicopters, motor vehicles, construction or demolition noise and sporting events or concerts.

5. *Light pollution* means brightening of the night sky by improper lighting that affects human health and sleep cycle.

6. *Thermal pollution* means an increase in temperature caused by human activity. The examples are warmer lake water from nearby manufacturing plant, increase in temperature in areas with a lot of concrete construction or vehicles especially in urban areas or cities.

7. *Visual pollution* refers to anything unattractive or visually damaging to the nearby landscape. Examples are skyscrapers that block a natural view, graffiti or carving on trees, rocks or other natural landscapes, billboards, litter, abandoned homes, junkyards and so forth.

Out of these seven types of pollution, the most common ones are air pollution and water pollution in the case of offshore exploration and production. Although most environmental regulatory attention, as well as that of the popular media, focuses on water pollution related impacts of extraction, less well known is the fact that offshore platforms comprise a significant source of certain air pollutants, for example, in 2008 in U.S. Gulf Coast total emissions of nitrogen oxides (No_x) exceeded 60,000 tons while total release of volatile organic compounds (VOC) approached 50,000 tons.

So, the emissions of air pollutants like VOC_s, SO_x, NO_x, fine particulate matter and greenhouse gases like CO_2 and CH_4 are the major causes of air pollution in offshore E & P activities whereas discharges to water mainly oil in muds and cuttings during drilling and oil in produced water besides waste spills are the major causes of water pollution. The main effects of such emissions or discharges are as follows.

The principal effect of VOC_s is their local ambient ozone-forming potential in combination with nitrogen oxides and sunlight. Ozone can affect the respiratory system in humans and affect plant growth. Some VOCs, such as benzene and 1,3-butadiene, are also potentially hazardous to health if high concentrations occur.

Sulphur oxides (SOx) lead to acid rain which may affect higher life forms, such as fish. Acid rain can also lead to forest damage such as defoliation.

Along with VOC_s and sunlight, nitrogen oxides (NO_x) can combine to increase ambient ozone and cause photochemical smog, particularly where there is no air dispersion. NO_x can also cause acid rain. Inhalation of NO and NO_2 can affect the respiratory system directly.

Carbon dioxide is the predominant greenhouse gas which could bring about global climate change. Carbon monoxide increases the lifetime of VOC_s by atmospheric chemistry and can also produce ozone in its own right, although slowly. Methane (CH_4) has a major impact on global warming potential.

Most of the air emissions in E & P arise from the use of fuel or from controlled flaring and venting, which are necessary for safe operation. Almost half of the emissions are hydrocarbons, consisting predominantly of methane. The remaining emissions, principally NO_x, SO_x and CO, are produced during fuel combustion.

Oil contamination of the sea from exploration and production activities arises nowadays primarily from the discharge of 'produced water', which itself comes from the reservoir and is cleaned typically to 30 ppm before discharge. Historically, some oil was discharged in oil-based drilling 'muds', but this has greatly diminished recently. Oil spills to the sea from exploration and production typically account for less than 5% of the total oil discharged from all sources. For example, natural seepage in the North Sea is four times as great as oil spills from E & P activities.

Pollution can be seen as a waste product and environmental management has become a major part of the oil industry. Historically, environmental management has been predominantly 'end-of-pipe' pollution control but recently the focus has been shifting toward pollution prevention.

Pollution Control in Drilling and Production

All pollution control techniques are process specific. So, briefly these have been outlined against the respective cases.

Drilling mud – Traditionally, oil-based muds have been used. The main types of pollution control technologies are substitution by biodegradable synthetic muds and water-based muds, treatment of drill cuttings, for example, solvent extraction and thermal treatment process and reinjection of the ground-up cuttings into an impermeable formation. Also, ship to shore for waste treatment and disposal can be an option, which is discussed in the next section.

Produced water – Historically, efforts have been concentrated on the separation of oil and water and the key technologies are separators, hydrocyclones and produced water reinjection. The treatment of produced water has been discussed in detail in Chapter 6.

Air – Reduction of venting and flaring together with improved operational procedures and leakage minimisation are some of the most cost-effective technologies. Purge substitution or management, flare gas recovery, compression and reuse are other control measures. Flaring has been also discussed in Chapter 6.

Spill during transportation – On sea-going tankers, double-skin vessels are being used and commissioned and procedures are continually improving. Also, ballast is segregated to avoid discharge of oily ballast water. This has been dealt with in a subsequent section.

Vapour recovery – Some of the main sources of VOC_s come from tanker loading and unloading; the major control technologies are closed-loop systems and vapour recovery units, liquefaction by refrigerated cooling and membrane systems.

10.4 Handling, Treatment and Disposal of Wastes

The E & P activities have a great impact on the environment and the main impact comes from the release of wastes into the environment. These wastes include hydrocarbons, solids contaminated with hydrocarbons and water contaminated with a variety of chemicals. To minimise the adverse impact due to those wastes it requires an understanding of the complex issues that an upstream industry faces.

These issues concern operations that generate wastes, their potential influence on the environment, mechanisms and pathways for waste migration, effective ways to handle or manage wastes, treatment methods to reduce their volume and/or toxicity and disposal methods.

Wastes are generated from a variety of activities associated with petroleum production. These wastes fall into the general categories of drilling wastes, produced water and associated wastes. In general, produced water accounts for about 98% of the total waste stream with drilling fluids and cuttings accounting for the remaining 2%. Other associated wastes combined contribute a few tenths of a percent to the total waste volume.

Drilling wastes include formation cuttings and drilling fluids. Water-based drilling fluids may contain viscosity control agents (e.g., clays), density control agents (e.g., barium sulphate, or barite), deflocculants (e.g., chrome-lignosulfonate or lignite), caustic (sodium hydroxide), corrosion inhibitors, biocides, lubricants, lost circulation materials and formation compatibility agents. Oil-based drilling fluids also contain a base hydrocarbon and chemicals to maintain its water-in-oil emulsion. The most commonly used base hydrocarbon is diesel, followed by less toxic mineral and synthetic oils. Drilling fluids typically contain heavy metals like barium, chromium, cadmium, mercury and lead. These metals can enter the system from materials added to the fluid or from naturally occurring minerals in the formations being drilled through. These metals, however, are not typically bioavailable.

Produced water virtually always contains impurities, and if present in sufficient concentrations, these impurities can adversely impact the environment. These impurities include dissolved solids (primarily salt and heavy metals), suspended and dissolved organic materials, formation solids, hydrogen sulphide and carbon dioxide and have a deficiency of oxygen. Produced water may also contain low levels of naturally occurring radioactive materials (NORM). In addition to naturally occurring impurities, chemical additives like coagulants, corrosion inhibitors, emulsion breakers, biocides, dispersants, paraffin control agents and scale inhibitors are often added to alter the chemistry of produced water. Water produced from waterflood projects may also contain acids, oxygen scavengers, surfactants, friction reducers and scale dissolvers that were initially injected into the formation.

Associated wastes are those other than drilling wastes and produced water. Associated wastes include the sludges and solids that collect in surface equipment and tank bottoms, water softener wastes, scrubber wastes, stimulation wastes from fracturing and acidising, wastes from dehydration and sweetening of natural gas, transportation wastes and accidental spills and releases.

One common, but incorrect, perception of the petroleum exploration and production industry is that it is responsible for large-scale hydrocarbon contamination of the sea. The primary source of hydrocarbon release into the ocean is from transportation by tankers. Oil production from offshore platforms contributes less than 2% of the total amount of oil entering the sea.

The influence or the impact of these wastes is primarily reported in terms of toxicity to the exposed organisms. LC_{50} is the measure of the toxicity and is defined as the lethal concentration required to kill 50% of the population.

It is measured in micrograms (or milligrams) of the material per litre or parts per million (ppm) of air or water. Understandably, the lower the value of LC_{50}, the more toxic is the material. High values of LC_{50} mean that high concentrations of the substance are required for lethal effects to be observed and naturally it indicates low toxicity.

The aromatic hydrocarbons like benzene, toluene, xylene and ethyl benzene which are commonly found in the petroleum industry exhibit higher toxicity on the order of 10 ppm, whereas high molecular weight paraffins are essentially nontoxic. Toxicities (LC_{50}) of water-based mud can be a few thousand ppm whereas that of polymer-based mud exceeds 1 million, which indicates less than 50% of the test species would have died during the test.

The heavy metals as such can cause damage to the liver, kidney or reproductive, blood forming or nervous systems but such metals present in offshore drilling fluid discharges combine quickly with the naturally abundant sulphates in seawater to form insoluble sulphates and precipitates that settle to the sea floor. This process renders the heavy metals inaccessible for bioaccumulation or consumption.

Various chemicals used during production activities also affect the environment, their toxicities vary considerably from highly toxic to essentially nontoxic, however, the concentration of chemicals actually encountered in the field are below toxic levels.

In most cases, the environmental impact of released wastes would be minimal if the wastes stayed at the point of release; unfortunately, most wastes migrate from their release points to affect a wider area. Because migration spreads the wastes over a wider area, the local concentrations and toxicities at any location will be reduced by dilution.

10.4.1 Handling

As far as handling or managing the wastes are concerned, it is done stepwise as follows:

1. To minimise the amount and/or toxicity of the waste that must be handled, which is done by maintaining careful control on chemical inventories, changing operations to minimise losses and leaks, modifying or replacing equipment to generate less waste, and changing the processes used to reduce or eliminate the generation of toxic wastes.

2. To reuse or recycle wastes. If wastes contain valuable components, those components can be segregated or separated from the remainder of the waste stream and recovered for use.

3. Wastes that cannot be reused or recycled must then be treated and disposed of.

10.4.2 Minimisation of Toxicity and Amount

10.4.2.1 Chemical Inventory

One aspect of waste minimisation is to carefully monitor inventories of all materials at a site. Accurate, written records of all raw and processed materials and their volumes should be kept for every stage of handling and production. A detailed material balance can help identify where unwanted losses and waste may be occurring.

10.4.2.2 Changing Operations

Another important method for minimising the amount of potentially toxic wastes generated is to change the operating procedures at the various sites. Many changes can be made to improve operations at a relatively low cost, particularly if planned in advance.

A very important step in improving operations is to keep different types of wastes segregated. Waste streams should never be mixed. Because the toxicities and regulations vary for different wastes, keeping the waste streams segregated allows the best disposal options to be selected for each waste. This minimises the volume of toxic wastes that must be handled under the most stringent and expensive regulations.

Optimised drilling operations provide a significant opportunity for minimising wastes. Because the total volume of drilling wastes is controlled primarily by the hole size and well depth, the smallest diameter hole will minimise the volume of cuttings generated and drilling mud used.

Many operational changes during production can also be implemented to minimise the total volume of waste generated. Routine inspection and/or pressure testing of all tanks, vessels, gathering lines and flow lines should be scheduled. Routine inspection and/or automatic pumps should be installed in all sumps.

Unfortunately, the largest volume of production waste is produced water and little can be done to minimise its production. In some formations, coning of water can be minimised by dually completing a well in both the water and oil zone. This can limit water coning and reduce the amount of water produced with the oil.

10.4.2.3 Modifying Equipment

Another important method for minimising the volume of potentially toxic wastes generated is to ensure that all equipment is properly operated and maintained. Inefficient equipment should be replaced with newer, more efficient equipment.

One of the first steps to be taken is to eliminate all leaks and spills from equipment. Drip pans can be used beneath the drilling rig floor to catch all water or mud drained from it. Leaking stuffing box seals should be replaced or new stuffing boxes installed. Fugitive emissions from leaking valves, flanges and such fittings can be minimised by replacing leaking equipment.

Internal combustion engines should be properly tuned and the proper fuel should be used.

If the volume of waste generated cannot be sufficiently reduced with the existing equipment, newer equipment should be installed. Important environmental features of newer equipment should be how easy they are to monitor and clean up, as well as how they facilitate waste recovery and recycling.

Automatic shutoff nozzles and low-volume, high-pressure nozzles should be installed on all hoses on the rig floor and wash racks to minimise wastewater.

More efficient separations equipment should be used to separate solids, hydrocarbons and water. Newer shale shakers can be used that are better at filtering out small solids than older equipment. Low shear pumps should be used for produced water to prevent hydrocarbon droplets from decreasing in size, because small droplets are more difficult to remove. Improved backwash equipment and better procedures can be used to extend filter life.

10.4.2.4 Changing the Process

Another important method for minimising the amount of potentially toxic wastes generated is to use less toxic materials for the various operational processes. Many studies of material substitutions have been made in oil-based muds whereby the toxicity has reduced below that of traditional muds.

10.4.3 Reuse or Recycle

Many of the materials in drilling and production waste streams can be used more than once. If materials are intended for future use, they are not wastes. The following materials have a potential for reuse: acids, amines, antifreeze, batteries, catalysts, caustics, coolants, gases, glycols, metals, oils, plastics, solvents, water, wax and some hazardous wastes.

Water has a considerable potential for reuse. For example, produced water, after treatment, can be reinjected for pressure maintenance during water floods or for steam injection in heavy oil recovery.

Material reuse can be facilitated by installing equipment that allows reuse. For example, closed-loop systems can be installed so that solvents and other materials can be collected and reused in plant processes. Reusable lube oil filters can be installed in some applications instead of throwaway filters. Flared natural gas can be reinjected for pressure control, or an alternate use for it can be found. Flaring should be restricted to emergency conditions only.

10.4.4 Treatment

Most waste treatment processes involve separating a waste stream into its individual components, that is,

1. Removing suspended or dissolved hydrocarbons and solids from water (i.e., treatment of water)
2. Removing hydrocarbons

As far as treatment of water is concerned, the basic principles of removing suspended oil are gravity separation, chemical treatment, heating, centrifugal force, filtration, or gas flotation which are discussed in Chapter 6. To remove the dissolved hydrocarbons, processes like adsorption in a bed of activated charcoal, volatilisation by lowering the partial pressure of the dissolved volatile organic carbon (VOC), biological treatment, precipitation, UV radiation or oxidation can be adopted.

Consequent to the separation of suspended or dissolved hydrocarbons, the solids either suspended or dissolved are also separated by the same common methods. The general methods adopted for separating suspended solids are gravity separation, filtration, coagulation, or flocculation and those for dissolved solids like salts and heavy metals are ion exchange process precipitation, reverse osmosis, evaporation/distillation or biological processes.

The solids like drill cuttings produced solids which get contaminated with hydrocarbons need to be cleaned before disposal. So, there are a variety of treatment methods available which are briefly discussed below.

10.4.4.1 Washing

One of the least expensive ways to remove most of the hydrocarbons from solids is to wash them. The solids can be entrained in a fluidised bed of upward-flowing, high-velocity water. This stream agitates the solids and opens the pore system to release the oil. The efficiency of this process can be enhanced by adding a surfactant (soap) to the water to lower the interfacial tension holding the oil to the solids. Washing is more effective in sandy soils containing low amounts of clay.

10.4.4.2 Adsorption

Another relatively low-cost method of removing some of the hydrocarbons contaminating solids is to mix the soil with a material that is strongly oil-wet, like coal or activated carbon.

A suspension of contaminated soil and the carbon can be tumbled in water at elevated temperatures to allow the oil to be absorbed by the carbon. The oily carbon is then separated from the water and clean sand by flotation.

10.4.4.3 Heating

Heating cuttings contaminated with hydrocarbons can help separate the hydrocarbons from the solids, particularly when being washed in water. This procedure is similar to using heat to break emulsions and separate hydrocarbons and water. Heating can also be used for hydrocarbon sludges.

10.4.4.4 Distillation/Pyrolysis

A more expensive method for removing light- and intermediate-weight hydrocarbon compounds is to distil them from the solids in a retort furnace. The solid/hydrocarbon mixture is heated to vaporise the light and medium molecular weight hydrocarbons and water. The gases are removed from the high-temperature chamber by either a nitrogen or steam sweep.

Distillation systems, however, have certain limitations like a fire hazard, corrosion problems and emission of pollutants.

10.4.4.5 Incineration

Another way to remove hydrocarbons from solids is to burn the mixture in an incinerator. Incinerators are specially designed burners that can burn the relatively small volume of combustible materials found in oily solids. Following combustion, the resulting ash, including any salts and heavy metals, is solidified to prevent leaching of any hazardous residue. Incineration typically removes over 99% of the hydrocarbons in the soil.

A significant limitation to incinerators is that they emit air pollutants, particularly metal compounds like barium, cadmium, chromium, copper, lead, mercury, nickel, vanadium and zinc.

10.4.4.6 Solvent Extraction

Solvent processes can also be used to separate hydrocarbons from solids. In these processes, a solvent with a low boiling point is mixed with the oily solids to wash the oil from the solids and dilute what remains trapped. The solvent is then separated from the hydrocarbons and solids by low-temperature distillation and reused. This method is expensive.

10.4.4.7 Biological Processes

Most hydrocarbons encountered in the upstream petroleum industry can be biologically converted to carbon dioxide and water by microbes like bacteria and fungi. During biological degradation, the hydrocarbons are eaten as food by the bacteria. This biological degradation can be enhanced by providing the optimum conditions for microbe growth. The deliberate enhancement of biological degradation is called *bioremediation*.

10.4.5 Disposal

Despite various processes like reuse or recycling of wastes and their treatments, wastes have to be disposed of. In some cases, the final disposal may be made on-site whereas in other cases, the wastes may be shipped for disposal off-site. In either case, these wastes can be disposed of above or below the surface of land or water. The choice of locations will vary depending on the type of wastes to be disposed. Here mainly the waste generated in offshore has been dealt with.

10.4.5.1 Disposal of Liquids

Wastewater can be discharged directly into the ocean as long as its quality meets regulatory standards, that is, its concentration of suspended and dissolved solids, chemicals and hydrocarbons is sufficiently low. Surface discharge is regulated in most areas, however, and permits for such discharge are required.

When wastewater is discharged offshore, the water is typically treated to remove only the hydrocarbon. The disposal pile is discussed in Chapter 6. Although the dissolved solids (salt) concentrations of most produced waters are high enough to be toxic to even marine life, the rapid mixing and dilution of the discharged water makes the resulting environmental impact negligible.

For near-shore discharges in shallow water, there is less opportunity for mixing and dilution of the discharged water, and a toxic plume can exist for some distance away from the discharge point. Such toxic plumes are of particular concern when discharging a dense, high-saline, oxygen-deficient brine because it can be trapped in subsurface topographic low areas. Because this trapped brine can significantly impact the local marine life, permits to discharge high-salinity brines near the shore may be difficult to obtain, even if the hydrocarbon content is low.

The other disposal methods for waste liquids, such as produced water, is to inject them into a subsurface formation through injection wells after mixing with treated water for pressure maintenance purposes.

10.4.5.2 Disposal of Solids

Waste solids can be discharged directly or into the ocean as long as their quality meets regulatory standards, that is, the concentration of contaminants like hydrocarbons and heavy metals is sufficiently low. Because such discharges are regulated, permits are required in most areas.

Offshore discharges of treated solids, such as drill cuttings and produced solids, are permitted in some areas. Offshore discharges, however, are prohibited within three miles of shore in the United States, and the discharge of oil-based drilling mud wastes are prohibited in all U.S. waters. Where offshore discharges are prohibited, waste solids must be transported to shore for disposal. This is generally more expensive than offshore treating and discharge.

10.5 Oil Spill – Containment and Recovery

In the offshore drilling and production operations, one of the major impacts to the neighbouring environment results from offshore releases of oil and the common cases are blowouts, riser or pipeline failure and tanker accidents. The oil slicks get carried to large distances and finally reach the shoreline if not properly contained in time. Though over time, by the natural processes oil slicks get dispersed and destroyed yet an early response strategy will minimise the amount of oil that reaches the shoreline.

When oil is spilled on open water of a sea or ocean, there are many mechanisms that play a role to determine the fate and behaviour of the spilled oil. These mechanisms are described below.

10.5.1 Spreading

The principal forces influencing the lateral spreading of oil on a calm sea are

1. Gravitational force (causing decrease in film thickness)
2. Surface tension
3. Inertial forces
4. Frictional forces

An example of spreading is best illustrated by an experiment carried out by Warren Spring Laboratory. In this, 120 tons of light Arabian crude oil was discharged at sea and its appearance and dimensions were observed for 4 days. Figure 10.1 shows the outline development and subsequent breakup of the oil slick, particularly interesting is the fact that they show the difference between major and minor dimensions of an oil spill.

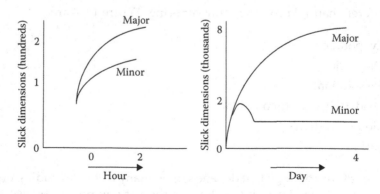

FIGURE 10.1
Slick development and breakup with time in major and minor dimensions.

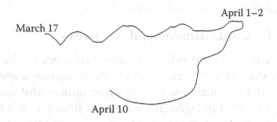

FIGURE 10.2
Drift perturbation of an oil spill.

10.5.2 Movement and Drift

Drift is a large-scale phenomenon and is a measure of the movement of the centre of mass of an oil slick. Nevertheless, wind direction and duration can have a dramatic effect on oil spill drift and trajectory and an example of wind-induced drift perturbations of an extensive near shore spill occurred after the wreck of Amoco Cadiz on 17th March 1978, which is illustrated in Figure 10.2.

Wind shift pattern which occurred on 2nd April caused an extensive oiling of previously clean coastal areas which had not been touched during the first 8 to 10 days after the spill.

Drift is also strongly influenced by waves and tidal currents. After the Amoco Cadiz spill, waves of 1.5–2 m and 6–7 m tides further affected the overall drift and the extent of the oiling of area. At the end of the first two weeks, a total of 72 km of coastline had been covered with oil, then following the dramatic wind shift change of 2nd April a total of `213 km were highly oiled and 107 km were heavily oiled.

10.5.3 Weathering

The different natural processes of weathering (Figure 10.3) are

1. Evaporation
2. Dispersion
3. Dissolution
4. Oxidation and photo-oxidation
5. Biodegradation
6. Sinking

The rate of weathering of oil depends upon many factors including oil type and ambient climatic conditions. Rate of the various processes also varies throughout the duration of an oil spill with many being greater in the first few hours.

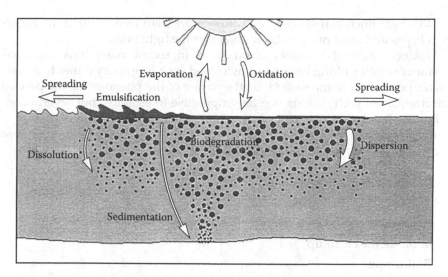

FIGURE 10.3
Different natural processes of weathering.

Evaporation – The most significant initial weathering process for oil spills on the sea is evaporation. Thus, the more volatile fractions of an oil (the light ends) are lost within the first few hours. Therefore, afterward the rate decreases and less volatile components will form a residue which has a higher specific gravity and viscosity than the original oil. The rate of evaporation is influenced by factors such as air and water temperatures, state of the sea, wind and rate of spreading.

Dispersion – Dispersion is of two types: (1) oil in water and (2) water in oil. Another important mechanism affecting the natural fate of oil on the sea surface is dispersion on the formation of oil in water emulsions. The rate of dispersion is a function of sea state and of the nature of the oil.

Dissolution – Some components of spilled oil are also lost through the process of dissolution. Whilst dissolution of the most suitable components in the oil will occur in the first few hours, other processes such as oxidation and biodegradation will constantly produce additional compounds that are water soluble.

Oxidation and photo-oxidation – Oxidation of certain hydrocarbons can also produce compounds that may act as emulsifiers and increase the natural dispersion of the spilled oil. The decomposition of oil can also take place by photo-oxidation when exposed to sunlight. As the process requires oxygen it tends to be a surface phenomenon and occurs most rapidly when the oil is spread into a thin film. The precise mechanisms are not well understood but it is thought that the process may be enhanced by trace metals such as vanadium but retarded by high sulphur content.

Biodegradation – Biodegradation has a significant effect on the removal of oil from the marine environment. Many species of marine bacteria, fungi and yeasts oxidise petroleum hydrocarbons by utilising them as a food source.

No single microbial species can utilise more than two or three hydrocarbon types and most preferentially consume the light ends.

Sinking – Several accounts of oil spills in recent years have reported instances of oil sinking because of increased specific gravity either to a midwater position or to the seabed. By the nature of the phenomena, the precise mechanism of such sinking is often impossible to establish but several contributory factors have been cited.

Beside natural processes of slick degradations, various other response strategies need to be adopted. Some of these are

1. Monitoring and evaluation
2. Use of chemical dispersants
3. Containment and recovery
4. Shoreline clean-up
5. Burning

All oil spills must be monitored, preferably by air from helicopter to determine their location and evaluate their potential threat. Oil spills which enter sensitive shoreline must be properly monitored to prevent environmental damage.

Chemical dispersants can be used to supplement the natural energy and help in the breakup of oil spills by breaking down the interfacial tension of the oil layer. Dispersants can be applied in a number of ways like 'boat mounted spray sets' or 'helicopter systems' or 'fixed wing systems'.

10.5.4 Containment and Recovery

This is the ideal strategy for oil spill response. To remove the pollutants from the environment it must be the most optimum response but unfortunately it is the most difficult strategy to achieve. The main problem is that the strategy is in direct conflict with the natural forces driving the oil spill. The sensitive locations threatened by advancing oil slicks can be protected with various kinds of equipment which are described below.

10.5.4.1 Booms

Booms are floating physical barriers to oil, made of plastic, metal, or other materials which slow the spread of oil and keep it contained. The efficiency of booms is limited to the conditions of sea, current speeds and so on. They are very useful in the near shore environment to protect sensitive areas (Figure 10.4).

Floating booms have four basic components:

1. A floatation chamber filled with air or some buoyant material.
2. A free board to prevent waves from washing oil from the top.

FIGURE 10.4
Floating boom. (U.S. Navy photo courtesy of Journalist 2nd Class Cassandra Thompson/ Wikimedia Commons/Public domain.)

3. A skirt to prevent oil from being swept underneath and a longitudinal support member to allow the boom to withstand the forces of winds, waves and current.

4. Weight to keep them perpendicular to the water surface.

There are three main types of boom. *Hard boom* is like a floating piece of plastic that has a cylindrical float at the top and is weighted at the bottom so that it has a 'skirt' under the water. If the currents or winds are not too strong, booms can also be used to make the oil go in a different direction (this is called 'deflection booming'). *Sorbent boom* looks like a long sausage made from a material that absorbs oil. If the inside of a disposable diaper is taken out and rolled into strips, it would act much like a sorbent boom. Sorbent booms don't have the 'skirt' that hard booms have, so they can't contain oil for very long. *Fire boom* is not used very much. It looks like metal plates with a floating metal cylinder at the top and thin metal plates that make the 'skirt' in the water. This type of boom is made to contain oil long enough that it can be lit on fire and burned up.

After containing the oil spill, the next job is to recover it. This is done by a *skimmer*.

An oil skimmer is a machine that separates a liquid from particles floating on it or from another liquid. A common application is removing oil floating on the sea surface. Skimmers can be towed, self-propelled, moored, or even used from shore. Many types of skimmers are available for use, depending on the kind of oil spilled and the weather conditions. Skimmers can be of four types.

1. Weir skimmers
2. Oleophilic skimmers
3. Suction/vacuum skimmers
4. Nonoleophilic/mechanical skimmers

10.5.4.1.1 Weir Skimmer

Weir skimmers function by allowing the oil floating on the surface of the water to flow over a weir. The height of the weir may be adjustable. Figure 10.5 shows how the funnel-shaped head of the weir allows the oil to flow into it and then the oil collected below is sucked into a storage tank.

10.5.4.1.2 Oleophilic Skimmer

Oleophilic (oil attracting) skimmers function by using a rotating element such as a drum, to which the oil adheres. The oil is wiped from the surface of the drum and collected. They are very efficient and do not pick up any appreciable amounts of water even when oil is not present. Oleophilic skimmers are distinguished not by their operation but by the component used to collect the oil. Ropes, discs, or drums are treated with a substance or otherwise manufactured to adhere to oil. The common one is a disk type (Figure 10.6).

10.5.4.1.3 Suction/Vacuum Skimmer

Vacuum skimmers (Figure 10.7) provide one of the most readily available spill response tools in the industry. An air pump is used to reduce the pressure in a storage tank, a suction hose is connected to the tank which is then used to collect the oil. The suction/vacuum skimmers work like a household vacuum cleaner. They suck the oil up from the water surface and store it in a tank like a vacuum bag.

10.5.4.1.4 Nonoleophilic/Mechanical Skimmer

Nonoleophilic skimmers are distinguished by the component used to collect the oil. A metal disc, belt, or drum is used in applications where an oleophilic material is inappropriate. Mechanical skimmers rely on physically picking up the oil from the surface of water by some mechanical means. Operation of a belt skimmer is shown in Figure 10.8. A comparative performance in terms of overall oil recovery efficiency and utility of these four types of skimmers are shown in Figure 10.9 against oils of different viscosity.

Shoreline clean-up is a very labour intensive, low technology process. There are generally three stages to the clean-up operations. Stage 1 deals with the

FIGURE 10.5
Weir skimmer. (Courtesy of Jason Radcliffe/Wikimedia Commons/Public domain.)

free floating or gross contamination at the shoreline. Stage 2 deals with the intermediate moderate contamination which involves techniques ranging from flushing to manual collection and Stage 3 deals with final clean-up, only when there is no danger of re-oiling.

The *burning* of oil spills has long been considered a response strategy from the earliest days of oil spill accidents. Weathering of the oil tends to make it more difficult to ignite. If local authorities and response experts agree then it may be considered to burn in-situ the oil slick or part of a slick before it

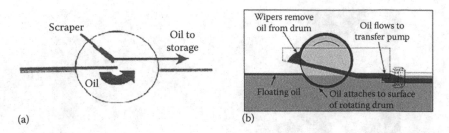

FIGURE 10.6
Oleophilic skimmer (a) disc skimmer (b) drum skimmer.

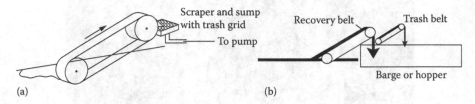

FIGURE 10.7
Suction/vacuum skimmer.

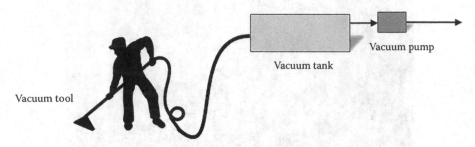

FIGURE 10.8
Belt skimmer (a) with trash grid (b) with trash belt.

reaches the coast. To do this, responders corral some of the oil from the slick in a fire-proof boom, then ignite it. This technique works best when the oil is fresh and the weather relatively calm.

Thus, it can be seen that there are a whole host of options available to the responder in dealing with a spill situation. It is important to fully understand and anticipate the future fate of the oil when selecting the response strategy. Once the desired option has been selected, the next stage is to ensure that the logistical support is available to sustain the strategy to enable it to be carried out along with the management structure to organise the response in such a way that objectives are clearly defined. Trained and experienced personnel are an absolute requirement for any response activity.

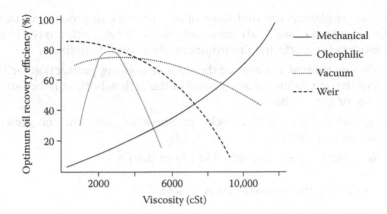

FIGURE 10.9
Performance of different skimmer.

10.6 Abandonment of Offshore Structures and Facilities

Nowadays, the problem regarding the abandonment of offshore oil and gas platforms is a matter for consideration not only to Greenpeace and other environmental institutions, but also a growing concern in the present policies of governments and oil and gas companies. This issue began in the summer of 1995 following the decision by Shell Brent Spar to abandon a floating storage tank located in the North Sea. The situation will worsen in the near future, when most of the major oil reservoirs in the world become depleted. It is estimated that in the near future approximately 500 large, steel and cement platforms will have to be removed. The cost of removal and abandonment is significant and agreements between the oil companies and governments will have to be reached, with the ever-present influence of environmental organisations, on the safest and most economical way of doing this. In fact, although it is overall an environmental issue, it is nonetheless also an economic issue. Here the technical aspects of the process of abandonment are discussed.

The entire abandonment process can be divided into seven discrete activities:

1. *Well abandonment*: the permanent plugging and abandonment of non-productive well bores.
2. *Preabandonment surveys/data gathering*: information-gathering phase to gain knowledge about the existing platform and its condition. Governing ministries or standards organisations should be contacted to determine permit and environmental requirements.
3. *Engineering*: development of an abandonment plan based on information gathered during preabandonment surveys.

4. *Decommissioning*: the shutdown of all process equipment and facilities, removal of waste streams and associated activities to ready the platform for a safe and environmentally sound demolition.

5. *Structure removal*: removal of the deck or floating production facility from the site, followed by removal of the jacket, bottom tether structures, or gravity base.

6. *Disposal*: the disposal, recycle, or reuse of platform components onshore or offshore.

7. *Site clearance*: final clean-up of sea-floor debris.

Each of these activities is discussed below.

10.6.1 Well Abandonment

Before the abandonment process starts, firstly all wells on the platform are to be permanently abandoned according to the recommended procedures. This means isolating productive zones of the well with cement, removing some or all of the production tubing and setting a surface cement plug in the well with the top of the plug approximately 30–50 m below the mudline. The inner casing string should be checked to ensure that adequate diameter and depths are available for the lowering of explosives or cutting tools.

To ensure no delays in structure removal, all well plug and abandonments should be completed several months prior to commencement of offshore decommissioning.

10.6.2 Preabandonment Surveys/Data Gathering

The preabandonment survey assesses the condition of the platform facilities and structure prior to beginning the abandonment. The survey should include the following:

1. *File surveys*. All available documentation concerning the platform design, fabrication, installation, commissioning, start-up and continuing operations are investigated. The file survey will familiarise the project engineer with the other appurtenances to the platform facility such as living quarters, process equipment, piping, flare system and pipelines and any additions/deletions or structural repairs to the jacket or the topside since the original installation.

2. *Geophysical survey*. Depending on the results of the file survey, the engineer may choose to have additional data gathered by means of side-scan sonar. This survey will indicate the amount of debris on the sea floor. In the case of deep-sea disposal, the sonar can determine if there are any obstructions at the dump site. Proximity of an

available dump site or 'rigs to reef' site, water depths and obstructions along the tow route are investigated as part of the geophysical survey.

3. *Environmental survey.* This consists of an environmental audit of the offshore platform to identify waste streams or other government controlled materials. At this time, items such as naturally occurring radioactive materials (NORM), asbestos, sludges, slop oils and hazardous/toxic wastes are identified and quantified.

4. *Structural survey.* A structural engineer can use observation and non-destructive ultrasonic testing techniques to evaluate the structural integrity. Items inspected will include condition and accessibility of lifting eyes, obstructions on the deck which may require removal and interfaces cutting for disassembly. The platform legs should be checked for damage that may obstruct explosives or cutting tools from accessing the proper cutting depth.

10.6.3 Engineering

Upon completion of preabandonment surveys, a strategy for decommissioning and abandonment can be developed. The engineering phase takes all of the data previously gathered and pieces it together to form a logical, planned approach to a safe abandonment. Of major concern during the development of this strategy is the safety of the operations. As with all offshore operations, there exists a high potential for accidents involving bodily injury or loss of life and the accidental discharge of soil and flammable, corrosive or toxic material into the environment; a risk analysis for all phases of the decommissioning should be performed.

After all the safety and environmental aspects of the project have been considered, details of the salvage process need to be identified. The sequence of process equipment and structure decommissioning and the salvage and disposal methods needs to be determined. Any required government permits should be submitted for approval.

A major determination for an effective and efficient abandonment program is proper selection of the salvage equipment. Normally derrick barges of higher capacity (135–6500 tonnes) are preferred for lifting maximum weights of components.

10.6.4 Decommissioning

A primary objective during the decommissioning is to protect the marine environment and the ecosystem by proper collection, control, transport and disposal of various waste streams. Decommissioning is a dangerous phase of the abandonment operation and creates the possibility of environmental pollution. Decommissioning and removal or abandonment in place should be carried out by personnel who have specific knowledge and experience in safety, process flows, platform operations, marine transportation, structural systems and pipeline operations.

The sequence of decommissioning the process system, utilities, power supplies and life support systems is important. The platform's power, communications and life support systems should be maintained for as long as practicable to support the decommissioning effort. Process systems throughout the platform will have to be flushed, purged and degassed in order to remove any trapped hydrocarbons.

Platform decommissioning will result in large amounts of waste liquids and solids. Where possible, waste liquids can be dealt with most cost-effectively by placing them in existing pipelines and sending them to existing operating facilities. If no ongoing operations are available, then the waste streams will have to be pumped into storage containers and transported onshore for disposal or recycling. The constituents of the waste stream will dictate the cost of disposal.

After the process piping and vessels have been cleaned and it has been determined that there is no future utility for the pipelines, pipeline decommissioning should commence. Pipelines departing the platform will either board another platform or commingle with another pipeline via a sub-sea tie-in. A surface to surface decommissioning is the least costly to perform. This requires pigging the line to vacate any residual hydrocarbons followed by flushing with one-line volume of detergent water followed by final rinsing with one-line volume of sea water. Upon completion of the pipeline purging operation, pipeline ends should be cut, plugs inserted and the ends buried below the sea-bed. In the case of a sub-sea tie-in, details of the sub-sea tap will have to be obtained so that pipeline decommissioning plans can be developed. The flowline can be pigged, flushed and disconnected if the receiving platform can accept the fluids; otherwise, the pipeline segment will have to be isolated from the adjoining trunk line and then decommissioned.

10.6.5 Structure Removal

The method of a structure removal will be determined by the structure design, availability of removal equipment, method of disposal and the legal requirements governing the jurisdiction in which the abandonment is to take place.

The majority of structures in moderate environments will be totally removed. Most regulatory bodies throughout the world require that the structure be removed anywhere from the mudline to 5 m below. The chief consideration when developing a removal procedure is to determine if the piles or well bores will be severed using explosive or nonexplosive methods.

(a) *Removals using explosives.* Severing platform piles and well bores with explosives is relatively effective compared with using nonexplosive methods, as multiple cuts can be made in a short period of time. This limits the amount

of time that removal support equipment must be on the site and limits personnel exposure to unsafe working conditions. Generally, explosives are the least
expensive and the method of choice for structure removal. However, when
explosives are used, more stringent regulations may become effective, including consultations with the local fishery or natural resource agencies.

Numerous studies are ongoing to reduce the harmful effects on local fish
populations during detonations. A technique to reduce the effects of explosives on habitat fisheries is to evacuate the platform piles of all water. This
reduces the resistance of the shock wave from the charge to the target.

(b) *Nonexplosive removals.* Use of nonexplosive removal techniques eliminates the impact due to shock waves. Consequently, costs and time associated with observers and additional permit conditions may be eliminated.
However, salvages using nonexplosive methods can be costlier since only
one pile or well bore can in practice be severed at one time. Each nonexplosive cut will typically take several hours to perform. Some of the nonexplosive severing technique are as follows.

High-pressure water/abrasive cutters. This system uses a high-pressure water
jet operating at anywhere from 200 to 4000 bar to perform the cut.

Mechanical cutters. Mechanical cutters use tungsten bit cutters that are
extended from a housing tool with hydraulic rams.

Diver cut. Internal or external pile or well bore cuts can be made with divers
using underwater burning equipment. This type of cut can be made internally if there is access for the diver into a large-diameter casing or piling.

Cryogenics. Cryogenics is a little used technology that consists of freezing
the platform pile in the area of a cut with CO_2. A relatively small explosive
charge is then placed at the elevation to be cut and detonated.

Plasma arc cutting. Plasma arc cutting is achieved by an extremely high-
velocity plasma gas jet formed by an arc and an inert gas flowing from a
small-diameter orifice.

Whether using explosives or nonexplosive methods of severing, obstructions in the pile can hinder the proper placement of charges or cutting tools
in the well bore or pile. Examples of obstructions include scale build-up,
damaged piling, mud, or pile stabbing guides. The removal of mud from the
pile is generally accomplished with the use of a combination of a water jet
and air lifting tools.

(c) *Alternative removal techniques.* Most structures are removed with heavy
lift equipment such as oceangoing derrick barges. Innovative methods of
decommissioning, removal and disposal must be proposed to offset the lack
of available salvage equipment and the high cost of equipment mobilisation to remote areas. An alternative approach is cutting the platform into
small, manageable components that lighter, more cost-effective equipment
work spreads can handle. The equipment that may be used includes crawler
cranes, A-frames and portable hydraulic cranes mounted on a cargo barge
and these methods use readily available equipment that can be rigged up
inexpensively.

Other forms of less expensive salvage support equipment include barge-mounted 'stiff legs' and converted jack-up drilling rigs. Stiff legs have the capability to handle large lifts, but generally have limited hook height and are not easily manoeuvrable during the lifting and setting of components on transport barges. Stiff legs are generally built to work in protected waters and are affected by rough seas.

Converted jack-up drilling rigs are becoming more common in the abandonment industry. Companies are converting obsolete rigs to lift vessels to take advantage of the increased need to supply salvage support equipment. This type of equipment can work in heavy seas when in the jacked-up position, but in the floating condition manoeuvrability is limited.

Another technique that can be used for the lifting of platform topsides is the Versatruss system (Figure 10.10). The method uses a series of A-frames mounted on tandem cargo barges. The combination of the A-frames, tension slings and the topside deck create a catamaran and truss effect for lift stability. This lift method also uses available equipment and requires relatively low-cost preparation.

(d) *Alternative structure uses.* If due to some reason or other structure removal does not become feasible, then alternative uses for the platform should be explored. The benefit of the alternative use should offset the costs to maintain the structure in place.

Some alternative uses may be fish farm, marine laboratory, military radar support structure, weather station, oil loading station, spur for deep-water developments, aviation/navigation beacon, tourism/recreational and power generation, that is, wind/wave.

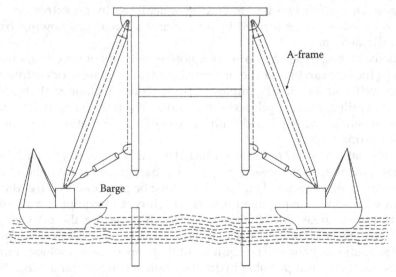

FIGURE 10.10
Versatruss system of lifting platform topside.

Leaving the structure in place should not create a hazard to local fishing industries or to navigation in the area.

(e) *Platform reuse.* Reuse is another option. If a potential development can finance the removal of a structure, this relieves the nonrevenue producing property from absorbing the salvage costs. Platform reuse can reduce the cycle time to get the new development in production, generating cash.

(f) *Partial removals.* This is a cost-saving proposal. Estimates show that total removal (Figure 10.11a) of structures will cost about $6.6 billion whereas partial removal will cost about $4.5 billion. These partial removal methods will consist of the following:

 i. Partial removal of jacket component (Figure 10.11b)

 ii. Toppling in place (Figure 10.11c)

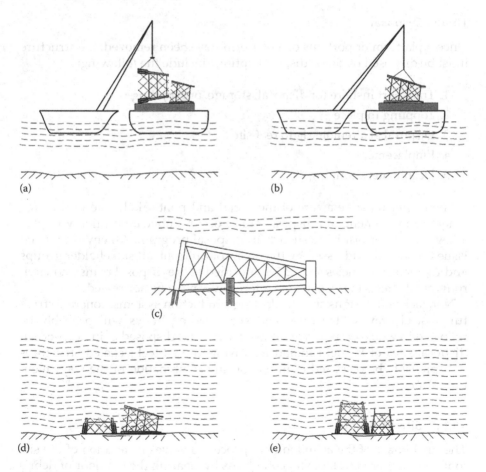

FIGURE 10.11
Total and partial removal methods of jacket components. (a) Total removal, (b) partial removal, (c) toppling in place, (d) removal of topside and toppling in place and (e) emplacement.

iii. Total removal of topside and toppling in place of the jacket only (Figure 10.11d)

iv. Emplacement (Figure 10.11e)

v. Transport to rigs to reef site

vi. Deep-water dumping

The choice of removal method will depend on cost, proximity to disposal sites, availability of removal equipment, location of the removal relative to shipping lanes and fishing interests and safety and environmental issues. In addition, the disposal method will play a key role in the decision on the removal method. The next section summarises the alternatives and key issues concerned with structure disposal.

10.6.6 Disposal

Once a platform or portions of a platform have been removed, the structure must be disposed of. Some disposal options include the following:

1. Transport inshore for disposal, storage, or recycling

2. Toppling in place

3. Disposal at a remote rig to reef site

4. Emplacement

5. Deep-water dumping

The owner must be aware of the social and political climate in the area where abandonment and disposal are to occur. Public perception will play a key role in performing a successful disposal program. All environmental issues should be addressed by the operator up front, all stakeholder groups and regulatory agencies should be informed of the disposal plans and environmental effects of the plan and alternatives must be addressed.

Non-jacketed designs such as floating production systems, concrete structures, steel gravity structures and spar loading buoys will probably be refloated in whole or in part and towed away and disposed of in deep-ocean disposal sites or brought inland for dismantling. Steel-jacketed structures will probably be disposed of in one or any combination of the ways mentioned above.

10.6.7 Site Clearance

The final phase of the abandonment process involves restoration of the site to its original predevelopment conditions by clearing the sea floor of debris and obstructions after platform removal. If the abandonment was a partial removal, site clearance procedures may vary from a total removal. In the case of total removal, debris should be removed, leaving the site trawlable and safe for fishing or other maritime uses.

Once the structure has been removed, the site is ready for a final clean-up if required. In shallow waters, a trawling vessel can be used to simulate typical trawling activities that may occur in the area after the platform removal.

Deeper water sites may not require trawling simulations to clear the area. Proper planning prior to the removal of debris can make a significant difference in controlling the costs. The geophysical survey performed with the side scan sonar during the pre-abandonment survey phase should identify the major debris, and this information will provide the basis for selecting the most effective equipment, personnel and timing. Upon completion of the bottom clean-up, job completion summaries should finally be submitted to the proper authorities.

Bibliography

American Petroleum Institute – Recommended Practice 75 (RP75), Development of a Safety and Environmental Management Program for Outer Continental Shelf (OCS) Operations and Facilities, 1993.

Enron Oil & Gas Company Safe Practices Manual, 1998.

Exposure to hazardous substances offshore, *The Annals of Occupational Hygiene*, http://www.annhyg.oxfordjournal.org.

Gomes, J. S. and F. B. Alves, *The Universe of the Oil and Gas Industry – From Exploration to Refining*, Partex Oil and Gas, 2013.

Oil Industry Advisory Committee, Management of Occupational Health Risks in the Offshore Oil & Gas Industry, Health and Safety Executive, 1996, www.hsebooks .co.uk.

Orszulik, S. T. *Environmental Technology in the Oil Industry*, Blackie Academic & Professional, Chapman and Hall, 1997.

Pant, P. K., S. S. Lamba, A. B. Chakrabarty, A. M. Pandey, and B. C. Kapala, *Oil Spill Response '99*, Proceedings of Conference & Exhibition in India on Oil Spill Response, Compiled by Oil Asia Journal, April 19–20, 1999.

Recommended Code of Practice, Safety & Environment Management, ONGC, Dehradun, India, 1985.

Reis, J. C. *Environmental Control in Petroleum Engineering*, Gulf Publishing Company, 1996.

Safety and Health Management System in Oil and Gas Industry, 2015, www.wipro .com.

Skimmer, 2015 http://en.wikipedia.org/wiki/skimmer-(machine).

Spill Containment Methods, 2015, http://response.restoration.noaa.gov/oil-and -chemical-spills/oil-spills/spill-containment.

Wardley-Smith, J. *The Prevention of Oil Pollution*, Graham & Trotman Limited, 1979.

Index

Printed in the United States
by Baker & Taylor Publisher Services

Printed in the United States
by Baker & Taylor Publisher Services